Photoelectrochemical Water Splitting

Materials, Processes and Architectures

RSC Energy and Environment Series

Editor-in-Chief:
Professor Laurence Peter, *University of Bath, UK*

Series Editors:
Professor Heinz Frei, *Lawrence Berkeley National Laboratory, USA*
Professor Ferdi Schüth, *Max Planck Institute for Coal Research, Germany*
Professor Tim S. Zhao, *The Hong Kong University of Science and Technology, Hong Kong*

Titles in the Series:
1: Thermochemical Conversion of Biomass to Liquid Fuels and Chemicals
2: Innovations in Fuel Cell Technologies
3: Energy Crops
4: Chemical and Biochemical Catalysis for Next Generation Biofuels
5: Molecular Solar Fuels
6: Catalysts for Alcohol-Fuelled Direct Oxidation Fuel Cells
7: Solid Oxide Fuel Cells: From Materials to System Modeling
8: Solar Energy Conversion: Dynamics of Interfacial Electron and Excitation Transfer
9: Photoelectrochemical Water Splitting: Materials, Processes and Architectures

How to obtain future titles on publication:
A standing order plan is available for this series. A standing order will bring delivery of each new volume immediately on publication.

For further information please contact:
Book Sales Department, Royal Society of Chemistry, Thomas Graham House, Science Park, Milton Road, Cambridge, CB4 0WF, UK
Telephone: +44 (0)1223 420066, Fax: +44 (0)1223 420247
Email: booksales@rsc.org
Visit our website at www.rsc.org/books

Photoelectrochemical Water Splitting
Materials, Processes and Architectures

Edited by

Hans-Joachim Lewerenz
Helmholtz-Zentrum Berlin and
Joint Center for Artificial Photosynthesis
California Institute of Technology, Pasadena, USA
Email: lewerenz@helmholtz-berlin.de

and

Laurence Peter
University of Bath, UK
Email: l.m.peter@bath.ac.uk

RSC Publishing

RSC Energy and Environment Series No. 9

ISBN: 978-1-84973-647-3
ISSN: 2044-0774

A catalogue record for this book is available from the British Library

Published by The Royal Society of Chemistry,
Thomas Graham House, Science Park, Milton Road,
Cambridge CB4 0WF, UK

Registered Charity Number 207890

For further information see our web site at www.rsc.org

Preface

World demand for energy continues to increase. Based on figures in BP's 2012 Statistical Review of World Energy,[1] global primary energy consumption in 2011 was equivalent to a thermal power output of 16.35 TW, an increase of 2.5% on the previous year and around 30% compared with a decade earlier. 87% of this energy was generated from carbon-based fuels. BP's World Energy Outlook 2030[2] predicts that global power output will rise to over 22 TW by 2030, and looking further, other growth models predict that energy consumption will at least double by 2050.[3,4] In the short term, shale gas will fill the gap in terms of carbon-based energy resources, but renewable energy resources will have to play an increased role if there is to be any hope of pegging global CO_2 emissions at a level that will reduce the impact of global climate change.

At present, renewables (including biofuels) account for only 2% of global primary energy consumption, but in reference 2 they are predicted to expand their share to around 6% by 2030. Since the potential for increases in the contribution from hydro and nuclear may be limited, this still leaves a huge increase in the consumption of oil, gas and coal. In the absence of viable carbon capture and storage technologies, this implies a massive increase in CO_2 emissions, even if the replacement of coal and oil by gas leads to lower CO_2 emissions per unit of energy generated.

Rapid expansion of terrestrial photovoltaics will go some way to addressing CO_2 emissions from electricity generation. Scenarios considered by the Intergovernmental Panel on Climate Change estimate the potential for power generation by photovoltaics (PV) at around 600-800 GW in 2050, but still this represents only around 2% of the total primary power required.[5] The main problem with photovoltaic power generation is intermittency. Large-scale deployment of PV will require the development of suitable electrical and chemical storage methods. Transport, which accounts for around 30% of primary

RSC Energy and Environment Series No. 9
Photoelectrochemical Water Splitting: Materials, Processes and Architectures
Edited by Hans-Joachim Lewerenz and Laurence Peter
© The Royal Society of Chemistry 2013
Published by the Royal Society of Chemistry, www.rsc.org

energy consumption, is likely to remain based on liquid (or increasingly gas) fuels, although electric vehicles will of course have some impact.

The development of methods of storing solar energy in chemical fuels has therefore become an important research priority, and countries are beginning to react to the problem by establishing large programmes of research into solar fuels. In the United States, the Joint Center for Artificial Photosynthesis (JCAP) was established in 2010. It is the world's largest research programme devoted to the development of an artificial solar fuel generation technology. Other centres in the United States include the Center for Bio-Inspired Solar Fuel Production at Arizona State University and the Research Triangle Solar Fuels Institute involving Duke, NC State and UNC Chapel Hill. Europe has been slower to address the issues, and the scale of funding is smaller than in the US. However, a new Solar Fuels programme has been started at the Helmholtz Centre in Berlin, and there are several initiatives elsewhere in Europe including the Nordic initiative for solar fuel development and The European Science Foundation's EuroSolarFuels programme. At the same time, Japan, Korea and Singapore are starting advanced artificial photosynthesis centres. In the UK, the Royal Society of Chemistry has published a helpful booklet that introduces the topic of solar fuels to a non-specialist audience and identifies some of the key strategic issues.[6]

The Editors felt that the recent rapid expansion of light-driven generation of solar fuels provided the raison d'être for a new book to reflect current progress and to highlight some of the key issues that need to be addressed by the research community. Although work on light-driven water splitting has continued since the much-cited work of Fujishima and Honda,[7] the recent upsurge of activity has brought new people and new ideas and methodologies. In this volume, we have tried to capture some of the energy and enthusiasm that is revitalizing this important research area. The chapters in the book cover a wide range of experimental and theoretical aspects that relate to the light-induced splitting of water and reduction of CO_2, such as materials science, interfaces, heterogeneous catalysts and (photo)electrochemical processes. In addition, new developments related to photonics, light management, excitation energy transfer and third generation approaches have been included in order to emphasize the potential for innovation in the field. We are grateful to the contributing authors, who are all experts in their respective fields and scientific disciplines, and we hope that the book will not only provide an authoritative overview of some of the most important current research directions but also stimulate debate and critical assessment of research priorities.

Hans-Joachim Lewerenz
Pasadena, California, USA

Laurie Peter
Bath, UK

April 2013

References

1. BP Statistical Review of World Energy June 2012. Available on the web at bp.com/statisticalreview.
2. BP Energy Outlook 2030. Available on the web at http://www.bp.com.
3. World Energy Technology Outlook 2050: WETO-H$_2$. 2006, European Commission, Brussels. Available on the web at http://www.ec.europa.eu/research/energy/pdf/weto-h2_en.pdf.
4. World Energy Consumption - What might the Future Look Like? 2008. Shell International BV. Available on the web at http://www.shell.com.
5. IPCC, 2011: IPCC Special Report on Renewable Energy Sources and Climate Change Mitigation. Prepared by Working Group III of the Inter-governmental Panel on Climate Change: [O. Edenhofer, R. Pichs-Madruga, Y. Sokona, K. Seyboth, P. Matschoss, S. Kadner, T. Zwickel, P. Eickemeier, G. Hansen, S. Schlömer, C. von Stechow (eds)]. Cambridge University Press, Cambridge, United Kingdom and New York, NY, USA, p. 1075.
6. Solar Fuels and Artificial Photosynthesis. Science and innovation to change our future energy options. Royal Society of Chemistry, Cambridge 2012. Available on the web at http://www.rsc.org/ScienceAndTechnology/Policy/Documents/solar-fuels.asp.
7. A. Fujishima and K. Honda, Electrochemical Photolysis of Water at Semiconductor Electrode, *Nature*, 1972, **238**, 37–38.

Contents

RSC Energy and Environment Series No. 9
Photoelectrochemical Water Splitting: Materials, Processes and Architectures
Edited by Hans-Joachim Lewerenz and Laurence Peter
© The Royal Society of Chemistry 2013
Published by the Royal Society of Chemistry, www.rsc.org

**Chapter 3   Structured Materials for Photoelectrochemical Water
               Splitting   52**
James McKone and Nathan Lewis

Author Biographies

Allen J. Bard was born in New York City on December 18, 1933. He attended The City College of the College of New York (B.S., 1955) and Harvard University (M.A., 1956, PhD., 1958). Dr. Bard joined the faculty at The University of Texas at Austin in 1958. He has been the Hackerman-Welch Regents Chair in Chemistry at UT since 1985. He has worked as mentor and collaborator with over 80 PhD students, 17 M.S. students, over 150 postdoctoral associates, and numerous visiting scientists. He has published over 900 peer-reviewed research papers, 75 book chapters and other publications, authored 3 books and has received over 23 patents. He served as Editor-in-Chief of the *Journal of the American Chemical Society* 1982–2001. His research interests involve the application of electrochemical methods to the study of chemical problems and include investigations in scanning electrochemical microscopy, electro-generated chemiluminescence and photoelectrochemistry.

Peter Bogdanoff received his PhD in physical chemistry at the Max-Volmer-Institute for Biophysical and Physical Chemistry of the Technical University Berlin. During his PhD work, he investigated vectorial electron transfer processes and the proton-stoichiometry during the photosynthesis at thylakoid membranes. In 1991 he joined the Department Solar Energetics at the former Hahn-Meitner-Institute in Berlin as a postdoctoral research fellow, and since 1998 he has been the head of the Electrochemistry and Photoelectrocatalysis research group, which is working on new materials and material combinations for energy conversion in (photo)electrocatalytic processes. One his main tasks has been the development and investigation of novel electro-catalysts for oxygen reduction, especially those based on transition-metal N_4 chelate centres, and their application in PEM fuel cells. Following foundation of Institute for Solar Fuels at the Helmholtz-Centre Berlin for Materials and

RSC Energy and Environment Series No. 9
Photoelectrochemical Water Splitting: Materials, Processes and Architectures
Edited by Hans-Joachim Lewerenz and Laurence Peter
© The Royal Society of Chemistry 2013
Published by the Royal Society of Chemistry, www.rsc.org

Energy in 2010, his research focus has changed to non-noble metal oxides and sulfides as electrocatalysts for light-driven splitting.

David J. Boston is a senior PhD student at the University of Texas at Arlington and works with Fred MacDonnell on the photochemical reduction of carbon dioxide using ruthenium polypyridyl complexes. He received his B.S. in chemistry from Iowa State University in 2006, where he received the Plagen's Research Stipend.

Ramón Collazo is Assistant Professor of Materials Science and Engineering at North Carolina State University in Raleigh. He has worked on the growth and characterization of wide band gap semiconductor thin films, especially nitrides and diamond, for the past 15 years. He has been particularly involved in developing a process for controlling the polarity in III-nitrides to develop lateral polarity homojunctions along with their application to the first lateral p/n junction. Additionally, he has been involved in the development of AlN bulk single crystal substrates, their surface preparation, and further epitaxial thin film deposition for optoelectronics and power device applications. Some of his current research interests are: III-N wide band gap semiconductors and control of their point defects, polar materials, optical characterization and non-linear optics. He was awarded the Facundo Bueso Medal for Physics and has authored over 100 publications in peer-reviewed journals. He has also been awarded several patents and given presentations at national and international conferences.

Nikolaus Dietz is a Professor of Physics at Georgia State University in Atlanta. His areas of expertise include: radiation interactions with matter; the growth, materials property analysis and defect characterization of photovoltaic group II-VI and I-III-VI$_2$ compound semiconductors, *e.g.* CdMnTe, CuInS$_2$, ZnGeP$_2$; the epitaxial thin film growth of group III-phosphide and group III-nitride compound semiconductors by chemical beam epitaxy, low-pressure and superatmospheric CVD and plasma-assisted MOCVD – and their physical properties characterization; the real-time optical thin-film growth diagnostic and process control; and the characterization of linear/nonlinear optical materials properties. Present research focuses on the development and exploration of new approaches for the fabrication of ternary and quaternary group III-nitrides/phosphides heterostructures of relevance in nanophotonics, opto-electronics, high-efficient photovoltaics, and photocatalytic devices. Dr. Dietz holds several patents and has published more than 140 papers in peer-reviewed journals as well a number of book chapters.

Kazunari Domen received his BSc (1976), MSc (1979), and PhD (1982) in chemistry from the University of Tokyo. He joined the Chemical Resources Laboratory, Tokyo Institute of Technology in 1982 as Assistant Professor and was subsequently promoted to Associate Professor in 1990 and Professor in 1996 before moving to the University of Tokyo as Professor in 2004. Professor

Domen works on heterogeneous water-splitting photocatalysts for generation of clean and recyclable hydrogen. In 1980, he reported the discovery of a NiO-SrTiO3 photocatalyst for the overall water splitting reaction – one of the earliest examples of stoichiometric H_2 and O_2 evolution on a particulate system. In 2005, he succeeded in achieving overall water splitting under visible light ($400\,nm < \lambda < 500\,nm$) using a GaN:ZnO solid solution photocatalyst. His research interests now include heterogeneous catalysis and materials chemistry, with particular focus on surface chemical reaction dynamics, photocatalysis, solid acid catalysis, and mesoporous materials.

Sebastian Fiechter studied materials science at the Crystallographic Institute of the Albert-Ludwig-University Freiburg im Breisgau (Germany). In 1983 he joined the group of Helmut Tributsch at the Hahn-Meitner-Institute in Berlin, where he developed new photoactive and photoelectrocatalytic materials. In the past 15 years he has been involved in the preparation and electrochemical characterization of electrocatalysts for fuel cells and photovoltaic hybrid electrolysers. Together with Peter Bogdanoff, he developed carbon-supported selenium-modified ruthenium catalysts and later porphyrin-based oxygen reduction catalysts, where catalytically-active $Fe-N_x$ and $Co-N_x$ centres are embedded in graphene nano-sheets. In 2011, he was appointed honorary Professor at the Technical University Berlin. Between 2008 and 2012, he was acting director of the newly-founded Institute for Solar Fuels at the Helmholtz-Centre Berlin for Materials and Energy and was appointed Deputy Director in 2012. As a principle investigator of the project "Nanostructured Materials for the Light-induced water Splitting", his activities are now focused on the realization of water splitting membranes where non-noble metal oxides and sulfides act as electrocatalysts integrated in PV thin film structures to split water into hydrogen and oxygen.

Michael Grätzel is Professor at the Ecole Polytechnique de Lausanne Switzerland, where he directs the Laboratory of Photonics and Interfaces. He pioneered the use of mesoscopic materials in energy conversion systems, in particular photovoltaic cells, lithium ion batteries and photo-electrochemical devices for the splitting of water into hydrogen and oxygen by sunlight. He also discovered a new type of solar cell based on dye sensitized nanocrystalline oxide films. Author of over 900 publications, two books and inventor of more than 50 patents, his work has been cited over 88,000 times (h-index 138) making him one of the 10 most cited chemists in the world.

Thomas Hannappel was appointed in 2011 a W3 full professor (physics) in the Photovoltaics Department of the University of Technology in Ilmenau. He is also scientific director of the solar center at the CiS research institute in Erfurt, Germany. Before, he was provisional head of the Materials for Photovoltaics Institute at the Helmholtz-Zentrum Berlin and lecturer at the Free University Berlin, where he received his doctorate in 2005. He obtained his PhD in Physics at the Technical University Berlin with studies on ultrafast dynamics of photo-induced charge carrier separation in dye solar cells, work performed at Fritz-Haber-Institute Berlin. In 2003/04 he conducted research on silicon/III-V-interfaces at NREL. His current investigations are focused on high-performance solar cells and critical interfaces and he is a key player in the fields of solar energy conversion and reactions of critical semiconductor interfaces including silicon/ and germanium/III-V-interfaces, and nano- and quantum-structures.

Anders Hellman received his PhD in theoretical physics in 2003 from University of Gothenburg, Sweden. Since then he has worked on various topics related to surface science, heterogeneous catalysis and materials for energy harvesting at research centers, such as, Haldor Topsøe A/S and the Center of Individual Nanoparticle Functionality (CINF), Denmark. He is currently an associate professor at Applied Physics, Chalmers University of Technology, Sweden, where he is also associated with the Competence Centre for Catalysis. His research interests range from automotive catalysis to plasmon-assisted water oxidation, where he has used several different computational methods, such as, density functional theory calculations, molecular dynamics, Monte-Carlo techniques, micro-kinetic models and finite difference time domain methods to understand the fundamental physics behind the observed phenomena. His research portfolio includes charge transfer and non-adiabaticity in surface reaction, ammonia synthesis, CO and methane oxidation, thin oxides supported on metals, and photoelectrochemical studies of water oxidation.

Kai-Ling Huang is currently a doctoral student at the University of Texas at Arlington since 2010. She works with Fred MacDonnell on the development of ruthenium polypyridyl complexes as potential photocatalysts for carbon dioxide reduction. Kai-Ling completed a Bachelor's Degree in Chemistry from National Taiwan University, Taiwan and continued her education with a Master's Degree in Catalysis, Molecular and Green Chemistry at Université de Rennes, France.

Heung Chan Lee obtained his BSc and MS degree at Korea University under the supervision of Keon Kim. He synthesized and modified polymer electrolyte membranes, and performed electrochemical evaluations for fuel cell application. He carried out his PhD studies at The University of Iowa under the supervision of Johna Leddy. Heung Chan Lee investigated magnetic field effects in electrochemical systems including heterogeneous and homogeneous electron transfer rates in magnetite composite modified semiconductor electrodes. During his PhD, he won three awards including the Jakobsen conference award in 2010. In 2011, he joined Allen J. Bard's group at the University of Texas at Austin and worked on new metal oxide semiconductors for photoelectrochemical energy conversion systems using scanning electrochemical microscopy (SECM). In 2011 he was awarded an Oronzio and Niccolò De Nora Postdoctoral Fellowship in Electrochemistry for his research proposal, "Developing new oxide semiconductors for solar energy conversion systems".

Chanelle Jumper is a current PhD student in the Scholes group at the University of Toronto, Canada. She is studying ultrafast dynamics of energy transfer in photosynthetic proteins, as well as model molecular systems. This work is aimed at understanding the physics involved in energy transfer processes for applications related to solar cell development.

Marc T.M. Koper is Professor of Surface Chemistry and Catalysis at Leiden University, The Netherlands. He received his PhD degree (1994) from Utrecht University (The Netherlands) in the field of electrochemistry. He was an EU Marie Curie postdoctoral fellow at the University of Ulm (Germany) and a Fellow of Royal Netherlands Academy of Arts and Sciences (KNAW) at Eindhoven University of Technology, before moving to Leiden in 2005. He has also been a visiting professor at Hokkaido University (Japan). His research interests are in fundamental studies of electrochemical and electrocatalytic processes through a combination of experimental and theoretical investigations. His group combines well-defined often single-crystalline electrodes with spectroscopic techniques to study electrocatalytic reactions of importance for energy and environmental issues, such as hydrogen evolution, oxygen reduction and oxygen evolution, carbon dioxide reduction, and nitrate reduction. His theoretical work includes theories of charge transfer reactions in condensed media and first-principles density-functional theory calculations of electrode surfaces.

Kevin C. Leonard is currently a postdoctoral research fellow in the Chemistry and Biochemistry Department at the University of Texas at Austin. He has a BS in Chemical Engineering and Applied Mathematics and an MS and PhD in Materials Science from the University of Wisconsin, Madison. Dr. Leonard's research interests include materials for renewable energy storage and characterization of these materials utilizing scanning electrochemical microscopy. He was awarded the 2012 Oronzio and Niccolò De Nora Foundation Fellowship in Electrochemistry.

Hans-Joachim Lewerenz is currently Department Head of the Joint Center for Artificial Photosynthesis at the California Institute of Technology and Deputy Director of the Solar Fuel Institute at the Helmholtz Zentrum Berlin. His research interests encompass photoelectrochemistry, solar energy conversion and surface science of semiconductors. He received his doctorate from the Technical University of Berlin (TUB) in 1978 with a thesis on photoemission into electrolytes, performed at the Fritz-Haber-Institute. After a two year postdoctoral stay at Bell Labs working on photoelectrochemical solar cells, he moved to the Brown Boveri Research Center in Switzerland before returning to Berlin to where he habilitated in physics at TUB and become group leader at the Hahn-Meitner-Institute. He was appointed as professor for physics at TUB and Head of the Department Interfaces at HMI in 1994. He has been guest professor at the Brandenburgisch Technical University Cottbus and visiting/ adjunct professor at North Carolina State University. Professor Lewerenz has published over 260 articles and 4 books and has authored 20 patents. He recently became editor of *Springer Briefs for Physics*.

Nathan S. Lewis is Professor of Chemistry at the California Institute of Technology since 1991. Since 2010 he has served as Principal Investigator of the Joint Center for Artificial Photosynthesis, the DOE's Energy Innovation Hub in Fuels from Sunlight, and since 1992 the Beckman Institute Molecular Materials Resource Center. His research interests include artificial photosynthesis and electronic noses. He continues to study ways to harness sunlight and generate chemical fuel by splitting water to generate hydrogen. He is developing the electronic nose, which consists of chemically sensitive conducting polymer film capable of detecting and quantifying a broad variety of analytes. Technical details focus on light-induced electron transfer reactions, both at surfaces and in transition metal complexes, surface chemistry and photochemistry of semiconductor/liquid interfaces, novel uses of conducting organic polymers and polymer/conductor composites, and development of sensor arrays that use pattern recognition algorithms to identify odorants, mimicking the mammalian olfaction process.

Fred MacDonnell is Professor of Inorganic Chemistry at the University of Texas at Arlington. He has worked for many years on topics related to solar energy conversion, most specifically with the development of ruthenium-based photocatalysts for multi-electron collection and catalysis. He has published 54 papers in peer-reviewed journals as well as several book chapters.

Kazuhiko Maeda received his BSc from Tokyo University of Science (2003), his MSc from the Tokyo Institute of Technology (2005), and his PhD from the University of Tokyo (2007) under the supervision of Professor Kazunari Domen. From 2008 to 2009, he was a postdoctoral fellow at Pennsylvania State University, where he worked with Professor Thomas E. Mallouk. He then joined The University of Tokyo as an Assistant Professor in 2009. Moving to Tokyo Institute of Technology in 2012, he was promoted to an Associate Professor. His research interests are photocatalytic and photoelectrochemical water splitting using semiconductor particles of (oxy)nitrides, inorganic metal oxide nanosheets, and polymeric carbon nitride, combining nanotechnology and materials chemistry. He has published more than 80 peer-reviewed original papers as well as several review papers and book chapters.

Matthias M. May is a PhD student at Humboldt-Universität in Berlin on a scholarship of the German National Academic Foundation. He studied physics in Stuttgart, Grenoble, and Berlin with focus on condensed matter and computational physics. In his diploma thesis, he investigated the electronic structure of charge-density waves in transition-metal dichalcogenides and began work on the water-semiconductor interface. His strongest research tools are angle-resolved photoelectron spectroscopy and in situ reflection anisotropy spectroscopy.

James McKone is in his fifth year of graduate studies at the California Institute of Technology in the research groups of Harry B. Gray and Nathan S. Lewis. A native of northern Iowa, he developed a passion for renewable energy as an undergraduate at Saint Olaf College, where he received a Bachelor of Arts degree in 2008, majoring in chemistry and music. His research at Caltech has been focused on development of earth-abundant semiconductor and catalyst materials for photoelectrochemical hydrogen evolution. James has also been actively involved in the design and rollout of several successful outreach efforts affiliated with the NSF CCI Solar program based at Caltech.

Noseung Myung obtained his BS degree at Yonsei University in Korea and his PhD degree at the University of Texas at Arlington with Professor Krishnan Rajeshwar as mentor. After a postdoctoral fellowship at Clark University, MA, USA, he has been a Professor of Applied Chemistry at Konkuk University Chungju Campus in Korea since 1996. His research interests include electro-deposition of semiconductor, photoelectrochemistry and method development for the analysis of semiconductor thin films using electrochemical quartz crystal microgravimetry and voltammetry.

Arthur J. Nozik is Professor Adjoint in the Department of Chemistry at the University of Colorado, Boulder; Senior Research Fellow Emeritus at the US National Renewable Energy Laboratory (NREL); and a Founding Fellow of the NREL/University of Colorado Renewable and Sustainable Energy Institute. Nozik has been Associate Director of the Los Alamos/NREL Energy Frontier Research Center and founding Scientific Director of the Colorado Center for Revolutionary Solar Photoconversion. His research interests include size quantization effects in semiconductor structures, including multiple exciton generation and the applications of unique effects in nanostructures and nanoscience to advanced approaches for solar photon conversion. He has published over 250 papers and book chapters, written or edited 6 books, holds 11 US patents and serves on many Scientific Advisory Boards and Committees. He received the 2011 ACS Esselen Award at Harvard University, the 2009 Science Award from the UN IREO, the 2008 Eni Award in Science and Technology, and the 2002 Research Award of the Electrochemical Society. He was Senior Editor of *The Journal of Physical Chemistry*, and is on the editorial advisory board of several other journals. He is a Fellow of the American Physical Society and the AAAS.

Evgeny E. Ostroumov is Postdoctoral Fellow in University of Toronto. He received his PhD in Physics from University of Düsseldorf and Max-Planck-Institute for Bioinorganic Chemistry, where he worked with Alfred Holzwarth exploring electronic properties of isolated chromophores as well as photo-protection mechanisms in high plants. During his MSc studies in Moscow State University he worked on laser diagnostics and remote sensing of phyto-plankton. His current research interests include application of nonlinear ultrafast spectroscopy to study energy transfer processes in natural complex systems.

Hyun S. Park received his BS degree (2006) and MS degree (2008) in Chemical Engineering from Seoul National University, Korea. He was a graduate student in Professor Allen J. Bard's group from 2010–2012, and just received his PhD degree from the Chemistry and Biochemistry Department at The University of Texas at Austin. His studies encompassed electrochemistry and photochemistry of metal oxide semiconductors. His current research interests include photo-electrochemistry for photon-energy conversion in water splitting system.

Bruce Parkinson received his BS in chemistry at Iowa State University in 1972 and his PhD from Caltech in 1977 under the guidance of Professor Fred Anson. After a year of post-doctoral studies at Bell Laboratories with Adam Heller he was a staff scientist at the Ames Laboratory. He then became a senior scientist at the Solar Energy Research Institute (now known as the National Renewable Energy Laboratory) in Golden, Colorado. He then joined the Central Research and Development Department of the DuPont Company in 1985. In 1991 he became Professor of Chemistry at Colorado State University until his departure to join the Department of Chemistry and the School of Energy Resources at the University of Wyoming in 2008. His current research covers a wide range of areas including materials chemistry, photochemistry on Mars and photoelectrochemical energy conversion. He has more than 200 publications in peer-reviewed journals and holds 5 US patents. His other interests include photography and swimming.

Laurie Peter is Professor of Physical Chemistry at the University of Bath. He has worked for many years on topics related to solar energy conversion, including dye-sensitized solar cells, earth-abundant materials for thin film PV, semiconductor photoelectrochemistry and photobiological systems for energy conversion. He has published around 290 papers in peer-reviewed journals as well as several book chapters and has been awarded a number of international prizes for his work. Much of his recent research has involved collaboration with other groups in the UK within the framework of two EPSRC Supergen Consortia: PV materials for the 21st Century and Excitonic Solar Cells. He currently has a number of international collaborations and splits his time between the University of Bath and the Ludwig Maximillian University in Munich, where he is working on a range of topics including light-driven water splitting and in-situ microwave measurements on solar cells.

Evgeny Ostroumov is a Postdoctoral Fellow at the University of Toronto. He received his PhD in Physics from University of Düsseldorf and the Max-Planck-Institute for Bioinorganic Chemistry, where he worked with Professor Alfred Holzwarth exploring the electronic properties of isolated chromophores as well as photo-protection mechanisms in high plants. During his MSc studies in Moscow State University, he worked on laser diagnostics and remote sensing of phytoplankton. His current research interests include the application of nonlinear ultrafast spectroscopy to study energy transfer processes in natural complex systems.

Krishnan Rajeshwar is a Distinguished University Professor and Associate Vice President for Research at the University of Texas at Arlington. He is the Editor of Electrochemical Society *Interface* since 1999. His research interests include photoelectrochemistry; solar energy conversion; renewable energy; materials chemistry; semiconductor electrochemistry; and environmental chemistry. He has edited books, special issues of journals, and conference proceedings and is the author of over 350 refereed publications.

Greg Scholes is the D.J. LeRoy Distinguished Professor at the University of Toronto in the Department of Chemistry. His research group examines photophysics in systems ranging from semiconductor nanocrystals to conjugated polymers to photosynthetic light-harvesting proteins. He is especially interested in uncovering microscopic details of light-induced energy capture and conversion processes in photosynthesis and organic photovoltaics using a combination of femtosecond laser experiments and theory. He serves as an Editorial Advisor for *New Journal of Physics* and Senior Editor for the *Journal of Physical Chemistry Letters*. He serves on several editorial advisory boards.

Klaas Jan Schouten was born in Gouda in 1986 and grew up in Moordrecht, the Netherlands. He obtained his MSc degree (*cum laude*) from Leiden University in 2009, doing research on the dissociation of hydrogen on stepped platinum surfaces under UHV conditions with Ludo Juurlink. He also investigated the adsorption of H^+ and OH^- on stepped platinum surfaces under electrochemical conditions. After that he continued with Marc Koper for a PhD project, focused on the mechanistic aspects of the electrochemical reduction of carbon dioxide on copper electrodes.

Kevin Sivula received a PhD degree from the Department of Chemical Engineering at the University of California, Berkeley in 2007 and is currently an Assistant Professor of Chemical Engineering at the École Polytechnique Fédérale de Lausanne. His research is directed towards engineering new, inexpensive and solution-processable semiconductor materials and implementing them in high performance devices – especially for solar energy conversion. He has published over 40 peer-reviewed papers on diverse subjects such as controlling interfacial electronics, nanostructured morphology, and self-assembly to gain insight into charge carrier transport, energy transfer, and optoelectronic performance of transistors, sensors, photovoltaic and photoelectrochemical devices.

Norma Tacconi is a Research Associate Professor at the University of Texas at Arlington. After stints at The Institute of Physical Chemistry (INIFTA, Argentina), University of Poitiers, CNRS Bellevue Laboratory, and the University of Geneva, she joined the group of Professor Rajeshwar at UT Arlington in the early 90s. She has co-authored over 160 papers in peer-reviewed journals as well as several book chapters. Her research interests include semiconductor nanostructures and nanocomposites for photovoltaic energy conversion, molecular catalysts for photo- and electroreduction of CO_2, semiconductor nanoparticles for environmental remediation, and photocatalytically-generated metal nanoclusters on carbon black for fuel cell applications.

John A. Turner is a Research Fellow at the National Renewable Energy Laboratory, and has been with the Laboratory since 1979 when was then the Solar Energy Research Institute. He received his BS degree in Chemistry from Idaho State University, his PhD in Analytical Chemistry from Colorado State University, and completed a postdoctoral appointment at the California Institute of Technology before joining the Laboratory. His work is primarily concerned with enabling technologies for the implementation of hydrogen-based systems into the energy infrastructure. His main research interests are photoelectrochemistry for the production of hydrogen from sunlight and water and the study of fundamental processes of charge transfer at semiconductor electrodes. He is a Fellow of the Renewable and Sustainable Energy Institute, a joint institute between NREL and the University of Colorado, Boulder and co-Editor of the *Journal of Renewable and Sustainable Energy*, an AIP journal.

Shijun Wang is currently a postdoctoral fellow working in Allen J. Bard's group at the University of Texas at Austin. He earned his PhD from the University of Southern Mississippi in 2010. He received his BS degree from the Anhui University of Science and Technology (China) in 1995 and MS degree from the Anhui University (China) in 2005. His research interests include semiconductor photoelectrochemistry, design and understanding novel solid state inorganic photocatalysis and electrocatalysis material for solar energy conversion.

CHAPTER 1

The Potential Contribution of Photoelectrochemistry in the Global Energy Future

BRUCE PARKINSON*[a] AND JOHN TURNER*[b]

[a] Department of Chemistry and School of Energy Resources, University of Wyoming, Laramie, WY 82071, USA; [b] National Renewable Energy Laboratory, Golden CO 8041, USA
*Email: bparkin1@uwyo.edu; John.Turner@nrel.gov

1.1 History

This chapter is directed at a realistic assessment of the role of photoelectrochemistry in the future global energy scenario. We start with a brief history of the development of photoelectrochemical solar energy conversion devices and then extrapolate to the future to offer an opinion about the contribution that photoelectrochemical devices and processes might provide.

We limit this discussion to devices that employ semiconducting materials to absorb solar energy and convert it to photoexcited charge carriers that are harnessed to produce either electrical power or chemical fuels. Photoelectrochemistry implies that the semiconductor is immersed in a solution and its properties are investigated in the dark and under illumination. The modern era of photoelectrochemistry began at Bell Laboratories during the early development of semiconducting materials for use in electronic devices. Bell Laboratories' researchers immersed various semiconductors such as Ge[1] and TiO$_2$[2] and measured their electrochemical response in the light and dark and

RSC Energy and Environment Series No. 9
Photoelectrochemical Water Splitting: Materials, Processes and Architectures
Edited by Hans-Joachim Lewerenz and Laurence Peter
© The Royal Society of Chemistry 2013
Published by the Royal Society of Chemistry, www.rsc.org

reported on their photocorrosion and even, in the case of TiO$_2$, the ability to evolve oxygen from water when illuminated with light of greater energy than the band gap. Other researchers, most notably Heinz Gerischer, began to publish experiments, models and theories to explain the energetics and kinetics of dark and photoinduced charge transfer at semiconductor electrolyte interfaces.[3–5] However it was not until the energy crisis of the early 1970s, and the paper by Fujishima and Honda,[6] that the connection between semiconductor photoelectrochemistry and solar energy received wide attention. Although oxygen evolution was observed as early as 1968 when illuminating a rutile electrode in solution,[2] the application of this concept to water photoelectrolysis was first pointed out by Fujishima and Honda in a series of experiments that used the *n*-type semiconductor rutile form of TiO$_2$.[7] While rutile is stable under illumination in aqueous electrolytes, its large band gap (3.0 eV) restricts its utilization to the UV portion of the solar spectrum and thus limits its ultimate efficiency. It should also be pointed out that the conduction band of rutile is not negative enough to reduce water, and so a "pH bias" was used in early work, where the oxygen-producing side of the cell was basic with respect to the hydrogen-producing electrolyte.[8] The simplicity of the Fujishima and Honda experiment, illuminating a rutile crystal electrode with UV light in an electrolyte to produce hydrogen and oxygen directly, and the energy crisis of the early 1970s, set off a flurry of further research in photoelectrochemistry aimed at solar energy conversion. Later work using a device with heterojunctions of III-V materials as photoelectrodes considerably increased the visible light conversion efficiency of direct water photoelectrolysis but at increased cost and decreased lifetime due to corrosion in aqueous electrolytes.[9]

In terms of solar energy conversion efficiency, the most successful devices that followed were not water-splitting devices but rather photoelectrochemical photovoltaic cells. Here, illumination of the semiconductor/electrolyte junction drove a reversible redox reaction such as sulfur/polysulfide with no net change in the chemical composition of the electrolyte, instead of photoelectrolysis where chemicals are consumed. These early cells used various semiconducting materials, such as GaAs,[10] CdSe,[11] Si,[12] and MoSe$_2$,[13] and reached efficiencies as high as 14% in laboratory cells and in some cases achieved respectable efficiencies with polycrystalline materials due in part to the spontaneous production of a conformal junction by the redox electrolyte.[14,15] The stability of some of these devices was quite good, especially for MoSe$_2$ and related materials,[13] but due to both the unproven long-term stability and issues related to encapsulation of the often corrosive liquid electrolytes employed, they never became serious alternatives to solid-state solar cells. These issues, along with the increased supply and lower cost of oil in the period between 1985 and 2000, reduced both the interest and the funding for research and development of photoelectrochemical energy conversion devices. A photoelectrochemical photovoltaic device did emerge in 1991, the nanocrystalline TiO$_2$ dye sensitized solar cell,[16] which promises to become a contender in low-cost thin-film photovoltaic solar cell market. This cell exploited some of the main advantages of semiconductor liquid junctions; the junction forms spontaneously and is conformal even when high-aspect ratio or porous nanocrystalline networks are involved. This cell has spawned an enormous amount of research, partly because it is rather easy to construct an inexpensive device with a respectable efficiency without sophisticated equipment. However, even if this device is improved and becomes a large commercial success, it will not solve the major problem of transitioning the world to a renewable energy economy. The problem is that over 80% of current energy use is fuels,[17] and so a method to convert solar energy, by far the most abundant renewable energy resource, to storable chemical energy is sorely needed. Therefore, there has been renewed interest in photoelectrolysis, culminating in the recent establishment of a "solar hub" by the United States Department of Energy to develop a device that can perform the photoelectrolysis of water within five years of its foundation in 2010.[18]

1.2   Solar Hydrogen

The photoelectrolysis of water to form molecular hydrogen and oxygen is an obvious and direct way to store solar energy as fuel.[19–21] Hydrogen represents stored energy in the form of chemical bonds, bonds that can directly or indirectly react with oxygen to release energy. The hydrogen can also be used to upgrade biofuels or coal to pure hydrocarbons and for the production of ammonia. At present, hydrogen is mainly produced by steam methane reforming using natural gas as the feed stock. This releases a large amount of CO_2 as a by-product and is clearly a non-sustainable process.

$$CH_4 + 2\,H_2O \rightarrow 4\,H_2 + CO_2 \tag{1.1}$$

By contrast, hydrogen produced from photoelectrolysis of water would be a sustainable process using two of our most abundant resources, water and sunlight, to produce hydrogen and then regenerating the water when the hydrogen is consumed, releasing the stored energy.

Hydrogen can also be obtained, along with molecular halogen, by the photoelectrolysis of haloacids. The advantage of this process is that the electrochemical oxidation of halides is kinetically much faster than the oxidation of water, resulting in less overpotential loss in the reaction. Indeed, in the 1970s, Texas Instruments spent tens of millions of dollars on the development of a very clever photoelectrolysis device to store solar energy by splitting hydrogen bromide.[22] The energy in the stored hydrogen gas and bromide/tribromide electrolyte could be recovered as electrical energy by recombining the hydrogen and bromine in a hybrid fuel cell/redox battery. While, strictly speaking, this device was not a photoelectrochemical system, since the semiconductor p/n junctions were insulated from the solution by a thin metal film, the Texas Instruments device showed nonetheless that the direct photoelectrolysis concept is scalable.

A large-scale operating water photoelectrolysis facility producing meaningful amounts of hydrogen would rapidly saturate the commercial demand for pure oxygen. Oxidation reactions, other than water oxidation, should be considered if there is a market for them, since there is much less demand for pure oxygen than there is for hydrogen. The only other commodity chemical currently produced on a very large scale by aqueous electrolysis is chlorine via the chlor-alkali process,[23] but again the demand for chlorine would be quickly met by any system producing enough hydrogen fuel to make an impact on world energy demand. However, photoelectrolysis technology could get an initial revenue boost from the sale of chlorine and hydroxide obtained from the photoelectrolysis of brine.

Approximately 44% of the world's population live within 150 km of coastal areas, thus the use of seawater as a feedstock in the photoelectrolysis process is an attractive option. However, the bulk of the anode reaction in seawater electrolysis produces chlorine even though oxygen is thermodynamically preferred. To avoid chlorine evolution from seawater, photoelectrolysis would then require desalination, an expensive and energy intensive process. The alternative is to identify an electrode material that would evolve oxygen from seawater without the concurrent evolution of chlorine. Manganese oxide-based anodes with a high current efficiency for oxygen evolution in preference to chlorine evolution have been reported.[24] However, the application of these MnO_x-based electrodes is limited due to their lack of stability, low voltage efficiency and an unknown mechanism for the preferential evolution of oxygen. Integrating such an electrode material into a photoelectrochemical cell would be a challenge.

Any efficient photoelectrolysis material must utilize the solar spectrum effectively and generate sufficient photovoltage to drive the water splitting reaction. Either a single or tandem semiconductor electrode system may be used.[6] The photovoltage must be greater than the thermodynamic value for the difference between the water oxidation and water reduction potentials (1.23 V at room temperature). Ideally, the semiconducting material(s) should have some catalytic activity for hydrogen or oxygen production so as to minimize the photovoltage in excess of 1.23 V that is required to overcome the electro-chemical overpotentials needed to drive the water oxidation and reduction reactions at the desired rate. Current densities (photoelectrolysis rates) of the order of 15–25 mA cm^{-2} at illumination intensities around one sun are needed for both single and tandem photoelectrode systems. In addition, a single semiconductor photoelectrode must have a valence band edge that is more positive than the water oxidation potential and a conduction band edge that is more negative than the water reduction potential so that the photogenerated holes and electrons have the necessary electrochemical driving force for the water-splitting reaction without needing to supply additional bias voltage. If a dual photoelectrode system is used, the conduction band of the p-type material must be negative of the water reduction potential and the valence band of the n-type material must be positive of the water oxidation potential. Stability of photoelectrolysis electrodes is crucial for a viable system since the significant capital investment needed for the support structures, electrolyte handling and gas handling systems requires that the system last for many years. Since the photoelectrode will be continually immersed in an electrolyte and illuminated with direct solar radiation, a long lifetime is a daunting challenge that must be adequately addressed even in the basic research stages. It is primarily for their thermodynamic stability that metal oxide semiconductors arguably hold the most promise for constructing a stable photoelectrolysis system.

For standard photovoltaic devices, the efficiency is determined by the product of the open circuit photovoltage, the short circuit current and the fill factor. However, for PEC devices, the only variable in the efficiency equation is the current.

$$\frac{\text{chemical potential} \times \text{rate}}{\text{light intensity}} = \frac{1.23\,\text{V} \times \text{current}\,(\text{mA cm}^{-2})}{100\,\text{mW cm}^{-2}} \tag{1.2}$$

Unlike solar cells, for photoelectrolysis cells (PEC) the potential is fixed to the chemical potential of the product, in this case hydrogen. The only variable in the efficiency calculation is the rate of the production of hydrogen – the current. Other methods including the "energy saved efficiency" have been used to inflate reported water splitting efficiencies, and these methods are reviewed in a prior publication.[25]

As stated above, either a single semiconductor electrode or a two-semiconductor electrode photoelectrolysis system can be envisioned. The single illuminated electrode system will need a band gap greater than about 1.7 eV in order to supply the photovoltage needed for water photoelectrolysis, given that photovoltage of about 2/3 the band gap is often the maximum obtainable from a good semiconductor material operating at its maximum power point. A single photoelectrode system would need to be attached to a catalytic electrode (Pt-catalyzed or a platinum mimic) that can accomplish the complementary water splitting half-reaction. A two-semiconductor system, with both p-type and an n-type semiconductor photoelectrodes, can be configured in several ways. In one configuration, the electrodes could be placed with the larger gap material absorbing the higher energy portion of the solar spectrum in front of the smaller band gap material of the opposite majority carrier type in which the wavelengths transmitted through the large band gap material would be absorbed in the smaller gap material, as shown in Figure 1.1. The two materials must absorb nearly equal numbers of solar photons since the current through each semiconductor must be matched for maximum efficiency. Recombination at the ohmic contacts sums the two photovoltages. The current would be less than a single band gap cell, but the efficiency could be considerably higher because the device will absorb more than twice the number of solar photons since there is considerable photon flux in the red and near IR region of the solar spectrum. This configuration has the advantage that one glass substrate, with a transparent conducting layer on both sides, could be used for both materials, reducing the total system costs. One must also design the device to have low ohmic losses as well as to have minimal gas cross-over between the two compartments. This will require either an ion selective

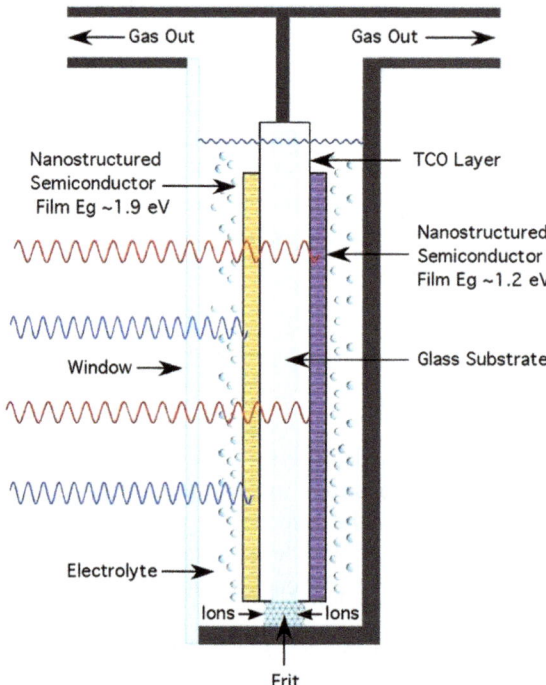

Gas Out ←

Gas Out →

Nanostructured
Semiconductor →
Film Eg ~1.9 eV

TCO Layer

Nanostructured
Semiconductor
Film Eg ~1.2 eV

Window →

Glass Substrate

Electrolyte →

Ions → ← Ions

Frit

Figure 1.1   Side view of a design for a tandem photoelectrolysis system that uses nanostructured semiconductor films of opposite carrier type that are deposited on opposite sides of a single substrate that is covered on both sides with a continuous transparent conducting oxide film. The larger band gap material is transparent to the lower energy radiation that is absorbed by the smaller gap film on the backside of the substrate.

membrane or frits that have high ionic conductivity to reduce ohmic losses. A schematic of a cross section of such a device is shown in Figure 1.1 and a front view of the electrode is shown in Figure 1.2. Another configuration would be to place the n- and p-type materials side by side, rather than putting a larger band gap material in front, allowing full-spectrum sunlight illumination of both electrodes. In this case, an n-type and p-type electrode of the same material or two different oppositely doped smaller band gap materials could be paired as long as the sum of their photovoltages at maximum power was greater than the ~1.8 eV needed for water photoelectrolysis and the photocurrents in both electrodes were matched. If two materials were used, the band gaps would have to be close to each other, since they need to absorb nearly the same flux of solar photons in order for matched photocurrents and maximum efficiency. Two stable materials would be needed for this configuration to work; a p-type material with a conduction band at least 0.8 eV negative of both the n-type material conduction band and the hydrogen potential, and valence band offsets of about 0.8 V. Two substrates would also be required, but they need not be transparent (as in the case of the stacked

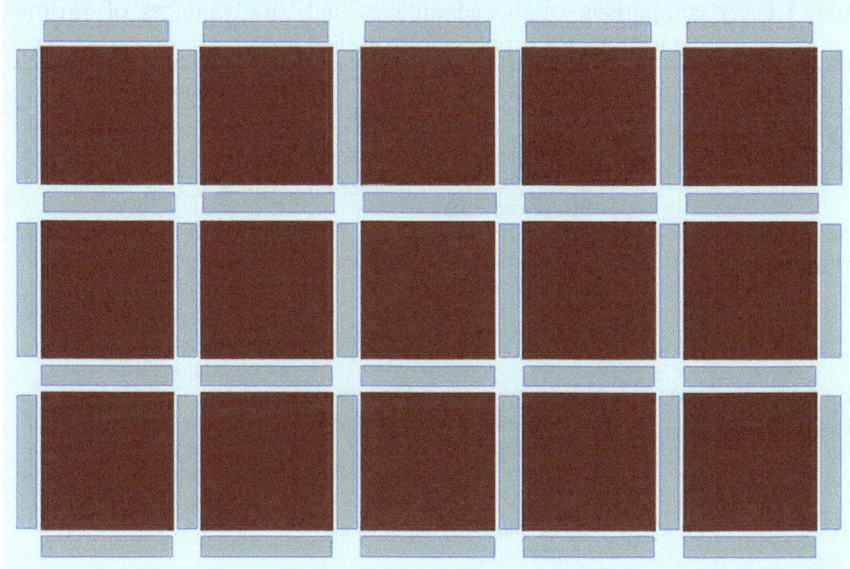

Figure 1.2   Front view of a possible configuration for the tandem photoelectrode from Figure 1.1 showing patches of nanostructured semiconducting thin film on a conducting glass substrate (maroon squares) surrounded by areas of a frit or membrane (blue rectangles) to allow for ion flow, prevent gas cross-over and to reduce resistance losses across the electrode compartments.

system discussed above and shown in Figures 1.1 and 1.2). A separator that conducts ions and inhibits gas mixing such as a membrane or a frit would also be required. Transparent substrates could be used in this case, since back illumination of both gas-evolving electrodes would eliminate all of the back-scattering of light from bubbles that prevents it from reaching the photoelectrodes. The result is that the number of possible material systems is broadened and cheaper or more abundant materials may be utilized. Table 1.1 summarizes some of the advantages and disadvantages of the different photoelectrolysis configurations discussed above. Systems that contain buried junctions, where charge separation is remote from the electrolyte, are really solid state solar cells in series with metal electrodes, and they are not considered here since they are essentially equivalent to hooking up photovoltaic cells to an electrolysis cell, albeit without the wires.

Additional losses are associated with the non-ideal placement of the band edge energies. Poor or no overlap of band edge energies with the water redox half reactions can significantly increase the energy losses or prevent the reactions from occurring. An approach to deal with the band-edge mismatch is to modify the surface with charged species to control the band edge energetics. Some work has been done on modifying the surface,[26,27] but this area is still relatively unexplored.

Table 1.1 A comparison of the advantages and disadvantages of various configurations for a photoelectrolysis device.

Configuration	Advantages	Disadvantages
Single photoelectrode	Only one substrate and one p or n type semiconductor needed. Dark electrode can be conventional.	Poor utilization of the solar spectrum since large band gap ($> \sim 1.7\,eV$) is needed.
Tandem photoelectrodes (Figures 1.2 and 1.3)	Needs only one substrate. More efficient utilization of the solar spectrum.	Need to identify one p and one n type semiconductor and needs a transparent conducting substrate. Two photons and current matching needed.
Separately illuminated photoelectrodes	Non-transparent substrates can be used. More efficient utilization of the solar spectrum. If both electrodes are back illuminated, scattering from bubbles is eliminated.	Need two semiconductors. Two photons and current matching needed. Twice the collector area of single or tandem configuration.

1.3   New Materials

At this point it would be useful to discuss the possible band gaps necessary for the as yet unknown materials to be used in an efficient solar photoelectrolysis device. The band gap values should be used as a guide for researchers' efforts on the discovery and optimization of these materials. Previous publications have considered limiting efficiencies for PEC water-splitting devices, and these are well covered in a recent review.[21] Calculations of theoretical efficiencies all hinge on assumptions of efficiency losses, which are primarily the overpotential losses for the hydrogen and oxygen half reactions. The efficiency loss estimates are typically based on the losses observed in electrolysis cells. However, electrolyzers operate at relatively high current densities, whereas at solar intensities, the maximum current at AM1.5 would range from around $30\,mA\,cm^{-2}$ to perhaps up to $100\,mA\,cm^{-2}$ for a single gap photoelectrolysis cell operating under mild solar concentration of around 3–5 suns (or half these values for a tandem cell). Figure 1.3 details the two-electrode voltage required to split water at current densities up to $130\,mA\,cm^{-2}$.[28] Even with these low current densities, high catalytic activity is still a necessity. Poor catalysis can allow charges to build up at the semiconductor/electrolyte interface, causing the band edges to unpin, producing a situation where the photogenerated carriers at the band edges are unable to accomplish the water splitting reaction and instead recombine. Bansal and Turner showed the impact of catalysis on band edge movement for the hydrogen evolution reaction in acid.[29] At $30\,mA\,cm^{-2}$, the overvoltage loss (H_2 and O_2), under this rather ideal condition, is only about $100\,mV$. Even with the current for a 3–5× sunlight concentration system, the overvoltage losses shown here are less than $250\,mV$. This data is for smooth

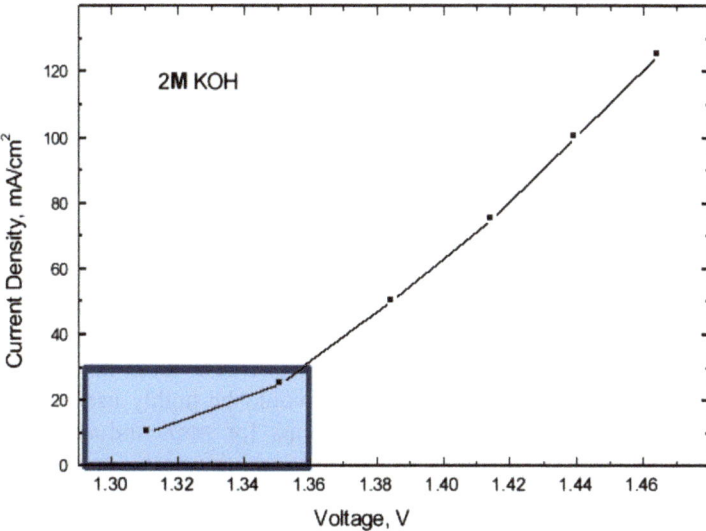

Figure 1.3 Steady-state current density *vs.* voltage for electrolysis of water at two platinum electrodes of equal area in 2M KOH. The box indicates the voltage needed for a typical solar generated current density of $30 \, mA/cm^2$.[28]

platinum electrodes, whereas nanostructuring or dispersing the Pt or other catalysts can reduce the local current density and thereby further reduce the overvoltage losses.

Clearly, it would be useful to know the approximate band gaps of materials that would be needed to provide high efficiency photoelectrolysis. Weber and Dignam analyzed this problem and came up with an optimum efficiency for a tandem PEC system of about 22% if a high fill factor and overpotential losses of 250 mV are assumed with band gaps near 1 eV and 1.8 eV for the two semiconductors.[30] In his chapter in this book, Nozik gives a series of calculations showing the solar-to-hydrogen efficiencies as a function of catalytic activity assuming an ideal semiconductor system (single junction or tandem). An assumed overvoltage loss of 0.4 volts is reasonable and results in a possible efficiency of 30% with a bottom cell at ~0.8 eV and a top cell at ~1.6 eV. Nozik and Hanna have also calculated the increased photoelectrolysis efficiencies possible from carrier multiplication (more than one electron per high energy photon) if these effects can be exploited in a practical device.[31] Carrier multiplication has been demonstrated in a quantum dot (QD) sensitized and a QD bulk absorber device.[32,33] However, new semiconducting materials must be discovered to make this a reality. Systems that convert any extra photovoltage to electrical power can also be designed, but the electrical power produced would most likely be more expensive than if produced from a conventional photovoltaic device, and given that this is the case, squeezing out enough photovoltage to overcome the overpotentials needed to photoelectrolyze water will be challenging enough.

To achieve an efficient, stable and affordable photoelectrochemical photo-electrolysis system, new semiconducting materials with band gaps much smaller than have currently been explored need to be discovered, developed and optimized.[34] Semiconducting oxides have the best chance of fulfilling the stability criteria since they can be thermodynamically stable, especially to the valence band holes that need to have the potential to oxidize water. Rocks are a good example of materials that are quite stable over many years in the presence of an electrolyte and sunlight, and most rocks are oxides. Given the choice, it will be far easier prevent corrosion in a thermodynamically stable photoelectrode than to kinetically stabilize it with a thin corrosion barrier that carriers can tunnel through. Since there are many millions of possible metal oxides that could be produced from the 60 or so metals in the periodic table mixed in various stoichiometries, combinatorial techniques would be highly useful for quickly producing and screening these combinations for semiconducting behavior. Several combinatorial techniques for accomplishing this task have been reported[35–37] and reviewed,[38] and so they are not be discussed in detail here. However, it is worthwhile to mention that the scaling up of a combinatorial search has been accomplished at the Joint Center for Artificial Photosynthesis (JCAP),[18] and due to the high throughput that they have achieved, the focus may soon shift from the discovery of the new oxide semiconductors to understanding and optimizing the properties as well as configuring the newly discovered materials for photoelectrolysis reactions. The latter task will be more challenging and time consuming than the discovery phase, and many researchers will be needed to investigate these many potential materials. Combinatorial techniques may still be useful for the optimization of material growth morphologies and doping densities, as was recently demonstrated for improving iron oxide photoanaodes.[39]

Another approach is to use density functional theory with high performance computing to calculate the electronic properties of candidate materials and to suggest alloys that should have the necessary properties.[40,41] While a number of possible alloy systems have been suggested using this approach, none have yet shown success. Solving this so-called inverse design problem is still in its early stages, but theory will still be important in helping to understand the role of defects and dopants in the new materials.[42]

Often materials are synthesized and tested for photochemical water splitting activity either as colloidal solutions or as powdered slurries using a sacrificial donor or acceptor. While this method may be useful for screening materials for water splitting activity, in our view it is unlikely that homogeneous colloidal solutions or slurries will be useful for a practical water splitting system. Reasons for this include: (i) the products are not produced in separate compartments, resulting in highly explosive mixtures of hydrogen and oxygen (an accident resulting in an explosion would likely terminate any large water splitting project); (ii) even if the explosive mixture can be handled with complete safety, energy is required to separate hydrogen and oxygen, reducing the overall efficiency of the water splitting process; (iii) since catalysts for hydrogen and oxygen production from water are also generally catalysts for the recombination of hydrogen and oxygen, illumination of such systems will eventually result in a photostationary state where forward and back reactions have equal rates and no more net water splitting can occur. In most publications reporting the use of semiconducting particles, this photostationary state is evident as a fall-off in the rate of hydrogen and oxygen production, whereupon the researchers need to purge the system and restart to obtain the initial faster gas production rate. The exchange current in the photostationary state will be related to the solubility of the gases in the electrolyte and the rate at which they are removed from the reactor.

In systems where the electrodes are not or cannot be separated (*e.g.* semiconductor particle systems), sacrificial reagents are often used in the photoreactors to scavenge either holes or electrons in order to avoid undesirable back reactions and poor efficiency for hydrogen or oxygen production. The most common examples of hole-scavenging agents introduced into solution are alcohols (usually methanol), amines (usually triethanolamine or EDTA) or sulfite salts. Electron scavengers such as the easily reduced Ag^+ have also been added. Electron or hole scavengers can be useful to study *one* of the water splitting half reactions without complications associated with the kinetics of the other half reaction (although using a three electrode potentiostat to study a single photoelectrode achieves the same end), but these additives are not viable for any practical system for sustainable energy production since they are not available in the large quantities that would be needed and/or they are much more valuable than the hydrogen produced.

1.4   Commercial Viability of Photoelectrolysis as a Route to Hydrogen

Having been through the basics and some efficiency calculations and various device configurations, one can ask at this point: what is the minimum solar-to-hydrogen (STH) conversion efficiency for a commercially viable PEC water-splitting device? Since the system must cover land area to collect sunlight, it is clear that the size of the plant will be directly related to the commercial viability, since a lower efficiency system will cover significant land area, meaning higher costs for land acquisition, longer piping for collecting the hydrogen and for distribution of the water feedstock. All these factors will increase the balance of plant costs and result in a commensurate increase in the cost of the produced hydrogen. Ultimately it is the price per kilogram of the produced hydrogen that will determine which PEC system can be used to produce sustainable hydrogen and, similar to PV, for photoelectrolysis the cost of that hydrogen is determined by the solar insolation, the STH efficiency, the lifetime, the cost of the photoconverters and the balance of plant. In an attempt to put realistic numbers on the cost of PEC produced hydrogen and how that relates to conversion efficiency, cell costs and lifetime, a technoeconomic analysis of PEC systems is needed. One recent chemical engineering analysis commissioned by DOE looked at four different engineering designs and calculated the costs of the produced hydrogen. The designs were (i) single-bed particle systems, (ii) dual-bed particle systems, (ii) flat plate PEC and (iv) low solar light concentration systems ($<10\times$).[43] The projected price of hydrogen was the lowest for the single bed system and highest for the flat plate systems, ranging from $\sim\$1/kg$ to $\sim\$10/kg$. However, the single bed system would produce a stoichiometric mixture of hydrogen and oxygen, which was clearly unacceptable to the chemical engineers doing the analysis since one accident would endanger operators, produce a negative public backlash and destroy a capital investment. The dual bed system requires a selective redox relay with redox properties that would be almost impossible to obtain, leaving the flat plate and the concentration systems as the only realistic options.

The main takeaway from this analysis is that the solar-to-hydrogen conversion efficiency has the largest impact on the hydrogen price, and that the efficiency required is much higher than previously thought acceptable for a viable device. A 10% system has generally been considered as an acceptable efficiency, but the technoeconomic analysis shows that the STH efficiency needs to be greater than 15% and in some cases, depending on the systems costs and lifetime, greater than 20%. (US Department of Energy website) One caveat though is that the hydrogen cost goal has been set by the US Department of Energy match that of hydrogen from steam methane reforming, $\sim\$2/kg$. A more realistic metric would be affordability, factoring in the costs of the impacts of greenhouse gases emitted by processes that use fossil fuels. A simple comparison would be gasoline at \$4.00/gallon and a 30 mpg fuel economy. The equivalent price for the same distance per kg H_2 would be $\sim\$8/kg$, since fuel cell vehicles have much more efficient fuel utilization. The \$8/kg cost must include $\sim\$1/kg$ for compression and delivery, so the price of the produced hydrogen should be closer to \$6/kg. Nonetheless, it is apparent that the efficiency of a viable PEC water splitting system must be better than 15% in order to achieve an acceptable price for the hydrogen. Another lesson form this analysis was that low solar light concentration ($<10\times$) could produce economic benefits, given that it may lower the costs of other components such as collector area and piping. Another attractive aspect of PEC hydrogen production is that the gases can be produced at higher than atmospheric pressure, reducing or eliminating the capital costs for compressors to compress the gas for pipeline distribution. Thermodynamically a ten-fold pressure increase is possible with a 59 mV addition to the overvoltage, which should be factored in to the band gap analysis presented above.

1.5   Photoelectrochemical Reduction of Carbon Dioxide

Carbon dioxide reduction using solar energy is often touted as a route to renewable solar fuels with no net carbon footprint, where any carbon that is either removed from the atmosphere or captured and reduced before entering the atmosphere will not add to the atmospheric CO_2 levels. Although the multi-electron, multi-proton chemistry involved in CO_2 reduction provides an interesting and challenging chemical problem and at first glance appears to be "carbon neutral", in our view it does not currently make sense as a strategy for reducing carbon dioxide emissions in the context of climate change when the big picture is considered. We examine this premise in the following section.

The first consideration for CO_2 reduction to impact climate change is the scale. Coal production in Wyoming alone is 425 million tons/year, resulting in annual emission of nearly a billion tons of CO_2 after burning in coal-fired plants (*cf.* Figure 1.4 – total US production is about 3 times this). An industrial scale process to reverse the impact of just Wyoming's coal production would dwarf even the current largest industrial chemical process, the production of sulfuric acid, by a factor of 25. If one considers reducing CO_2 to a liquid fuel using renewable energy thinking that there would be value added, the logical source of CO_2 would be a fossil fuel burning power plant since it is very concentrated in the stack, whereas it takes energy to extract and concentrate it from the air. The largest source of CO_2 in the US (as well as many other countries in the world) is from coal burning plants, and so we qualitatively examine the energy balance of this approach.

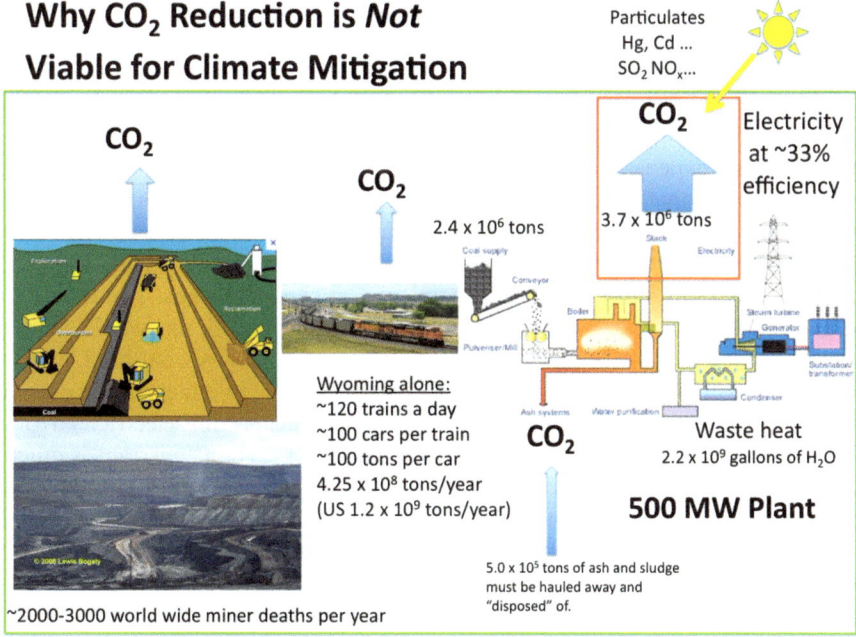

Figure 1.4 A system to be evaluated using thermodynamics must be defined by placing it in a box and examining all the inputs and outputs to the system. Often when CO_2 reduction is considered the box is placed only around the emission of CO_2 from the stack of the plant (red box) and the impression is that the energy from the sun is free and the CO_2 is free and needs to be mitigated. However, when the system is defined as the whole coal mining to disposal of ash (green box), the value of replacing the coal-fired plant up front with renewable energy becomes even more compelling.

At first sight, photoelectrochemical CO_2 reduction looks attractive. The raw materials needed (CO_2 and sunlight) are free, and one might even get a subsidy to utilize or prevent the release of the CO_2. However, this thinking only considers a small part of the entire coal burning system (the stack) as shown with the red box in the upper right of Figure 1.4. In order to do a proper analysis as in thermodynamics, one must put the box around the whole system, as in the larger green box in Figure 1.4. It is then apparent that there is a considerable release of CO_2 from burning fossil fuels in machines for mining the coal, shipping it to the power plants in the locations far from the mine (*e.g.* Wyoming to the Midwest in the US or from Australia to China) and then pulverizing the coal before combustion. The coal is then burned to produce heat that is converted to electricity at an efficiency of about 33% in a modern coal-fired plant (the efficiency of older plants can be as low as 20%). Along with CO_2, burning coal also results in the emission of sulfur dioxide, mercury, other heavy metals and radioactive metals. Huge volumes of coal ash, which contains toxic metals, must be disposed of by shipping it to off-site (and out of sight) locations using more fossil fuel (although some of the ash can be used in building materials). These wastes have resulted in widespread water pollution in the US since their disposal is not federally regulated. However, focusing on CO_2, we see that at least 3 equivalents of CO_2 are emitted to the atmosphere to produce one equivalent of electrical energy. However it is much more than three when considering the previously mentioned fossil fuel-based energy inputs into the whole coal burning to electricity cycle. Therefore, it appears to make no sense to use diffuse, expensive to harvest sunlight to reverse the combustion reaction even to produce methanol, the easiest liquid fuel to synthesize. This is especially true when considering that the solar converter will be at most 20% efficient and that it takes six electrons per CO_2 molecule to convert it to methanol with efficiency losses in all of the electrochemical and chemical steps. The cost and stability of the catalysts also needs to be considered. Instead, renewable energy should be installed to replace the electricity from as many coal-fired power plants as possible "up front" before any attempt is made to undertake any large scale CO_2 reduction. In other words, the cost of generating electricity from solar photovoltaics and preventing a given amount of carbon emissions is much lower than the cost of capturing and reducing CO_2 after burning a fossil fuel. Implementation of photovoltaic (PV) arrays to replace the electrical output of coal-fired plants would pay off by reducing CO_2 emissions by probably a factor of four or more per unit of electric energy produced as well as by eliminating toxic emissions and reducing the number of deaths from coal mining that averages several thousand per year world-wide. Stated another way, to deploy an artificial photosynthetic system to totally reverse the CO_2 output of a typical 300 MW coal-fired plant, the land and solar collector area needed would be at least three or four times that needed to simply replace the electrical output of a 300 MW coal plant with photovoltaics. In practice, the area would be much more since it would be difficult to drive the complex chemistry at the same efficiency as direct PV electricity production (10% would be impressive). Solar conversion economics is driven by the capital costs of producing large areas of collectors, so the capital costs for conventional PV would then probably be at least six times less than needed to construct a system to convert the CO_2 to liquid fuel when considering the both the efficiency and the complexity of a complete CO_2 reduction installation that needs investment in gas handling, purification systems and catalysts. Further efficiency improvements of the solar PV systems will be gained if they are decentralized, unlike the conventional power plants, to be near the consumption of the power, resulting in consider-able reductions of the very high transmission losses. Of course solar PV can only replace the peak output of a coal plant during daylight hours, and other renewable power or energy storage would be needed on the grid to replace the coal-fired electricity during the night where wind energy may be more available. This discussion is most pertinent to the many coal-fired plants that operate at high capacity during daylight hours and scale back output during darkness.

If a liquid fuel source with a minimal carbon footprint is required, it makes more sense to use the abundant coal resources, gasify the coal and do well-established high-temperature reactions using hydrogen produced from sunlight by the photoelectrolysis of water to reduce the gasified coal to methanol by the reaction sequence:

$$C+\frac{1}{2}O_2 \rightarrow CO \quad \text{followed by} \quad CO + 2H_2 \rightarrow CH_3OH \tag{1.3}$$

Despite the fact that it still produces net carbon dioxide emission, this approach has several advantages. First, the photoelectrolysis of water is much easier and more fully developed than the multielectron reduction of CO_2. Secondly, the high-temperature gas-phase catalytic chemistry is well known and fairly well optimized. Also, only one equivalent of CO_2 is emitted into the atmosphere from burning or using the methanol in a direct methanol fuel cell, as opposed to much three to as much as six when one tries to reverse the combustion in a coal burning plant. The gasification route may still be a pre-ferred method of liquid fuel production even if water is used as the reductant by employing water gas shift chemistry after gasification.

$$2C + O_2 \rightarrow 2CO$$

$$2CO + 2H_2O \rightarrow 2H_2 + 2CO_2 \tag{1.4}$$

$$CO + 2H_2 \rightarrow CH_3OH$$

In this approach, a minimum of 3 CO_2 molecules are released (considering that even these well-studied catalytic processes are not 100% efficient) after use of the liquid fuel, and ideally the gasification/fuel producing plant could be located near a geological formation that is acceptable for CO_2 sequestration. Therefore, any large-scale CO_2 reduction scheme for mitigation of atmospheric CO_2 emissions makes absolutely no sense until virtually all coal-burning plants are replaced with carbon-free power such as wind, nuclear or solar.

Considering possible sources of CO_2, while the reserves of fossil fuels are large, they are finite and there are climate change issues to be considered and so ultimately, to be sustainable we must get our CO_2 from the air. While > 400 ppm CO_2 has a potent impact as a global warming gas, it still is rather dilute and would require great expenditures of energy and capital to collect and concentrate it for reaction. Biomass is nature's way of collecting CO_2 from the atmosphere and generating reduced carbon that can be converted into fuels. However, typical bio-fuels have significant oxygen atom content, and pure hydrocarbons are preferred as fuels. Again, a better approach would be to pyrolyze biomass and use the pyrolysis oil as feed stock combined with PEC hydrogen for hydro-cracking/hydroforming to produce carbon-neutral fuels of the sort currently used in the transportation infrastructure. Huge amounts of hydrogen produced from methane are already used in the petroleum refining industry, and photoelec-tochemically produced hydrogen from water would also eliminate CO_2 emissions associated with this industry. Of course eliminating deforestation and replanting of forests is the most near term and effective way of CO_2 mitigation since photosynthesis by self-replicating trees is low tech and effective.

Photoelectrochemically generated hydrogen from water could also be useful in another large industrial process that releases CO_2, nitrogen fixation via the Haber Bosch process. This process already consumes 3–5% of the world's natural gas and 1–2% of the world's energy supply, and therefore renewable hydrogen would have large global impact.

$$N_2 + 3H_2 \rightarrow 2NH_3 \tag{1.5}$$

1.6 Summary and Conclusions

In summary, stable photoelectrochemical water splitting systems are the pref-erable target for photoelectrochemical energy conversion and, if made cheaply and with a high enough efficiency to be a cost-effective renewable hydrogen source, could realistically impact large-scale future energy conversion and storage. In order for this to happen, researchers must concentrate their efforts on discovering and improving stable materials that are capable of efficient solar photoelectrolysis. Since the hydrogen economy is the only long-term sustain-able renewable fuel-based energy alternative, the question is not if but when will the world begin its transition to this energy carrier. In addition to research and development, barriers, to this conversion include the availability and low cost of fossil fuels, especially coal and recently natural gas, and the lack of a hydrogen infrastructure as well as political and sociological issues. Perhaps some of the latter issues can be addressed by educating people about the true costs of continued exploitation of non-renewable reduced carbon resources, such as climate change, land and ecosystem destruction, air and water pollution and their detrimental impact on human health and wellbeing. One hopes that this awareness will inspire countries to shift their subsidies from non-renewable to sustainable energy systems such as solar driven hydrogen production. To quote Sinclair Lewis, "It is difficult to get a man to understand something when his pay check depends upon him not understanding it."

Some first steps that have been taken are exemplified by the recent increases in government funding of solar energy research and by the establishment of several centers focused on solar hydrogen production, including the Joint Center for Artificial Photosynthesis (JCAP) in California and the Korean Center for Artificial Photosynthesis (KCAP) in Korea as well as solar fuel research projects in the European Union, China and Japan. We can hope that this investment will result in research breakthroughs that will put us on the path to a world economy based on clean and sustainable energy and that photo-electrochemically produced hydrogen will be a large or at least contributing part of that economy. In Winston Churchill's words, "People and countries can do the right thing once they have exhausted all other alternatives."

Acknowledgement

Funding from the Division of Chemical Sciences, Geosciences, and Biosciences, Office of Basic Energy Sciences of the U.S. Department of Energy through Grant #DE-FG02-05ER15750 is acknowledged.

References

1. W. H. Brittain and P. J. Boddy, *J. Electrochem Soc.*, 1962, **109**, 574–582.
2. P. J. Boddy, *J. Electrochem. Soc.*, 1968, **115**, 199–203.
3. H. Gerischer, *J. Electrochem. Soc.*, 1966, **113**, 1174–1182.
4. H. Gerischer, Semiconductor photoelectrochemistry, in *Physical Chemistry: An Advanced Treatise*, ed. H. Eyring, D. Henderson, W. Yost, Academic Press, New York, 1970, pp. 463–542.
5. H. Gerischer, Solar photoelectrolysis with semiconductor electrodes, in *Solar Energy Conversion: Solid State Physics Aspects*, ed. B.O. Seraphin, pp. 115–172.
6. A. Fujishima and K. Honda, *Nature*, 1972, **238**, 37–38.
7. A. Fujishima and K. Honda, *Bull. Chem. Soc. Japan*, 1971, **44**, 1148–1150.
8. A. J. Nozik, *Nature*, 1975, **257**, 383–386.
9. O. Khaselev and J. A. Turner, *Science*, 1998, **280**, 425–427.
10. B. A. Parkinson, A. Heller and B. Miller, *Appl. Phys. Lett.*, 1978, **33**, 521–523.
11. C. Levy-Clement, R. Triboulet, J. Rioux, A. Etcheberry, S. Licht and R. Tenne, *J. Appl. Phys.*, 1985, **58**, 4703–4709.
12. J. F. Gibbons, G. W. Cogan, C. M. Gronet and N. Lewis, *Appl. Phys. Lett.*, 1984, **45**, 1095–1098.
13. G. Kline, K. Kam, D. Canfield and B. A. Parkinson, *Solar Energy Materials*, 1981, **4**, 301–308.
14. G. Hodes, D. Cahen, J. Manassen and M. David, *J. Electrochem. Soc.*, 1980, **127**, 2252–2254.
15. W. D. Johnston, Jr., H. J. Leamy, B. A. Parkinson, A. Heller and B. Miller, *J. Electrochem. Soc.*, 1980, **127**, 90–95.
16. B. O. Regan and M. Grätzel, *Nature*, 1991, **353**, 737–740.

17. Key World Energy Statistics 2012, International Energy Agency, www.iea.org.
18. http://solarfuelshub.org.
19. J. Turner, *Science*, 1991, **285**, 687–689.
20. K. Rajeshwar, *J. Appl. Electrochem.*, 2007, **37**, 765–787.
21. M. G. Walter, E. L. Warren, J. R. McKone, S. W. Boettche, Q. Mi, E. A. Santori and N. S. Lewis, *Chem. Rev.*, 2010, **110**, 6446–6473.
22. J. R. White, F. Fan and A. J. Bard, *J. Electrochem. Soc.*, 1985, **132**, 544–550.
23. T. F. O'Brien, V. Bommaraju, F. Hine, *Handbook of Chlor-Alkali Technology*, Springer Science, Boston, MA, 2005.
24. K. Izumiya, E. Akiyama, H. Habazaki, N. Kumagai, A. Kawashima and K. Hashimoto, *Electrochimica Acta*, 1998, **43**, 3303–3312.
25. B. A. Parkinson, *Acc. Chem. Res.*, 1984, **17**, 431–437.
26. S. S. Kocha and J. A. Turner, *J. Electrochem. Soc.*, 1995, **142**, 2625–2630.
27. H. S. Hilal and J. A. Turner, *Electrochim. Acta*, 2006, **51**, 6487–6497.
28. O. Khaselev, A. Bansal and J. A. Turner, *Int. J. Hydrogen Energy*, 2001, **26**, 127–132.
29. A. Bansal and J. A. Turner, *J. Phys. Chem. B*, 2000, **104**, 6591–6598.
30. M. Weber and M. Dignam, *Int. J. Hydrogen Energy*, 1986, **11**, 225–271.
31. M. Hanna and A. J. Nozik, *J. App. Phys.*, 2006, **100**, 74510.
32. J. B. Sambur, T. Novet and B. A. Parkinson, *Science*, 2010, **330**, 63–66.
33. O. E. Semonin, J. M. Luther, S. Choi, H.-Y. Chen, J. Gao, A. J. Nozik and M. C. Beard, *Science*, 2011, **334**, 1530–1533.
34. E. Osterloh, *Chem. Mater.*, 2008, **20**, 35–54.
35. M. Woodhouse, G. Herman and B. A. Parkinson, *Chem. Mater.*, 2005, **17**, 4318–4324.
36. T. Arai, Y. Konishi, Y. Iwasaki, H. Sugihara and K. Sayama, *J. Comb. Chem.*, 2007, **9**, 574–581.
37. T. F. Jaramillo, S.-H. Baeck, A. Kleiman-Shwarsctein, K.-S. Choi, G. D. Stucky and E. W. McFarland, *J. Comb. Chem.*, 2005, **7**, 264–271.
38. M. Woodhouse and B. A. Parkinson, *Chemistry of Materials*, 2008, **20**, 2495–2502.
39. J. He and B. A. Parkinson, *ACS Comb. Sci.*, 2011, **13**, 399–404.
40. W.-J. Yin, H. Tang, S.-H. Wei, M. M. Al-Jassim, J. A. Turner and Yanfa Yan, *Phys. Rev. B*, 2010, **82**, 045106.
41. C. B. Feng, W. J Yin, J. L. Nie, X. T. Zu, M. N. Huda, S. H. Wei, M. M. Al-Jassim, J. A. Turner and Y. F. Yan, *Applied Phys. Lett.*, 2012, **100**, 023901.
42. J. D. Perkins, T. R. Paudel, A. Zakutayev, P. F. Ndione, P. A. Parilla, D. L. Young, S. Lany, D. S. Ginley, A. Zunger, N. H. Perry, Y. Tang, M. Grayson, T. O. Mason, J. S. Bettinger, Y. Shi and M. F. Toney, *Phys. Rev. B.*, 2011, **84**, 205207.
43. B. D. James, G. N. Baum, J. Perez, K. N. Baum, *Technoeconomic Analysis of Photoelectrochemical (PEC) Hydrogen Production*, www1.eere.energy.gov/hydrogenandfuelcells/pdfs/pec_technoeconomic_analysis.pdf (December 2009).

CHAPTER 2

Kinetics and Mechanisms of Light-Driven Reactions at Semiconductor Electrodes: Principles and Techniques

LAURENCE PETER

Department of Chemistry, University of Bath, Bath BA2 7AY,
United Kingdom
Email: l.m.peter@bath.ac.uk

2.1 Introduction

This chapter examines a range of electrochemical, optical and other experimental methods that can be used to obtain information about the kinetics and mechanisms of electrode reactions driven by the capture of photogenerated minority carriers at the interface between an electrolyte and an illuminated semiconductor electrode. The main emphasis is on systems in which the semiconductor electrode is part of an electrochemical cell rather than isolated as a colloidal particle at which balanced anodic and cathodic reactions occur under illumination. The treatment is selective, with particular attention being paid to the reactions that take place during light-driven water splitting. A more general overview of techniques used in semiconductor photoelectrochemistry can be found in reference 1.[1] The present chapter begins by contrasting the differences between electrode reactions at metal electrodes and illuminated semiconductor electrodes, before going on to clarify the confusing range of

RSC Energy and Environment Series No. 9
Photoelectrochemical Water Splitting: Materials, Processes and Architectures
Edited by Hans-Joachim Lewerenz and Laurence Peter
Published by the Royal Society of Chemistry, www.rsc.org

different rate constants that are used in the literature to describe the kinetics of electrode reactions involving photogenerated minority carriers. A relatively simple approach is then used to show how information about rate constants for minority carrier reactions at illuminated semiconductor electrodes can be derived from measurements involving perturbing either the illumination or the electrode potential. This approach is illustrated with examples of light-driven oxygen evolution at n-type hematite electrodes and hydrogen evolution at p-type silicon electrodes. The chapter finishes with a discussion of recent optical measurements on thin hematite films, which provide information about reaction mechanisms, intermediates and kinetics that is difficult to obtain from electrical measurements alone.

2.2   Conventional and Nanostructured Photoelectrodes

2.2.1   Conventional Semiconductor Electrodes

Pioneering studies in the photoelectrochemistry of semiconductors were carried out in the 1960s using single crystal samples with well-defined oriented surfaces. Early work on germanium and silicon was complicated by the high reactivity of these elements in water, and a clearer picture began to emerge only when single crystal ZnO electrodes[2,3] were used, since samples could be cleaved and etched to produce well-defined polar surfaces with a low density of surface defects. CdS[4,5] was also widely studied at this time, and interest moved later from these II-VI semiconductors to III-V materials such as GaAs, GaP and InP,[6,7] which are used in solid-state electronic devices. Whereas these II-VI and III-V materials were readily available in semiconductor quality, transition metal oxide single crystals of interest for water splitting such as rutile (TiO_2)[8,9] and hematite $(\alpha\text{-}Fe_2O_3)$[10,11] proved much more difficult to prepare in pure form with controlled doping levels. The photoelectrochemical behaviour of single crystals of both of these oxides is strongly influenced by bulk and surface states that introduce non-ideality. More recent work on water splitting has focussed on compact or nanostructured polycrystalline films of oxides such as TiO_2, WO_3, Fe_2O_3 and $BiVO_4$. Understanding these more complicated systems involves consideration of the influence of additional factors such as surface morphology, grain boundaries and electrolyte screening in porous structures.

2.2.2   Potential and Charge Distribution at the Semiconductor-Electrolyte Junction

The fundamental properties of the semiconductor-electrolyte junction have been reviewed in a number of text books.[12–14] This section is therefore restricted to a very brief overview of some essential concepts that have been developed from work on well-defined single crystal semiconductor electrodes.

A convenient reference point for describing the potential and charge distribution at the junction between a semiconductor electrode and the solution is the flatband potential, U_{fb}. This is the electrode potential at which there is no electrical field in the semiconductor side of the junction, so that the conduction and valence band energies are constant. If a potential is applied to the electrode so as to withdraw majority carriers from the junction, a depletion or space charge region is formed. For n-type semiconductors, a positive space charge region is formed when the applied potential is made more positive than U_{fb}: the converse is true for p-type semiconductors. Since the space charge is due to immobile ionized donor (for n-type) or acceptor (for p-type) species, the width of the space charge region, W_{sc}, depends on the doping density N, the relative permittivity, ε, and the potential drop across the space charge region, $\Delta\phi_{sc}$, which corresponds to a band bending of $q\Delta\phi_{sc}$.

$$W_{sc} = \left(\frac{2\Delta\phi_{sc}\varepsilon\varepsilon_0}{qN}\right)^{1/2} \tag{2.1}$$

The charge in the space charge region in the semiconductor is balanced by an ionic charge of opposite sign in the electrolyte. For concentrated electrolytes, this charge is effectively located at the outer Helmholtz plane of the electrical double layer. The semiconductor-electrolyte junction can therefore be modelled as a space charge capacitance C_{sc} in series with a Helmholtz capacitance, C_H. The dependence of C_{sc} on the potential drop across the space charge region ($\Delta\phi_{sc}$) is given by the Mott–Schottky equation (here for an n-type electrode)

$$\frac{1}{C_{sc}^2} = \frac{2}{N_d\varepsilon\varepsilon_0}\left[\Delta\phi_{sc} - \frac{k_BT}{q}\right] \tag{2.2}$$

For low or moderately doped semiconductors under depletion conditions, $C_H \gg C_{sc}$, so that it is often assumed that changes in electrode potential appear mainly across the space charge region. In this case, $\Delta\phi_{sc}$ can be replaced by $U - U_{fb}$, where U is the applied potential. However, for many of the materials that are studied in the context of water splitting, this approximation may not be valid for several reasons. Firstly, C_{sc} may be comparable with C_H if the material is doped to levels above 10^{24} m^{-3}, as is often the case for non-stoichiometric oxides. Secondly, the derivation of eqn (2.2) involves an assumption that no electronic charge is stored at the surface of the semiconductor. However, many semiconductor electrodes have imperfect surfaces with high densities of surface states that can be charged by electron exchange with the semiconductor, giving rise to changes in the potential drop, $\Delta\phi_H$, across the Helmholtz layer. In extreme cases, the semiconductor may behave more like a metal, with nearly all of the change in potential drop appearing across the Helmholtz layer as the potential is varied. This phenomenon is referred to as Fermi level pinning.[15] Even in the case of hydrogen terminated-silicon surfaces, which have very low densities of surface states, the build-up of photogenerated minority carriers near the surface as a consequence of slow electron transfer kinetics results in changes in $\Delta\phi_H$ that can be detected by photocapacitance measurements.[16] In this chapter, this phenomenon is referred to as light-induced Fermi level pinning, to distinguish it from Fermi level pinning in the dark arising from charging of surface states by majority carriers.

2.2.3 Collection of Minority Carriers at the Illuminated Semiconductor-Electrolyte Junction

In the ideal case where minority carriers reaching the edge of the space charge region are transferred via the interface to the solution redox species without loss by recombination, the generation-collection problem can be formulated in terms of three characteristic lengths. These are the width of the space charge region, W_{sc}, the penetration depth, $1/\alpha(\lambda)$, of the light incident on the surface of the electrode and L, the minority carrier diffusion length. Here, α is the absorption coefficient of the semiconductor at the wavelength, λ. The minority carrier diffusion length (L_p in the case of holes in an n-type semiconductor) is determined by the mobility, μ, and bulk lifetime, τ, of minority carriers.

$$L_p = \sqrt{\frac{k_B T}{q} \mu_p \tau_p} \qquad (2.3)$$

It is important to realize that the minority carrier lifetime in eqn (2.3) is defined for electron-hole recombination in the bulk semiconductor, where there is a large excess of majority carriers. The diffusion length concept cannot be used when recombination occurs at the semiconductor surface or in the space charge region. The diffusion length is also difficult to define in the case of mesoporous semiconductors (see section 2.2.4), where the majority carrier density (and hence the rate of recombination) is controlled by the external bias voltage. The three characteristic lengths are illustrated in Figure 2.1.

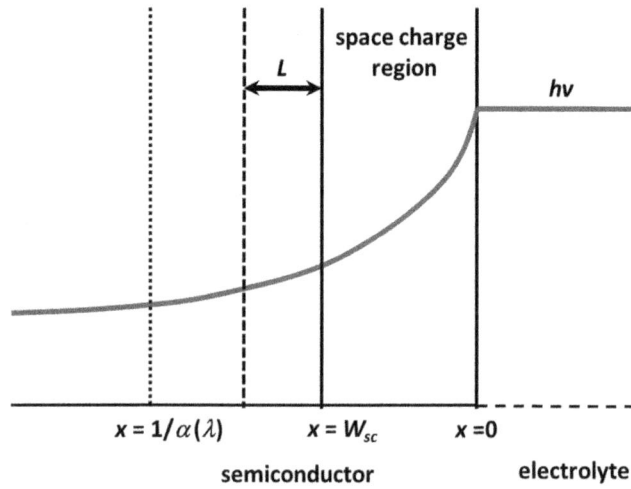

Figure 2.1   The three characteristic lengths appearing in the derivation of the Gärtner equation. W_{sc} – the width of the space charge region; L – the minority carrier diffusion length; $1/\alpha(\lambda)$ – the penetration depth of the light.

The solution of the generation-collection problem is known as the Gärtner equation,[17,18] which gives the external quantum efficiency (or IPCE: incident photon to current conversion efficiency) as

$$IPCE = \frac{j_{photo}}{qI_0} = \frac{g}{I_0} = 1 - \exp - \left(\frac{\alpha W_{sc}}{1 + L}\right) \qquad (2.4)$$

where I_0 is the incident photon flux, g is the minority carrier flux, W_{sc} is given by eqn (2.1) and L by eqn (2.3). The ideal photocurrent-voltage behaviour predicted by the Gärtner equation is only seen in semiconductor/electrolyte systems involving fast outer sphere electron transfer processes. If electron transfer is slower, as is the case for the multistep reactions involved in oxygen or hydrogen evolution, the concentration of minority carriers builds up as carriers 'queue up' close to the interface. As a consequence, recombination of electron hole pairs at the surface and/or in the space charge region competes with charge transfer, reducing the external quantum efficiency, particularly for low values of band bending. Recombination, which is a major limiting factor in water splitting reactions at semiconductor electrodes, is considered in section 2.3.4.

2.2.4 Nanostructured and Mesoporous Electrodes

Nanostructured and mesoporous electrodes differ in several important respects from the flat surfaces described in the preceding two sections. Firstly, the characteristic length scale of the nanostructured electrode may be substantially smaller than the Debye length L_D, which is given by

$$L_D = \left(\frac{\varepsilon \varepsilon_0 k_B T}{2q^2 N}\right)^{1/2} \qquad (2.5)$$

where N is the dopant concentration. This means that the size of the elements of the nanostructure is too small to allow development of substantial band bending. For example, if we consider a spherical particle of a doped semi-conductor, solution of the Maxwell Boltzmann equation gives the maximum potential drop between the centre of the particle and the surface as

$$\Delta\phi_{max} = \frac{k_B T}{6q} \left[\frac{r_0}{L_D}\right]^2 = \frac{1}{3} \frac{r_0^2 q N}{\varepsilon \varepsilon_0} \qquad (2.6)$$

To illustrate the effect of size, we can consider a 20 nm diameter fully depleted anatase particle with a doping density of $10^{23}\,\text{m}^{-3}$ and a relative permittivity of 40. Eqn (2.6) indicates that the maximum band bending in this case is only 4.5 meV. This is smaller than the average thermal energy $k_B T$, so the effects of band bending can be neglected. In this case, photoexcited charge carriers reach the surface by diffusion. Often the minority carrier diffusion length (eqn (2.3)) is cited as a relevant parameter for nanostructured oxide electrodes, leading to the suggestion that recombination losses can be min-imized by ensuring that the characteristic length scale of the nanostructured electrode is smaller than the minority carrier diffusion length. However, this line of reasoning neglects the fact that the pseudo first order minority carrier lifetime depends linearly on the majority carrier density, which changes when the Fermi level is altered by changing the applied potential. In the case of a mesoporous n-type semiconductor, the free electron density falls rapidly as the Fermi level moves away from the conduction band with increasingly positive potential, and for large applied bias voltages recombination may no longer be pseudo first order, since the electron and hole concentrations become similar. Furthermore, if interfacial minority carrier transfer is a slow process – as it is in the case of light-driven oxygen evolution, for example – minority carriers may be scattered back many times from the interface before being transferred or lost by recombination. Unfortunately, no satisfactory description of electron-hole recombination in nanostructured oxide electrodes is available currently. The development of theoretical descriptions will need to take into account that recombination may involve trapped carriers; for example, mobile holes and trapped electrons or vice versa.

Similar considerations regarding band bending apply to other nanostructured electrodes such as nanorods or nanotubes: in this case, it is the smallest characteristic dimension that is relevant. If the nanostructured electrode has a more complex structure, then small features may be almost field-free, whereas significant electrical fields will be present in the space charge region that can be developed in larger features.[19,20] Figure 2.2 illustrates this point. The spherical nanoparticle (a) and the nanorods (b) can both be characterised by their radius. Whether they have a significant electrical field normal to the surface will depend on their size and doping density. In the case of highly doped nanorods, for example, separation of electron hole pairs may be assisted by the electrical field perpendicular to the long axis provided that the rod radius is larger than W_{sc}. By contrast the structure shown in (c) is more complicated. The larger parts could allow development of a space charge layer in which the electrical field assists electron-hole separation, whereas the smaller features

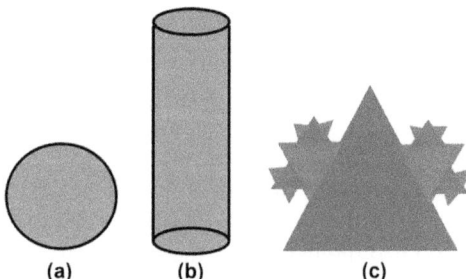

(a)   (b)   (c)

Figure 2.2   Schematic illustration of nanostructures: (a) spherical nanoparticles in which band bending can often be ignored; (b) larger nanorods in which space charge effects may be important in highly doped materials such as ZnO, assisting separation of electron hole pairs; (c) complex structures with different length scales where space charge effects may be confined to the largest feature dimensions, with collection of carriers taking place by diffusion in the smallest features.

could be field-free, so that diffusion is the dominant transport mechanism. The schematic structure in Figure 2.2c highlights the fact that nanostructured systems with different characteristic length scales are likely to show complex transport and electron transfer behaviour, making fundamental studies difficult. For this reason, it may be better to use plane surfaces for fundamental studies, in spite of the fact that nanostructured electrodes often perform much better.[21]

2.3   Rate Constants and Reaction Orders

2.3.1   Electrode Kinetics at Metal Electrodes

A large range of electrochemical techniques has been developed to study of the kinetics of electrode processes at metal electrodes.[22] Transient and periodic methods rely on perturbation of the electrode potential with different time-dependent functions (*e.g.* step, sweep, and sinusoidal modulation). In the case of metal electrodes, the potential perturbation affects the rate constants for electron transfer directly because the free energy of activation for electron transfer depends on electrode potential. The reason for this is the fact that changes in applied potential appear across the Helmholtz layer. The potential dependence of the rate constants for electron transfer at metal electrodes are given by the well-known expressions

$$\overrightarrow{k} = k^0 \exp\left[\frac{-\alpha nq(E - E^0)}{k_B T}\right] \quad \overleftarrow{k} = k^0 \exp\left[\frac{(1 - \alpha)nq(E - E^0)}{k_B T}\right] \quad (2.7)$$

Here $\overrightarrow{k}$ and $\overleftarrow{k}$ are the rate constants for the transfer of electrons from the electrode (cathodic reaction) and to the electrode (anodic reaction), respectively, and k^0 is the standard heterogeneous rate constant. E is the applied potential, E^0 is the standard reduction potential of the electrode process and α is the cathodic transfer coefficient. For a redox reaction

$$O + ne^- \rightleftarrows R \quad (2.8)$$

the net current density is given by

$$j = \overleftarrow{j} + \overrightarrow{j} = nq\left(\overleftarrow{k}R - \overrightarrow{k}O\right) = nqk^0\left[R\exp\left[\frac{(1 - \alpha)nq(E - E^0)}{k_B T}\right]\right.$$

$$\left. -O\exp\left[\frac{-\alpha nq(E - E^0)}{k_B T}\right]\right] \quad (2.9)$$

where R and O are the concentrations of electron donor (reduced) and electron acceptor (oxidized) species at the electrode surface (these local values may be determined by mass transport of reacting species).

An important concept in electrode kinetics is the overpotential, η, which is the difference between the applied potential and the equilibrium electrode potential defined by the Nernst equation. Eqn (2.9) – rewritten in terms of the overpotential – is the well-known Butler Volmer equation. It is important to realize that in the case of electrode reactions at metal electrodes, the additional free energy associated with the overpotential directly influences the rate constants for forward and reverse electron transfer. As we shall see, the concept of overpotential can also be applied to electrode processes such as oxygen evolution taking place at illuminated semiconductors, but in this case it is mainly the concentration term in the rate equation that is affected and not the rate constant.

2.3.2 Kinetics of Minority Carrier Reactions at Semiconductor Electrodes

It can be seen that the standard heterogeneous rate constant k^0 in eqn (2.7) has units of m s^{-1}. This reflects the fact that the concentration of electrons in the metal does not appear in the rate expression because it is assumed to be independent of electrode potential, *i.e.* changes in potential only affect the rate constants. The situation in the case of electrode reactions at wide band gap semiconductor electrodes is different. In the dark, the semiconductor/ electrolyte junction is rectifying, and current only flows under conditions in which there is a significant concentration of majority carriers at the interface (accumulation conditions). Both the concentration of majority carriers and the potential drop in the Helmholtz layer depend on potential, so that eqn (2.7) does not provide a proper description of the rate of an electrode process at a semiconductor electrode in the dark. The kinetics of dark reactions at semiconductor electrodes is not considered further here; instead the chapter focusses on reactions at illuminated semiconductor electrodes under conditions in which photogenerated minority carriers are responsible for the electrode reaction. As shown below, we can distinguish between reactions of majority and minority carriers in this case, and the rate expressions need to take into account the fact that the concentrations of these reactant species depend on applied potential and illumination conditions.

For any minority carrier reaction at an illuminated semiconductor photoelectrode, the net rate of electron or hole transfer can be inferred directly from the current flowing in the external circuit. However, the current gives no information about the rate constants for electron or hole transfer. In the most general case, the kinetics of a light-driven process at an illuminated n-type semiconductor electrode in the absence of surface recombination can be described by the general expression

$$j = nqk_{et}p^a R^b \tag{2.10}$$

Here, j is the current density, n is the number of electrons transferred in the electrode reaction. k_{et} is the rate constant for the electron (hole) transfer process, p is the hole density at the surface of the electrode and R is the concentration of electron donor species in solution. a and b are the reaction orders with respect to holes and electron donor species, respectively. For p-type photoelectrodes, p^a is replaced by n^a, where n is the concentration of electrons at the surface, and R^b is replaced by O^b, where O is the concentration of electron acceptor species in solution. In the case of water splitting, water (or OH^-) is the electron donor for light-driven oxygen evolution, and water (or H^+) is the electron acceptor (for light-driven hydrogen evolution). If the reaction orders a and b are both unity and the concentrations p and R are expressed in m^{-3}, the rate constant k_{et} in eqn (2.10) has units $m^4\,s^{-1}$. In the case of water splitting, the concentration R can be treated as constant to a good approximation, so that the concentration of minority carriers at the surface is the main variable. It is usually assumed that k_{et} is independent of electrode potential provided that the semiconductor is not highly doped and that there is not a high density of surface states. The reasoning behind this assumption is that the majority of any changes in electrode potential appear across the semiconductor and not across the Helmholtz layer because the space charge capacitance C_{sc} is much smaller than the Helmholtz capacitance C_H. However, the approximation will not be valid if the semiconductor is so highly doped that C_{sc} becomes comparable with C_H or if charging of surface states or minority carrier build-up causes changes in the potential drop in the Helmholtz layer (Fermi level pinning). In this case, changes in potential influence not only the surface concentration of minority carriers, but also the activation energy for electron or hole transfer.

For simple outer sphere redox reactions, k_{et} is given by[23]

$$k_{et} = \nu_n \kappa_n \kappa_{el} \tag{2.11}$$

Here, ν_n is a vibrational frequency along the reaction coordinate (ca. $10^{13}\,s^{-1}$) and κ_n, κ_{el} are nuclear and electronic coupling terms, respectively. The nuclear term depends of the driving force ΔG^0 for the electron transfer reaction, and on the reorganization energy, λ.

$$\kappa_n = \exp-\left[\frac{(\Delta G^0 + \lambda)^2}{4\lambda k_B T}\right] \tag{2.12}$$

The electronic coupling term κ_{el} is estimated to be of the order of $10^{-38}\,m^4$ for a typical semiconductor/electrolyte system.[24] The maximum value for κ_n (1) is obtained when $\Delta G^0 = -\lambda$. This corresponds to a maximum value of $k_{et} = 10^{-25}\,m^4\,s^{-1}$. However, the reactions involved in water splitting involve bond breaking/formation and surface-bound intermediates. The rate-determining steps leading to molecular oxygen formation have high activation energies, so that the rate constants will be many orders of magnitude lower than for outer sphere redox systems.

The kinetics of light-driven minority carrier reactions at semiconductor electrodes can also be formulated in a way that is consistent with the approach used for solid state junctions. In this case, the electrolyte species is treated as a state that can capture minority carriers arriving at the surface. The rate of the hole capture process (and the corresponding current density) for an n-type electrode is given by

$$j = q v_{th} p N_R \sigma_p \tag{2.13}$$

Here, $v_{th} = 3k_B T/m^*$ is the thermal velocity ($\approx 10^5$ m s^{-1} for $m^* = m_e$) of holes with effective mass m^*, p is the concentration of holes at the surface (m^{-3}), N_R is the surface concentration (m^{-2}) of the electron donor species and σ_p is the hole capture cross section (m^2). N_R is related to R in eqn (2.10) by $N_R = R\delta_H$, where δ_H is a reaction layer thickness of the order of 1 nm. The rate constant k_{et} in eqn (2.10) is therefore related to the thermal velocity and capture cross section by

$$k_{et} = \delta_H v_{th} \sigma_p \tag{2.14}$$

It follows from eqn (2.14) that the capture cross section is of the order of 10^{-19} m^2 for fast outer sphere electron transfer reactions ($k_{et} \approx 10^{-25}$ m^4 s^{-1}). The capture cross sections for reactions involving large re-organization energies will be many orders of magnitude smaller than this.

A useful approximation involves formulating the rate of minority carrier reactions in terms of the surface concentration p_{surf} (m^{-2}) of carriers rather than the volume concentration (m^{-3}) at the surface.[25] At the same time, the reaction order is reduced by considering that the concentration of solution species R is constant. So, for an n-type semiconductor under illumination

$$j = q k' p_{surf} R = q k p_{surf} \tag{2.15}$$

where $p_{surf} = p\delta_{sc}$, δ_{sc} is a reaction length in the semiconductor (ca.1 nm) and $k = k'R$. If $n = a = b = 1$, the pseudo-first order rate constant k (units s^{-1}) in eqn (2.15) is related to the rate constant k_{et} by

$$k = \frac{k_{et}}{\delta_{sc}} R \tag{2.16}$$

As discussed below, this rate constant defines a time constant $\tau = 1/k$ that appears in the time and frequency dependent response of the photoelectrode. If we assume a concentration of redox species equal to 10^{-2} M ($\equiv 6 \times 10^{24}$ m^{-3}), for example, and the upper limit of $k_{et} = 10^{-25}$ m^4 s^{-1}, the upper limit for k in eqn (2.16) is 6×10^8 s^{-1}, which corresponds to a characteristic relaxation time constant of around 1 ns. The fact that the relaxation times observed for water splitting reactions are many orders of magnitude longer than this (milliseconds to seconds) reflects the large activation energies for individual steps in the multistep reactions.

2.3.3 The Concept of Overpotential for Minority Carrier Reactions

The free energies of electrons and holes in an illuminated semiconductor are defined by the respective quasi Fermi levels, $_nE_F$ and $_pE_F$. Illumination of the photoelectrode with light with energy $hv > E_g$, creates non-equilibrium steady-state populations of electrons (n) and holes (p). The electron and hole concentrations are given by

$$n = N_c \left(\frac{1}{1 + \exp\left(\frac{E_c - {_nE_F}}{k_B T}\right)} \right) \quad p = N_v \left(\frac{1}{1 + \left(\frac{_pE_F - E_v}{k_B T}\right)} \right) \quad (2.17)$$

where E_c, E_v are the conduction and valence band energies, and N_c, N_v are the conduction and valence band densities of states. If we consider an n-type semiconductor, the equilibrium hole density in the dark is very low since $n_{eq} \gg p_{eq}$. Illumination creates equal excess concentrations (Δn and Δp) of electrons in the conduction band and holes in the valence band. Even for moderately-doped n-type semiconductors, $n_{eq} + \Delta n \approx n_{eq}$, so that $_nE_F$ remains close the equilibrium (dark) Fermi level E_F. However, since $p_{eq} + \Delta p \gg p_{eq}$, the quasi Fermi level for holes moves towards the valence band, increasing the driving force for oxidation reactions. The driving force for water oxidation is determined by the free energy difference between $_pE_F$ at the surface and the equilibrium redox Fermi level for the O_2/H_2O couple. This situation is illustrated in Figure 2.3 for the case of an n-type electrode that has been biased into depletion and where slow electron transfer kinetics leads to a build-up of holes at the interface, driving $_pE_F$ below $E_F(O_2/H_2O)$.

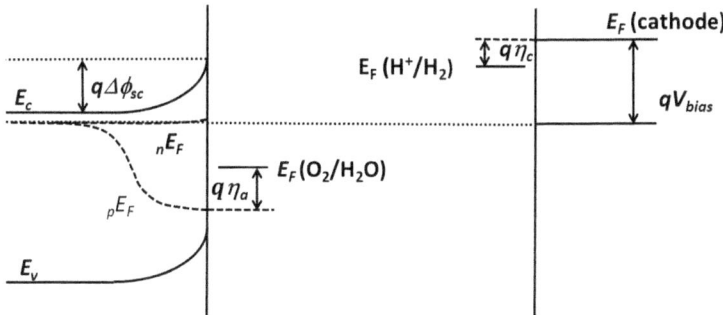

Figure 2.3 Energy diagram for a photoelectrolysis cell operating with an external bias voltage V_{bias}. The n-type semiconductor photoanode is biased into depletion, causing band bending $q\Delta\phi_{sc}$. The splitting of the electron and hole Fermi levels under illumination is shown along with the resulting overpotential, η_a, for oxygen evolution. The cathodic overpotential for hydrogen evolution at the metal cathode has been exaggerated for clarity.

The difference in free energies ΔE_F, which is a measure of the driving force, defines the overpotential $\eta_a = \Delta E_F / q$. Figure 2.3 also illustrates the energetics of the cathode, where application of a bias voltage has raised the Fermi level of the metal cathode sufficiently far above the equilibrium Fermi level for the H^+/H_2 couple to generate hydrogen.

It should be noted at this point that the concept of 'overpotential' is different at the two electrodes in the photoelectrolysis cell shown in Figure 2.3. Provided that the photoanode is not too highly doped and behaves ideally, the change in free energy of holes ($_pE_F$) is associated with the entropic (*i.e.* concentration) part of the free energy: the internal energy contribution remains constant, provided that the potential drop across the Helmholtz layer does not change. By contrast, the change in free energy of electrons (Fermi level shift) in the metal cathode is associated with the change in potential drop across the Helmholtz layer, which affects the internal energy of the system.

2.3.4 Competition between Charge Transfer and Recombination

So far, only interfacial electron transfer reactions involving photogenerated minority carriers have been considered. Minority carrier lifetimes in bulk semiconductors range from milliseconds for very pure low-doped silicon wafers to picoseconds for impure and polycrystalline materials. Bulk recombination is generally a (pseudo) first order process because the concentration of majority carriers is much higher than the concentration of minority carriers. This means that we can define a minority carrier lifetime, τ_{min}. The low τ_{min} values for many practical water-splitting photoelectrodes means that minority carriers move only a short distance before recombining with majority carriers.

In the neutral bulk region of an n-type semiconductor electrode, electron-hole recombination determines the minority carrier diffusion length defined by eqn (2.3). However, recombination in the space charge and surface regions requires a different approach. The model of depletion layer photocurrents developed originally by Gärtner[17] for solid state junctions and extended to semiconductor/electrolyte junctions by Wilson[26] and by Butler[18] assumes that all minority carriers holes that reach the edge of the space charge region at $x = W$ are transferred to the contacting phase (*i.e.* no recombination occurs in the space charge region or at the surface). This assumption gives rise to the boundary condition that the excess hole concentration Δp at $x = W$ is zero. However, this approximation will no longer be valid if electron transfer at the interface is slow, leading to a build-up of holes. A more realistic approach that includes the effects of recombination of electrons and holes in the space charge region as well as the influence of the rate of interfacial charge transfer has been given by Reichmann,[27] and El Guibaly and Colbow[28] have extended this treatment to include surface recombination. The increasing complexity of these models and the absence of reliable values for the variables probably explain why these extensions of the basic Gärtner model have been almost entirely ignored in the later literature. More recently, numerical methods have been used to model the semiconductor/electrolyte junction under steady-state conditions.[29,30] In this chapter, a simpler phenomenological approach[25] is described that includes surface recombination, but not space charge recombination. In spite of this simplification, the analysis leads to a useful semi-quantitative understanding of the factors that control the kinetics of light-driven electrode reactions.

Surface recombination occurs at free semiconductor surfaces as well as in different types of semiconductor junctions. In solid state physics, recombination is normally described in terms of a surface recombination velocity and equilibrium electron (Hall Shockley Read) statistics.[31] The process involves a state that can capture both holes and electrons, promoting recombination. By contrast, non-equilibrium kinetic descriptions of the competition between surface recombination and interfacial electron transfer have been used extensively for semiconductor electrodes. The basic processes considered are illustrated in Figure 2.4 for a n-type photoelectrode, which shows the capture of holes (minority carriers) and electrons (majority carriers) as well as hole transfer from to solution redox species.

The model is a simplified version of an earlier version in which hole transfer from the valence band and from a surface state were considered separately.[32] k_{tr} is the rate constant for hole transfer and k_{rec} is the rate constant for electron/hole recombination at the surface. Initially, we assume that k_{tr} is independent of applied potential, whereas k_{rec} decreases with increased band bending as a

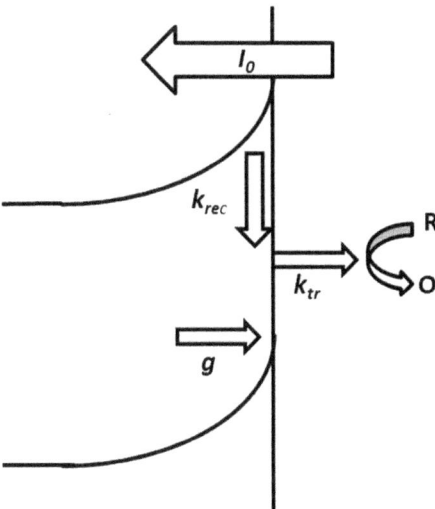

Figure 2.4   Simplified kinetic scheme showing the competition between hole transfer (rate constant k_{tr}) and recombination (rate constant k_{rec}). I_0 is the incident photon flux and g is the flux of holes into the surface predicted by the Gärtner equation (eqn (2.4)).

consequence of the potential dependence of the electron concentration at the surface.

$$k_{rec} = \sigma_n v_{th} n_{surf} = \sigma v_{th} N_d \exp - \left[\frac{\Delta\phi_{sc}}{k_B T} \right] \qquad (2.18)$$

where σ_n is the capture cross section for electrons (majority carriers). The surface concentration of trapped holes is given by

$$p_{surf} = \frac{g}{k_{tr} + k_{rec}} \qquad (2.19)$$

where g is the flux of holes into the surface defined by the Gärtner equation (cf. eqn (2.4)). The recombination current is

$$j_{rec} = qk_{rec}p_{surf} = qg\frac{k_{rec}}{k_{tr} + k_{rec}} \qquad (2.20)$$

It follows that the normalized photocurrent density (IPCE: cf. eqn (2.4)) is given by

$$IPCE = \frac{j_{photo}}{qI_0} = \frac{qg - j_{rec}}{qI_0} = \frac{g}{I_0}\left(\frac{k_{tr}}{k_{tr} + k_{rec}} \right) \qquad (2.21)$$

The accumulation of holes at the surface of the semiconductor alters the potential drop across the Helmholtz layer. In this chapter, this phenomenon is referred to as light-induced Fermi level pinning. If we assume that holes are located (or trapped) at the surface, the light-induced change in potential drop across the Helmholtz layer is given by

$$\Delta V_H = \frac{qp_{surf}}{C_H + C_{sc}} \qquad (2.22)$$

At constant applied potential, this increase in potential drop across the Helmholtz layer has to be balanced by a corresponding decrease in $\Delta\phi_{sc}$.

Figure 2.5 illustrates the effects of surface recombination and minority carrier build-up on the current voltage characteristics for a case in which hole transfer at the interface is slow ($k_{tr} = 10 \text{ s}^{-1}$). Figure 2.5a compares the ideal IPCE *vs.* applied potential plot predicted by the Gärtner equation with the displaced plot expected when surface recombination and hole accumulation occur. It can be seen that surface recombination consumes all of the photo-generated holes until the applied potential is increased sufficiently to allow hole transfer to begin to compete effectively (photocurrent onset). The build-up of holes near the semiconductor surface changes in the potential drop in the Helmholtz layer, and Figure 2.5b shows the flattening of in the variation of $\Delta\phi_{sc}$ with applied potential that is characteristic of light-induced Fermi level pinning.

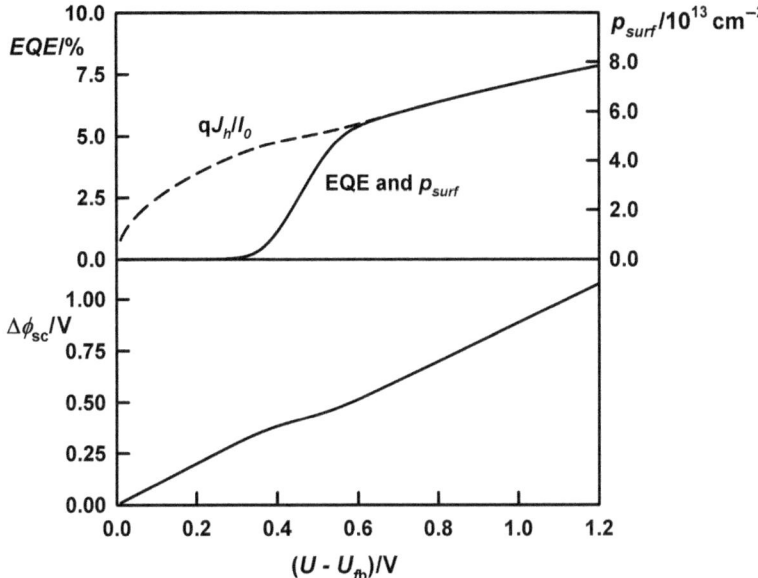

Figure 2.5 Comparison of normalized photocurrent (EQW) voltage plot calculated from the Gärtner equation (no surface recombination) with plot calculated for slow hole transfer and surface recombination using the model described in the text. The lower plot shows how the potential drop across the space charge region varies with applied potential. Note the inflection around 0.5 V (light-induced Fermi level pinning) due to the build-up of holes at the interface in the photocurrent onset region. Values used in the calculation: $k_{tr} = 10\,s^{-1}$, $\alpha = 1.5 \times 10^5\,cm^{-1}$, $N_d = 10^{19}\,cm^{-3}$, $\varepsilon = 25$, $C_H = 100\,\mu F\,cm^{-2}$, $\sigma_n = 10^{-16}\,cm^2$, $v_{th} = 10^5\,cm\,s^{-1}$, $I_0 = 10^{16}\,cm^{-2}\,s^{-1}$.

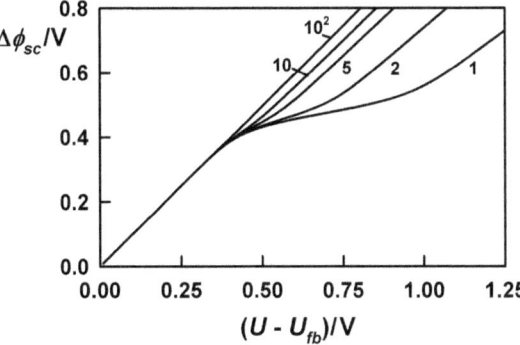

Figure 2.6 Plot showing the influence of the rate constant for hole transfer on light-induced Fermi level pinning. Values of k_{tr} (s^{-1}) are shown on the plot. Other values used in the calculation are the same as for Figure 2.5.

It is clear from the preceding discussion that enhanced surface recombination and light-induced Fermi level pinning are a direct consequence of slow minority carrier transfer. This is illustrated by Figure 2.6, which shows that increasing k_{tr} by only a factor of 10 is predicted to remove light-induced Fermi level pinning.

2.4   Determination of Rate Constants from Transient Photocurrents

As shown above, surface recombination displaces the onset of the photocurrent relative to the flatband potential. The effects of surface recombination are also evident in the photocurrent response to chopped illumination. In the photocurrent onset region, the photocurrent response shows a characteristic decay from a 'spike' to a steady state during the illumination period, followed by an overshoot and decay back to zero during the dark period. An example of this kind of response in the case of a hematite photoanode is shown in Figure 2.7.

The explanation of the spikes is as follows.[33] When the light is switched on, photogenerated electron/hole pairs are separated rapidly, charging the space charge capacitance. However, the corresponding 'instantaneous' displacement or charging current is not a measure of charge transfer across the interface. Under continued illumination during the 'on' period, the concentration of holes at the surface builds up towards a steady state value, which is reached when the rate of arrival of holes is balanced by the combined rates of interfacial transfer and recombination. As the hole concentration builds up, increasing numbers of electrons begin to flow to the surface, where recombination takes place. This flux of electrons corresponds to a current of opposite sign to the hole current. In the steady state, the measured photocurrent is the sum of the hole and electron currents. When the light is switched off, the photo-induced hole displacement current falls abruptly to zero, and the current in the external circuit changes sign as electrons continue to move towards the surface to recombine with the remaining holes. Since the recombination current is linearly proportional to the surface hole concentration, the cathodic current overshoot

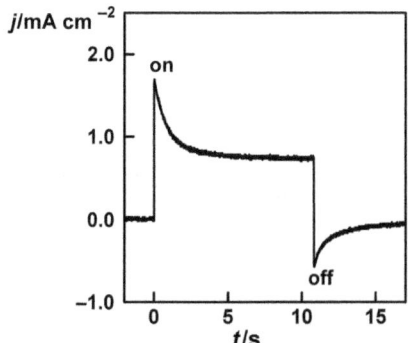

Figure 2.7   Normalized transient photocurrent response of thin film α-Fe$_2$O$_3$ electrode at 0 V *vs.* Ag|AgCl showing the decay and overshoot characteristic of surface electron-hole recombination. Electrolyte 1.0 M NaOH. Illumination 455 nm ca. 11 mW cm^{-2}.

reflects the decay of the surface hole concentration by recombination and interfacial transfer.

The time-dependent solution of the problem for the 'on' transient is given by

$$\frac{j(t) - j(\infty)}{j(0) - j(\infty)} = e^{-t/\tau} \tag{2.23a}$$

Here, $\tau = (k_{tr} + k_{rec})^{-1}$, $j(0)$ is the instantaneous photocurrent and $j(\infty)$ is the steady-state photocurrent given by

$$\frac{j(\infty)}{j(0)} = \frac{k_{tr}}{(k_{tr} + k_{rec})} \tag{2.23b}$$

The steady state current is expected to increase as the potential is made more positive since the surface concentration of electrons – and hence k_{rec} – decrease as the band bending increases (cf. eqn (2.8)). When the light is switched off at time t_0, the overshoot decays exponentially back to zero.

$$\frac{-j(t > t_0)}{j(0) - j(t_0)} = e^{-(t-t_0)/\tau} \tag{2.23c}$$

Typical decay and overshoot transients are illustrated in in Figure 2.8a, which has been calculated for $k_{tr} = k_{rec} = 1\,\text{s}^{-1}$. The corresponding time dependence of the surface hole concentration is plotted in Figure 2.8b.

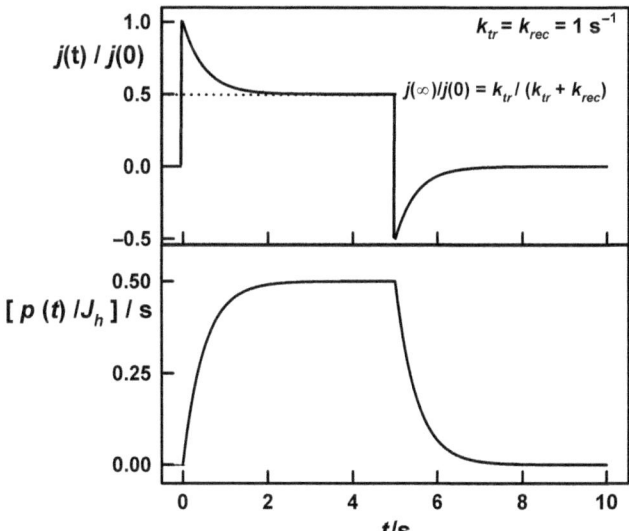

Figure 2.8   The upper plot shows a normalized photocurrent transient calculated for $k_{tr} = k_{rec} = 1\,\text{s}^{-1}$. Note that the steady state photocurrent $j(\infty)$ is determined by the competition between charge transfer and recombination, and the decay time constant is determined by $(k_{tr} + k_{rec})^{-1}$. Values of k_{tr} and k_{rec} can be obtained by analyzing experimental photocurrent transients. The lower plot shows the corresponding build up and decay of the surface hole concentration $p(t)$ normalized by dividing by the hole flux J_h into the surface.

Inspection of eqn (2.23) shows that the ratio $k_{tr}/(k_{tr}+k_{rec})$ and the sum $(k_{tr}+k_{rec})$ can be derived from experimental transients, so that the rate constants for charge transfer and recombination can be obtained. In practice, large amplitude changes in light intensity can lead to other effects such as changes in band bending, and a better approach is to use small amplitude light pulses superimposed on a dc background illumination or small amplitude sinusoidal modulation of the light intensity, as described in the next section.

2.5 Intensity-Modulated Photocurrent Spectroscopy

Intensity-modulated photocurrent spectroscopy (IMPS)[33–37] is a convenient way of measuring the rate constants for charge transfer and recombination. The method involves using a small sinusoidal modulation of light intensity (typically less than 10% of the dc illumination level) and measuring the phase and magnitude of the photocurrent response as a function of frequency. The result, which is usually plotted in the complex plane, can be analyzed to find obtain k_{tr} and k_{rec}. The relaxation time constant $(k_{tr}+k_{rec})^{-1}$ seen in the exponential decay and overshoot of photocurrent transients gives rise to a semi-circular response that is described by the normalized transfer function

$$\frac{j_{photo}(\omega)}{q\,\tilde{J}_h} = \frac{k_{tr}+i\omega}{k_{tr}+k_{rec}+i\omega} \tag{2.24a}$$

Here, the tilde over J_h draws attention to the fact that we are dealing with the modulated hole current, not the dc flux. The real and imaginary parts of the IMPS response are given by

$$\mathrm{Re}\left[\frac{j_{photo}(\omega)}{q\,\tilde{J}_h}\right] = \frac{k_{tr}(k_{tr}+k_{rec})+\omega^2}{(k_{tr}+k_{rec})+\omega^2} \tag{2.24b}$$

$$\mathrm{Im}\left[\frac{j_{photo}(\omega)}{q\,\tilde{J}_h}\right] = \frac{k_{rec}\omega}{(k_{tr}+k_{rec})+\omega^2} \tag{2.24c}$$

This response corresponds to a semicircle with a high frequency (HF) intercept on the real axis at 1 and a low-frequency intercept equal to $k_{tr}/(k_{tr}+k_{rec})$. The maximum of the semicircle occurs at a radial frequency $\omega_{max}=k_{tr}+k_{rec}$. Comparison with the transient response described in the preceding section shows that the HF limit corresponds to the instantaneous photocurrent and the LF limit to the steady state photocurrent. The time constant of the transient response is $1/\omega_{max}$.

Both the transient and IMPS responses are affected by the RC time constant of the electrochemical system,[37] which is determined by the series resistance, R, and the combination of C_{sc} and C_H. In the case of transient photocurrents, the effect of the RC time constant is to determine the rise time of the photocurrent. In the case of the IMPS response, the RC time constant gives rise to an

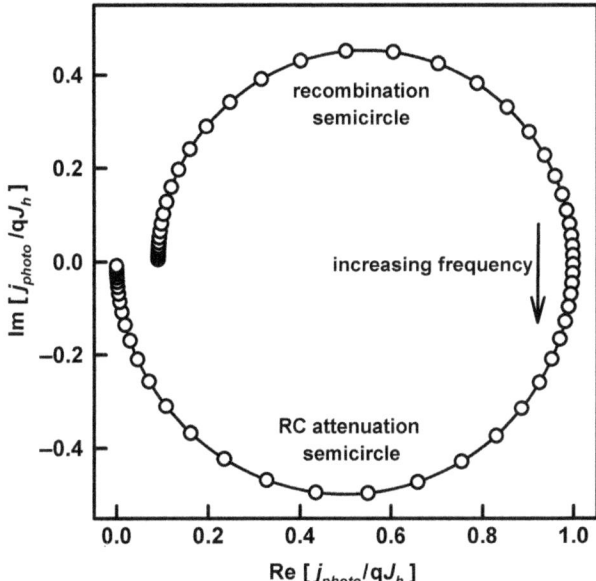

Figure 2.9 Generalized IMPS response calculated from eqns (2.24) and (2.25) (RC attenuation). Provided that the RC time constant is much shorter than $(k_{tr} + k_{rec})^{-1}$, the upper semicircle can be used to derived the two rate constants.

additional semicircle that is located in the lower quadrant of the complex plane as illustrated in Figure 2.9. The radial frequency at the minimum of this lower semicircle is given by

$$\omega_{\min} = \frac{C_{sc} + C_H}{RC_{sc}C_H} \tag{2.25}$$

Provided that ω_{max} and ω_{min} are separated by at least an order of magnitude, then the upper semicircle can still be used to determine values of k_{tr} and k_{rec}. However, if the time constants are closer, deconvolution becomes necessary using a value for the *RC* time constant determined, for example, by impedance measurements.

Figure 2.10 is an example of an IMPS plot for a thin film hematite photoanode in the photocurrent onset region where spikes appear in the chopped photocurrent. In this example, the hematite electrode had been treated with a dilute Co(II) solution, which substantially improves the current-voltage characteristics[38] (the effect of this cobalt treatment is discussed below). The values of the rate constants derived from the normalized low-frequency intercept (ca. 0.6) and ω_{max} (ca. 4 s^{-1}) are $k_{tr} \approx 2.4$ s^{-1} and $k_{rec} \approx 1.6$ s^{-1}. These extraordinarily low values give a clear indication of how slow the oxidation of water by photogenerated holes is at hematite electrodes.

The potential dependence of k_{tr} and k_{rec} has been derived from IMPS measurements on untreated and Co(II)-treated hematite film electrodes.[39,40]

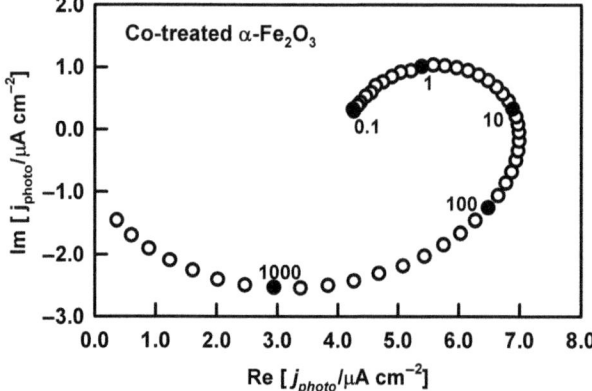

Figure 2.10   IMPS response of α-Fe$_2$O$_3$ film electrode in the photocurrent onset
region showing the recombination and RC attenuation semicircles. The
electrode was treated with a drop of 10 mM cobalt nitrate solution.
Electrolyte 1.0 M NaOH. Electrode potential -0.3 V *vs.* Ag|AgCl.
Illumination 455 nm, dc photon flux 1.1×10^{17} cm^{-2} s^{-1}.

The results show that the semiconductor/electrode interface is very non-ideal in
the case of untreated electrodes, with a substantial fraction of the applied
potential change appearing across the Helmholtz layer. Treatment of the
electrode with a drop of dilute Co(II) solution evidently supresses surface
recombination substantially, but the electrode behaviour is still far from ideal.

Both the upper and lower semicircles in Figure 2.10 are flattened. The
flattening of the recombination semicircle can be interpreted as evidence for a
distribution of the rate constants k_{tr} and k_{rec} as a consequence of surface
heterogeneity,[41] whereas the flattening of the *RC* semicircle is attributed to the
frequency-dependent permittivity of hematite.[40]

Figure 2.11 contrasts the potential dependence of k_{tr} and k_{rec} for untreated
hematite electrodes and hematite electrodes that have been treated with Co(II).
The remarkable S-shaped potential dependence of the two rate constants for
the untreated hematite electrode is similar to the behaviour observed in an
IMPS study of n-GaAs electrodes.[41] The S-shape of the plot for k_{rec} is un-
expected. Since k_{rec} depends on the surface concentration of electrons, it should
decrease monotonically as the band bending increases. Instead, k_{rec} increases
above 0 V *vs.* Ag|AgCl, before falling at higher potentials. Fermi level pinning
due to the build-up of holes at the surface (or in surface states) is expected to
produce a flat region in the k_{rec} plot, but it cannot explain the observed S-shape
of the plot. It seems likely that changes in the chemical composition of the
surface associated with the trapping of holes at surface sites may be responsible,
since this could alter the surface dipole potential, effectively changing the
flatband potential. The k_{tr} plot tracks the k_{rec} plot and therefore probably has
the same explanation: the decrease in band bending that gives rise to the
maximum in the k_{tr} plot corresponds to an increase in $\Delta\phi_H$, which increases the
rate constant for hole transfer by changing the activation energy.

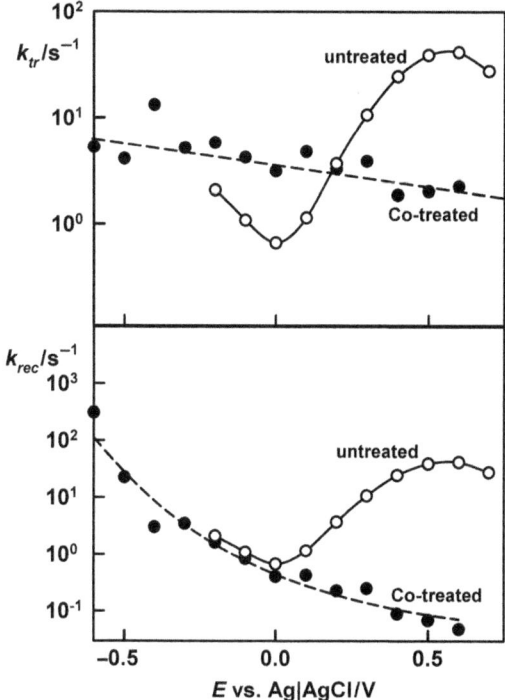

Figure 2.11   Plots showing the effect of Co(II) treatment on the rate constants for charge transfer and recombination at hematite film electrodes. The Co(II) treatment involved placing a drop of 10 mM cobalt nitrate solution on the electrode surface for 60 s, followed by rinsing with pure water for 30 s. Electrolyte 1.0 M NaOH. Illumination 455 nm, 11 mW cm^{-2}.

In the case of the Co(II)-treated electrode, k_{tr} decreases smoothly over the entire potential range, showing that the band bending increases monotonically with applied potential. However, the change in $\Delta\phi_{sc}$ estimated from the k_{tr} plot using eqn (2.18) is only 200 mV for a change in applied potential of 1 volt. This highly non-ideal behaviour appears to suggest that most of the change in applied potential must appear across the Helmholtz layer, even in the case of the Co(II)-treated electrode. However, this conclusion is not supported by the fact that k_{tr} appears to fall slightly over the same potential range rather than increasing as would be expected if the increase in $\Delta\phi_H$ increases k_{tr}. Clearly the kinetics of recombination and charge transfer are more complicated than the simple treatment given here. This is not surprising, since the photo-oxidation of water to produce oxygen is a four-electron process that involves intermediate species and a series of elementary steps. Further indications of this complexity are provided by the observation that k_{tr} increases linearly with light intensity, whereas k_{rec} exhibits a square root dependency.[42] To date, the most detailed treatment of the IMPS response expected for multi-step electron transfer reactions considers a two-step oxidation process of the type involved in the photoanodic decomposition of II-VI semiconductors such as CdS.[25] This treatment shows that the phenomenological rate constants k_{tr} and k_{rec} need to be interpreted in terms of particular reaction schemes. In addition, the effects of the modulation of the potential distribution across the interface caused by changes in surface hole concentration need to be considered. Such a detailed treatment has not been attempted yet for the four-electron oxidation of water by photogenerated holes.

2.6 Photoelectrochemical Impedance Spectroscopy

Photoelectrochemical impedance spectroscopy (PEIS) involves measurement of the impedance of photoelectrodes under steady-state illumination conditions. The method has been widely used to study single-crystal semiconductor electrodes,[43–50] and several authors have discussed the equivalence between PEIS and IMPS.[51–53] In this chapter, we discuss expressions for the PEIS response a semiconductor electrode derived using the phenomenological approach outlined in the previous sections. As before, we consider the competition between interfacial charge transfer and surface recombination. Small amplitude sinusoidal modulation of the electrode potential perturbs the electron density and hence the recombination rate constant (*cf.* eqn (2.18)). This perturbation results in a modulation of the recombination current (*cf.* eqn (2.20) for the steady-state expression). The potential perturbation also gives rise to a charging current. As a first approximation, it is assumed that the flux of holes, J_h, into the surface is not perturbed by the potential modulation. It is also assumed that the potential modulation does not affect k_{tr}. The analysis predicts that the PEIS response should exhibit two semicircles in the complex plane.[53,54] In the case that $Csc \ll C_H$, the low frequency semicircle corresponds to a parallel RC network with the values

$$R_{LF} = \frac{k_B T}{q^2 J_h} \left(\frac{k_{tr} + k_{rec}}{k_{tr}} \right) \tag{2.26a}$$

$$C_{LF} = \frac{q^2 J_h}{k_B T} \frac{1}{(k_{tr} + k_{rec})} \tag{2.26b}$$

$$\omega_{max,LF} = \frac{1}{R_{LF} C_{LF}} = k_{tr} \tag{2.26c}$$

The high frequency semicircle corresponds to a second parallel RC network with values

$$R_{HF} = \frac{k_B T}{q^2 J_h} \left(\frac{k_{tr} + k_{rec}}{k_{rec}} \right) \tag{2.27a}$$

$$C_{HF} = C_{sc} \tag{2.27b}$$

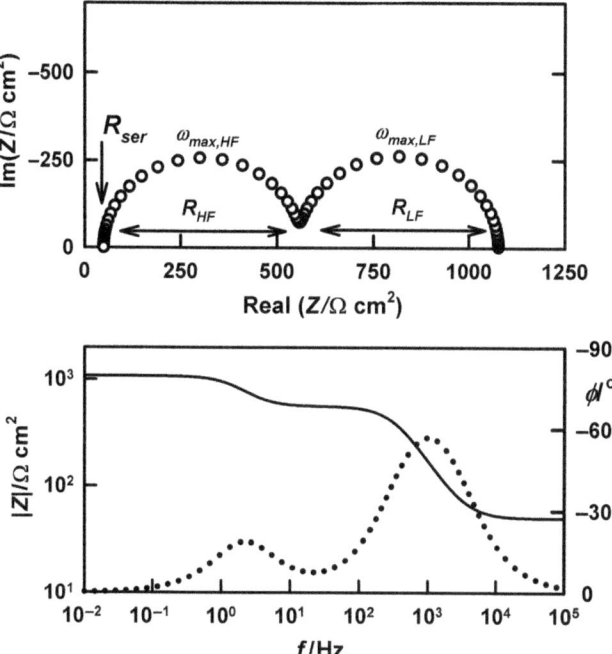

Figure 2.12   Complex plane and Bode plots of the PEIS response calculated from eqn
(2.26) for $k_{tr} = k_{rec} = 10^2\,s^{-1}$, $J_h = 10^{15}\,cm^{-2}\,s^{-1}$, $C_{sc} = 1\,\mu F\,cm^{-2}$,
$C_H = 100\,\mu F\,cm^{-2}$. The complex plane plot shows the two semicircles,
and the Bode plot shows the separation of the low and high frequency
RC time constants (continuous line: magnitude; broken line: phase).

$$\omega_{\max,HF} = \frac{q^2 J_h}{k_B T C_{sc}} \left[\frac{k_{rec}}{k_{rec} + k_{tr}} \right] \tag{2.27c}$$

Figure 2.12 illustrates the main features of the PEIS response predicted by the
theory.

Leng *et al.*[55] have re-derived these expressions and have extended them for
cases in which surface state charging occurs and in which partial Fermi level
pinning introduces a potential dependence into k_{tr} given by the Butler Volmer
equation. The interested reader is referred to the original paper for details.
Alternative approaches to the analysis of PEIS spectra have also been proposed
that are based on equivalent circuits in which the resistors and capacitances
correspond to distinct physical processes.[52,56–58] Figure 2.13 shows the calculated potential dependence of R_{LF} and C_{LF} in the photocurrent onset region for
the ideal case where the potential dependence of the surface recombination rate
constant k_{rec} is given by eqn (2.18) and k_{tr} is independent of potential.

Figure 2.14 shows a typical PEIS response for an untreated hematite thin film
photoanode. Two semicircles can be seen, but they are both flattened compared
with the ideal response. In principle, the flattened semicircles can be fitted using

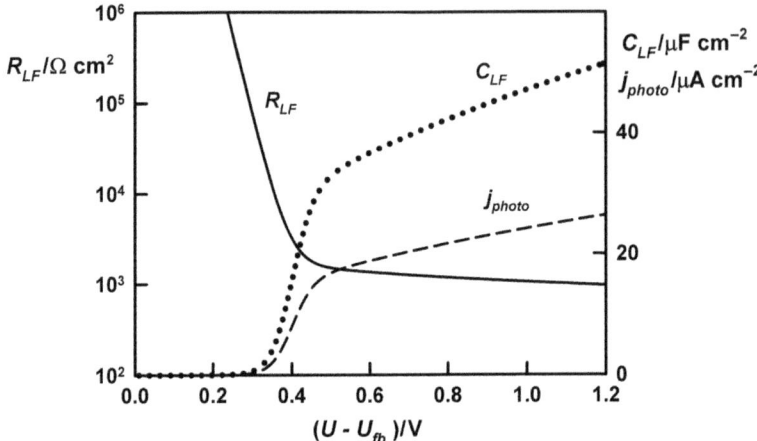

Figure 2.13   Theoretical plots of R_{LF} and C_{LF} obtained from eqn (2.26) using values of J_h and k_{rec} calculated using eqn (2.4) and eqn (2.18), respectively. Other values: $k_{tr} = 20\,\text{s}^{-1}$, $\alpha = 1.5 \times 10^5\,\text{cm}^{-1}$, $N_d = 10^{19}\,\text{cm}^{-3}$, $\varepsilon = 25$, $C_H = 100\,\mu\text{F cm}^{-2}$, $\sigma_n = 10^{-16}\,\text{cm}^2$, $v_{th} = 10^5\,\text{cm s}^{-1}$, $I_0 = 2 \times 10^{15}\,\text{cm}^{-2}\,\text{s}^{-1}$.

constant phase shift elements, but this approach is inadvisable since it introduces additional arbitrary fitting parameters that have no physical meaning. Figure 2.14 also shows that a reasonably satisfactory fit can be obtained to two parallel RC elements connected in series with the ohmic resistance. The low frequency semicircle is the most important since it can be used to obtain k_{tr} and k_{rec} values.

The experimental dependence of R_{LF} and C_{LF} on potential for untreated hematite electrodes is shown in Figure 2.15, along with the derived values of the rate constants k_{tr} and k_{rec}. It can be seen that the potential dependence of k_{rec} has the same sinusoidal form seen in Figure 2.11, although in this case k_{tr} does not show such a pronounced maximum.

2.7   Following Interfacial Electron Transfer using Microwave Reflectivity Measurements

In some cases, the build-up of minority carriers at the illuminated semiconductor/electrolyte interface can be detected directly as a change in conductivity. *In situ* microwave reflectance measurements on semiconductor/electrolyte junctions were pioneered by Tributsch and co-workers.[59,60] The method involves terminating a microwave waveguide with a semiconductor sample that is in contact on its other face with an electrolyte in an electrochemical cell. The microwave reflectivity of the sample can be perturbed by changing the potential or by illumination. Low-doped samples are commonly used in order to obtain good sensitivity. Microwaves reflected back from the sample undergo a phase change and are deflected via a circulator to a detector.

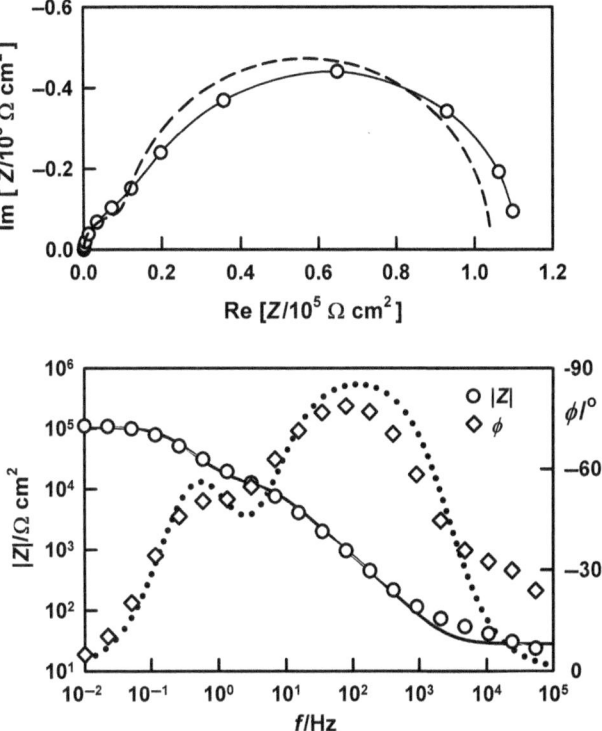

Figure 2.14   Example of the PEIS response of a hematite film electrode. Electrolyte 1.0 M NaOH. Potential 0.1 V *vs.* Ag|AgCl. Illumination 455 nm, ca.1 mW cm^{-2}.

The normalised change in microwave reflectance is related to the change in conductivity, $\Delta\sigma$, of the sample brought about by the perturbation.

$$\Delta R_M = \frac{\Delta P_r}{P_i} = R_M \frac{\Delta P_r}{P_r} = S < \Delta\sigma > = \frac{Sq}{d} \int_0^d \left[\mu_n \Delta n(x) + \mu_p \Delta p(x)\right] dx \quad (2.28)$$

Here P_i and P_r are the incident and reflected microwave power respectively, ΔP_r is the change in reflected microwave power, R_M is the unperturbed microwave reflectivity, S is a sensitivity factor, d is the sample thickness, q is the elementary charge, $\Delta n(x)$ and $\Delta p(x)$ are the position dependent excess electron and hole densities resulting from illumination and μ_n, μ_p are the electron and hole mobilities. The density profiles of photogenerated electrons and holes can be calculated numerically for given conditions (intensity and wavelength of illumination and rate constant for interfacial electron transfer). Integration of the profiles then gives the change in mean conductivity and hence the microwave response. S can be measured[16] or it can be calculated using a filter stack model to represent the carrier profiles.[16] Numerical calculations show that a measurable light-induced microwave response will only be observed if charge

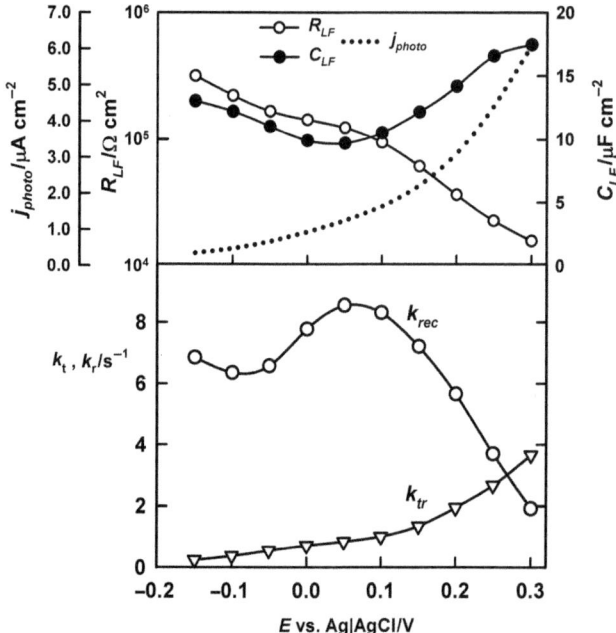

Figure 2.15 (a) Experimental values of R_{LF} and C_{LF} for an untreated hematite electrode. Electrolyte 1.0 M NaOH. Illumination 455 nm, ca. 1 mW cm^{-2}. The steady state photocurrent is also shown. The lower plot shows the corresponding values of k_{tr} and k_{rec} derived by analysing the PEIS responses.

transfer is slow, so that the concentration of minority carriers build up at the interface, changing the conductivity of the sample.[29] In principle, the magnitude of the response can be used to drive the rate constant for minority carrier reaction at the surface.

Microwave reflectance measurements have been used to investigate light-driven hydrogen evolution at low-doped p-Si(111) electrodes in buffered fluoride solutions.[16,29] The silicon surfaces are hydrogen terminated in this electrolyte, and the density of surface states is known to be exceptionally small. Figure 2.16 compares the microwave and photocurrent responses seen at a potential that is in the photocurrents saturation region. Whereas the photocurrent response follows the square waveform of the illumination waveform, the microwave response shows a much slower rise, corresponding to the build-up of the concentration of electrons (minority carriers) at the interface.

The study in ref. 16 found that the magnitude of the steady-state light-induced microwave response and the rate constant for electron transfer depended on the square root of the illumination intensity. It was shown that this is consistent with the two step mechanism

$$H^+ + e^- \rightarrow H^\circ \tag{2.29a}$$

$$H^0 + H^0 \rightarrow H_2 \tag{2.29b}$$

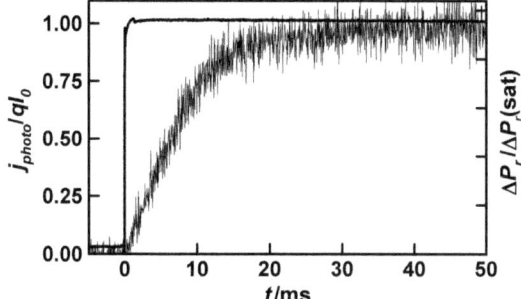

Figure 2.16   Comparison of normalized photocurrent and microwave reflectance responses of a p-Si(111) electrode in the photocurrent saturation region (-0.8V *vs.* Ag|AgCl: see photocurrent plot in Figure 2.17). Electrolyte 1.0 M NH_4F (pH 3.0). Note that the photocurrent response is instantaneous, whereas the microwave signal shows a characteristic rise time that is determined by k_{tr} (note that k_{rec} is negligible in the photocurrent saturation region). The microwave response shows directly the build-up of the electron concentration to a steady state value.

Here, H° represents a hydrogen atom intermediate that must be stabilized in some way by interaction with the silicon.

In a further development of the microwave reflectance method known as light-modulated microwave reflectivity: LMMR, sinusoidally modulated illumination is used and the phase and magnitude of the microwave reflectivity is measured as a function of modulation frequency.[16,61] The normalized microwave response takes the form

$$\text{Re}\left[\frac{\Delta R_M}{R_M}\right] = \frac{k_{tr}(k_{tr} + k_{rec})}{(k_{tr} + k_{rec})^2 + \omega^2} \qquad (2.30a)$$

$$\text{Im}\left[\frac{\Delta R_M}{R_M}\right] = \frac{k_{tr}\omega}{(k_{tr} + k_{rec})^2 + \omega^2} \qquad (2.30b)$$

The intensity-modulated microwave reflectance response corresponds to a semicircle in the complex plane with a high-frequency intercept at the origin and a low frequency intercept on the real axis at $k_{tr}/(k_{tr} + k_{rec})$. It follows that the low-frequency intercept will be 1 in the photocurrent saturation region when k_{rec} is negligible. By comparison, the semicircular IMPS response (see section 2.5, eqn (2.24)) has a high frequency intercept of 1 and a low-frequency intercept at $k_{tr}/(k_{tr} + k_{rec})$. This means that the semicircle becomes smaller as k_{rec} decreases, until it condenses to a point at unity in the photocurrent saturation region, so that no kinetic information can be derived. This difference highlights the power of the microwave method: it can still be used even in cases where surface recombination is absent.

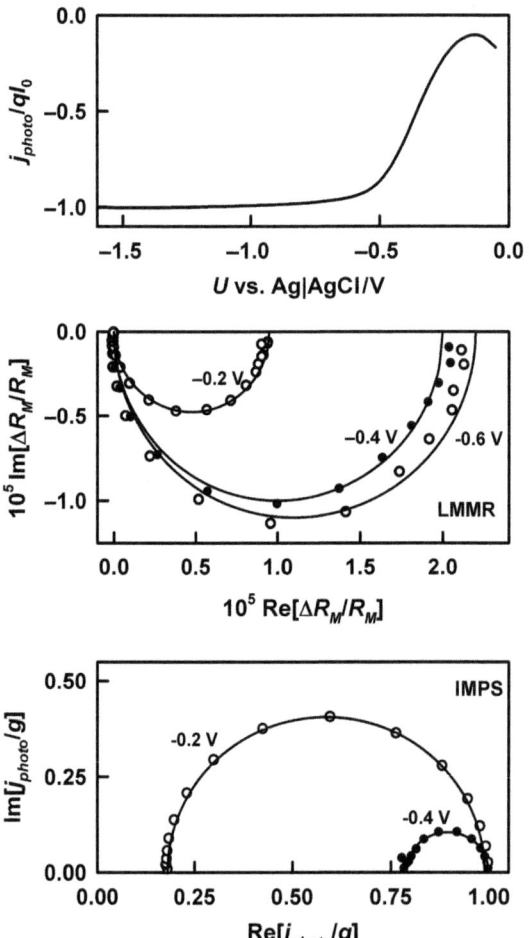

Figure 2.17 Comparison of LMMR response of p-Si(111) electrode in 1.0 M NH$_4$F (pH 3.0) with the normalized IMPS responses. The normalized photo-current plot is also shown. Note that the LMMR response becomes larger as the band bending is increased, whereas the diameter of the IMPS semicircle decreases (the response at −0.6 V, which is not shown, is simply a point on the real axis). This means that kinetic information can be obtained by LMMR under conditions where the IMPS response no longer provides information about k_{tr} and k_{rec}.

Figure 2.17 contrasts the way in which the LMMR and IMPS responses of a p-Si(111) electrode change as the potential is made more negative, increasing the band bending and decreasing k_{rec}. It can be seen that the LMMR semicircle become larger, whereas the IMPS semicircles become smaller. Analysis of the IMPS and LMMR responses gives consistent values of k_{tr} and k_{rec} as a function of potential: the interested reader is referred to the original paper[16] for details. Although clearly a powerful method, LMMR has not been widely used. One of the reasons is that the measurements need to be made on low-doped materials in order to obtain adequate sensitivity.

2.8 Potential and Light-Modulated Absorption Spectroscopy

One of the main problems with the (photo)electrical measurements discussed so far is their lack of chemical specificity. The emphasis on charge carriers and the simplifications inherent in the kinetic modelling obscure the fact that reactions involving molecular species are taking place. If we consider the four-electron oxidation of water, for example, it must involve a series of surface-bound intermediates. Identification of reaction intermediates holds the key to understanding the mechanisms of water splitting reactions. Time-resolved transient absorbance measurements have been used by several groups to follow charge carrier dynamics in systems in which charge transfer is slow. In the case of hematite electrodes, a long-lived 'hole' species has been identified by its characteristic absorbance in the visible region.[62,63] The lifetime of this species is consistent with the low values of k_{tr} measured by IMPS and PEIS. It is not clear whether the species is in fact a free or trapped hole or whether it is a long-lived intermediate in the water oxidation process. Transient absorbance methods have also been used to investigate the role of cobalt-based catalysts in light driven water splitting at hematite electrodes.[64]

The detection of surface-bound intermediates in light-driven water splitting requires an optical system with high sensitivity. The required sensitivity is achieved in time-resolved absorbance measurements by averaging a large number of transients. An alternative approach is to use modulation techniques in which the system is perturbed with a small amplitude electrical or optical signal and the change in absorbance (or reflectance) is detected with a lock-in amplifier or frequency response analyzer. Potential-modulated absorbance spectroscopy (PMAS) and potential-modulated reflectance spectroscopy (PMRS) have been applied extensively in electrochemistry. In the case of light-driven water splitting, the periodic perturbation can be light rather than voltage. Light-modulated absorbance spectroscopy (LMAS) has been used together with PMAS to study thin semi-transparent hematite electrodes.[65] In the dark, water oxidation at hematite electrodes takes place at potentials considerably more positive than the equilibrium O_2/H_2O potential. A quasi-reversible redox couple observed in the onset region has been attributed to the Fe(IV)/Fe(III) system.[66] PMAS spectra recorded in this region are similar to the spectra attributed to long-lived holes.[62] Oxygen evolution at illuminated hematite electrodes takes place at more negative potentials, where the dark current is negligible. As Figure 2.18 shows, LMAS spectra recorded in this region appear to be identical with the PMAS spectra recorded at the onset of oxygen evolution in the dark. This points to a common species being involved in water oxidation in the dark and under illumination. It seems likely that the

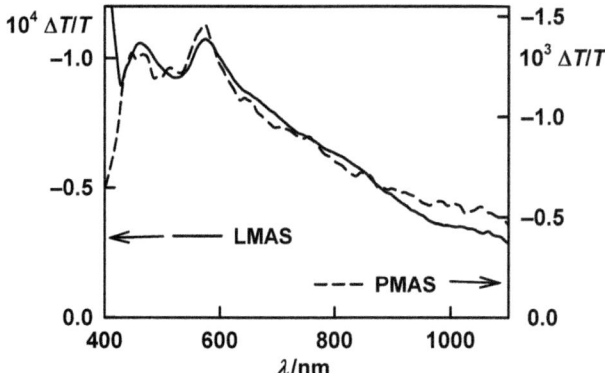

Figure 2.18   Comparison of LMAS and PMAS spectra of a mesoporous Fe_2O_3 film (thickness *ca.* 80 nm). The LMAS spectrum was obtained at 0.4 V *vs.* Ag|AgCl using modulated UV illumination (370 nm) to create electron hole pairs. The PMAS spectrum was obtained at 0.7 V *vs.* Ag|AgCl using 100 mV potential modulation. Modulation frequency 2.7 Hz. Electrolyte 1.0 M NaOH.

species detected is a long-lived intermediate in the four-electron oxidation process: candidates include surface bound Fe(IV) or a Fe(III)-peroxo species.

If a surface species is formed by reversible oxidation of a surface site, *e.g.*

$$Fe(III)_{surf} \rightleftharpoons Fe(IV)_{surf} + e^- \qquad (2.31)$$

modulation of the potential or illumination changes the number densities (concentrations) of oxidized and reduced species (Δn_O, Δn_R) by an amount that depends on the Faradaic charge passed, *i.e.* on the integral of the Faradaic component of the total ac current.

$$\tilde{Q}_F = zq\Delta n_O = -zq\Delta n_R = \int j_F(\omega t)dt \qquad (2.32)$$

where z is the number of electrons transferred and q is the elementary charge. The corresponding normalized change in absorbance depends on the difference in (wavelength dependent) optical absorption cross sections, $\sigma_O(\lambda)$ and $\sigma_R(\lambda)$ of the two species.[67,68]

$$\frac{\Delta T}{T} = \Delta n[\sigma_O(\lambda) - \sigma_R(\lambda)] = \frac{\tilde{Q}_F}{zq}[\sigma_O(\lambda) - \sigma_R(\lambda)] \qquad (2.33)$$

The modulated charge can be derived from impedance measurements, which allows quantitative evaluation of the difference in optical cross sections. This in turn leads to information about the molar absorption coefficients of the species. A full discussion is beyond the scope of this chapter, and the interested reader is referred to the original paper[65] for further details.

2.9 Conclusions and Outlook

This chapter presents a selection of experimental methods rather than an exhaustive overview. The discussion highlights the fact that the consideration of kinetics and mechanisms for multiple electron transfer reactions at illuminated semiconductors is a subject that is still in its infancy. Progress towards the realization of efficient light-driven water splitting requires a better understanding of the chemical nature and energy landscape of the solid/electrolyte interface under illumination. A key issue that has not been considered here is catalysis. Further work is needed to identify catalysts that promote hole transfer to the electrolyte without increasing surface recombination. Identification of intermediates and the study of their temporal evolution under non-steady-state conditions using advanced spectroscopic methods will be an essential step towards relating the phenomenological rate constants derived from the methods described here to the elementary processes taking place during oxygen evolution.

Acknowledgements

The author thanks Upul Wijayantha and Asif Tahir (Loughborough University) as well as Frank Marken and Charles Cummings (University of Bath) for their collaborative work on hematite that provided the data shown in Figures 2.7, 2.10, 2.11, 2.14 and 2.15.

References

1. L. M. Peter, H. Tributsch, in *Nanostructured And Photoelectrochemical Systems For Solar Photon Conversion*, ed. M. D. Archer, A. J. Nozik, Imperial College Press, London, 2008, pp. 675–736.
2. J. F. Dewald, *J. Electrochem. Soc.*, 1958, **105**, C49–C49.
3. J. F. Dewald, *Bell System Technical Journal*, 1960, **39**, 615–639.
4. P. A. Kohl and A. J. Bard, *J. Am. Chem. Soc.*, 1977, **99**, 7531–7539.
5. R. H. Wilson, *J. Electrochem. Soc.*, 1979, **126**, 1187–1188.
6. P. A. Kohl and A. J. Bard, *J. Electrochem. Soc.*, 1978, **125**, C159–C159.
7. L. S. R. Yeh and N. Hackerman, *J. Phys. Chem.*, 1978, **82**, 2719–2726.
8. E. C. Dutoit, F. Cardon and W. P. Gomes, *Ber.Bunsenges. Phys. Chem*, 1976, **80**, 475–481.
9. R. H. Wilson, L. A. Harris and M. E. Gerstner, *J. Electrochem. Soc.*, 1979, **126**, 844–850.
10. M. P. Dare-Edwards, J. B. Goodenough, A. Hamnett and P. R. Trevellick, *J. Chem. Soc. Faraday Trans. I*, 1983, **79**, 2027–2041.
11. S. Sahami and J. H. Kennedy, *J. Electrochem. Soc.*, 1985, **132**, 1116–1120.
12. S. R. Morrison, *Electrochemistry of semiconductor and metal electrodes*, Plenum Press, New York, 1980.
13. N. Sato, *Electrochemistry at metal and semiconductor electrodes*, Elsevier, Amsterdam, 1998.

14. Y. V. Pleskov and Y. Y. Gurevich, *Semiconductor Photoelectrochemistry*, Consultants Bureau, New York, 1985.
15. A. J. Bard, A. B. Bocarsly, F. R. F. Fan, E. G. Walton and M. S. Wrighton, *J. Am. Chem. Soc.*, 1980, **102**, 3671–3677.
16. M. J. Cass, N. W. Duffy, L. M. Peter, S. R. Pennock, S. Ushiroda and A. B. Walker, *J. Phys. Chem. B*, 2003, **107**, 5864–5870.
17. W. W. Gärtner, *Phys. Rev*, 1959, **116**, 84.
18. M. A. Butler and D. S. Ginley, *J. Mat. Sci*, 1980, **15**, 1–19.
19. I. Mora-Sero, F. Fabregat-Santiago, B. Denier, J. Bisquert, R. Tena-Zaera, J. Elias and C. Levy-Clement, *Appl. Phys. Lett.*, 2006, **89**, 203117.
20. J. Bisquert, *PCCP*, 2008, **10**, 49–72.
21. K. Sivula, F. Le Formal and M. Grätzel, *Chemsuschem*, 2011, **4**, 432–449.
22. A. J. Bard and L. R. Faulkner, *Electrochemical Methods: Fundamentals and Applications*, 2nd edn., John Wiley and Sons Inc, New York, 2001.
23. K. E. Pomykal, A. M. Fajardo and N. S. Lewis, *J. Phys. Chem.*, 1996, **100**, 3652–3664.
24. W. J. Royea, A. M. Fajardo and N. S. Lewis, *J. Phys. Chem. B*, 1997, **101**, 11152–11159.
25. L. M. Peter, E. A. Ponomarev and D. J. Fermin, *J. Electroanal. Chem.*, 1997, **427**, 79–96.
26. R. H. Wilson, *J. Appl. Phys.*, 1977, **48**, 4292–4297.
27. J. Reichman, *Appl. Phys. Lett.*, 1980, **36**, 574–577.
28. F. El Guibaly and K. Colbow, *J. Appl. Phys.*, 1982, **53**, 1737–1740.
29. M. J. Cass, N. W. Duffy, L. M. Peter, S. R. Pennock, S. Ushiroda and A. B. Walker, *J. Phys. Chem. B*, 2003, **107**, 5857–5863.
30. S. J. Anz and N. S. Lewis, *J. Phys. Chem. B*, 1999, **103**, 3908–3915.
31. P. T. Landsberg, *Recombination in Semiconductors*, Cambridge University Press, Cambridge, 1991.
32. E. A. Ponomarev and L. M. Peter, *J. Electroanal. Chem.*, 1995, **396**, 219–226.
33. L. M. Peter, *Chem. Rev.*, 1990, **90**, 753–769.
34. J. Li and L. M. Peter, *J. Electroanal. Chem..*, 1985, **193**, 27–47.
35. J. Li and L. M. Peter, *J. Electroanal. Chem.*, 1986, **199**, 1–26.
36. L. M. Peter, *Chem. Rev.*, 1990, **90**, 753–769.
37. L. M. Peter and D. Vanmaekelbergh, in *Adv. Electrochem. Sci. Eng.*, R. C. Alkire, D. M. Kolb, Weinheim, 1999, vol. 6 , pp. 77–163.
38. A. Kay, I. Cesar and M. Graetzel, *J. Am. Chem. Soc.*, 2006, **128**, 15714–15721.
39. C. Y. Cummings, F. Marken, L. M. Peter, A. A. Tahir and K. G. U. Wijayantha, *Chem. Commun.*, 2012, **48**, 2027–2029.
40. L. M. Peter, K. G. U. Wijayantha and A. A. Tahir, *Faraday Discuss.*, 2012, **155**.
41. R. Peat and L. M. Peter, *Ber. Bunsen-Ges. Phys. Chem*, 1987, **91**, 381–386.
42. K. G. U. Wijayantha, S. Saremi-Yarahmadi and L. M. Peter, *PCCP*, 2011, **13**, 5264–5270.
43. W. P. Gomes and F. Cardon, *J. Electrochem. Soc.*, 1982, **129**, 2874.

44. W. P. Gomes, D. Vanmaekelbergh and F. Cardon, *J. Electrochem. Soc.*, 1986, **133**, C334–C334.
45. D. Vanmaekelbergh and F. Cardon, *Journal of Physics D-Applied Physics*, 1986, **19**, 643–656.
46. D. Vanmaekelbergh, W. P. Gomes and F. Cardon, *J. Electrochem. Soc.*, 1987, **134**, 891–894.
47. D. Vanmaekelbergh and F. Cardon, *Semicond. Sci. Technol.*, 1988, **3**, 124–133.
48. D. Vanmaekelbergh and F. Cardon, *Electrochim. Acta*, 1992, **37**, 837–846.
49. J. Schefold and H. M. Kühne, *J. Electroanal. Chem.*, 1991, **300**, 211–233.
50. J. Schefold, *J. Electroanal. Chem.*, 1993, **362**, 97–108.
51. D. Vanmaekelbergh, A. R. Dewit and F. Cardon, *J. Appl. Phys.*, 1993, **73**, 5049–5057.
52. J. Schefold, *J. Electroanal. Chem.*, 1992, **341**, 111–136.
53. E. A. Ponomarev and L. M. Peter, *J. Electroanal. Chem.*, 1995, **397**, 45–52.
54. D. J. Fermin, E. A. Ponomarev and L. M. Peter, *J. Electroanal. Chem.*, 1999, **473**, 192–203.
55. W. H. Leng, Z. Zhang, J. Q. Zhang and C. N. Cao, *J. Phys. Chem. B*, 2005, **109**, 15008–15023.
56. B. Klahr, S. Gimenez, F. Fabregat-Santiago, J. Bisquert and T. W. Hamann, *Energy Environ. Sci*, 2012, **5**, 7626–7636.
57. B. Klahr, S. Gimenez, F. Fabregat-Santiago, T. Hamann and J. Bisquert, *J. Am. Chem. Soc.*, 2012, **134**, 4294–4302.
58. J. Schefold, *J. Electroanal. Chem.*, 1995, **394**, 35–48.
59. G. Schlichthörl and H. Tributsch, *Electrochim. Acta*, 1992, **37**, 919–931.
60. H. Tributsch, G. Schlichthörl and L. Elstner, *Electrochim. Acta*, 1993, **38**, 141–152.
61. G. Schlichthörl, E. A. Ponomarev and L. M. Peter, *J. Electrochem. Soc.*, 1995, **142**, 3062–3067.
62. S. R. Pendlebury, M. Barroso, A. J. Cowan, K. Sivula, J. W. Tang, M. Gratzel, D. Klug and J. R. Durrant, *Chem. Commun.*, 2011, **47**, 716–718.
63. S. R. Pendlebury, A. J. Cowan, M. Barroso, K. Sivula, J. Ye, M. Graetzel, D. R. Klug, J. Tang and J. R. Durrant, *Energy Environ. Sci*, 2012, **5**, 6304–6312.
64. M. Barroso, A. J. Cowan, S. R. Pendlebury, M. Grätzel, D. R. Klug and J. R. Durrant, *J. Am. Chem. Soc.*, 2011, **133**, 14868–14871.
65. C. Y. Cummings, F. Marken, L. M. Peter, K. G. U. Wijayantha and A. A. Tahir, *J. Am. Chem. Soc.*, 2012, **134**, 1228–1234.
66. C. Y. Cummings, M. J. Bonne, K. J. Edler, M. Helton, A. McKee and F. Marken, *Electrochem. Commun.*, 2008, **10**, 1773–1776.
67. R. S. Hutton, M. Kalaji and L. M. Peter, *J. Electroanal. Chem.*, 1989, **270**, 429–436.
68. M. Kalaji and L. M. Peter, *J. Chem. Soc. Faraday Trans.*, 1991, **87**, 853–860.

CHAPTER 3

Structured Materials for Photoelectrochemical Water Splitting

JAMES McKONE AND NATHAN LEWIS*

California Institute of Technology, Division of Chemistry and Chemical Engineering, 1200 E. California Blvd, Pasadena, CA 91125
*Email: nslewis@caltech.edu

3.1 Introduction

Efficient and economical photoelectrochemical water splitting requires innovation on several fronts. Tandem solar absorbers could increase the overall efficiency of a water splitting device, but economic considerations motivate research that employs cheap materials combinations. The need to manage simultaneously light absorption, photogenerated carrier collection, ion transport, catalysis, and gas collection drives efforts toward structuring solar absorber and catalyst materials.

This chapter divides the subject of structured solar materials into two principal sections. The first section investigates the motivations, benefits, and drawbacks of structuring materials for photoelectrochemical water splitting. We introduce the fundamental elements of light absorption, photogenerated carrier collection, photovoltage, electrochemical transport, and catalytic behavior. For each of these elements, we discuss the figures of merit, the critical length scales associated with each process and the way in which these length scales must be balanced for efficient generation of solar fuels. This discussion

RSC Energy and Environment Series No. 9
Photoelectrochemical Water Splitting: Materials, Processes and Architectures
Edited by Hans-Joachim Lewerenz and Laurence Peter
Published by the Royal Society of Chemistry, www.rsc.org

assumes a working knowledge of the fundamentals of semiconductor-liquid junctions; for more details the reader is encouraged to consult review articles.[1–3] The second section of this chapter reviews recent approaches for generating structured semiconductor light absorbers and structured absorber-catalyst composites. This literature review emphasizes the insights gained in the last six years that are specifically related to photoelectrochemical water splitting, rather than to general photoelectrochemistry or photovoltaic applications. This chapter concludes with perspectives and an outlook for future efforts aimed at solar water splitting using structured materials. The realization of a practical, efficient, and useful water splitting device requires significant new developments in materials synthesis as well as deeper understanding of the relevant chemistry and physics. This chapter is intended to motivate such developments.

3.2 Interplay between Materials Properties and Device Characteristics

Practical solar water splitting requires light absorption, separation and collection of photogenerated charge carriers, and charge-carrier transport to catalytic sites to produce gases that must be safely and economically separated and stored. The overall process must occur at economically relevant efficiencies (> 10 mA cm^{-2} under 1 sun illumination) with negligible degradation in performance over a multi-year timescale.

One model device for photoelectrochemical water splitting is shown in Figure 3.1. Practical, economical devices may not resemble this model, but it remains useful for highlighting the interplay and design tradeoffs that exist when effectively balancing light management, catalysis, and mass transport. The device in Figure 3.1 utilizes a tandem water-splitting structure in which each solar absorber operates as an electrode for redox chemistry. The use of a tandem solar absorber configuration enables better matching with the solar spectrum,[4,5] and separation of the solar absorber electrodes by embedding each component in a selectively permeable membrane enables the separation of the evolved gaseous products. For solar water splitting, the structured absorber on the top of the device absorbs short wavelength, high-energy photons and transmits longer wavelength photons. This first structured absorber serves as the photoanode, where holes oxidize water to produce oxygen gas and protons with the help of an oxygen-evolving catalyst. At the structured solar absorbing photocathode at the bottom of the device, electrons reduce protons to produce hydrogen gas, assisted by a hydrogen-evolution catalyst. A proton-permeable membrane separates photoanodic oxygen evolution and photocathodic hydrogen evolution to prevent formation of a combustible mixture.

Irrespective of the final device configuration, solar absorbers in photoelectrochemical solar fuels devices must accomplish several tasks. They must convert energetic photons into mobile charge carriers and must facilitate the separation of photogenerated electrons and holes. Either a built-in electric field

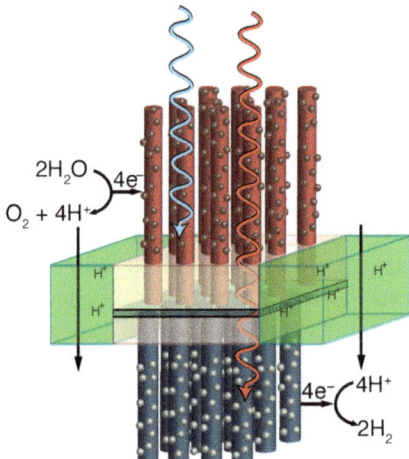

$2H_2O$

$O_2 + 4H^+$

$4e^-$

H^+ H^+ H^+

H^+ H^+ H^+

$4e^-$ $4H^+$

$2H_2$

Figure 3.1   Schematic of a proposed water splitting device utilizing tandem, structured solar absorbers and a proton-permeable membrane for ion transport. Image copyright 2013, Elizabeth A. Santori; used with permission.

from a crystalline semiconductor,[3] or kinetic separation as in a dye-sensitized solar cell can accomplish the necessary charge separation.[6] Following charge separation, minority carriers must travel to an interface and perform catalyzed redox chemistry with minimal deleterious carrier recombination. Beyond light absorption and electron/hole collection, the device in Figure 3.1 highlights one of the principal driving forces of structured solar absorbers. In a device that physically separates reduction and oxidation reactions, increasing the distance between the photoelectrodes severely attenuates the device performance due to mass transport effects. A device with structured solar absorber photoelectrodes similar to the configuration in Figure 3.1 offers the potential to reduce performance-attenuating mass transport losses. This chapter discusses several of the aforementioned issues as well as the way in which the structuring of solar absorbers favorably or adversely affects light management, photovoltage, mass transport, and catalysis within a photoelectrochemical cell.

3.2.1   Light Absorption and Collection of Photogenerated Carriers in Crystalline Semiconductors

The nature of the electronic transition that accompanies photon absorption determines the light-absorption properties in semiconductor materials. The requirement for efficient light absorption guides a significant portion of the motivation for structuring solar absorbers, and absorption phenomena contribute to several of the aforementioned figures of merit.

Electronic transitions associated with light absorption in semiconductor materials can be classified according to the alignment of the crystal energy bands with respect to momentum wave vectors in k-space.[7] Figure 3.2 highlights this alignment for energy *vs.* crystal momentum diagrams for silicon (left)

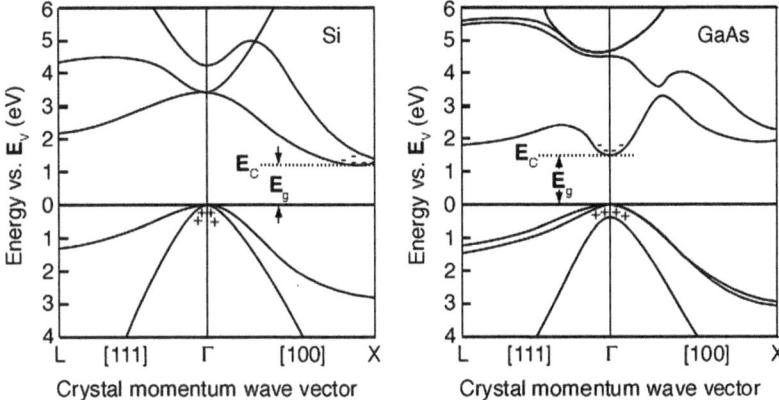

Figure 3.2 Energy versus crystal momentum diagram for the indirect band gap silicon (left) and direct band gap gallium arsenide (right). The differences between the band alignments within these materials yield significantly different photon absorption properties from each other.
Data from Cohen and Chelikowsky.[8]

and gallium arsenide (right).[7,8] For materials such as silicon, in which the maximum of the valence-band energy, E_v, and the minimum of the conduction-band energy, E_c, do not align, optical absorption requires the absorption or emission of a phonon. Thus silicon has an *indirect* band gap, E_g. In contrast to silicon, the conduction band minimum for gallium arsenide is aligned with the valence-band maximum, so photon absorption in GaAs does not require concomitant phonon processes. Gallium arsenide – and all materials with a similar energy *vs.* crystal momentum relationship – have *direct* band gaps.

The nature of the band gap has profound implications for the length scales associated with light absorption. As phonon processes occur on $\sim 10^{-12}$ s timescales, photons must travel significantly longer distances within an indirect band-gap material relative to distances traveled in a direct band-gap material that is limited by the $\sim 10^{-15}$ s timescale for purely electronic interband transitions. Absorption coefficients, α, for direct gap materials are generally much larger than absorption coefficients for indirect-gap materials. As a result, the characteristic absorption length, α^{-1}, is significantly smaller for direct-gap materials relative to α^{-1} for materials with an indirect energy band gap. The plot in Figure 3.3 depicts the differences between α^{-1} for the indirect band-gap material, silicon,[9] relative to α^{-1} for the direct band-gap material, gallium arsenide.[10] For example, at 800 nm, $\alpha_{Si} = 850$ cm^{-1}, so silicon requires 11.8 µm to absorb 63% (or $1 - e^{-1}$) of the incoming light.[9] In contrast, gallium arsenide requires only 1.1 µm to absorb 63% of 800 nm radiation.[10] An understanding of these length scales has profound implications for photoelectrochemistry with semiconductor absorbers.

In a traditional inorganic semiconductor-based photoelectrochemical cell, the collection of photogenerated carriers is linked to the light absorption properties because light absorption and photogenerated minority-carrier collection occur

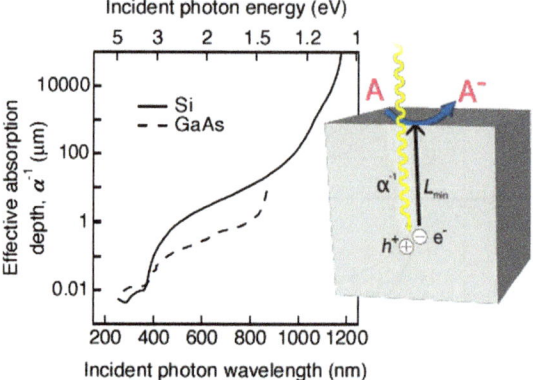

Figure 3.3   The characteristic absorption length, α^{-1}, *vs.* incident photon wavelength for silicon, an indirect absorber material as well as for gallium arsenide, a direct absorber. Over a broad wavelength range, absorption of $1 - e^{-1}$ or 63% of the incoming light requires significantly longer path lengths in silicon relative to gallium arsenide.

along the same spatial axis. Among other factors, efficient carrier collection from bulk crystalline semiconductors requires that the effective minority-carrier diffusion length, L_{min}, is comparable to the penetration depth of the light, α^{-1}. The inset in Figure 3.3 demonstrates the relationship between L_{min} and α^{-1}, which are two critical figures of merit for solar absorbers for photoelectrochemical water splitting. As L_{min} depends on material purity and quality,[11] photoelectrochemical cells that employ this geometry often require highly purified materials, resulting in high overall cost for the active absorber materials. Thus, decoupling L_{min} from α^{-1} is desirable and motivates significant efforts into structured solar absorber materials. The coupling between L_{min} and α^{-1} may not place such a significant restriction on material purity in a direct band-gap solar absorber, but structuring direct band-gap absorbers might still be advantageous for water splitting applications, as discussed in Section 3.2.3.

Long absorption path lengths require highly pure, highly crystalline materials or a modification of light-absorption properties within a semiconductor absorber to achieve optimal performance. Such modifications can reduce the quantity of material required to absorb a significant fraction of above-band-gap photons. In photovoltaic modules, nanoscale interfacial features ideally increase light trapping and decrease the required semiconductor thickness by as much as $4n^2$, where n is the refractive index.[12,13] This enhancement is known as the Lambertian limit, and such nanoscale structuring effectively increases α and decreases α^{-1} but does not fundamentally affect the minority-carrier collection length of the material (techniques for leveraging plasmonic properties for increased light absorption are discussed elsewhere in this volume).

In contrast to increasing light absorption *via* scattering, spatially decoupling light absorption from the collection of photogenerated carriers has the potential to reduce the constraints on purity. Highly-ordered microwire arrays are

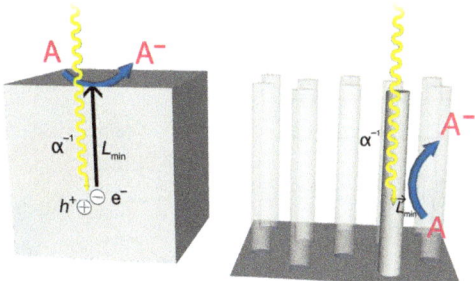

Figure 3.4 Traditional bulk semiconductor geometries (left) require high-quality materials such that the photogenerated minority-carrier diffusion length, L_{min}, is comparable to the characteristic light-absorption length, α^{-1}. In contrast, structured devices such as wire arrays (right) enable the orthogonalization of light absorption and carrier collection. Decoupling L_{min} from α^{-1} reduces certain materials quality constraints.

a prototypical example of a device geometry that decouples light absorption from carrier collection.[14,15] Figure 3.4 contrasts the critical length scales for minority-carrier collection, L_{min}, and α^{-1} for a bulk semiconductor (left) and for a single microwire in a highly ordered array (right). In contrast to the planar, bulk semiconductor case, the microwire-array geometry enables light absorption along the longitudinal wire axis, while photogenerated carrier collection occurs along the radial axis. Thus the microwires can be made sufficiently long for high photon absorption, as implied by Figure 3.3, while maintaining a small radius, permitting the use of semiconductors with low L_{min}. This orthogonalization of light absorption relative to carrier collection is a principal benefit of structured solar absorbers.

In addition to highly ordered microwire arrays, several other structured solar absorber geometries successfully decouple L_{min} from α^{-1}. Section 3.3 of this chapter details recent work on structured absorbers that involve porous morphologies, ordered and un-ordered high-aspect ratio assemblies, and modified dye-sensitized solar cells for photoelectrochemical solar water splitting devices. Each of these geometries presents its own set of benefits and challenges for solar water splitting. However, the remainder of this analysis will focus on highly ordered microwire arrays. We intend this discussion to highlight some of the advantages and disadvantages of structured absorbers and to encourage critical thinking towards nontraditional materials, new geometries, and new characterization techniques for use in photoelectrochemical solar fuels devices.

Experimental, theoretical, and computational modeling studies have indicated that efficient photovoltaic and photoelectrochemical performance can be achieved from highly ordered silicon microwire-array electrodes. This research not only validates the microwire geometry, but also illustrates more generally the benefits provided by structuring solar absorbers. Figure 3.5 presents a summary of light absorption experiments on vapor-liquid-solid (VLS)-grown

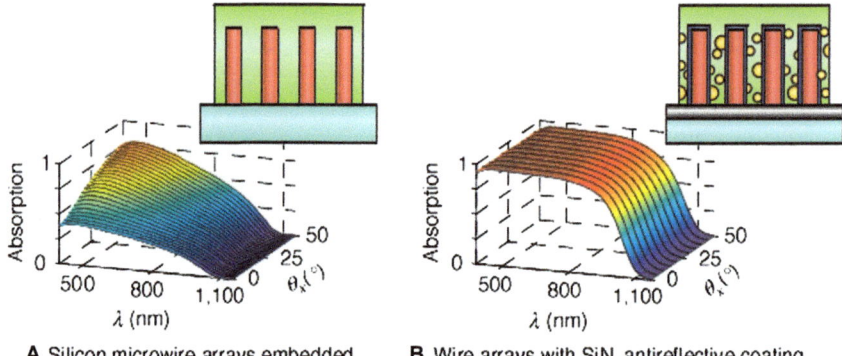

A Silicon microwire arrays embedded
in PDMS on a quartz substrate

B Wire arrays with SiN$_x$ antireflective coating,
Ag back reflector, and Al$_2$O$_3$ scattering particles

Figure 3.5 Optical absorption results demonstrate that PDMS-embedded silicon
microwire arrays have poor absorption at normal incidence but increasing
absorption at off-normal angles (frame A). The addition of scattering
elements, including a silver back reflector, silicon nitride antireflective
coating, and alumina particles, significantly increases absorption at all
angles and all wavelengths of incoming light (frame B).
Image adapted from Kelzenberg *et al.*[16]

silicon microwire arrays that were impregnated with poly(dimethylsiloxane)
(PDMS) and then cleaved from the bulk Si(111) growth substrate. In this
configuration, the wire arrays demonstrated poor optical absorbance when illuminated at normal incidence but increasing absorbance as the incoming light
was rotated away from the substrate normal (Figure 3.5A).

Light-scattering techniques including a silver back reflector, a silicon nitride
antireflective coating, and 900 nm alumina particles embedded in the PDMS
each increased the overall light absorption of the microwire arrays. The combination of these three light-trapping methods increased the maximum absorption to 0.96 and increased the normal incidence absorption to 0.92
(Figure 3.5B). The exceptional light-trapping ability of the silicon microwire
arrays exceeded the planar light-trapping limit for $\lambda > 800$ nm and also exceeded
the simulated day-integrated absorption for planar light-trapping silicon.
Interestingly, this light absorption occurred in a structure that had a fractional
silicon volume of 4.2%, and a total quantity of silicon equal to a 2.8 µm-thick
wafer of identical area. This exceptional light-absorption behavior is a major
benefit of structuring solar absorbers.[14,16,17]

In addition to effective light trapping and a reduction in materials purity
requirements, structured solar absorbers would not provide viable photovoltaic
or photoelectrochemical devices without facilitating effective carrier collection.
Several tools for quantifying the carrier properties of bulk semiconductors are
also useful for structured solar absorbers. In photoelectrochemical and
photovoltaic systems, analyses of the surface recombination velocity, external
quantum yield, and internal quantum yield can help elucidate the details of
photogenerated carrier transport and recombination within, and at the surface

of, structured semiconductors. Additionally, quantification of the open-circuit photovoltage at an illuminated semiconductor-liquid interface is a primary tool for ascertaining the overall device quality and performance.[18]

3.2.2 Open-Circuit Photovoltage at Structured Semiconductors

Models of charge transfer at illuminated semiconductor-liquid junctions balance the current density of electron-hole pair photogeneration with the current densities for deleterious processes.[1,2,18] Such deleterious currents include carrier recombination in the depletion region, tunneling through the interfacial potential barrier, thermionic emission at the interfacial potential barrier, recombination *via* bulk trap states, recombination *via* surface states, and interfacial charge transfer. Ideally, all of the deleterious currents should be suppressed relative to the current due to carrier recombination in the bulk, which is an intrinsic property of the semiconductor material.[18] Solving the ideal diode equation for zero net current yields equation (3.1), the open–circuit photovoltage, V_{oc}, at illuminated semiconductor-liquid junctions. Since current flow is a kinetic phenomenon, the V_{oc} for an illuminated semiconductor-liquid junction is a kinetic parameter, not a thermodynamic parameter, and represents a fundamental figure of merit for photoelectrochemical water splitting.

$$V_{oc} = n \frac{k_B T}{q} \ln \frac{J_{ph} N_{D,A} L_{min}}{q D_{min} n_i^2} \tag{3.1}$$

Here n is the diode ideality factor, J_{ph} is the short circuit photocurrent density (photocurrent per area) under illumination, N_D (N_A) is the donor (acceptor) density, L_{min} is the minority-carrier diffusion length, D_{min} is the minority-carrier diffusion coefficient and n_i is the intrinsic carrier density. The term $N_{D,A}^{-1} L_{min}^{-1} q D_{min} n_i^{-2}$ is frequently simplified as J_s, the saturation current density, equation (3.2).

$$V_{oc} = n \frac{k_B T}{q} \ln \frac{J_{ph}}{J_s} \tag{3.2}$$

For bulk, planar semiconductors, the current density ratio on the right hand side of equation 3.2 values reduces to a ratio of the photocurrent to saturation current because the areas in J_{ph} and J_s are identical. However, current density comparisons are more complicated for structured semiconductors. The area defined for photocurrent density refers to the rectilinear parallel projection of the structured semiconductor onto a plane that is illuminated by the incoming light. Such a parallel projection defines the semiconductor projected surface area. Conversely, the area defined for the saturation current density includes the entire contact area of the rectifying junction. For a rectifying junction formed conformally at the interface of a structured semiconductor-liquid junction, the contact area is the geometric surface area of the semiconductor.

The dimensionless figure of merit γ defined in equation (3.3) reconciles the discrepancy between these two areas.[2]

$$\gamma = \frac{\text{geometric surface area}}{\text{projected surface area}} \qquad (3.3)$$

Incorporation of γ into equation (3.1) enables a straightforward comparison of the currents and current densities in equation 3.4.

$$V_{oc} = n\frac{k_B T}{q}\ln\frac{J_{ph}}{\gamma J_s} \qquad (3.4)$$

Equation (3.4) demonstrates a potential disadvantage of structured solar absorbers. When the semiconductor-liquid junction forms the rectifying contact, increasing the surface area of the solar absorber will decrease the attainable V_{oc} due to the increase in γ. This decrease may be obviated by employing small contact area heterojunctions,[19,20] optical concentrators, or a nanoemitter-style photoelectrochemical cell.[21] The nanoemitter solution could be particularly interesting, as the nanopatterened contact may further function as a scattering element and/or as a solar fuels catalyst. Indeed, the ideally designed nanoemitter solar fuels device may ultimately possess a geometric surface area that is smaller than the projected surface area, yielding $\gamma < 1$ and $V_{oc} > V_{oc,planar}$.

Returning to the example of highly ordered silicon microwire array photoelectrodes, initial studies yielded an open circuit photovoltage, $V_{oc} = 389\,\text{mV}$ and a short-circuit current density, $J_{sc} = 1.43\,\text{mA cm}^{-2}$ for n-Si wire electrodes in contact with dimethylferrocene$^{+/0}$ in methanol under ELH-simulated 1-Sun illumination (halogen lamp).[15] Although this silicon microwire performance was well below the $V_{oc} = 670\,\text{mV}$ and $J_{sc} = 20\,\text{mA cm}^{-2}$ values obtained using planar n-type silicon,[22] these initial studies demonstrated the viability of highly ordered silicon microwire arrays for photoelectrochemistry.

Subsequent experiments on Si microwire arrays established more efficient performance. Spectral response measurements on polymer-embedded, $67\,\mu\text{m}$-long wire arrays in methyl viologen$^{2+/+\bullet}_{(aq)}$ demonstrated nearly unity internal quantum efficiencies between 400 nm and 900 nm.[17] Scanning photocurrent measurements revealed a minority-carrier effective diffusion length, $L_{eff} = 10\,\mu\text{m}$ in individual $0.9\,\mu\text{m}$ radius silicon microwires,[23] which achieved the $L_{eff} > r$ required to produce efficient photocarrier collection in a microwire configuration.[14] Arrays of p-Si microwires with an n$^+$ emitter layer achieved $V_{oc} = 540\,\text{mV}$ and energy-conversion efficiencies exceeding 5% for photoelectrochemical hydrogen production under $100\,\text{mW cm}^{-2}$ ELH-simulated 1 Sun illumination using a Pt co-catalyst.[20] These results illustrate the usefulness of the microwire morphology as well as the methodology of employing structured semiconductors for photoelectrochemical water splitting.

Light absorption and photovoltage considerations begin to highlight the complexity involved in designing a highly efficient photoelectrochemical water-splitting device. Next we consider electrochemical transport processes, which incorporate yet another layer of complexity involving larger size scales and are especially relevant for structured solar absorbers and catalysts for water splitting.

3.2.3 Electrochemical Transport at Structured Semiconductors

The understanding and control of electrochemical transport within a photoelectrochemical cell are critical to the construction of practical solar water-splitting devices. Non-optimized mass transport will result in inefficient use of catalyst material and will also produce overpotentials that degrade the device performance.

Based on comparisons between α^{-1} and L_{min}, indirect-gap semiconductors likely benefit more from structuring than direct-gap semiconductors, because direct absorbers (with larger α values) intrinsically require less material to achieve complete light absorption. However, solar absorbers play roles that are more complicated than converting light to oxidizing and reducing equivalents. In a hydrogen-producing system, balancing the proton flow requires proton-permeable channels between the photoanode and photocathode. The use of two planar, direct-gap semiconductors in the geometry of Figure 3.1 might result in excellent light absorption at the expense of proton flow. Therefore, even direct-gap semiconductors may require a degree of structuring to obtain efficient photoelectrochemical solar fuels devices.

Electrochemical mass transport can be categorized according to three driving forces: convection, migration, and diffusion. Mechanical movement of the bulk solution generates convective transport. Migration refers to the motion of charged species under the influence of electric fields. Concentration gradients produce diffusive transport. A combination of these driving forces determines the mass transport of reactants to an electrode surface and the transport of products away from the electrode surface. Excess supporting electrolyte, or operation in either highly acidic or highly alkaline media, effectively suppresses ion migration forces. Additionally, structured photoelectrodes are likely to suppress bulk solution motion arising from convective forces. Thus, diffusion is likely to dominate mass transport in the vicinity of structured photoelectrodes.

In addition to laminar-flow mass transport phenomena, photoelectrochemical production of gaseous H_2 will affect several aspects of device operation. Gas bubbles will modify light scattering and generate some local solution convection, but it remains unclear whether these factors will produce a net increase or decrease in device performance. Gas bubbles may form at nucleation sites on the photoelectrode surface, decreasing the overall performance by blocking catalytic sites from access to the solution. However, the increased convective flow due to bubble surface detachment and motion may facilitate mass transport.

Mass transport in electrochemical systems has been the focus of significant research for over a century. Such transport studies include mass transport at bulk, planar electrodes,[24] diffusion-limited mass transport during metal deposition at planar microelectrodes,[25,26] and transport near conical and hemispherical ultramicroelectrodes.[27] Several research groups have characterized mass transport, diffusion, and migration phenomena in dye-sensitized solar cells.[28–32] However, mass transport effects at structured solar fuels photoelectrodes remain relatively unexplored.

In contrast to the rigorous mathematical treatments available for planar electrodes and certain microelectrode configurations, models of mass transport do not provide straightforward equations that describe the solution concentration profiles in the vicinity of highly structured electrodes. Recently, simulations in COMSOL Multiphysics have elucidated the mass transport in highly ordered p-type silicon microwire arrays in contact with a solution of cobaltocene$^{+/0}$ in acetonitrile.[33] Figure 3.6 highlights relationships demonstrated by these studies. In simulations where the wire diameter, d, is comparable to the wire pitch spacing, the concentration profile of reactant species resembles a quasi-planar boundary layer (frame A). In contrast, wire array geometries in which $d \ll$ pitch yield a conformal boundary layer (frame B). Simulations that modeled the experimental arrangements demonstrated that the solution-phase reactant concentration decreased to zero over the lower $\sim 70\%$ of the wire length relative to the bulk concentration above the microwire (frame C). In this example, the entire microwire length may be necessary for effective light absorption, but less than one third of the microwire length is an effective redox electrode.

Depletion of solution-phase reactants will adversely affect the photoelectrochemical device performance.[33] A critical adverse effect is a concentration overpotential, η_c, that represents a Nernstian potential shift due to changing concentrations of reactants and products relative to bulk concentrations (equation (3.5)).

$$\eta_c = \frac{RT}{nF} \left(\ln \frac{C_{ox}^{\nu_{ox}}}{C_{red}^{\nu_{red}}} - \ln \frac{C_{ox}^{*\nu_{ox}}}{C_{red}^{*\nu_{red}}} \right) \qquad (3.5)$$

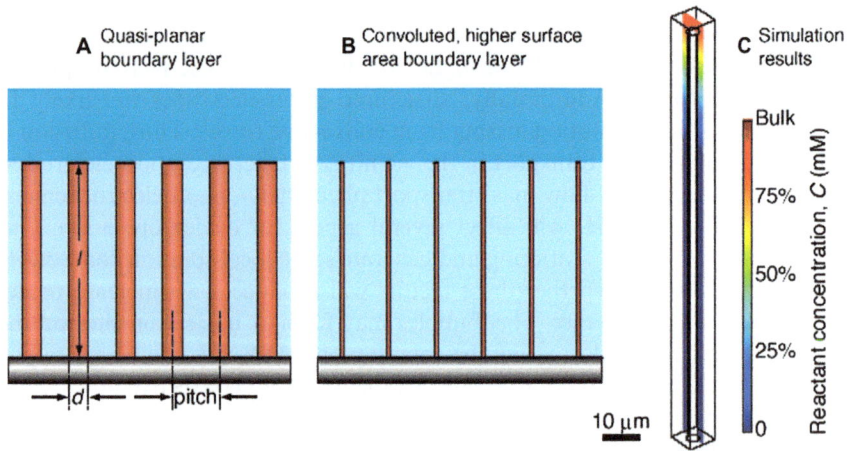

Figure 3.6 The geometries, including size and spacing of solar absorber photoelectrodes, affect solution-phase mass transport. In highly ordered microwire arrays in which the wire diameter is comparable to the spacing, solution phase reactants deplete, forming a quasi-planar boundary layer to the outside solution (frame A). In contrast, the volume between wires is less likely to deplete of reactants when the spacing is significantly greater than the wire diameter (frame B). COMSOL Multiphysics simulation results (frame C) demonstrate the complete depletion of solution-phase reactants over $\sim 70\%$ of the $100\,\mu m$ wire length.

In equation (3.5), n is the stoichiometric number of electrons transferred, R is the gas constant, F is Faraday's constant C_{Ox} and C_{Red} are the concentrations of oxidized and reduced species at the electrode surface, respectively, C_{Ox}^* and C_{Red}^* are the corresponding concentrations in the bulk solution, and ν_{Ox} and ν_{Red} are the stoichiometric numbers of the oxidized and reduced species, respectively. This concentration overpotential adds to the kinetic catalyst overpotential that is required to sustain a desired current density from an ideal catalyst, such that $\eta_{total} = \eta_{catalyst} + \eta_c$. Thus, if the surface concentrations or concentration profiles are known, equation (3.5) quantifies the voltage penalty that a structured electrode will incur to overcome mass transport effects associated with its structure.

Simulations that elucidate mass transport effects will be useful when designing and characterizing structured solar absorber photoelectrodes. The characterization of structured photoelectrodes in non-aqueous photoelectrochemical cells with redox species that exhibit facile, 1-electron transfer reactions is also helpful in understanding the interplay between electrode structure and concentration overpotential.

3.2.4 Catalysis at Structured Semiconductors

An understanding of the catalytic phenomena at structured solar absorber materials is not straightforward. In the simple case, a structured solar absorber with a ratio γ of its geometric surface area to its projected surface area should decrease the turnover demands for a conformally deposited catalyst by the same ratio. For example, a highly ordered microwire array containing 100 μm long, 1.8 μm diameter microwires on a 7 μm square pitch exhibits $\gamma = 5.2$. For this microwire array, a catalyst that is uniformly distributed along the structured solar absorber need only support an actual current density of 1.9 mA cm^{-2} for the device to achieve a 10 mA cm^{-2} current density on the basis of projected (i.e. illuminated) area. The implication of this simple case predicts that structuring solar absorbers places a reduced demand on the catalytic rates, and might thus enable the use of non-noble metal catalysts for efficient, economical photoelectrochemical water splitting.

While structured catalysts may indeed reduce the demands for high catalytic activity, several factors complicate the behaviors of structured catalysts. Continuous metallic catalyst layers will behave differently from a discontinuous and/or a non-metallic catalyst. A sufficiently conductive, continuous, metallic catalyst layer deposited on a structured electrode will equilibrate to a uniform potential along the layer surface. Despite being equipotential, this continuous catalyst layer will not necessarily sustain a uniform current density. As Figure 3.6 demonstrates, the volume between adjacent microwires within a highly ordered array quickly depletes of reactants, and the electrode process becomes limited by mass transport from the bulk. Such depletion would render catalyst near the base of the microwire arrays less productive than catalyst material closer to the wire tops. For a non-uniform catalyst layer, all of the catalyst material is not necessarily at an equipotential but rather is at an electrochemical potential that is determined by the minority-carrier quasi-Fermi level at the local semiconductor-catalyst interface. As a consequence of being poised at different potentials, a non-uniform catalyst layer will result in spatially non-uniform current densities, which may decrease device efficiency.

The catalytic performance is convolved into catalyst structure, photoelectrode structure, light absorption, carrier transport, and mass transport. Challenges will remain for localizing catalyst material in regions along a structured solar absorber to maximize the performance. Similarly, the quantification of catalytic performance and isolation from other processes at structured photoelectrodes will remain a challenge in the development of solar water splitting devices. Structured photoelectrodes, however, will certainly relax the constraints on exchange rates or turnover frequencies.

3.2.5   Outlook

Structuring of solar absorbers offers advantages and disadvantages for a photoelectrochemical solar fuels device. Disadvantages include decreased catalyst utilization efficiency and decreases in open-circuit photovoltage. Advantages include decoupling light absorption and charge-carrier collection, enabling the use of less pure material, and decreasing the turnover constraints on the associated catalyst system. These advantages promise to decrease device costs by permitting use of lower amounts of less pure conventional materials (*e.g.* Si, GaAs, Pt, Ru) or in enabling the use of otherwise less ideal materials (*e.g.* oxide absorbers, non-noble catalysts).

Considering the different optical, carrier generation and mass transport phenomena that occur simultaneously at structured photoelectrodes, it is important to note that very few research groups are incorporating structures and characterizing multiple key phenomena. As this chapter illustrates, photoelectrode structuring is often advantageous to one aspect of overall performance while disadvantageous to another. The most effective studies will therefore combine experimentation and computation to investigate simultaneously several characteristics of a putative photoelectrochemical water-splitting device.

3.3   Review of Recently Demonstrated Advantages of Structured Materials for Photoelectrochemical Water Splitting

This section highlights some features in the development of structured semiconductors and catalysts for the hydrogen evolution reaction (HER), the oxygen evolution reaction (OER), and overall water splitting. We restrict this discussion to developments from the last five years and focus on solar water splitting as opposed to photovoltaics or other photoelectrochemical reactions. Thus, Section 3.3 represents a targeted assessment of the "state of the art" in solar water splitting rather than an exhaustive catalogue of all of the work performed in the field over the last five years.

While research must ultimately investigate solar absorption alongside catalysis, mass transport, and light management, few research groups are addressing the multitude of challenges that exist in the production of a viable photoelectrochemical water splitting device. Therefore, this section presents separately recent experiments and results regarding (1) non-traditional photoelectrode materials with a range of morphologies, (2) the structuring and characterization of traditional semiconductor absorbers, and (3) structured catalyst materials. Finally, this chapter concludes with an outlook and broader considerations for research into practical and efficient photoelectrochemical water splitting.

3.3.1 Metal Oxide Photoelectrodes

Metal oxides have traditionally dominated the field of semiconductor-coupled oxygen evolution, due to the natural stability of the oxides under the highly oxidative conditions demanded by the OER. Accordingly, much recent work has focused on developing highly structured oxides for efficient oxygen evolution.

3.3.1.1 Hematite

Hematite (α-Fe$_2$O$_3$), is readily synthesized from inexpensive materials, can be doped n-type, and possesses a 2.2 eV band gap. These properties continue to drive extensive research by several groups. Recent studies have illuminated the fundamental challenge for water oxidation using hematite, namely poor transport properties and short minority-carrier lifetimes resulting in minority-carrier transport distances in the tens of nm.[34,35] Additionally, hematite exhibits strongly anisotropic behavior, with electrons and holes traveling more easily along the (001) crystal planes.[36] All of these difficulties point to the desirability of developing highly nanostructured materials that may permit facile charge carrier collection at the electrolyte interface.

The Grätzel group has successfully achieved control over both the structure and doping of hematite thin films and has demonstrated progressively greater efficiencies for conversion of incident white light to O$_2$ (g) by synthesizing nanostructured films using both physical and chemical deposition methods. Initial experiments utilized chemical vapor deposition to synthesize hematite films with nanoscale features, producing photocurrent densities of several mA cm^{-2} at the thermodynamic water oxidation potential.[37] Subsequent experiments employed solution-phase deposition and generated hematite thin films that achieved similarly large photocurrent densities by decoupling the chemical benefit of sintering from the concomitant increase in feature size.[38]

Several research groups have successfully coupled overlayers on Fe$_2$O$_3$ substrates for electronic or catalytic benefits. Zhong and Gamelin electrodeposited a cobalt oxide catalyst onto hematite photoelectrodes and observed enhanced oxygen-evolution activity at potentials less than the thermodynamic oxygen-evolution potential.[39] The Grätzel group noted a similar enhancement in catalytic activity by deposition of an iridium oxide co-catalyst or through the use of one of several group 13 oxides on the surface of nanostructured Fe_2O_3 films.[40,41]

Another recent area of interest in hematite photoelectrochemistry is the use of plasmonic nanoparticles to enhance light absorption in ultrathin films. Deposition of gold nanoparticles that were tens of nm in diameter increased the light absorption in Fe_2O_3 due to the coupling and scattering of the incoming light with plasmon modes within the nanoparticles.[42,43] Despite the increased light absorption due to plasmonic coupling, Thimsen et al. saw no enhancement in water splitting photocurrent.[42] Thomann and coworkers, however, saw an enhancement by more than a factor of ten for photocurrents in the spectral region corresponding to surface plasmon resonances.[43]

Hematite has very promising physical and optical properties, and has garnered much research interest as a water splitting photoanode. However, hematite photoelectrodes have yet to attain sufficient photovoltage (>1 V) or photocurrent density (>5 mA cm^{-2}) for efficient water oxidation. Hamann recently provided a thorough discussion of the reasons for the disparity between the expected and actual performance, concluding primarily that methods need to be developed to suppress, or out-compete, charge-carrier recombination at the hematite surface to allow for efficient, productive hole collection.[44]

3.3.1.2   Tungsten Oxide

Tungsten oxide has been the subject of active investigation as a semiconductor for water oxidation. Tungsten oxide does not suffer from the same unfavorable electronic properties as hematite, as in crystalline form it has reasonably long minority carrier lifetimes and isotropic electronic properties. Tungsten oxide is also stable in acidic media. A challenge for WO_3, however, is its indirect band gap of ~ 2.7 eV, which prevents absorption of a significant fraction of the solar spectrum. Thus, work on nanostructured WO_3 has focused on achieving high quantum efficiencies, on attempts to reduce the band gap, and on the development of nanostructures for more efficient charge capture at the semiconductor-solution interface.

Several recent studies have leveraged the morphology of nanocrystalline WO_3 to maximize light capture and carrier collection. One approach by Hong et al. involved the solution phase, hydrothermal synthesis of nanocrystals of WO_3.[45] Calcination at various temperatures afforded control over the grain size of the material. Films that were calcined at 600 °C gave the highest energy-conversion efficiencies, attributable to the best compromise between high crystallinity and high hole collection efficiency.

In another approach, researchers synthesized a unique "flake-wall" morphology in WO_3 nanoparticles that were prepared by a solvothermal method.[46] The technique allowed the deposition of structures directly onto conductive tin-oxide-coated glass plates by seeding with an underlayer of nanocrystalline WO_3. Annealed solvothermal films generated nearly an order of magnitude higher photocurrent densities than un-annealed solvothermal films or the electrodes that had only the nanocrystalline underlayer. In yet another approach, researchers synthesized WO_3 in a unique inverse opal morphology to enhance light absorption of the films.[47] This inverse-opal structure led to a doubling of photocurrent densities relative to those of conventional, compact films.

Several research groups have also increased the rate of oxygen evolution on WO_3 by the addition of catalysts to the surface of the material. Liu *et al.* used atomic layer deposition (ALD) to deposit an overlayer of Mn oxide on WO_3 and saw a small enhancement in the photocurrent, but a large enhancement of O_2 yield.[48] Seabold and Choi deposited a cobalt oxide catalyst by electrodeposition and observed greatly improved stability for water oxidation on WO_3.[49]

Recently the Lewis group reported that in the absence of catalyst, the faradaic efficiencies of oxygen evolution in acidic media are quite low on WO_3, as oxidation of the electrolyte counter ion (*e.g.* chloride, sulfate, phosphate) is favored over the OER.[50] Catalysts were added either to decompose the oxidized counter ion or to facilitate direct transfer of holes to water molecules, and oxygen evolution proceeded with high faradaic yield.

Given its rather large indirect band gap, WO_3 films will not be viable for efficient photoelectrochemical water splitting unless credible methods can be developed for significantly increasing their visible light absorption without disrupting carrier transport properties. Further details of techniques employed in synthesizing nanostructured and doped WO_3 can be found in a recent review by Liu *et al.*[51]

3.3.1.3 Bismuth Vanadate

The optical properties of the monoclinic form of $BiVO_4$ make it similar to, or even more attractive than, WO_3 for driving the OER using visible light irradiation. $BiVO_4$ has a direct band gap of approximately 2.4 eV, allowing efficient absorption of blue photons. Additionally, the band-edge alignment allows generation of a relatively large photovoltage for water oxidation compared to other metal oxides.[52,53]

Several research groups have improved photoelectrochemical water oxidation by the introduction of controlled structure into $BiVO_4$ films. Berglund *et al.* found that deposition of nanostructured, vanadium-rich $BiVO_4$ films resulted in OER activity under illumination that was several times higher than that of stoichiometric films, even after the excess vanadium had dissolved into the electrolyte.[54] Luo *et al.* synthesized $BiVO_4$ films by a chemical bath deposition, resulting in a variety of microstructured morphologies.[55] Interestingly, the highest photocatalytic activities were obtained for films that were relatively compact, although with smaller crystallite size. This result implies that diffusion of holes to the surface may be a limiting factor in the performance of crystalline $BiVO_4$ films.

The photoelectrochemical water oxidation activity of $BiVO_4$ films can also be enhanced by addition of the group VI metals Mo and W. The Bard group carried out a combinatorial study of the Bi-V-W oxide system using a scanning electrochemical microscopy (SECM) technique.[56] They found that a material consisting of the ratios $4.5:5:0.5$ of Bi, V, and W, respectively, gave the highest photoactivity, both in the combinatorial experiment and in bulk film studies. In another approach, Hong *et al.* synthesized thin film heterojunctions of $BiVO_4$ and WO_3 by sequential deposition of multilayers of the respective precursors.[57] They found that the most active heterojunction consisted of one layer of $BiVO_4$ atop three layers of WO_3. The enhancement was attributed to leveraging of the favorable charge-transfer properties of WO_3 alongside the high light absorption in $BiVO_4$.

Perhaps the most successful approach in the development of $BiVO_4$ photoanodes has been with films that incorporated overlayers of oxide co-catalysts for the OER. The Bard group leveraged their SECM approach to screen combinatorially a variety of catalytic materials deposited onto Bi-W-V oxide films.[58] Interestingly, they found that the highly active OER catalyst iridium oxide did not significantly enhance the photoactivity, whereas the less active catalyst materials Pt and Co_3O_4 did enhance the photocurrent densities by nearly an order of magnitude. They attributed this result to the fact that the interfacial properties of the semiconductor/catalyst junction are critically important in determining overall photoelectrode efficiency.

Several other research groups have found that deposition of oxide catalysts on the surface of $BiVO_4$ photoelectrodes significantly enhances its water oxidation efficiency. Pilli and coworkers observed an enhancement in water oxidation activity for $BiVO_4$ both upon doping with 2% Mo in place of V as well as upon deposition of a cobalt oxide catalyst onto the surface.[59] Zhong *et al.* saw a similar enhancement with a Co oxide co-catalyst on tungsten-substituted $BiVO_4$, which they attributed to efficient suppression of surface recombination on application of the co-catalyst.[60] Seabold and Choi obtained high photocurrent densities (on the order of 2 mA cm^{-2} short-circuit current density) for water oxidation under AM1.5 illumination for a $BiVO_4$ film synthesized by an electrodeposition/calcination technique and coated with an iron oxyhydroxide co-catalyst.[61] These composite films were stable for several hours under oxygen evolution conditions while being illuminated in neutral aqueous electrolytes.

Further efforts are warranted in the suppression of recombination losses in the $BiVO_4$ bulk, as well as coupling efficient OER co-catalysts to the surface for simultaneous enhancement in catalytic activity and suppression of surface recombination losses. Additionally, systematic efforts in the generation of micro- or nanostructured $BiVO_4$ may produce higher charge-carrier collection efficiencies, similar to what has been seen with WO_3 and Fe_2O_3. Also it is important to determine the stability limits of $BiVO_4$ in terms of pH and electrochemical potential in aqueous solutions. With success in these areas, $BiVO_4$ may emerge as a very promising metal oxide photoanode for water oxidation.

3.3.1.4   Other Oxide Systems

Several other systems that are composed of transition metal oxides have gained recent interest for photoelectrochemical water splitting. The Mallouk group demonstrated a variation on the dye-sensitized solar cell (DSSC) as an oxygen-evolution system.[62] As with a conventional DSSC, the system is comprised of titania nanoparticles that are functionalized with light-absorbing ruthenium bipyridine dyes. Conventional dye-sensitized cells use a reversible redox couple to regenerate the dye from its oxidized state following electron injection into a mesoporous TiO_2 electrode, but the Mallouk system transfers highly oxidizing holes from the dye to an IrO_2 co-catalyst, which then oxidizes water. The electrons injected into the titania layer produce hydrogen at the counter electrode when an additional bias is provided. This type of system allows for utilization of visible photons in oxide-based water-splitting systems, but the overall efficiencies need to be improved. Additionally, the long-term stability of the sensitizer complexes under highly oxidizing conditions needs elucidation.

Several research groups have explored copper (I) oxide (Cu_2O) as a p-type oxide semiconductor material. The Lewis group demonstrated stable photoelectrochemistry, and photovoltages of over 800 mV, from thermally prepared Cu_2O in contact with non-aqueous redox couples.[63] However, stability was lost in aqueous media due to reduction of the oxide to copper metal on the surface, resulting in the subsequent loss of photovoltage. The Grätzel group circumvented the problem of instability of Cu_2O by introducing protective layers of aluminum/zinc oxide and TiO_2 by atomic layer deposition.[64] Their system was able to evolve hydrogen stably for >1 hour using a Pt co-catalyst, albeit with low photovoltages. The Choi group explored the electrodeposition of Cu_2O, and with careful tuning of electrodeposition conditions grew the semiconductor in controllably branched, dendritic structures.[65] Further work on Cu_2O may yield structured materials that can perform the HER efficiently.

3.3.2 High Aspect-Ratio Structures

Over the last decade, significant research efforts have investigated semiconductor nanowires and microwires. The greatest proportion of these efforts has focused on use of wire structures as candidate materials for thin-film photovoltaics and other optoelectronic devices. Several research groups have also developed the wire geometry specifically for photoelectrochemistry and water splitting. The recent review literature contains extensive discussion of the history and progress of nanowire fabrication and solid state devices.[66–72] Here we provide a short overview of the progress on structured Si and III-V semiconductor developments with respect to their potential uses for water splitting.

3.3.2.1 Si Structures

Chemical etching procedures have been developed to generate rods and/or porous structures in silicon in a top-down manner *via* anisotropic metal-assisted etching.[73] A highly porous array of silicon nanowires is produced that has the electronic quality of the parent wafer.[74] This top-down nanostructuring allows for generation of high photocurrent densities, due to the significant antireflective properties of the etched Si; hence the nickname "black silicon" is given to such nanostructures.[75]

Researchers at the National Renewable Energy Laboratory have successfully used black silicon photocathodes for photoelectrochemical hydrogen generation.[76] Significant enhancements in photocurrent density were observed due to the antireflective properties of the nanoporous coating. Additionally, the onset of hydrogen evolution was shifted positive by several hundred mV for black silicon electrodes relative to planar Si controls. The catalytic shift was attributed to relaxed catalytic turnover requirements as a result of increased Si surface area, and was also likely due to the advantageous presence of trace Au remaining on the porous Si surface after the metal-assisted etching procedure. Chemically-etched silicon nanowire photoelectrochemical solar cells utilizing redox couples other than H^+/H_2 or O_2/H_2O have demonstrated remarkably high solar energy conversion efficiencies,[48,77,78] implying that the water splitting half reactions could also be driven efficiently with the proper electrode architectures and catalysts.

In addition to nanostructures, Si microstructures have also been fabricated by top-down etching procedures, and these systems have been utilized for solar hydrogen generation. Recently, Hou *et al.* demonstrated a photocathode based on p-type Si micropillars that were generated using a dry etching procedure.[79] These pillars were decorated with a molybdenum sulfide cubane cluster as an earth-abundant hydrogen evolution catalyst. The structured composite device generated a photocurrent density of ~ 10 mA cm^{-2} at the reversible potential for hydrogen-evolution. The observed current density was larger than the photocurrent density generated by a planar control sample. These photocathodes also evolved hydrogen stably for at least one hour.

Silicon nano- and microstructures have also been prepared using a bottom-up synthesis approach that takes advantage of a vapor-liquid-solid growth mechanism, in which a Si/metal eutectic selectively crystallizes silicon onto a substrate from a vapor-phase precursor such as $SiCl_4$ or various silanes.[80] Several research groups have demonstrated VLS-based Si nanowires and microwires in photoelectrochemical solar cells.[15,19,81–84] However, there are only a few examples of VLS-grown silicon structures for photoelectrochemical hydrogen evolution.[20,83] Si microwire arrays can also be embedded in a polymer and removed from the growth substrate while retaining their photoelectrochemical activity,[85–87] potentially allowing fabrication of water-splitting device architectures that utilize tandem solar absorbers on either side of an ionically conductive membrane.[79,88,89]

3.3.2.2 III-V Structures

The III-V semiconductors gallium arsenide and indium phosphide have both been utilized as photocathodes for the efficient generation of hydrogen from acidic electrolytes.[90–92] Since both of these materials have direct band gaps, highly structured morphologies might not be needed to improve light absorption or charge-carrier collection. Nevertheless, controlled structuring of the semiconductor or catalyst layers in III-V photoelectrodes may be desirable to enable novel device geometries or to relax catalyst turnover requirements.

Gallium phosphide is a III-V semiconductor with an indirect fundamental band gap at relatively high energy, making it an interesting candidate as a structured absorber in tandem water-splitting systems. Methods have been devised for top-down formation of wire or porous structures in GaP through anisotropic etching.[93] Recently, the Maldonado group used this electro-chemical etching technique to generate nanostructured n-type and p-type gallium phosphide with vertically oriented pores of varying depth.[94,95] The structured material showed much greater efficiency than planar controls for collecting excited charge carriers in both regenerative and fuel-forming modes.

Thus far, oxygen or hydrogen evolution has not been reported from highly structured III-V semiconductors. This is due in part to the low stability of III-V semiconductors under the reducing or oxidizing conditions required for the HER and OER, respectively. An illustrative example is the work of Khasalev and Turner on a full water-splitting system based on multi-junction, planar, III-V semiconductors.[96] Although this system generated hydrogen and oxygen with high energy-conversion efficiency, it was stable for only a few hours. With continued progress in nanoscale control over composition and morphology, water-splitting devices based on III-V semiconductors might be made stable under long-term operation.

3.3.3   Water Splitting by Colloidal Particles

Many research groups have attempted to split water using colloidal particles. A recent analysis has suggested that colloidal water splitting is the best approach for efficient, scalable solar hydrogen generation, provided that the colloidal species are composed of abundant elements and provided that that inexpensive and safe methods for separating the products from an explosive mixture can be developed.[97]

To date, there are very few examples of full water splitting that use colloidal particle suspensions in the absence of sacrificial reagents. The Domen group has reported the net generation of H_2 and O_2 gases in colloidal systems through careful suppression of the parasitic (and thermodynamically downhill) back-reactions.[98,99] The researchers relied on an overlayer of CrO_3 on Rh particles deposited onto colloidal particles of GaN/ZnO solid solutions. The chromia overlayer enabled net water splitting on these particles by affording selectivity of the Rh cores for the HER, due to selective permeability of CrO_3 to protons and H_2 but not water or oxygen. Work from the Domen group is detailed elsewhere in this volume, but this selective system is worth noting for its control over nanostructure, providing the necessary components for overall water splitting.

3.3.4   Water Splitting Catalysis by Structured Materials

Several research groups have advanced the development of nanostructured catalysts for the HER and the OER. Significant work has focused on the replacement of the noble metals Pt, Ru, and Ir that are commonly used in proton-exchange membrane electrolyzers, with non-noble alternatives, or on leveraging structured geometries to minimize the quantities of expensive elements. The following is a brief discussion of several research highlights in the development of heterogeneous HER and OER catalysts, with special emphasis on systems that have been developed with control of features at the nanoscale.

3.3.4.1   Hydrogen Evolution

Several research groups have recently demonstrated molybdenum sulfides as catalysts for the HER.[100] These sulfides have been widely studied for their use in hydrodesulfurization,[101] but have demonstrated viability as hydrogen-evolution catalysts based on DFT calculations from the Nørskov group that suggested the HER catalytic activity of such systems could approach that of pure Pt.[102] Subsequent experimental work in the Chorkendorff group demonstrated facile HER catalysis at the edge sites of nanocrystalline MoS_2 lamellae.[103]

Recent work from the Chorkendorff and Jaramillo groups sought to maximize the density of active edge sites of MoS_2. For example, a precursor was deposited onto high surface-area carbon paper and subsequently annealed under a sulfidizing atmosphere to yield a supported catalyst of nominally high surface area.[104] This carbon-supported metal sulfide catalyst produced exchange current densities on the order of 10^{-6} A cm^{-2} based on estimated total surface area. The activity was increased by addition of small amounts of Co salts to the precursor solutions. The Jaramillo group reported catalytic activity from a structured MoO_3-MoS_2 core-shell morphology.[47] Synthesized by sulfidizing the outer layer of a nanostructured MoO_3 layer, this morphology avoided ohmic losses due to high resistivity of the MoS_2, which is far less conductive than MoO_3. The high surface area MoO_3-MoS_2 composite attained high geometric activity and demonstrated extended stability under acidic conditions.

Transition metal sulfide electrocatalysts can be deposited from molecular precursors onto structured Si for photoelectrochemical hydrogen evolution. As discussed previously, Hou *et al.* observed efficient catalysis to yield a net energy conversion of incoming light energy to stored energy in $H_2(g)$.[79] A subsequent systematic study of transition metal sulfides derived from molecular precursors observed the highest energy conversion efficiency from Mo and Cu/Mo sulfides, but the highest stability was exhibited by pure Mo sulfide.[105]

Amorphous, rather than crystalline, molybdenum sulfide also is an efficient HER catalyst. This active material can be either electrodeposited from ammonium thiomolybdate under anaerobic conditions or can be chemically synthesized by precipitation of nanoparticles.[106,107] Interestingly, the electrodeposited material is formed by passing both anodic and cathodic current through the working electrode, which is unusual for a material intended only to catalyze a reduction reaction. Similar to the results of Chorkendorff *et al.*,

Merki *et al.* observed an enhancement in activity upon the addition of the first row metals Fe, Ni, and Co to the amorphous films.[107]

Several groups have characterized the catalytic activity of molybdenum-containing alloys. Rocheleau *et al.* studied a Co-Mo catalyst for solar water splitting,[108] and the Lewis group studied a Ni-Mo alloy for photoelectrochemical hydrogen evolution.[83] Previously studied for alkaline electrolyzer applications, Ni-Mo has demonstrated high activity over thousands of hours.[109,110] Lewis and coworkers showed that an alloy of Ni and Mo could be electrodeposited directly onto p-type Si substrates for efficient hydrogen evolution under mildly acidic conditions. The as-deposited films were nanoparticulate, and the apparent catalytic activity increased when the material was deposited onto Si microwire arrays, due to the multi-scale roughness enhancement afforded by the nanostructured catalyst on the microstructured semiconductor substrate. Researchers at Sun Catalytix have recently employed a related catalyst system, Ni-Mo-Zn, which was integrated into a water splitting cell that utilized a triple junction amorphous silicon solar cell as a substrate.[111] Three-component Ni-Mo-X catalysts have been previously studied for alkaline electrolysis,[112,113] whereas the Sun Catalytix researchers reported stable performance under buffered conditions at neutral pH.

3.3.4.2 Oxygen Evolution

Significant recent efforts have targeted understanding and development of cobalt oxide for the OER. The 2008 publication by Kanan and Nocera stimulated recent investigations into cobalt oxide catalysts in which amorphous Co oxide is electrodeposited at neutral pH from Co salts.[114] These recent studies follow from previous research on Co oxides for water oxidation in alkaline and neutral pH.[115–117] The initial experiments suggested that the phosphate buffer played some role in the formation of the catalyst, although subsequent work has shown that other buffers, or Co metal films, generate similar coatings.[118,119]

The reports on Co oxide catalysts have driven efforts to understand the mechanism of their operation. Nocera and coworkers utilized EPR and X-ray techniques to suggest a cubane structure for the active species, where the Co is proposed to undergo a redox transition from Co^{III} to Co^{IV} during catalytic turnover.[120] Studies of Co oxide OER catalysts over a range of pH suggest that the active catalytic mechanism transitions from primarily heterogeneous to primarily homogeneous at pH values below 3.[121]

The Nocera group has investigated Co oxide as a commercially viable system for efficient solar driven oxygen evolution. They generated current densities of tens of mA cm^{-2} for oxygen evolution at overpotentials below 300 mV in neutral pH at amorphous Co oxide deposited onto high surface area Ni foams.[122] Additionally, this electrodeposited Co oxide catalyst resisted poisoning by Ca^{2+} ions and other contaminants found in natural waters, and the catalyst exhibited a linear increase in catalytic activity with mass loading. This linear scaling of activity with mass loading implies that the electrodeposited Co oxide catalyst exhibits a large electrochemically active surface area for the

OER, either as a result of nanoscale features or due to three-dimensional porosity in the film. The same methodology for the deposition of Co oxide films has been used in several functional systems for solar water splitting incorporating amorphous and crystalline Si semiconductors.[111,123,124]

Several researchers have incorporated Co oxide catalysts as active materials for driving the OER on semiconductor absorber substrates. The Gamelin group deposited amorphous Co oxide onto nanocrystalline hematite, yielding an increase in the overall efficiency for oxygen evolution.[39,125] The Choi group deposited Co oxide onto ZnO and hematite nanostructures under illumination and found that the catalyst morphology and energy-conversion efficiency could be modulated by judicious control of the deposition conditions.[126,127] The Choi group also deposited Co oxide onto WO_3 photoelectrodes and observed a significant increase in the selectivity of the composite film toward oxygen evolution relative to formation of peroxo-species, which also enhanced the long-term stability of the oxygen-evolution system.[49]

In addition to studies of non-noble metal catalysts, a traditional water oxidation catalyst, IrO_2, is of interest with the goal of minimizing the iridium loading while maintaining high oxygen evolution activity. The Murray group reported a mesoporous IrO_2 film consisted of nanoscale oxide particles that were synthesized in the solution phase and then flocculated onto an electrode that was maintained at positive bias.[128] Significant oxygen-evolution activity with 100% Faradaic efficiency was observed at overpotentials as low as 250 mV. Additionally, the Mallouk group has published several deposition methods that produce nanoscale iridium oxide films for electrochemical oxygen evolution.[129]

Nanoscale noble metals and oxides have recently been explored for both of the water splitting half reactions on planar Si electrodes that are protected from deleterious interfacial reactions by thin oxide layers. Lewerenz and Muñoz have carried out extensive work on so-called "nanoemitter" junctions between planar Si and either Pt or Ir metals accompanied by surface Si oxide. They demonstrated stable, sustained electrochemical reactions on Si surfaces under conditions that normally result in silicon degradation.[130] McIntyre, Chidsey, and coworkers recently leveraged ALD-deposited TiO_2 for the protection of planar n-type Si electrodes for sustained oxygen evolution under alkaline conditions using an evaporated Ir co-catalyst.[131]

3.3.5 Advances in Modeling Heterogeneous Catalysis

Computational modeling of active redox catalysis for the water-splitting half reactions has produced notable, recent advances. Several research groups, led primarily by Nørskov and collaborators, have recently undertaken the challenge of developing DFT models that are sufficiently accurate to *predict* active materials for the efficient evolution of hydrogen and oxygen, as well as other fuel-forming reactions.[132,133]

Recent DFT modeling studies indicate a need for nanoscale control over the catalyst composition and morphology. Nørskov *et al.* predicted that the activity of MoS_2 for hydrogen evolution would stem primarily from active sites at the edges of the lamellar crystal structure.[102] This prediction was confirmed experimentally by the Chorkendorff lab.[103] These results imply that molybdenum sulfides must be nanostructured to maximize the proportion of step edge sites and thus obtain an optimum hydrogen evolution efficiency.

Other recent work showed that composites consisting of adlayers of one metal on another could attain higher catalytic activities for the HER than either of the constituent metals.[134,135] These results suggest that new structural/compositional motifs can be used for new, highly catalytic nanomaterials to be coupled with light absorbers for efficient water splitting.

DFT modeling results from Nørskov and Rossmeisl indicate conserved differences in energy between intermediates for the OER on transition metal oxides.[136,137] These relationships may result in an upper bound for the catalytic activity of any metal oxide OER catalyst that is modest in comparison with the high activities of noble metals for hydrogen evolution. The development of multifunctional catalysts that consist of chemically distinct "active sites" for various primary steps could circumvent this limitation. These sites would need to be located sufficiently closely to allow facile exchange of intermediate species, and would thus require control over composition and structure at the nanoscale. With several applications beyond photoelectrochemical water splitting, the successful demonstration of a rationally designed nanoscale, multifunctional electrocatalyst would be very significant.

3.3.6 Broader Considerations – Beyond Small

Recent efforts in the development of systems for photoelectrochemical water splitting demonstrate the need for control over composition and morphology at the micro, nano, and even atomic scale. In addition, successful photoelectrochemical water splitting systems based on nanostructured film morphologies require a proper understanding of mass transport in relation to nanostructured geometries. Alternatively, colloidal water-splitting systems may require some form of convection to maintain particles in the suspended form. Both of these characteristics demand an understanding of diffusion and convection of species from the nanoscale to the tens or hundreds of micron scale. Efficient generation of O_2 and H_2 gas implies copious bubble formation, the dynamics of which may significantly influence key characteristics such as light absorption and reactant transport. New work must address all such features of an overall water splitting process.

The potential need for gas separation and pressure management in water splitting systems also requires study. Systems with no physical barrier between the oxygen-evolving anode and the hydrogen-evolving cathode may require a means to minimize losses from the comparatively facile reverse reactions of oxygen reduction and hydrogen oxidation, respectively. Barrierless systems will also require schemes to safely manage and separate an explosive H_2/O_2 mixture. Systems that employ a separator between the anode and cathode must minimize ohmic losses due to ionic transport over distances between the two compartments. Another key concern for multi-compartment systems is the potential need for active pressure management, as the 2:1 stoichiometry of hydrogen and oxygen evolution, respectively, implies rapid buildup of differential pressures that may affect sustained operation of the system.

Many key insights involving distance scales larger than a few microns can be gained from previous experience as well as from collaborations between the fields of chemical engineering and systems design. If we are to have economical solar water splitting systems in the near future, both large scale and small scale developments must continually feed back to one another to efficiently move toward functional, scalable solutions.

Acknowledgements

This work was supported in part by the Joint Center for Artificial Photosynthesis, a DOE Energy Innovation Hub. The contribution from NSL was supported through the Office of Science of the U.S. Department of Energy under award No. DE-SC0004993; the contributions from JRM and RLG were supported by BP and by the U.S. Department of Energy under award No. DE-FG02-03ER15483. JRM additionally acknowledges the U.S. Department of Energy Office of Science for a graduate research fellowship.

References

1. M. X. Tan, P. E. Laibinis, S. T. Nguyen, J. M. Kesselman, C. E. Stanton and N. S. Lewis, *Prog. Inorg. Chem.*, 1994, **41**, 21–144.
2. M. G. Walter, E. L. Warren, J. R. McKone, S. W. Boettcher, Q. X. Mi, E. A. Santori and N. S. Lewis, *Chem. Rev.*, 2010, **110**, 6446–6473.
3. S. Maldonado, A. G. Fitch and N. S. Lewis, in *Nanostructured and photoelectrochemical systems for solar photon conversion*, ed. M. D. Archer and A. J. Nozik, Imperial College Press, London, 2008, vol. 3.
4. J. R. Bolton, S. J. Strickler and J. S. Connolly, *Nature*, 1985, **316**, 495–500.
5. A. J. Bard and M. A. Fox, *Acc. Chem. Res.*, 1995, **28**, 141–145.
6. M. Grätzel and J. R. Durrant, in *Nanostructured and photoelectrochemical systems for solar photon conversion*, ed. M. D. Archer and A. J. Nozik, Imperial College Press, London, 2008, vol. 3.
7. S. M. Sze and K. K. Ng, *Physics of Semiconductor Devices*, 3rd edn., Wiley-Interscience, New York, 2007.
8. M. L. Cohen and J. R. Chelikowsky, *Electronic structure and optical properties of semiconductors*, 2nd edn., Springer-Verlag, Berlin, 1988.
9. M. A. Green and M. J. Keevers, *Prog. Photovoltaics*, 1995, **3**, 189–192.
10. H. C. Casey, D. D. Sell and K. W. Wecht, *J. Appl. Phys.*, 1975, **46**, 250–257.
11. V. Schlosser, *IEEE Trans. Electron Dev.*, 1984, **31**, 610–613.
12. S. E. Han and G. Chen, *Nano Lett.*, 2010, **10**, 4692–4696.
13. E. Yablonovitch, *J. Opt. Soc. Am.*, 1982, **72**, 899–907.

14. B. M. Kayes, H. A. Atwater and N. S. Lewis, *J. Appl. Phys.*, 2005, **97**, 114302.
15. J. R. Maiolo, B. M. Kayes, M. A. Filler, M. C. Putnam, M. D. Kelzenberg, H. A. Atwater and N. S. Lewis, *J. Am. Chem. Soc.*, 2007, **129**, 12346–12347.
16. M. D. Kelzenberg, S. W. Boettcher, J. A. Petykiewicz, D. B. Turner-Evans, M. C. Putnam, E. L. Warren, J. M. Spurgeon, R. M. Briggs, N. S. Lewis and H. A. Atwater, *Nature Mater.*, 2010, **9**, 239–244.
17. M. D. Kelzenberg, PhD thesis, California Institute of Technology, 2010.
18. N. S. Lewis, *J. Electrochem. Soc.*, 1984, **131**, 2496–2503.
19. S. W. Boettcher, J. M. Spurgeon, M. C. Putnam, E. L. Warren, D. B. Turner-Evans, M. D. Kelzenberg, J. R. Maiolo, H. A. Atwater and N. S. Lewis, *Science*, 2010, **327**, 185–187.
20. S. W. Boettcher, E. L. Warren, M. C. Putnam, E. A. Santori, D. Turner-Evans, M. D. Kelzenberg, M. G. Walter, J. R. McKone, B. S. Brunschwig, H. A. Atwater and N. S. Lewis, *J. Am. Chem. Soc.*, 2011, **133**, 1216–1219.
21. T. Stempel, M. Aggour, K. Skorupska, A. Munoz and H. J. Lewerenz, *Electrochem. Commun.*, 2008, **10**, 1184–1186.
22. M. L. Rosenbluth and N. S. Lewis, *J. Am. Chem. Soc.*, 1986, **108**, 4689–4695.
23. M. C. Putnam, D. B. Turner-Evans, M. D. Kelzenberg, S. W. Boettcher, N. S. Lewis and H. A. Atwater, *Appl. Phys. Lett.*, 2009, **95**.
24. A. J. Bard and L. R. Faulkner, *Electrochemical Methods: Fundamentals and Applications*, 2nd. edn., Wiley, New York, 2001.
25. B. Scharifker and G. Hills, *J. Electroanal. Chem.*, 1981, **130**, 81–97.
26. J. Mostany, J. Mozota and B. R. Scharifker, *J. Electroanal. Chem.*, 1984, **177**, 25–37.
27. R. M. Penner, M. J. Heben and N. S. Lewis, *Anal. Chem.*, 1989, **61**, 1630–1636.
28. G. P. Kalaignan and Y. S. Kang, *J. Photoch. Photobio. C*, 2006, **7**, 17–22.
29. Y. Lin, Y. T. Ma, L. Yang, X. R. Xiao, X. W. Zhou and X. P. Li, *J. Electroanal. Chem.*, 2006, **588**, 51–58.
30. W. Hyk and J. Augustynski, *J. Electrochem. Soc.*, 2006, **153**, A2326–A2341.
31. N. Papageorgiou, M. Gratzel and P. P. Infelta, *Sol. Energ. Mat. Sol. C*, 1996, **44**, 405–438.
32. J. J. Lee, G. M. Coia and N. S. Lewis, *J. Phys. Chem. B*, 2004, **108**, 5269–5281.
33. C. Xiang, A. C. Meng and N. S. Lewis, *Proc. Natl. Acad. Sci. USA*, 2012, **109**, 15622–15627.
34. B. M. Klahr and T. W. Hamann, *J. Phys. Chem. C.*, 2011, **115**, 8393–8399.
35. K. Sivula, F. Le Formal and M. Grätzel, *ChemSusChem*, 2011, **4**, 432–449.
36. C. M. Eggleston, A. J. A. Shankle, A. J. Moyer, I. Cesar and M. Grätzel, *Aquat. Sci.*, 2009, **71**, 151–159.

37. A. Kay, I. Cesar and M. Gratzel, *J. Am. Chem. Soc.*, 2006, **128**, 15714–15721.
38. J. Brillet, M. Gra, K. Sivula and P. Fe, *Nano Lett.*, 2010, **10**, 4155–4160.
39. D. K. Zhong and D. R. Gamelin, *J. Am. Chem. Soc.*, 2010, **132**, 4202–4207.
40. T. Hisatomi, F. Le Formal, M. Cornuz, J. Brillet, N. Tétreault, K. Sivula and M. Grätzel, *Energy Environ. Sci.*, 2011, **4**, 2512–2515.
41. S. D. Tilley, M. Cornuz, K. Sivula and M. Grätzel, *Angew. Chem., Int. Ed.*, 2010, **122**, 6549–6552.
42. E. Thimsen, F. Le Formal, M. Grätzel and S. C. Warren, *Nano Lett.*, 2011, **11**, 35–43.
43. I. Thomann, B. A. Pinaud, Z. Chen, B. M. Clemens, T. F. Jaramillo and M. L. Brongersma, *Nano Lett.*, 2011, **11**, 3440–3446.
44. T. W. Hamann, *Dalton Trans.*, 2012, **41**, 7830–7834.
45. S. J. Hong, H. Jun, P. H. Borse and J. S. Lee, *Int. J. Hydrogen Energy*, 2009, **34**, 3234–3242.
46. F. Amano, D. Li and B. Ohtani, *Chem. Commun. (Cambridge, U.K.)*, 2010, **46**, 2769–2771.
47. X. Chen, J. Ye, S. Ouyang, T. Kako, Z. Li and Z. Zou, *ACS Nano*, 2011, **5**, 4310–4318.
48. R. Liu, Y. Lin, L.-Y. Chou, S. W. Sheehan, W. He, F. Zhang, H. J. M. Hou and D. Wang, *Angew. Chem., Int. Ed.*, 2011, **50**, 499–502.
49. J. A. Seabold and K.-S. Choi, *Chem. Mater.*, 2011, **23**, 1105–1112.
50. Q. Mi, A. Zhanaidarova, B. S. Brunschwig, H. B. Gray and N. S. Lewis, *Energy Environ. Sci*, 2012, **5**, 5694–5694.
51. X. Liu, F. Wang and Q. Wang, *Phys. Chem. Chem. Phys.*, 2012, **14**, 7894–7911.
52. K. Sayama, A. Nomura, T. Arai, T. Sugita, R. Abe, M. Yanagida, T. Oi, Y. Iwasaki, Y. Abe and H. Sugihara, *J. Phys. Chem. B*, 2006, **110**, 11352–11360.
53. A. Walsh, Y. Yan, M. N. Huda, M. M. Al-jassim and S.-H. Wei, *Chem. Mater.*, 2009, **21**, 547–551.
54. S. P. Berglund, D. W. Flaherty, N. T. Hahn, A. J. Bard and C. B. Mullins, *J. Phys. Chem. C.*, 2011, **115**, 3794–3802.
55. W. Luo, Z. Wang, L. Wan, Z. Li, T. Yu and Z. Zou, *J. Phys. D: Appl. Phys.*, 2010, **43**, 405402–405402.
56. H. Ye, J. Lee, J. S. Jang and A. J. Bard, *J. Phys. Chem. C*, 2010, **114**, 13322–13328.
57. S. J. Hong, S. Lee, J. S. Jang and J. S. Lee, *Energy Environ. Sci.*, 2011, **4**, 1781–1781.
58. H. Ye, H. S. Park and A. J. Bard, *J. Phys. Chem. C*, 2011, **115**, 12464–12470.
59. S. K. Pilli, T. E. Furtak, L. D. Brown, T. G. Deutsch, J. A. Turner and A. M. Herring, *Energy Environ. Sci.*, 2011, **4**, 5028–5028.
60. D. K. Zhong, S. Choi and D. R. Gamelin, *J. Am. Chem. Soc.*, 2011, **133**, 18370–18377.

61. J. A. Seabold and K.-S. Choi, *J. Am. Chem. Soc.*, 2012, **134**, 2186–2192.
62. W. J. Youngblood, S-H. A. Lee, K. Maeda and T. E. Mallouk, *Acc. Chem. Res.*, 2009, **42**, 1966–1973.
63. C. Xiang, G. M. Kimball, R. L. Grimm, B. S. Brunschwig, H. A. Atwater and N. S. Lewis, *Energy Environ. Sci.*, 2011, **4**, 1311–1311.
64. A. Paracchino, V. Laporte, K. Sivula, M. Grätzel and E. Thimsen, *Nat. Mater.*, 2011, **10**, 456–461.
65. C. M. McShane and K.-S. Choi, *J. Am. Chem. Soc.*, 2009, **131**, 2561–2569.
66. E. C. Garnett, M. L. Brongersma, Y. Cui and M. D. McGehee, *Annu. Rev. Mater. Res.*, 2011, **41**, 269–295.
67. A. I. Hochbaum and P. Yang, *Chem. Rev.*, 2010, **110**, 527–546.
68. X. Liu, Y. Z. Long, L. Liao, X. Duan and Z. Fan, *ACS Nano*, 2012, **6**, 1888–1900.
69. W. Lu and C. M. Lieber, *J. Phys. D: Appl. Phys.*, 2006, **39**, R387–R406.
70. W. Lu and C. M. Lieber, *Nat. Mater.*, 2007, **6**, 841–850.
71. P. J. Pauzauskie and P. Yang, *Mater. Today*, 2006, **9**, 36–45.
72. M. Yu, Y.-Z. Long, B. Sun and Z. Fan, *Nanoscale*, 2012, **4**, 2783–2796.
73. Z. Huang, N. Geyer, P. Werner, J. de Boor and U. Gösele, *Adv. Mater.*, 2011, **23**, 285–308.
74. K. Peng, X. Wang and S.-T. Lee, *Appl. Phys. Lett.*, 2008, **92**, 163103–163103.
75. H. M. Branz, V. E. Yost, S. Ward, K. M. Jones, B. To and P. Stradins, *Appl. Phys. Lett.*, 2009, **94**, 231121–231121.
76. J. Oh, T. G. Deutsch, H.-C. Yuan and H. M. Branz, *Energy Environ. Sci.*, 2011, **4**, 1690–1690.
77. K-Q. Peng, X. Wang, X.-L. Wu and S.-T. Lee, *Nano Lett.*, 2009, **9**, 3704–3709.
78. X. Shen, B. Sun, F. Yan, J. Zhao, F. Zhang, S. Wang, X. Zhu and S. Lee, *ACS Nano*, 2010, **4**, 5869–5876.
79. Y. Hou, B. L. Abrams, P. C. K. Vesborg, M. E. Björketun, K. Herbst, L. Bech, A. M. Setti, C. D. Damsgaard, T. Pedersen, O. Hansen, J. Rossmeisl, S. Dahl, J. K. Nørskov and I. Chorkendorff, *Nat. Mater.*, 2011, **10**, 434–438.
80. R. S. Wagner and W. C. Ellis, *Appl. Phys. Lett.*, 1964, **4**, 89–89.
81. A. P. Goodey, S. M. Eichfeld, K.-K. Lew, J. M. Redwing and T. E. Mallouk, *J. Am. Chem. Soc.*, 2007, **129**, 12344–12345.
82. E. A. Santori, J. R. Maiolo Iii, M. J. Bierman, N. C. Strandwitz, M. D. Kelzenberg, B. S. Brunschwig, H. A. Atwater and N. S. Lewis, *Energy Environ. Sci.*, 2012, **5**, 6867–6867.
83. J. R. McKone, E. L. Warren, M. J. Bierman, S. W. Boettcher, B. S. Brunschwig, N. S. Lewis and H. B. Gray, *Energy Environ. Sci.*, 2011, **4**, 3573–3583.
84. G. Yuan, H. Zhao, X. Liu, Z. S. Hasanali, Y. Zou, A. Levine and D. Wang, *Angew. Chem., Int. Ed.*, 2009, **48**, 9680–9684.

85. K. E. Plass, M. A. Filler, J. M. Spurgeon, B. M. Kayes, S. Maldonado, B. S. Brunschwig, H. A. Atwater and N. S. Lewis, *Adv. Mater.*, 2009, **21**, 325–328.

86. J. M. Spurgeon, S. W. Boettcher, M. D. Kelzenberg, B. S. Brunschwig, H. A. Atwater and N. S. Lewis, *Adv. Mater.*, 2010, **91125**, 3277–3281.

87. J. M. Spurgeon, K. E. Plass, B. M. Kayes, B. S. Brunschwig, H. A. Atwater and N. S. Lewis, *Appl. Phys. Lett.*, 2008, **93**, 032112–032112.

88. S. L. McFarlane, B. A. Day, K. McEleney, M. S. Freund and N. S. Lewis, *Energy Environ. Sci.*, 2011, **4**, 1700–1700.

89. J. M. Spurgeon, M. G. Walter, J. Zhou, P. A. Kohl and N. S. Lewis, *Energy Environ. Sci.*, 2011, **4**, 1772–1772.

90. E. Aharon-Shalom and A. Heller, *J Electrochem. Soc.*, 1982, **129**, 2865–2865.

91. A. Heller, *Solar Energy*, 1982, **29**, 153–162.

92. A. Heller and R. G. Vadimsky, *Phys. Rev. Lett.*, 1981, **46**, 1153–1153.

93. B. H. Erne, D. Vanmaekelbergh and J. J. Kelly, *J. Electrochem. Soc.*, 1996, **143**, 305–314.

94. K. Hagedorn, S. Collins and S. Maldonado, *J. Electrochem. Soc.*, 2010, **157**, D588–D588.

95. M. J. Price and S. Maldonado, *J. Phys. Chem. C.*, 2009, **113**, 11988–11994.

96. O. Khaselev and J. A. Turner, *Science*, 1998, **280**, 425–427.

97. B. D. James, G. N. Baum, J. Perez and K. N. Baum, *Technoeconomic Analysis of Photoelectrochemical (PEC) Hydrogen Production,* GS-10F-009, U.S. DOE, 2009.

98. K. Maeda, K. Teramura, D. Lu, N. Saito, Y. Inoue and K. Domen, *J. Phys. Chem. C.*, 2007, **111**, 7554–7560.

99. M. Yoshida, K. Takanabe, K. Maeda, A. Ishikawa, J. Kubota, Y. Sakata, Y. Ikezawa and K. Domen, *J. Phys. Chem. C*, 2009, **113**, 10151–10157.

100. A. B. Laursen, S. Kegnæs, S. Dahl and I. Chorkendorff, *Energy Environ. Sci.*, 2012, **5**, 5577–5577.

101. V. Chandra Srivastava, *RSC Adv.*, 2012, **2**, 759–759.

102. B. Hinnemann, P. G. Moses, J. Bonde, K. P. Jorgensen, J. H. Nielsen, S. Horch, I. Chorkendorff and J. K. Norskov, *J. Am. Chem. Soc.*, 2005, **127**, 5308–5309.

103. T. F. Jaramillo, K. P. Jorgensen, J. Bonde, J. H. Nielsen, S. Horch and I. Chorkendorff, *Science*, 2007, **317**, 100–102.

104. J. Bonde, P. G. Moses, T. F. Jaramillo, J. K. Norskov and I. Chorkendorff, *Faraday Discuss.*, 2008, **140**, 219–231.

105. Y. Hou, B. L. Abrams, P. C. K. Vesborg, M. r. E. Björketun, K. Herbst, L. Bech, B. Seger, T. Pedersen, O. Hansen, J. Rossmeisl, S. r. Dahl, J. K. Nórskov and I. Chorkendorff, *J. Photonics Energy*, 2012, **2**, 026001–026001.

106. D. Merki, S. Fierro, H. Vrubel and X. Hu, *Chem. Sci.*, 2011, **2**, 1262–1262.

107. D. Merki, H. Vrubel, L. Rovelli, S. Fierro and X. Hu, *Chem. Sci.*, 2012, **3**, 2515–2525.

108. R. E. Rocheleau, E. L. Miller and A. Misra, *Energy Fuels*, 1998, **12**, 3–10.
109. D. E. Brown, M. N. Mahmood, A. K. Turner, S. M. Hall and P. O. Fogarty, *Int. J. Hydrogen Energy*, 1982, **7**, 405–410.
110. D. E. Brown, M. N. Mahmood, M. C. M. Man and A. K. Turner, *Electrochim. Acta*, 1984, **29**, 1551–1556.
111. S. Y. Reece, J. A. Hamel, K. Sung, T. D. Jarvi, A. J. Esswein, J. J. H. Pijpers and D. G. Nocera, *Science*, 2011, **334**, 645–648.
112. B. E. Conway, H. Angersteinkozlowska, M. A. Sattar and B. V. Tilak, *J. Electrochem. Soc.*, 1983, **130**, 1825–1836.
113. B. E. Conway and L. Bai, *J. Chem. Soc., Faraday Trans. I*, 1985, **81**, 1841–1862.
114. M. W. Kanan and D. G. Nocera, *Science*, 2008, **321**, 1072–1075.
115. B. Wakkad and A. Hickling, *Trans. Faraday. Soc.*, 1950, **46**, 820–824.
116. *United States Pat.*, 3,399, **966**, 1968.
117. C. Iwakura, A. Honji and H. Tamura, *Electrochim. Acta*, 1981, **26**, 1319–1326.
118. Y. Surendranath, M. Dincă and D. G. Nocera, *J. Am. Chem. Soc.*, 2009, **131**, 2615–2620.
119. E. R. Young, D. G. Nocera and V. Bulović, *Energy Environ. Sci.*, 2010, **3**, 1726–1726.
120. J. G. McAlpin, Y. Surendranath, M. Dincă, T. A. Stich, S. A. Stoian, W. H. Casey, D. G. Nocera and R. D. Britt, *J. Am. Chem. Soc.*, 2010, **132**, 6882–6883.
121. J. B. Gerken, J. G. McAlpin, J. Y. C. Chen, M. L. Rigsby, W. H. Casey, R. D. Britt and S. S. Stahl, *J. Am. Chem. Soc.*, 2011, **133**, 14431–14442.
122. A. J. Esswein, Y. Surendranath, S. Y. Reece and D. G. Nocera, *Energy Environ. Sci.*, 2011, **4**, 499–499.
123. J. J. H. Pijpers, M. T. Winkler, Y. Surendranath, T. Buonassisi and D. G. Nocera, *Proc. Natl. Acad. Sci. U.S.A.*, 2011, **108**, 10056–10061.
124. E. R. Young, R. Costi, S. Paydavosi, D. G. Nocera and V. Bulović, *Energy Environ. Sci.*, 2011, **4**, 2058–2058.
125. D. K. Zhong, M. Cornuz, K. Sivula, M. Grätzel and D. R. Gamelin, *Energy Environ. Sci.*, 2011, **4**, 1759–1759.
126. E. M. P. Steinmiller and K.-S. Choi, *Proc. Natl. Acad. Sci. U.S.A.*, 2009, **106**, 20633–20636.
127. K. J. McDonald and K.-S. Choi, *Chem. Mater.*, 2011, **23**, 1686–1693.
128. T. Nakagawa, C. A. Beasley and R. W. Murray, *J. Phys. Chem. C*, 2009, **113**, 12958–12961.
129. Y. Zhao, N. M. Vargas-Barbosa, E. A. Hernandez-Pagan and T. E. Mallouk, *Small*, 2011, **7**, 2087–2093.
130. H. J. Lewerenz, *J. Electroanal. Chem.*, 2011, **662**, 184–195.
131. Y. W. Chen, J. D. Prange, S. Dühnen, Y. Park, M. Gunji, C. E. D. Chidsey and P. C. McIntyre, *Nature Mat.*, 2011, **10**, 539–544.
132. L. A. Kibler, *ChemPhysChem*, 2006, **7**, 985–991.
133. J. K. Nørskov, T. Bligaard, J. Rossmeisl and C. H. Christensen, *Nat. Chem.*, 2009, **1**, 37–46.

134. M. E. Björketun, A. S. Bondarenko, B. L. Abrams, I. Chorkendorff and J. Rossmeisl, *Phys. Chem. Chem. Phys.*, 2010, **12**, 10536–10541.

135. J. Greeley, T. F. Jaramillo, J. Bonde, I. Chorkendorff and J. K. Norskov, *Nat. Mater.*, 2006, **5**, 909–913.

136. J. Rossmeisl, A. Logadottir and J. K. Nørskov, *Chem. Phys.*, 2005, **319**, 178–184.

137. J. Rossmeisl, Z. W. Qu, H. Zhu, G. J. Kroes and J. K. Nørskov, *J. Electroanal. Chem.*, 2007, **607**, 83–89.

CHAPTER 4

Tandem Photoelectrochemical Cells for Water Splitting

KEVIN SIVULA* AND MICHAEL GRÄTZEL

Institute of Chemical Sciences and Engineering, École Polytechnique Fédérale de Lausanne, 1015-Lausanne, Switzerland
*Email: kevin.sivula@epfl.ch

4.1 Introduction and Motivation for Using Multiphoton Systems

The realization of an energy economy based on solar hydrogen is contingent on identifying a scalable system for solar H_2 production at costs commensurate with the price of H_2 generated from conventional sources (2–3 US\$ kg^{-1} for the steam reforming of natural gas).[1] Thus, a balance between minimizing system complexity and maximizing energy conversion efficiency and device longevity must be considered for any practical photoelectrochemical (PEC) water splitting system for solar H_2 production. The most straightforward approach for solar H_2 production using PEC water splitting employs a single semiconductor photoelectrode in combination with a metal counter electrode. Here, hydrogen can be produced from water upon the absorption of two photons (under the condition that one electron-hole pair is produced for each photon absorbed) since the water reduction reaction requires two electrons to make one molecule of H_2. This single-absorber, two-photon approach (more simply called S2) appears quite promising at first glance. Since only 1.23 V are needed for water splitting under standard conditions from a thermodynamic standpoint, one could believe that a semiconductor with a minimum band-gap, E_g, of 1.23 eV

RSC Energy and Environment Series No. 9
Photoelectrochemical Water Splitting: Materials, Processes and Architectures
Edited by Hans-Joachim Lewerenz and Laurence Peter
© The Royal Society of Chemistry 2013
Published by the Royal Society of Chemistry, www.rsc.org

(an absorption wavelength cut-off of 1008 nm) could be effective. Based on the standard AM1.5 G solar spectrum (1000 W m^{-2}), a semiconductor with such a band-gap would operate at maximum overall solar-to-hydrogen conversion efficiency, η_{STH}, of 47.4% assuming a unity quantum conversion efficiency ($\phi_{conv} = 1$) and no other losses.[2]

However, significant loss processes are involved and are unavoidable in any energy conversion process. Figure 4.1 shows an electron energy scheme of S2 PEC water splitting using a photoanode and highlights these loss processes.

Firstly, there is an entropic energy loss resulting from the amount of work that must be done to extract the excited states. The chemical potential of the excited state, $\Delta\mu_{ex}$, represents the maximum amount of energy available to do work,[3] and is necessarily significantly less than E_g. The difference between these quantities depends on the band-gap of the semiconductor and the illumination intensity; it has been calculated by Bolton *et al.*[4] to be about $E_g/4$ for a typical semiconductor at terrestrial solar intensities, meaning that $\Delta\mu_{ex} \approx 0.75\,E_g$. In addition, kinetic overpotentials, η_{ox} and η_{red}, exist at both electrodes resulting from the energies of the intermediate species involved in the complex water reduction and oxidation reaction mechanisms. Even using the best catalysts,[5] the overpotential of the water oxidation reaction, η_{ox}, is notoriously large (0.2–0.4 eV) for reasonable current densities (*ca.* 10 mA cm^{-2}) as a consequence of the four-electron nature of the oxygen evolution reaction (OER)

These loss processes imply that a semiconductor with E_g much greater than 1.23 eV must be used to realize S2 PEC water splitting. This is clearly illustrated in Figure 4.1. Since a semiconductor can only harvest photons with energy greater than E_g, the requirement to use a material with a larger band-gap implies a lower possible η_{STH}. Indeed, while the ideal (Schottky-Queisser) solar

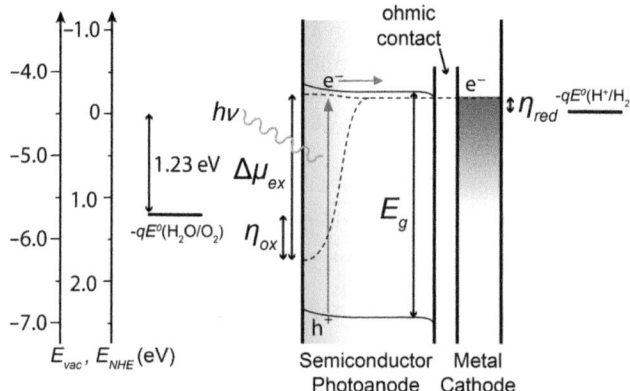

Figure 4.1 Electron energy scheme of S2 PEC water splitting using a photoanode. The absorption of a photon (hv) by the semiconductor with a band-gap (E_g) creates an electron-hole pair with free energy of $\Delta\mu_{ex}$. This free energy must be greater than the energy needed for water splitting (1.23 eV) plus the overpotential losses at both the anode and the cathode, η_{ox} and η_{red}, for the water splitting reaction to occur.

energy conversion limit using one absorber material is 34% for AM1.5G irradiation, Weber and Dignam[2] followed by Bolton *et al.*[6] analyzed the "upper limit" for η_{STH} using S2 water splitting with realistic losses and reported maximum values of 11.6% (using $E_g = 2.2$) and 17% (using $E_g = 2.03$), respectively. The latter report additionally considered photon absorption and collection losses and estimated that 10% should be entirely feasible with an S2 approach. In reality, the realization of a S2 PEC water splitting device with $\eta_{STH} = 10\%$ is clearly a difficult task. A review by Walter *et al.*[5] in 2010 summarizes decades of progress on using either a photocathode or a photoanode for S2 water splitting. While many promising S2 photocathodes exist (*e.g.* GaP and InP), their cost and stability in aqueous environments remain critical limitations. Semiconductor oxide photoanodes offer unmatched stability for water splitting, but, due to the large energy loss processes discussed above, materials with band-gaps greater than 3.0 eV (*e.g.* SrTiO$_3$ and KTaO$_3$) must be employed, limiting η_{STH} to less than 2%.

It remains possible that a single semiconductor material that possesses the optimum band-gap, conduction and valence band levels, stability and availability will be identified, and a device exploiting the S2 approach will realize over 10% η_{STH}. However, the failure to do so after decades of work has led researchers to consider alternative approaches. This chapter introduces an approach to PEC water splitting that increases system complexity but also energy conversion efficiency. By employing multiple light absorbers in tandem, PEC systems can both harvest a significant portion of the solar spectrum and provide enough $\Delta\mu_{ex}$ to afford water splitting at high η_{STH}. In the next sections, various approaches using multi-absorber systems will be presented, analyzed, and discussed using examples from literature. In addition, since the primary goal of PEC research is to develop a system that balances system complexity, cost, and efficiency, particular attention will be given to these aspects.

4.2 Strategies and Limitations of Multi-Absorber Systems

4.2.1 "Brute Force" Strategies

The most obvious approach to accumulate sufficient $\Delta\mu_{ex}$ for water splitting with semiconductor materials that can also absorb a large fraction of the solar spectrum is to connect multiple photovoltaic (PV) cells in series. For example, a traditional pn-junction silicon solar cell generates a voltage of 0.5–0.6 V at its maximum power point under standard conditions. Thus three of these cells connected electrically in series would create sufficient voltage to split water. This "brute force" PV + electrolysis approach is limited by the price and availability of PV devices and electrolyzers, which, at the time of this writing makes the price of the H$_2$ produced around 10 US\$ kg^{-1}.[1] Nevertheless, this path has generated fair amount of interest, and overall η_{STH}'s up to 9.3% have been demonstrated using single crystal silicon PV modules and high-pressure electrolysis.[7] It has also been suggested that a complete optimization of these systems could deliver H_2 at a price of 4 US\$ kg^{-1}.[8] This price may decrease further if novel thin-film PV technologies are employed. For example, Dhere and Jahagirdar[9,10] have reported the use of two high voltage $CuIn_{1-x}Ga_xS_2$ (GIGS) PV cells side-by-side to generate a η_{STH} up to 8.8%. A major drawback with the brute force approach in general comes from the voltage output of a pn-junction solar cell, which is strongly dependent on the illumination intensity. This necessitates complicated switching mechanisms to ensure that the optimum number of cells is connected in series during variations in light intensity caused by haze or cloud cover.

4.2.2 The Tandem Cell Concept

Besides using PV cells side-by-side to obtain sufficient voltage for water splitting, cells can be stacked on top of one another, in tandem, if they can individually harvest different portions of the solar spectrum. The first (top) cell, with band-gap E_{g1} absorbs photons from the sun with a wavelength smaller than $\lambda_1 = hc/E_{g1}$. Ignoring scattering events, photons with $\lambda > \lambda_1$ are transmitted through to the next cell with a band-gap $E_{g2} < E_{g1}$, and so on (see Figure 4.2). If the cells are connected electrically in series, the total photo-current density will be limited by the cell producing the least current, whereas the total voltage will be the addition of all the cells used.

This tandem cell strategy not only increases the voltage provided by the overall cell, but also significantly enhances the upper limit of solar power conversion (solar-to-electricity) efficiency from 34% in the one absorber case to 42% (2 cells), 49% (3 cells), or 68% (for an infinite number of cells) for unconcentrated AM1.5G sunlight.[11] Of course, using an infinite number of cells is not possible and in practice the addition of each layer brings technical difficulties that greatly increase the complexity and cost of these devices.

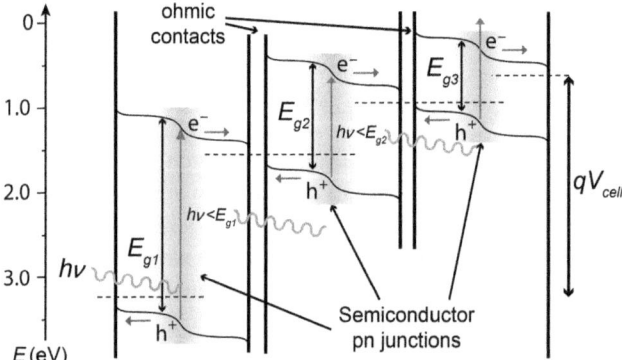

Figure 4.2 Electron energy scheme of a generic tri-level tandem cell for solar energy conversion. Here three semiconductors (E_{g1-3}) are employed as pn-junctions and connected by transparent ohmic contacts to give a final photopotential of V_{cell}.

Nevertheless, the precise control of materials used and device architecture (typically triple-junction epitaxially grown III-V semiconductor devices) has pushed the solar-to-electricity power conversion efficiency, η_{PC}, past 40%.[12] However, the high cost of producing these devices relegates their use to specific niches (*e.g.* extra-terrestrial application). The use of these types of "PV/PV" tandem cells for PEC water splitting is discussed in Section 4.3 and also other chapters of this book. First, since the operation requirements and theoretical maximum η_{STH} for a water splitting PEC tandem cell are distinct from its electricity producing (PV) counterpart, it is worthwhile to examine closely the benefits and limitations of this strategy for solar water splitting.

4.2.3 The D4 Strategy and its Potential

While very impressive photovoltaic efficiencies have been demonstrated using triple junction cells as mentioned above, the maximum possible η_{PC} does not change significantly going from 2 to 3 cells (from 42 to 49%). This brings into question the return on the device complexity added by the third cell. In addition, gaining the $\Delta\mu_{ex}$ necessary for water splitting should be feasible with only two light absorbers. In fact, a two-photosystem solution is precisely what evolved in plant photosynthesis, which uses sunlight to split water and stores the energy in carbohydrate vectors. Finally, since the water splitting reaction necessarily entails two separate half-reactions, it seems natural to use two light absorbers for PEC water splitting.

In a dual-absorber approach where each absorbed photon creates one excited electron-hole pair, four photons (two in each absorber) must be absorbed to create one molecule of H_2. Following Bolton's convention,[6] this is designated a D4 approach. A simple way to accomplish D4 water splitting is to use a separately illuminated photoanode and photocathode in a side-by-side (not tandem) configuration. Weber and Dignam[2] evaluated the potential η_{STH} of this approach and reported an upper limit of 16.6% with $E_{g1} = E_{g2} = 1.4\,eV$ – a modest increase over the 11.6% predicted with an S2 approach. However, using an integrated tandem D4 approach, 22% η_{STH} was predicted to be possible with $E_{g1} = 1.8\,eV$ and $E_{g2} = 1.15\,eV$. A more extensive analysis by Bolton plotted the maximum η_{STH} with respect to different absorber band-gaps and reasonable losses.[3,6] The formalism presented in that work can be extended to a generalized expression for the expected tandem D4 solar-to-hydrogen conversion efficiency as a function of the chosen semiconductors and the expected losses:

$$\eta_{STH} = \frac{j_{min}\Delta G^0_{H_2}\eta_{farad}}{2qE_S} \tag{4.1}$$

where $\Delta G^0_{H_2}$ represents the standard free energy of the hydrogen produced (lower heating value, 2.46 eV per molecule of H_2), η_{farad} is the Faradaic efficiency for the water splitting reactions, q is the elementary charge $(1.6022\times10^{-19}\,C = 1\,eV\,V^{-1})$, and E_S represents the power flux from the sun incident on the tandem cell. For standard conditions (AM1.5G), this latter value is $1000\,\mathrm{W\,m^{-2}}$. The factor of 2 in the denominator represents the stoichiometric relation between one molecule of H_2 and the number of electrons needed to produce it electrochemically from H_2O. Finally, j_{min} represents the photocurrent density ($\mathrm{A\,m^{-2}}$) of the D4 tandem device (*i.e.* the electrical current that is flowing between the two cells), which is zero if there is not sufficient $\Delta\mu_{ex}$ generated by the tandem cell and is otherwise limited by the individual cell producing the least amount of photocurrent, as the cells are assumed be connected in series:

$$j_{min} = \begin{cases} 0, & \Delta\mu_{ex} \leq 1.23\,eV + \eta_{ox} + \eta_{red} \\ \min(j_1, j_2), & \Delta\mu_{ex} > 1.23\,eV + \eta_{ox} + \eta_{red} \end{cases} \qquad (4.2)$$

Here, $\Delta\mu_{ex}$ is equal to the sum of the free energies produced by each cell, $\Delta\mu_{ex1} + \Delta\mu_{ex2}$. The individual current densities produced by each cell, j_1 and j_2, can be expressed with respect to the known spectral irradiance from the sun[13] as a function of photon wavelength ($\mathrm{W\,m^{-2}\,nm^{-1}}$), $E_S(\lambda)$, and the wavelength cut-offs for each semiconductor ($\lambda_i = hc/E_{gi}$, $i = 1,2$, again setting $E_{g1} > E_{g2}$) as

$$j = \int_{\lambda_{i-1}}^{\lambda_i} \frac{E_S(\lambda)\varphi_{conv,i}(\lambda)}{(hc/\lambda)} d\lambda, \text{ for } i = 1, 2 \qquad (4.3)$$

where $\phi_{conv,i}(\lambda)$ represents the quantum conversion efficiency for each semiconductor as a function of wavelength (*i.e.* the ratio of photocurrent density and absorbed photon flux), and λ_0 represents the smallest wavelength for which there is an appreciable photon flux from the sun (*ca.* 300 nm for AM1.5G solar irradiation).

For simplicity, the above formalisms assume that each semiconductor absorbs all photons with energy $h\nu > E_g$ and transmits all photons with energy $h\nu < E_g$. No reflection or scattering losses are included for this ideal case. Thus values for j_1 and j_2 calculated in this way represent upper bounds for real systems. In real tandem systems, reflection and scattering losses exist, but can be addressed with device engineering. Also, in real systems, both $\Delta\mu_{ex}$ and ϕ_{conv} are complicated functions of the light intensity and the semiconductor properties. Moreover, the overpotentials for the oxidation and reduction reactions, η_{ox} and η_{red}, are related to j_{min} by the Tafel relation. Thus, additional assumptions are needed to obtain reasonable values for η_{STH}. In principle, ϕ_{conv} can be made very close to unity even in real systems,[3] so $\phi_{conv} = 1$ will be used to evaluate equation (4.3). If we also assume that no corrosion or undesirable side reactions occur then $\eta_{farad} = 1$. Bolton selected some values for the remaining loss processes based on reasonable assumptions regarding the overpotentials and the free energy.[6] The minimum reasonable value for the energy loss, $U_{loss} = E_{g1} + E_{g2} - \Delta\mu_{ex} + \eta_{ox} + \eta_{red}$, was chosen to be 1.2 eV (or 0.6 eV for each

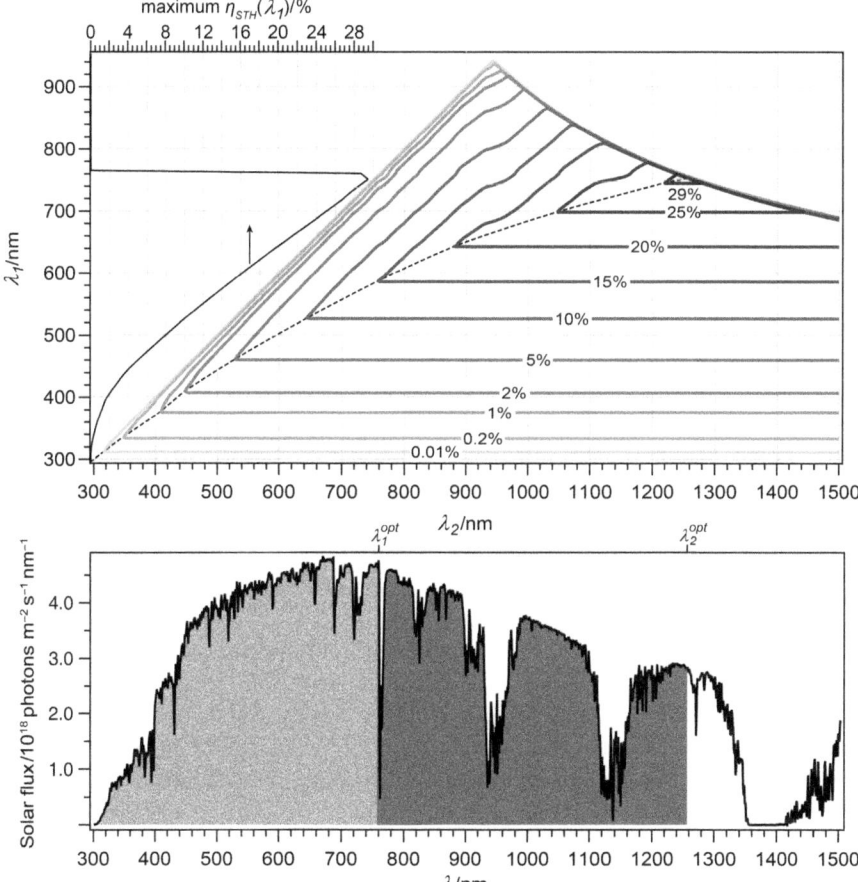

Figure 4.3 (Top) Contour plot (thick grey lines) showing the maximum reasonable η_{STH} with AM 1.5G incident radiation ($1000\,W\,m^{-2}$) and an total loss, U_{loss}, set at 1.4 eV as it depends on the chosen semiconductor cut-off wavelengths, λ_1 and λ_2 where ($\lambda_i = hc/E_{gi}$, $i = 1,2$, with $E_{g1} > E_{g2}$). The maximum η_{STH} as it depends on the choice of the wider band-gap absorber (λ_1) is also shown (thin black line). See text for full explanation. (Bottom) The AM 1.5 G solar photon flux as a function of wavelength. The optimum values of λ_i are indicated and the areas corresponding to the photons harvested by each absorber layer are shaded.

semiconductor). Figure 4.3 (top) shows a contour plot showing how a reasonable upper limit for η_{STH} depends on the chosen semiconductor cut-off wavelengths, λ_1 and λ_2. The plot was obtained using equations (4.1)–(4.3) and the above assumptions with slightly higher losses ($U_{loss} = 1.4$ eV total).

The shapes of the contours (thick grey lines in Figure 4.3), which represent values of λ_1 and λ_2 that result in the same η_{STH}, are easily rationalized. First, the contours remain to the right side of the 45° line because as this is where the set condition $E_{g1} > E_{g2}$ is satisfied. The upper-right region, where η_{STH} is undefined, represents semiconductor combinations that do not possess sufficient $\Delta\mu_{ex}$ for water splitting in a D4 configuration given U_{loss}. Calculating the upper limit of η_{STH} and optimum λ_1, λ_2 pairing for an arbitrary system can be simply done with the help of Figure 4.3 by first choosing λ_1. For example, take $\lambda_1 = 477$ nm ($E_{g1} = 2.6$ eV), which corresponds to WO$_3$, a material commonly used as a semiconductor photoelectrode for water splitting.[14] A horizontal line drawn at $\lambda_1 = 477$ nm would intercept first with the solid black maximum-η_{STH} line. Reading the value on the top scale gives the maximum value of η_{STH} for WO$_3$ as 6.3%. Continuing back on the imagined horizontal line at $\lambda_1 = 477$ nm further, it intercepts with a dashed black line running against the contour lines. This line represents the values of λ_1 and λ_2 that meet the current matching condition. Reading the value of λ_2 at that point gives $\lambda_2 = 561$ nm and thus the minimum value for $E_{g2} = hc/\lambda_2 = 2.2$ eV. Choosing larger λ_2 (smaller E_{g2}) will not result in a higher η_{STH} as the photocurrent will be limited by the top absorber, λ_1. For this reason the grey contour lines are all horizontal to the right of the dashed black "current matching line".

To identify the optimum values for both λ_1 and λ_2 with $U_{loss} = 1.4$ eV, the current matching line can be followed for increasing λ_1 and λ_2 (and η_{STH}) until the maximum η_{STH} of 29.9 % with $\lambda_1 = 755$ and $\lambda_2 = 1253$ nm. The distribution of the photons harvested by each absorber in this optimum case can be visualized in the bottom of Figure 4.3, where the AM 1.5G solar photon flux is plotted as a function of the wavelength. The optimum values for λ_1 and λ_2 are indicated, and shaded areas indicate photons harvested (lighter shading for λ_1). The areas of the two shaded regions in this plot are identical, as the optimized circumstances exactly satisfy the current matching condition. By coincidence, the cut off for absorber 1, λ_1, coincides with the sharp dip in the solar spectrum at 762 nm, which is due to the weak spin-forbidden electronic absorption of molecular oxygen.

Higher assumed values for U_{loss} result in lower values for the maximum η_{STH} at the optimum conditions. For example, with U_{loss} (total) $= 1.6$ eV, the optimum conditions give $\eta_{STH} = 27.1$ % using $\lambda_1 = 720$ and $\lambda_2 = 1120$ nm, and with $U_{loss} = 2.0$ eV, $\eta_{STH} = 21.6$ % using $\lambda_1 = 655$ and $\lambda_2 = 925$ nm.[6] The maximum predicted η_{STH} for arbitrarily chosen λ_1, λ_2 does not change when assuming higher values for U_{loss}, as long as sufficient $\Delta\mu_{ex}$ can still be generated. For example, with $U_{loss} = 2.0$, the maximum η_{STH} for WO$_3$ remains 6.3%, with the current matching λ_2 remaining as 561 nm. However, in this case, λ_2 must also be greater than 1914 nm to ensure that sufficient $\Delta\mu_{ex}$ is generated.

Overall the model presented in this section only gives reasonable expected maximum values for η_{STH}, which can be of use in describing the basic capability (or limitations) of specific materials or materials combinations. This is certainly of use in determining the interest in developing specific systems. However, the model does not include any additional constraints on selecting the materials with respect to the conduction and valence band energy levels, or work function; factors that add to the difficulty of finding ideal and complementary materials for a water-splitting tandem cell.

4.2.4 D4 Device Architectures

Beyond band-gap selection, an important practical consideration for tandem cell construction is the types of junctions used. While photovoltaic (pn-junction) technology is well developed for solar electricity production, it has been reasoned that the direct photoelectrolysis of water at a semiconductor-liquid junction can potentially produce more efficient solar-to-hydrogen conversion compared PV-electrolysis device.[15] For this reason, many groups have investigated tandem cells where one or both of the absorbers are in direct contact with the aqueous electrolyte. In addition, alternative photoelectrochemical devices, like dye sensitized solar cells (DSSCs) can also be employed as one of absorbers in a tandem scheme. Thus, many D4 configurations are possible. Figure 4.4 shows, from an energy perspective, two classic examples in addition to the PV/PV combination previously mentioned.

The p-type photocathode/n-type photoanode device (Figure 4.4a) is the simplest in terms of junctions with only one interconnect between the two semiconductors together with two semiconductor liquid junctions (SCLJs). Here, materials must be chosen such that when the device is in contact with the electrolyte, $CB_2 < -qE^0(H+/H_2) < CB_1 < VB_2 < -qE^0(H_2O/O_2) < VB_1$, where VB_i and CB_i represent the valence band and conduction band energy levels of semiconductor i, respectively. A similar situation is necessary for a photoelectrode/PV configuration (Figure 4.4b) with the added complication of aligning an additional n-type energy level and the work function of the metal electrode used to perform the reduction reaction (this is usually necessary as the semiconductors employed for PV cells are not stable when performing the water splitting reaction).

Despite the constraints on both the material's band-gap and energy levels, various material combinations have been demonstrated in D4 configurations. Specific examples can be categorized as photocathode/photoanode, photoelectrode/PV, or PV/PV. Specific examples of these configurations are explored in the next sections.

4.3 PV/PV Strategies

Devices employing two or more pn-junctions in tandem have attained the highest reported η_{STH}.[16,17] Champion devices use the well-known band-gap engineering of III-V semiconductor systems to optimize light harvesting and $\Delta\mu_{ex}$. In particular, Licht et al. have employed an AlGaAs/Si (pn-pn) structure using Pt-black and RuO_2, as reduction and oxidation catalysts, respectively, and obtained solar-to-hydrogen conversion efficiencies as high as 18.3% under simulated AM0 sunlight (135 mW cm^{-2}).[18,19] Turner and co-workers have investigated monolithic GaAs/GaInP$_2$ (pn-p, pn-pn, or pn-pn-p) systems and attained η_{STH} up to 12.5%.[20–22] While the conversion efficiencies reported with these systems are quite impressive, major concerns exist about the price (due to the requirements of high material purity and costly fabrication methods) and stability of these devices when they are used in contact with aqueous

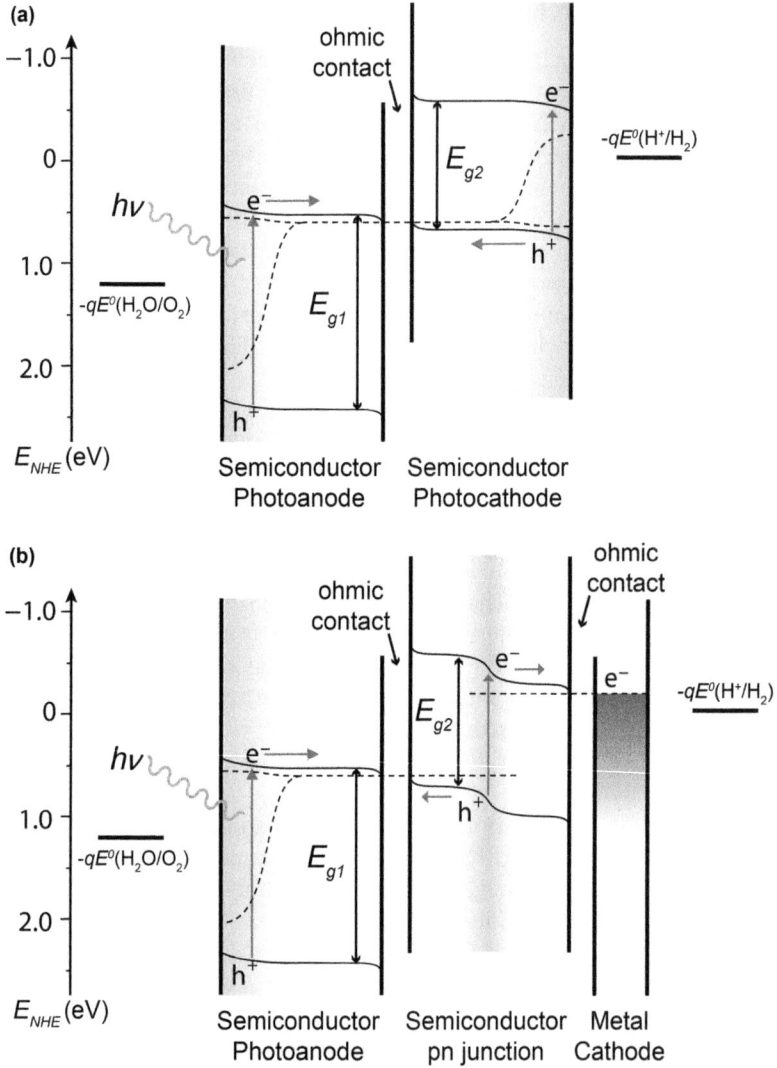

Figure 4.4   Electron energy schemes of two possible D4 tandem water splitting device
configurations: (a) shows an n-type photoanode connected to p-type
photocathode and (b) shows a n-type photoanode with a pn-junction.

electrolyte.[23] A comprehensive overview of these III-V systems will not be
presented here as other sections of this book are dedicated to that purpose.

Of course, it is not necessary to construct an all-PV tandem system from
III-V semiconductors; many other systems have also been reported. One system
of particular interest is based on amorphous silicon (a-Si). Bockris and co-
workers first showed that a triple stack of n-type/intrinsic/p-type (nip) a-Si (not
a D4 system, as 6 photons need to be absorbed for one H_2) on a Ti substrate
could give up to $\eta_{STH} = 7.5\%$ under AM1 (100 mW cm^{-2}) illumination when

islands of Pt and RuO_2 were used as a reduction and oxidation catalysts, and the device was directly submerged into aqueous electrolyte.[24] Since the band-gap of the a-Si was the same for each nip-junction, the layer thicknesses were optimized so each layer produced the same photocurrent (*i.e.* the top two layers were thin enough to transmit some photons with energy $hv > E_g$). This concept was later extended to increase the efficiency for electricity production by mixing some Ge in with the Si to modify the E_g of the bottom two layers.[25] Using these devices for water splitting, an η_{STH} of 7.8% under AM1.5 ($100\,mW\,cm^{-2}$) illumination was achieved in 1 M KOH electrolyte with CoMo (reduction) and $NiFe_yO_x$ (oxidation) catalysts.[26] The stability of these devices was improved to greater than 31 days while maintaining 5–6% η_{STH} using a fluorine-doped SnO_2 protection layer on the cathode.[27] More recent efforts with these devices have also shown that operation at near neutral pH is possible with $\eta_{STH} = 4.7\%$.[28] While these a-Si based devices are presumably less expensive than III-V semiconductor based tandem cells, their fabrication still requires relatively expensive vacuum processing. Despite this, at the time of this writing, the mass production by several companies of triple junction a-Si device modules for photovoltaic energy conversion is beginning. It is not yet clear what the price per kg H_2 would be if these cells were used for water splitting. Furthermore, the open-circuit voltages produced by such triple junction a-Si cells are around 2.4 V – much higher than is needed for water electrolysis. The optimization of this class of devices specifically for solar water splitting has yet to be accomplished.

4.4 Photoelectrode/PV Systems

The most widely recognized photoelectrode/PV device is a monolithic, epi-taxially grown III-V semiconductor system developed by Turner with a GaAs pn-junction coupled to a $GaInP_2$ photocathode.[22] While the performance of this system was impressive at $\eta_{STH} = 12.5\%$, the $GaInP_2$ photocathode quickly corrodes when in contact with the aqueous electrolyte.[23] In general, this is a major drawback of using direct semiconductor-liquid junctions. However, with certain semiconductors the photoelectrode/PV system has many advantages.

4.4.1 Advantages and Operation of a Photoanode/PV Tandem Cell

Identifying semiconductor materials that form stable junctions with water under illumination has been a major challenge in the field of PEC water splitting. Transition metal oxide semiconductors have, however, excelled in this aspect, with the prototype being TiO_2.[29,30] Unfortunately, a conundrum exists for transition metal oxides with respect to their band-edge energy levels and band-gap. The conduction band-edge energy is strongly influenced by the metal d-orbitals and varies from metal to metal. However, the valence band edge energy of a transition metal oxide is most strongly influenced by oxygen 2p orbitals, thus it does not change appreciably from a potential around 2.5–3.0 V *vs.* NHE when changing the metal.[31] In addition, the band-edge energies typically exhibit a Nertsian pH response. Thus, for a transition metal oxide to have a small enough E_g for adequate solar light harvesting its CB energy must be well below the water reduction potential. This excludes them from being used in a S2 water splitting scheme but essentially makes them ideal for use as photoanodes in a D4 tandem cell. Moreover, the difficulty in identifying a stable p-type cathode makes a photoanode/PV tandem device a good compromise, with device complexity and stability.

Since there are several examples of this type of tandem cell, it is useful to illustrate general device function. The operation conditions of a photoanode/PV tandem device can be predicted by comparing the current density-voltage characteristics of each individual cell under the actual illumination conditions of that separate part (given its position in the particular tandem configuration).[2] Figure 4.5 shows the current density with respect to the applied potential for a realistic but arbitrary photovoltaic cell (solid black curve) and a photoanode (solid grey curve) under the appropriate illumination conditions. The operating potential U_{op} is defined as the potential measured at a common node between the PV and the PEC cell (*i.e.* where the curves overlap). The current density at this point, j_{op}, can be used directly in equation (4.1) as j_{min} to predict the η_{STH} of the tandem cell. In this way the properties of the individual cell can be independently varied to maximize j_{op} and, accordingly, the η_{STH} of the device.

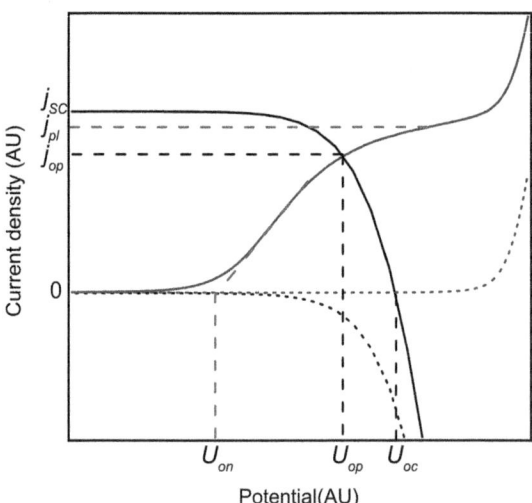

Figure 4.5   The operation conditions of a photoanode/PV tandem device as shown by the idealized current-voltage curves for a PV cell (black) on top of a n-type photoanode (grey) in the dark (broken lines) and under illumination (solid lines). Relevant parameters of the curves are indicated and described in the text.
Reprinted with permission from J. Brillet, *et al., J. Mater. Res.*, 2010, **25**, 17–24. Copyright 2010 Materials Research Society.

It should be noted that, in a real device, the maximum j_{op} is not necessarily j_{min} as expressed in equation (4.2). That is to say that the maximum j_{op} is not found by merely maximizing the current output from each cell (denoted as the short circuit current density, j_{sc}, for the PV cell and the plateau photocurrent density, j_{pl}, for the photoanode). U_{loss} (as defined in section 4.2.3) and other device non-idealities (such as high series and low shunt resistances) necessitate that the characteristic cell potentials (denoted as the open circuit photo-potential, U_{oc}, and the photocurrent onset potential, U_{on}, for the PV and the photoanode, respectively) be also considered in device optimization.

4.4.2 Photoanode/PV Examples

The possibility of a photoanode/PV tandem device to employ a stable transition metal oxide photoanode and a complimentary PV device has attracted many research groups to investigate various promising systems. For example, a TiO_2 photoanode in tandem with a thin film PV device based on $Cu(In,Ga)Se_2$/CdS produced hydrogen at a rate of $0.052\,\mu L\,cm^{-2}\,s^{-1}$ during unassisted solar water splitting (corresponding to an external quantum efficiency of 1.02%).[32] While less interest has been paid to using the prototypical TiO_2 as a photoanode due to its large band-gap ($E_g = 3.2\,eV$, max $\eta_{STH} = 2\%$), this work demonstrated the importance of using optimized protective layers ($Nb_{0.03}Ti_{0.97}O_{1.84}$ in this case) to eliminate corrosion of the PV cell in the aqueous conditions.

Further research efforts have focused on using more promising transition metal oxides such as WO_3, and Fe_2O_3. As mentioned in section 4.2, WO_3 can potentially convert up to 6.3% of the energy of AM1.5G sunlight into hydrogen. Hematite ($\alpha\text{-}Fe_2O_3$) is a more promising transition metal oxide with a band gap of 2.1 eV. This corresponds to a maximum η_{STH} of 15.4% according to Figure 4.3, if a second absorber with $\lambda_2 > 774\,nm$ ($E_{g2} < 1.6\,eV$) is used. Besides its relatively small band-gap, $\alpha\text{-}Fe_2O_3$ is the most stable form of iron oxide at ambient conditions, and both iron and oxygen are ubiquitous atoms in the earth's upper crust, making hematite an outstanding candidate for solar energy conversion on a scale commensurate with the global energy demand. Furthermore, both WO_3 and Fe_2O_3 photoanodes can be prepared by inexpensive solution-based techniques.[33]

Miller and co-workers have investigated combining tungsten[34] or iron[35] oxide photoanodes in tandem with a-Si:Ge PV. As a drawback, the large overpotential for water oxidation (η_{ox}) and the relatively low U_{oc} of a a-Si device requires a double junction PV in tandem to provide sufficient potential to drive the overall water splitting reaction (similar to the PV only case where 3 pn-junctions were needed with a-Si). Despite this, 3% η_{STH} was obtained with the WO_3/a-Si:Ge/a-Si:Ge device.[34] A similar device for iron oxide with only one PV (Fe_2O_3/a-Si:Ge) did not split water without an external bias, but it was shown that a 0.65 V bias "savings" was earned under AM1.5G illumination.[35] This result could reasonably be improved with the recent advances in $\alpha\text{-}Fe_2O_3$ photoanode performance.[36]

4.4.3 The Photoanode/DSSC Tandem Cell

From a practical perspective, the attractive aspects of using a widely available, highly stable and inexpensively produced photoanode are diminished when using a tandem component that requires relatively expensive processing techniques (*e.g.* the chemical vapour deposition of a-Si). Thus, more recently, investigations have focused on using next-generation photovoltaics that can also be fabricated with inexpensive, solution processed methods. The dye sensitized solar cell (DSSC) is the prototype example[37] of this class of photovoltaic device and thus has attracted significant attention for use in solar water splitting tandem cells with a stable photoanodes.[38] The photoanode/DSSC combination was first suggested by Augustynski and Grätzel[39] with WO_3 as the photoanode – suggesting that device would be capable of 4.5% η_{STH} given the performance of the two devices.[40] In practice, tandem devices with WO_3 have been constructed by Park and Bard[41] and Arakawa *et al.*[42] giving η_{STH}'s up to 2.8% at AM1.5G (100 mW cm^{-2}) using Pt as the cathode. However, similar to the a-Si based devices, two DSSCs connected in series to the photoanode were necessary to achieve overall water splitting. This was accomplished by positioning these two DSSCs side-by-side behind the WO_3 photoanode.

This photoanode/2×DSSC architecture does not fundamentally provide a limitation to the possible solar-to-hydrogen conversion efficiency for WO_3 or even the more promising Fe_2O_3, as less than one third of the available solar photons have wavelengths shorter than 600 nm, and pan-chromatic dyes with high quantum efficiency extending beyond 900 nm are being developed.[43] However, the architecture does present a challenge to device construction as the two DSCs need to be constructed each with half of the active area of the photoanode in order to normalize the total area of the device.

The development of new dyes for the DSSC during the past decade has initiated work to explore alternative device architectures for optimizing light harvesting in water splitting tandem cells. For example, all-organic dyes, such as the squaraine dyes, which have a narrow absorption bandwidth extending into the far red region of the visible, have demonstrated solar power conversion efficiencies over 6% under AM 1.5G illumination.[44,45] Specifically, the dye coded SQ1 only absorbs light with wavelengths between 550 and 700 nm. In addition to reducing costs considerably compared to a ruthenium-based dye, Brillet *et al.* pointed out that two new distinct tandem cell configurations become accessible when using this type of dye.[46] The possible configurations are shown in Figure 4.6. Besides the benchmark photoanode/2×DSSC design (Figure 4.6a), a true tri-level device (photoanode/DSSC-squaraine dye/DSSC-panchomatic dye) would also be effective (Figure 4.6b). This configuration would eliminate the need to construct two DSSCs side-by-side. A second possibility is to arrange two side-by-side DSSCs in front of a hematite photoanode (Figure 4.6c). This "front DSSC" configuration is particularly attractive in view of light harvesting, as transition metal oxides (in particular hematite) have high indexes of refraction, which increase reflection. Moreover, the front DSSC approach eliminates the need to deposit the photoanode on a

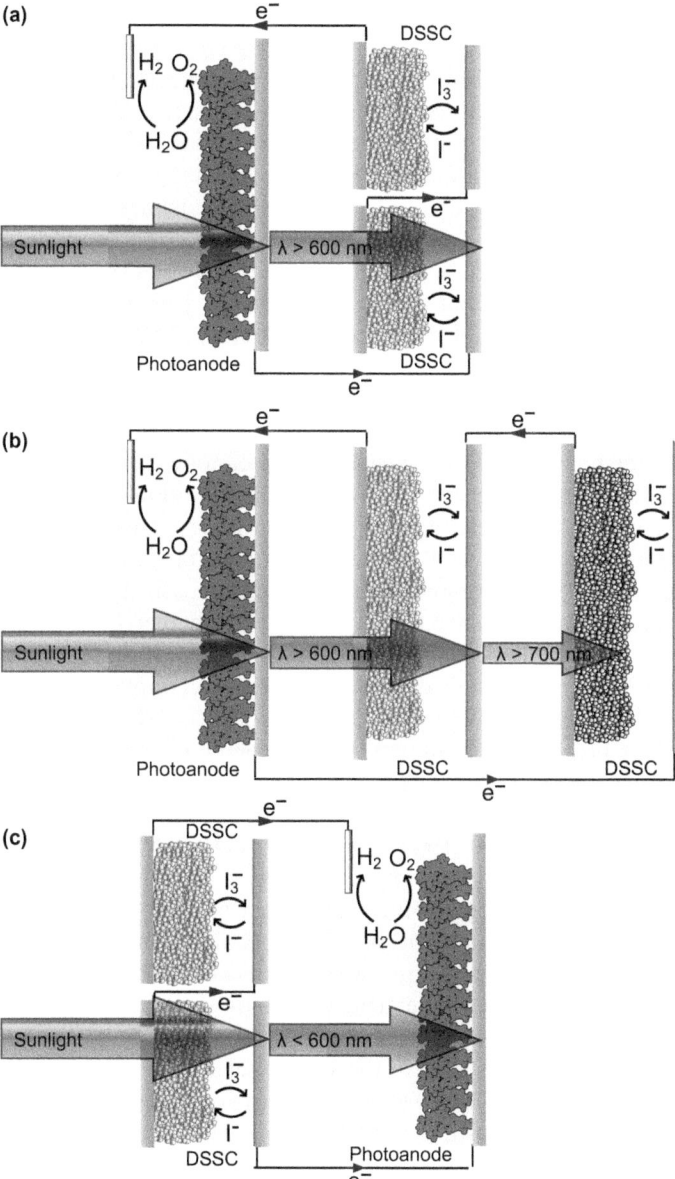

Figure 4.6 The layouts of three architectures for water splitting tandem cells using a n-type photoanode and two dye sensitized solar cells in series. (a) the conventional "back DSSC" configuration, (b) a "trilevel" configuration, and (c) the "front DSSC" configuration.
Reprinted with permission from J. Brillet, *et al., J. Mater. Res.*, 2010, **25**, 17–24. Copyright 2010 Materials Research Society.

transparent conducting glass and an inexpensive metal foil support could instead be used.

Brillet *et al.* determined how these two new photoanode/DSC tandem concepts performed compared to standard architecture using state-of-the-art nanostructured hematite[47] photoanodes. The tri-level tandem architecture (photoanode/DSSC-squaraine dye/DSSC-panchromatic dye) produced the highest operating current density and thus the highest expected solar-to-hydrogen efficiency of 1.36%. The conventional photoanode/2×DSSC architecture gave an efficiency of 1.16% and the front DSSC configuration 0.76%. It should be noted that these values were far below the expected 3.3% that should have been possible with the nanostructured hematite photoanodes used. Evidently, reduced light harvesting caused by scattering and reflection were limiting the overall conversion efficiency. This is illustrated by the optical and electronic characteristics of the tri-level tandem cell presented in Figure 4.7. Here it is shown that while the hematite photoanode only produces photocurrent for $\lambda > 600$ nm, the light transmitted to the middle and back cells is severely attenuated by scattering and reflection losses. This causes the DSSC photocurrent to limit the overall performance of the tandem cell. While device engineering should improve the operating current density by reducing reflection, scattering and resistive losses, DSSCs with higher U_{oc}'s would enable construction of a true D4 photoanode/DSSC water splitting tandem cell.

Indeed, in 2012, a specifically designed cobalt redox couple combined with an all-organic dye (coded Y123) gave DSSCs with $U_{oc} > 1.0$ V at 1 sun conditions.[48] This breakthrough allowed the first demonstration of a D4 photoanode/DSSC water splitting tandem cell.[49] Devices were assembled with both nanostructured WO_3 and Fe_2O_3 photoanodes, and j_{op} was measured to give η_{STH} values of 3.10% and 1.17% with the WO_3/DSSC and Fe_2O_3/DSSC combinations, respectively. An optical analysis also compared the predicted photocurrent from the integration of IPCE measurements and the actual *j-U* behavior of the device measured *in situ*. This optical analysis and the *j-U* curves for each device are shown in Figure 4.8. In the case of the Fe_2O_3/DSSC tandem cell, the j_{op} is far from the plateau region of the hematite electrode photocurrent. This results in a performance far from the maximum obtainable. The limitation of this system is clearly the late onset of the photocurrent in the photoelectrode, despite the use of state-of-the-art strategies to shift the onset of the photocurrent to less positive potentials by means of surface catalysis and passivation. However, in the case of the WO_3/DSSC tandem cell, the photocurrent onset is not a limitation and the j_{op} is very close to the plateau region of the photoanode (approx. 1000 mV). The limiting factor in this case is the low photocurrent obtainable by the photoanode due to less than ideal absorption capability of tungsten trioxide in the visible region of the solar spectrum. Despite this, the Fe_2O_3/DSSC device still exhibited a near unity faradaic efficiency and good stability over an eight-hour testing period. Overall this work suggests that the low η_{STH} of the D4 hematite/DSSC tandem cell offers the larger room for improvement, in particular, further reduction of the overpotential for water oxidation.

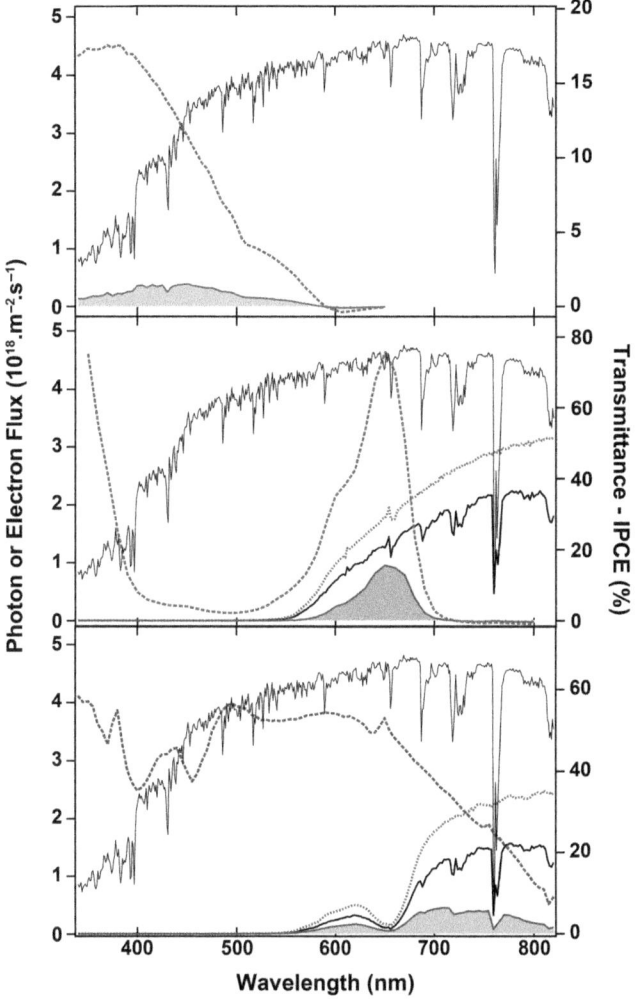

Figure 4.7 The optical and electronic characteristics of the "trilevel" tandem cell. Top panel: The solid line shows the incident solar flux. The IPCE of the hematite photoanode employed (dashed line) is convoluted with the solar flux to give the electron flux in the photoanode (shaded area). Middle panel: The solar flux is convoluted with the transmittance of the hematite (dotted line) to give the incident irradiance on the squaraine DSC (solid bold line). The irradiance is convoluted with the IPCE of the squaraine DSC (dashed line) to give the electron flux in the intermediate photovoltaic device (shaded area). Bottom panel: The solar flux is convoluted with the transmittance of the squaraine dye plus hematite (dotted line) to give the incident irradiance on the black DSC (solid bold line). The irradiance is convoluted with the IPCE of the black DSC (dashed line) to give the electron flux in the bottom photovoltaic device (shaded area). Reprinted with permission from J. Brillet, *et al., J. Mater. Res.*, 2010, **25**, 17–24. Copyright 2010 Materials Research Society.

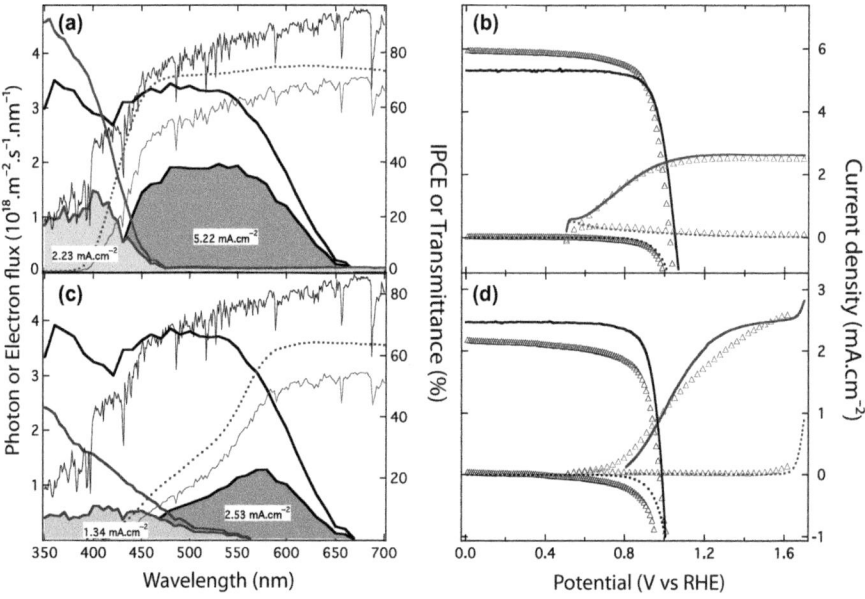

Figure 4.8 Spectral response study of the (a) WO$_3$ (light grey)/DSC (dark grey) and (c) Fe$_2$O$_3$ (light grey)/DSC (dark grey) D4 tandem cells. The transmittance of the photoanode (dashed lines) convoluted to the photon flux on the photoanode (thin black line) allows the calculation of the photon flux on the DSC (thin grey line). The shaded areas represent the electron density calculated from the IPCE (solid lines) convoluted to the photon flux on each device. *j-U* curves of (b) WO$_3$ (light grey)/DSC (dark grey) and (d) Fe$_2$O$_3$ (light grey)/DSC (dark grey) tandem cells under AM 1.5 G irradiation. The solid lines refer to the linear voltage sweep predicted from the IPCE integration and the triangles refer to *in situ* measurements. Reprinted from reference 49.

4.5 Photoanode/Photocathode Systems

As mentioned in Section 4.2, the most straightforward and simple way to construct a D4 tandem water splitting cell is to use an n-type semiconductor photoanode and a p-type semiconductor cathode. However, this approach is the least developed of the water splitting systems found in the literature. The reason for this is clearly due to a lack of suitable photocathode materials for water reduction. Many materials, for example GaInP$_2$,[50] Si,[51,52] SiC,[53] WS$_2$,[54] Cu(In,Ga)Se$_2$,[55] Cu$_2$O,[56] CuYO$_2$,[57] CaFe$_2$O$_4$,[58] and even p-type (Mg^{2+} doped) Fe$_2$O$_3$[59] have been investigated as water reducing p-type electrodes, but the magnitude of the photocurrent or the stability in aqueous solutions have remained limiting factors. Despite this a few efforts have been made to demonstrate D4 photoanode/photocathode tandem cells.

Early work by Nozik[60] established the general theory for combining photoanodes and photocathodes into tandem cells and introduced the n-TiO$_2$/p-GaP system. Ohmic contacts between single crystals of n-TiO$_2$ and p-GaP

($E_g = 2.26$ eV) gave a tandem cell that was found to evolve both hydrogen and oxygen without an applied potential. A high internal resistance limited the conversion efficiency of the cell for H_2 evolution, which was calculated to be 0.25% at zero bias (based on a total electrode area of 1.9 cm^2 and 85 mW cm^{-2} of net incident simulated sunlight). An oxide layer forming on the surface of the p-GaP was the likely cause of the device instability.[61]

Very little research attention was given to constructing photoanode/photocathode tandem devices for many years after this seminal demonstration. More recently, following the advances in oxide photoanode performance, additional demonstrations of photoanode/photocathode tandem devices have appeared. For example, while also known to be unstable in aqueous systems,[23] GaInP$_2$ ($E_g = 1.83$ eV) photocathodes have been combined with either WO$_3$[50] or Fe$_2$O$_3$[62] photoanodes by Wang and Turner. Under intense white light illumination (>200 mW cm^{-2}) the WO$_3$/GaInP$_2$ combination produced a detectable photocurrent that rose linearly with light intensity to reach $j_{op} = 20$ μA cm^{-2} at 1000 mW cm^{-2}. Due to the insufficient potential difference, the system did not function at illumination intensities below 200 mW/cm^2 (recall that the minority carrier quasi Fermi level and thus the $\Delta\mu_{ex}$ of each electrode should change proportionally with the logarithm of the illumination intensity). For the case of Fe$_2$O$_3$/GaInP$_2$, negligible photocurrent was observed even at 10 sun illumination due to the mismatch of the conduction band minimum of the spray-pyrolyzed Fe$_2$O$_3$ thin film and the valence band maximum in the GaInP$_2$. Employing surface dipoles to raise the conduction band[63] of Fe$_2$O$_3$ may be useful for this combination of materials, which is clearly not ideal. Moreover the limited availability of indium in the earth's crust prevents the application of this material on a global scale.

A more novel p-type material made from abundant elements, CaFe$_2$O$_4$ ($E_g = 1.9$ eV), has been paired with n-TiO$_2$ (in a side-by-side configuration) to give a device operating at $j_{op} = 110$ μA cm^{-2} in 0.1 M NaOH with the light from a 500 W Xe lamp.[64] However, the Faradaic efficiency for water splitting was found to be only 12%, and Fe and Ca were detected in the electrolyte after the device test.

Nanostructuring techniques have also been employed to enhance the performance of inexpensive electrode materials for photoanode/photocathode tandem cells. Grimes and co-workers used an anodization technique to oxidize Cu-Ti films to obtain p-type nanotubular Cu-Ti-O films (1000 nm length, 65 nm pore diameter, 35 nm wall thickness).[65] The films were primarily CuO ($E_g = 1.4$ eV) with a Cu$_4$Ti$_3$ impurity phase detected. These films (on transparent F:SnO$_2$ substrates) were placed in tandem with nanotubular TiO$_2$ to give a working tandem device (see Figure 4.9). A $j_{op} = 0.25$ mA cm^{-2} (η_{STH} around 0.30%) under standard illumination conditions and a reasonable stability in the minutes time scale were observed (however, photocurrent was negligible after 5 hours). It should be noted that the TiO$_2$ was exposed to 1 M KOH while the Cu-Ti-O was exposed to 0.1 M Na$_2$HPO$_4$ and the electrolyte compartments were connected by a salt bridge. The dissimilar pH values lead to a favourable chemical bias equivalent to about 0.4 V in this case. Exposing the

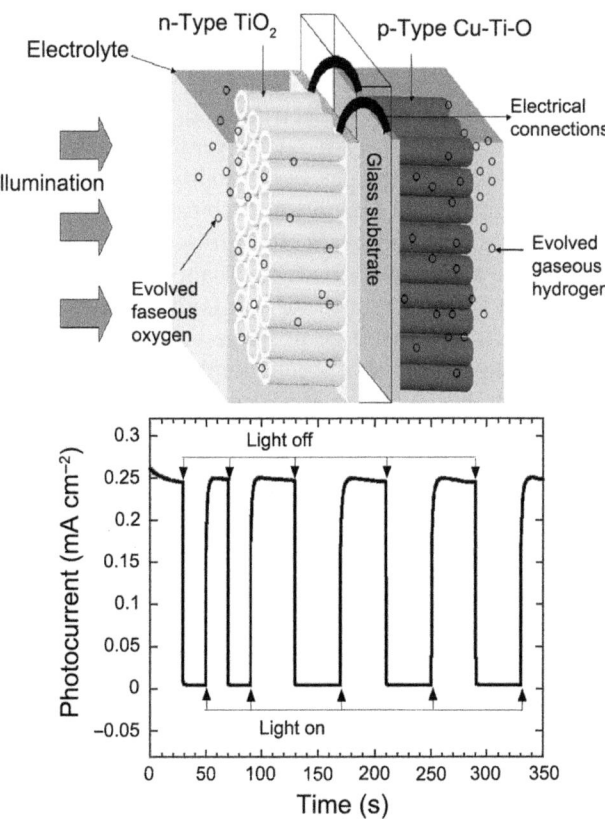

Figure 4.9   (Top) Illustration of a D4 tandem cell comprised of n-type TiO_2 and p-type Cu-Ti-O nanotube array films, with their substrates connected through an ohmic contact. (Bottom) Photocurrent from the D4 tandem cell under global AM 1.5 illumination. Light is incident upon the oxygen evolving TiO_2 side of the diode, with the visible portion of the spectrum passing to the Cu-Ti-O side. The n-TiO_2 side of the device is kept immersed in a 1 M KOH aqueous solution, the p-Cu-Ti-O side is kept in 0.1 M Na_2HPO_4 with a salt bridge linking the two sides solution.
Reprinted with permission from G. K. Mor, *et al.*, *Nano Lett.*, 2008, **8**, 1906–1911. Copyright 2008 American Chemical Society.

Cu-Ti-O to the KOH caused the rapid decay of photocurrent as the CuO was reduced to copper. Recent efforts to stabilize p-type photocathodes using overlayers deposited *via* atomic layer deposition may improve the performance of this and many other photocathode materials.[66]

4.6   Practical Device Design Considerations

Regardless of the nature of the combination of devices chosen to make a tandem cell for solar water splitting, consideration must be given to how the device will be illuminated and how the evolved gases will be collected. This latter point is especially important, given that the stoichiometric combination of H_2 and O_2 produced during water splitting is highly explosive when mixed. Thus, effective gas separation is necessary. Fortunately, the intrinsic configuration of water splitting tandem cell – one side for water oxidation and another side for water reduction – facilitates the separate collection of gases. A simple device has been described which uses 2 electrolyte compartments with identical compositions separated by a glass frit or membrane for ionic equilibration and to prevent reverse water splitting reaction from occurring.[67] An idealized variation of this device is shown in Figure 4.10a. Here, gravity is the driving force for separating the bubbles of gas forming on the electrode's surface and the remaining electrolyte. However, this force is maximized when the photoelectrodes are vertical, which requires the illumination direction to be perpendicular to the gravitational force. Without using mirrors, this illumination condition is only satisfied at sunrise or sunset. Thus, additional device considerations are necessary. For example, the device could simply be placed with the photoanode facing in the direction of the Earth's equator and at an angle optimized to maximize insolation while still providing sufficient buoyancy

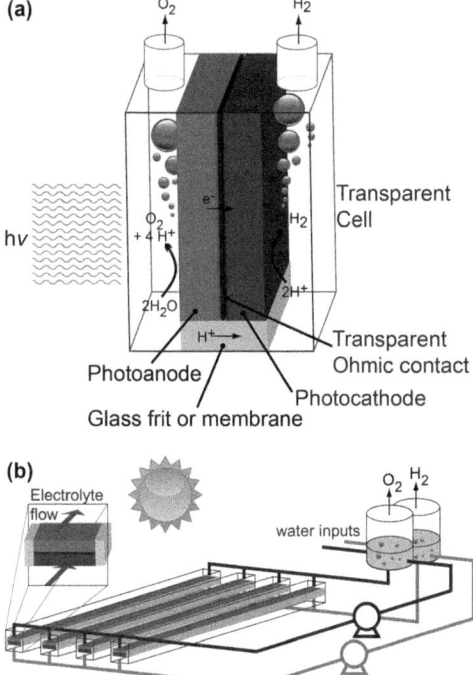

Figure 4.10   (Top) A simple schematic of a photoanode/photocathode tandem cell with gas collection. (Bottom) Conceptual design of a large-scale reactor with integrated tandem cells and flowing electrolyte to facilitate gas collection.
The bottom figure is adapted with permission from E. L. Miller, *et al.*, *Int. J. Hydrogen Energy*, 2003, **28**, 615–623. Copyright 2003 Elsevier.

force. At large latitudes (north or south), this would allow for near optimal illumination during the whole day while also permitting the gravity collection of the gases. Since the illumination direction near the equator is parallel to the gravitational force, the cells should be oriented flat with respect to the earth's surface to maximize insolation. This leads to difficulties with bubble accumulation at the surface of the bottom cell which would reduce the performance of the tandem cell by lowering the surface area of the electrode exposed to the electrolyte.

Relying on gravity to separate the evolved gases from the electrolyte also requires the bubbles to reach a certain size before buoyancy forces become larger than surface forces and they detach from the photoelectrode surface in any orientation. Thus to overcome any potential losses in the large scale implementation of the tandem cell, a trough-type flow design has been proposed (Figure 4.10b). Here two glass frits (or membranes) are used to prevent mixing of the two electrolyte compartments, and the electrolyte is pumped though the two compartments.[68] The flowing electrolyte adds an extra force to shear the bubbles off the surface, which are then carried along with the electrolyte to separating tanks where the gases are collected at ambient pressure. This type of system would allow the tandem cell to be at any angle, enabling the optimum insolation at any latitude or by using solar tracking without any concern of bubbles masking the active area of either electrode.

In addition to these relatively simple methods of tandem cell device implementation, many other systems have been proposed for different operation scales, and even with concentrated sunlight.[68,69] A common issue addressed in tandem cell device design is implementing separate functional layers for catalysis or protection. Of course these layers increase the cost of the system, but they are necessary when employing unstable materials or those with large overpotentials. The potential cost of producing the tandem device and its longevity are equally important as its overall solar-to-hydrogen efficiency. Is has been estimated that a device lifetime of 10 years, a cost of 100 US\$ m^{-2}, and a $\eta_{STH} = 10\%$ are necessary to produce hydrogen by PEC at a cost comparable to that of steam reforming of methanol (H$_2$ at a price of about 5 US\$ kg^{-1}).[70] Thus reducing device complexity and employing inexpensive fabrication methods while employing stable materials are the main issues for the practical implementation of the solar water splitting tandem cell.

4.7 Summary and Outlook

In this chapter, the motivation and theoretical framework for the PEC water splitting tandem cell were examined. While a single perfect material (in an S2 scheme) could reasonably convert 10% of the incident solar irradiation to chemical energy stored in hydrogen, that single perfect material has not been found. An integrated tandem approach (D4 scheme) could reasonably convert over 20% of the sun's energy (even with large assumed losses) and is more flexible regarding material choice. Many different systems have been investigated using various combinations of photovoltaic cells and photoelectrodes.

In order to be economically competitive with simple "brute force" strategies or the production of H_2 from fossil fuels, a practical water splitting tandem cell must optimize cost, longevity and performance. Due to the lack of stable photocathode materials, a promising approach to meet the cost and performance targets is to use a stable photoanode material in tandem with an inexpensive PV cell like the dye sensitized solar cell (DSSC). Promising photoanode materials are stable transition metal oxides exemplified by hematite (α-Fe_2O_3), which has vast potential given its band-gap, energy levels, abundance, and stability. The successful demonstration of a D4 Fe_2O_3 DSSC tandem cell with an solar-to-hydrogen efficiency of 1.17%, demonstrates the convincing promise of this device, but much more research is needed to realize a D4 tandem device with an $\eta_{STH} > 10\%$. Further reducing the overpotential for the oxygen evolution reaction (by passivating surface traps and adding catalysts)[71,72] and increasing the photocurrent (by nanostructuring)[73–75] are, at the time of writing, ongoing research topics for hematite and other encouraging photoanode materials. Device complexity and cost would also drop dramatically if stable photocathode materials with suitable band-gap energies for employment with hematite were developed. Other chapters of this book are dedicated to describing approaches to improving the performance of materials like hematite and to finding new materials that will someday be employed in an efficient, inexpensive and stable tandem device for the overall conversion and storage of the sun's energy into molecular hydrogen.

Acknowledgement

The authors acknowledge the Swiss Federal Office of Energy (Project number 102326, PECHouse) for supporting our research and development of the PEC tandem cell.

References

1. N. R. C. Committee on Alternatives and Strategies for Future Hydrogen Production and Use, National Academy of Engineering, *The Hydrogen Economy:Opportunities, Costs, Barriers, and R&D Needs*, National Academies Press, Washington, 2004.
2. M. F. Weber and M. J. Dignam, *J. Electrochem. Soc.*, 1984, **131**, 1258–1265.
3. J. R. Bolton, *Solar Energy*, 1996, **57**, 37–50.
4. J. R. Bolton, A. F. Haught and R. T. Ross, in *Photoelectrochemical Conversion and Storage of Solar Energy*, J S Connolly, Academic New York, 1981, pp. 297–339.
5. M. G. Walter, E. L. Warren, I R. McKone, S. W. Boettcher, Q. Mi, E. A. Santori and N. S. Lewis, *Chem. Rev. (Washington, DC, U. S.)*, 2010, **110**, 6446–6473.
6. J. R. Bolton, S. J. Strickler and J. S. Connolly, *Nature*, 1985, **316**, 495–500.

7. N. Kelly, T. Gibson and D. Ouwerkerk, *Int. J. Hydrogen Energy*, 2008, **33**, 2747–2764.
8. T. L. Gibson and N. A. Kelly, *Int. J. Hydrogen Energy*, 2008, **33**, 5931–5940.
9. N. G. Dhere and A. H. Jahagirdar, *Thin Solid Films*, 2005, **480–481**, 462–465.
10. A. H. Jahagirdar and N. G. Dhere, *Sol. Energy Mater. Sol. Cells*, 2007, **91**, 1488–1491.
11. A. D. Vos, *J. Phys. D: Appl. Phys*, 1980, **13**, 839.
12. H. Cotal, C. Fetzer, J. Boisvert, G. Kinsey, R. King, P. Hebert, H. Yoon and N. Karam, *Energy Environ. Sci*, 2009, **2**, 174.
13. ASTM-International, West Conshohocken, PA, 2008.
14. C. Santato, M. Ulmann and J. Augustynski, *J. Phys. Chem. B*, 2001, **105**, 936–940.
15. A. J. Bard and M. A. Fox, *Acc. Chem. Res.*, 1995, **28**, 141–145.
16. S. Licht, *J. Phys. Chem. B*, 2001, **105**, 6281–6294.
17. O. Khaselev, A. Bansal and J. A. Turner, *Int. J. Hydrogen Energy*, 2001, **26**, 127–132.
18. S. Licht, B. Wang, S. Mukerji, T. Soga, M. Umeno and H. Tributsch, *J. Phys. Chem. B*, 2000, **104**, 8920–8924.
19. S. Licht, B. Wang, S. Mukerji, T. Soga, M. Umeno and H. Tributsch, *Int. J. Hydrogen Energy*, 2001, **26**, 653–659.
20. X. Gao, S. Kocha, A. J. Frank and J. A. Turner, *Int. J. Hydrogen Energy*, 1999, **24**, 319–325.
21. S. S. Kocha, D. Montgomery, M. W. Peterson and J. A. Turner, *Sol. Energy Mater. Sol. Cells*, 1998, **52**, 389–397.
22. O. Khaselev and J. A. Turner, *Science*, 1998, **280**, 425–427.
23. O. Khaselev and J. A. Turner, *J. Electrochem. Soc.*, 1998, **145**, 3335–3339.
24. G. H. Lin, M. Kapur, R. C. Kainthla and J. O. M. Bockris, *Appl. Phys. Lett.*, 1989, **55**, 386–387.
25. X. Deng, X. Liao, S. Han, H. Povolny and P. Agarwal, *Sol. Energy Mater. Sol. Cells*, 2000, **62**, 89–95.
26. R. E. Rocheleau, E. L. Miller and A. Misra, *Energy Fuels*, 1998, **12**, 3–10.
27. N. A. Kelly and T. L. Gibson, *Int. J. Hydrogen Energy*, 2006, **31**, 1658–1673.
28. S. Y. Reece, J. A. Hamel, K. Sung, T. D. Jarvi, A. J. Esswein, J. J. H. Pijpers and D. G. Nocera, *Science*, 2011, **334**, 563.
29. A. Fujishima and K. Honda, *Nature*, 1972, **238**, 37–38.
30. A. Fujishima, X. T. Zhang and D. A. Tryk, *Surf. Sci. Rep*, 2008, **63**, 515–582.
31. Y. Xu and M. A. A. Schoonen, *Am. Mineral.*, 2000, **85**, 543–556.
32. B. Neumann, P. Bogdanoff and H. Tributsch, *J. Phys. Chem. C*, 2009, **113**, 20980–20989.
33. B. D. Alexander, P. J. Kulesza, L. Rutkowska, R. Solarska and J. Augustynski, *J. Mater. Chem.*, 2008, **18**, 2298–2303.
34. A. Stavrides, *Proc. SPIE*, 2006, **6340**, 63400K.

35. E. Miller, R. E. Rocheleau and S. U. M. Khan, *Int. J. Hydrogen Energy*, 2004, **29**, 907–914.
36. K. Sivula, F. Le Formal and M. Grätzel, *Chemsuschem*, 2011, **4**, 432–449.
37. B. O'Regan and M. Gratzel, *Nature*, 1991, **353**, 737–740.
38. M. Gratzel, *Nature*, 2001, **414**, 338–344.
39. J. Augustynski, G. Calzaferri, J. C. Courvoisier and M. Gratzel, Hydrogen energy progress XI : proceedings of the 11th World Hydrogen Energy Conference, Stuttgart, Germany, 1996.
40. M. Gratzel, *Cattech*, 1999, **3**, 3–17.
41. J. H. Park and A. J. Bard, *Electrochem. Solid-State Lett*, 2006, **9**, E5–E8.
42. H. Arakawa, C. Shiraishi, M. Tatemoto, H. Kishida, D. Usui, A. Suma, A. Takamisawa and T. Yamaguchi, *Solar Hydrogen and Nanotechnology II*, 2007, **6650**, 65003–65003.
43. M. K. Nazeeruddin, P. Péchy, T. Renouard, S. M. Zakeeruddin, R. Humphry-Baker, P. Cointe, P. Liska, L. Cevey, E. Costa, V. Shklover, L. Spiccia, G. B. Deacon, C. A. Bignozzi and M. Grätzel, *J. Am. Chem. Soc.*, 2001, **123**, 1613–1624.
44. J. H. Yum, P. Walter, S. Huber, D. Rentsch, T. Geiger, F. Nuesch, F. De Angelis, M. Gratzel and M. K. Nazeeruddin, *J. Am. Chem. Soc.*, 2007, **129**, 10320.
45. Y. Shi, R. B. M. Hill, J.-H Yum, A. Dualeh, S. Barlow, M. Grätzel, S. R. Marder and M. K. Nazeeruddin, *Angew. Chem.*, 2011, **123**, 6749–6751.
46. J. Brillet, M. Cornuz, F. Le Formal, J. H. Yum, M. Graetzel and K. Sivula, *J. Mater. Res.*, 2010, **25**, 17–24.
47. I. Cesar, K. Sivula, A. Kay, R. Zboril and M. Graetzel, *J. Phys. Chem. C*, 2009, **113**, 772–782.
48. J.-H Yum, E. Baranoff, F. Kessler, T. Moehl, S. Ahmad, T. Bessho, A. Marchioro, E. Ghadiri, J-E Moser, C. Yi, M. K. Nazeeruddin and M. Grätzel, *Nat. Commun*, 2012, **3**, 631.
49. J. Brillet, J. H. Yum, M. Cornuz, T. Hisatomi, R. Solarska, J. Augustynski, M. Graetzel and K. Sivula, *Nat. Photonics*, 2012, **6**, 824.
50. H. Wang, T. Deutsch and J. A. Turner, *J. Electrochem. Soc.*, 2008, **155**, F91.
51. R. N. Dominey, N. S. Lewis, J. A. Bruce, D. C. Bookbinder and M. S. Wrighton, *J. Am. Chem. Soc.*, 1982, **104**, 467–482.
52. Y. Hou, B. L. Abrams, P. C. K. Vesborg, M. E. Björketun, K. Herbst, L. Bech, A. M. Setti, C. D. Damsgaard, T. Pedersen, O. Hansen, J. Rossmeisl, S. Dahl, J. K. Nørskov and I. Chorkendorff, *Nat. Mater.*, 2011, **10**, 434–438.
53. D. H. van Dorp, N. Hijnen, M. Di Vece and J. J. Kelly, *Angew. Chem. Int. Ed.*, 2009, **48**, 6085–6088.
54. J. A. Baglio, G. S. Calabrese, D. J. Harrison, E. Kamieniecki, A. J. Ricco, M. S. Wrighton and G. D. Zoski, *J. Am. Chem. Soc.*, 1983, **105**, 2246–2256.
55. D. Yokoyama, T. Minegishi, K. Maeda, M. Katayama, J. Kubota, A. Yamada, M. Konagai and K. Domen, *Electrochem. Commun.*, 2010, **12**, 851–853.

56. W. Siripala, A. Ivanovskaya, T. F. Jaramillo, S-H Baeck and E. W. McFarland, *Sol. Energy Mater. Sol. Cells*, 2003, **77**, 229–237.
57. M. Trari, A. Bouguelia and Y. Bessekhouad, *Sol. Energy Mater. Sol. Cells*, 2006, **90**, 190–202.
58. Y. Matsumoto, M. Omae, K. Sugiyama and E. Sato, *The Journal of Physical Chemistry*, 1987, **91**, 577–581.
59. C. Leygraf, M. Hendewerk and G. Somorjai, *J. Solid State Chem.*, 1983, **48**, 357–367.
60. A. J. Nozik, *Appl. Phys. Lett.*, 1976, **29**, 150.
61. D. S. Ginley and M. B. Chamberlain, *J. Electrochem. Soc.*, 1982, **129**, 2141–2145.
62. H. Wang and J. A. Turner, *J. Electrochem. Soc.*, 2010, **157**, F173–F178.
63. Y. S. Hu, A. Kleiman-Shwarsctein, G. D. Stucky and E. W. McFarland, *Chem. Commun. (Cambridge, U. K.)*, 2009, 2652–2654.
64. S. Ida, K. Yamada, T. Matsunaga, H. Hagiwara, Y. Matsumoto and T. Ishihara, *J. Am. Chem. Soc.*, 2010, **132**, 17343–17345.
65. G. K. Mor, O. K. Varghese, R. H. T. Wilke, S. Sharma, K. Shankar, T. J. Latempa, K.-S. Choi and C. A. Grimes, *Nano Lett.*, 2008, **8**, 1906–1911.
66. A. Paracchino, V. Laporte, K. Sivula, M. Grätzel and E. Thimsen, *Nat. Mater.*, 2011, **10**, 456–461.
67. M. Gratzel and J Augustynski, International Patent WO 2001002624, 2000.
68. E. L. Miller, R. E. Rocheleau and X. M. Deng, *Int. J. Hydrogen Energy*, 2003, **28**, 615–623.
69. L. J. Minggu, W. R. Wan Daud and M. B. Kassim, *Int. J. Hydrogen Energy*, 2010, **35**, 5233–5244.
70. B. D. James, G. N. Baum, J. Perez and K. N. Baum, *Technoeconomic Analysis of Photoelectrochemical (PEC) Hydrogen Production DOE Report* 2009. Contract # GS-10F-009J.
71. F. Le Formal, N. Tetreault, M. Cornuz, T. Moehl, M. Gratzel and K. Sivula, *Chem. Sci*, 2011, **2**, 737–743.
72. D. K. Zhong, M. Cornuz, K. Sivula, M. Gratzel and D. R. Gamelin, *Energy Environ. Sci*, 2011, **4**, 1759–1764.
73. K. Sivula, F. Le Formal and M. Gratzel, *Chem. Mater.*, 2009, **21**, 2862–2867.
74. J. Brillet, M. Grätzel and K. Sivula, *Nano Lett.,*, 2010, **10**, 4155–4160.
75. K. Sivula, R. Zboril, F. Le Formal, R. Robert, A. Weidenkaff, J. Tucek, J. Frydrych and M. Grätzel, *J. Am. Chem. Soc.*, 2010, **132**, 7436–7444.

CHAPTER 5

Particulate Oxynitrides for Photocatalytic Water Splitting Under Visible Light

KAZUHIKO MAEDA AND KAZUNURI DOMEN*

The University of Tokyo, Department of Chemical System Engineering,
7-3-1 Hongo, Bunkyo-ku, Tokyo, 113-8656, Japan
*Email: domen@chemsys.t.u-tokyo.ac.jp

5.1 Introduction

Catalytic splitting of pure water into hydrogen and oxygen in the presence of semiconductor powders with visible light is a promising approach from the viewpoint of converting solar energy into chemical energy. The reaction (eq. (1)) is typical of "uphill-type reaction" with a large positive change in the Gibbs free energy ($\Delta G^0 = 238$ kJ$\cdot$mol^{-1}).

$$2H_2O \rightarrow 2H_2 + O_2 \tag{1}$$

whose half reactions are described as follows:

$$2H^+ + 2e^- \rightarrow H_2 \tag{2}$$

$$2H_2O + 4h^+ \rightarrow O_2 + 4H^+ \tag{3}$$

The research was initially triggered by the potential of TiO_2-based photo-electrochemical reactions for the decomposition of water into H_2 and O_2.[1] Since then, a range of semiconductor materials have been examined as powder photocatalysts in view of the emphasis placed on large-scale

RSC Energy and Environment Series No. 9
Photoelectrochemical Water Splitting: Materials, Processes and Architectures
Edited by Hans-Joachim Lewerenz and Laurence Peter
© The Royal Society of Chemistry 2013
Published by the Royal Society of Chemistry, www.rsc.org

hydrogen production.[2-6] For efficient solar energy conversion, a photocatalytic material that can work under visible light ($\lambda > 400$ nm) – the main component of sunlight – needs to be devised.

Figure 5.1 shows a schematic illustration of the basic principle of overall water splitting on a semiconductor particle. Under irradiation with an energy equivalent to or greater than the band gap of the semiconductor photocatalyst, electrons in the valence band are excited into the conduction band, leaving holes in the valence band. These photogenerated electrons and holes cause reduction and oxidation reactions, respectively. To achieve overall water splitting, the potential equivalent to the bottom of the conduction band must be more negative than the reduction potential of H^+ to H_2 (0 V *vs.* NHE at pH 0), while the top of the corresponding potential for valence band must be more positive than the oxidation potential of H_2O to O_2 (1.23 V *vs.* NHE). Therefore, the minimum photon energy thermodynamically required to drive the reaction is 1.23 eV, which corresponds to a wavelength of *ca.* 1000 nm, in the near infrared region. Accordingly, it would appear possible to utilize the entire spectral range of visible light ($400 < \lambda < 800$ nm). However, there are activation barriers in the charge transfer processes between the photocatalyst and water molecules, necessitating a photon energy greater than the band gap of the photocatalyst to drive the overall water-splitting reaction at a reasonable rate. In addition, the back reaction; that is, water formation from H_2 and O_2, must be strictly inhibited, and the photocatalysts themselves must be stable in the reaction. Although there is a large number of materials that possess suitable band-gaps, materials that function as a photocatalyst for overall water splitting are limited due to other factors, as will be mentioned later.

Since 1980, many metal oxides have been examined as photocatalysts for water splitting. They include transition metal oxides containing metal ions of Ti^{4+}, Zr^{4+}, Nb^{5+}, Ta^{5+}, or W^{6+} with d^0 electronic configuration and typical metal oxides having metal ions of Ga^{3+}, In^{3+}, Ge^{4+}, Sn^{4+}, or Sb^{5+} with d^{10}

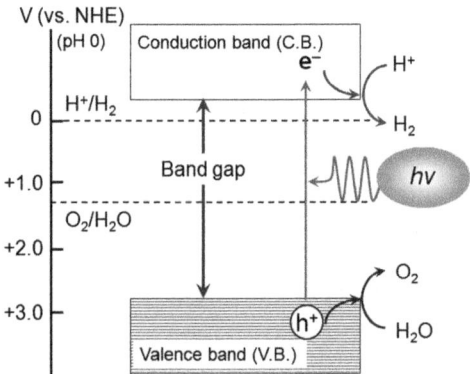

Figure 5.1 Basic principle of overall water splitting on a semiconductor particle. Reprinted with permission from Reference 3. Copyright 2007 American Chemical Society.

electronic configuration.[4,5] More than 100 metal oxides have been reported to exhibit water splitting activity, and some have achieved high quantum efficiencies of several tens of percent. However, metal oxide photocatalysts in principle work only under UV light ($\lambda < 400$ nm), as explained later. Although some non-oxide materials such as CdS and CdSe have been examined, particularly for visible-light catalysis, successful photocatalytic systems have yet to be achieved, primarily due to the lack of oxygen productivity and inherent instability of the materials.[7,8]

The two major obstacles to the development of powder photocatalysts are the discovery of new stable photocatalytic materials and the construction of a suitable visible light-driven photocatalytic system. The application of metal oxides for photocatalysis in the visible-light region is complicated by the deep valence band positions (O2p orbitals) of these materials, resulting in a band gap that is too large to harvest visible light.[9] In general, the conduction band of a metal oxide photocatalyst is formed by empty d orbitals of a transition metal or s,p orbitals of a typical metal, which lie above the water reduction potential of H$_2$O (0 V *vs.* NHE at pH 0). On the other hand, the potential of the valence band, which consists of O2p orbitals (*ca.* +3 V), is considerably more positive than the water oxidation potential (1.23 V). Therefore, the band gaps of metal oxide photocatalysts inevitably become wider than 3 eV, if they meet the thermodynamic requirement for water splitting.

Since the N2p orbital has a higher potential energy than the O2p orbital, it would be interesting to use a metal nitride or metal oxynitride as a photocatalyst. Figure 5.2 shows the schematic band structures of the metal oxide NaTaO$_3$ and (oxy)nitride BaTaO$_2$N, both of which have the same perovskite structure.[3] The top of the valence band (*i.e.* the highest occupied molecular orbital, HOMO) of the metal oxide consists of O2p orbitals. When N atoms are

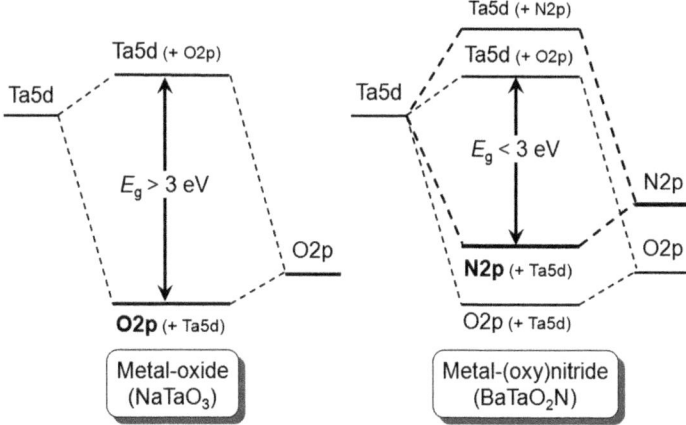

Figure 5.2   Schematic band structures of a metal oxide (NaTaO$_3$) and metal (oxy)nitride (BaTaO$_2$N).
Reprinted with permission from Reference 3. Copyright 2007 American Chemical Society.

partially or fully substituted for O atoms in a metal oxide, the HOMO of the material is expected to shift above that of the corresponding metal oxide without affecting the level of the bottom of the conduction band (*i.e.* the lowest unoccupied molecular orbital, LUMO). DFT band structure calculations for $BaTaO_2N$ indicated that the HOMO consists of hybridized N2p and O2p orbitals, whereas the LUMO is mainly composed of empty Ta5d orbitals. The HOMO for the (oxy)nitride is at a higher potential energy than that of the corresponding oxide due to the contribution of N2p orbitals, making the band gap small enough to respond to visible light ($<3\,eV$). Similar results were obtained in calculations for other (oxy)nitrides. In this chapter, (oxy)nitrides are presented as non-oxide type materials for photocatalytic water splitting under visible light, focusing on recent research progress made by the authors' group.

5.2 Oxynitrides Having Early-Transition Metal Cations

Some particulate (oxy)nitrides containing early transition-metal cations are stable and harmless materials, and can be readily obtained by nitriding a corresponding metal-oxide powder.[10-26] Note that this type of material differs from materials doped with nitrogen. Figure 5.3 shows the UV-visible diffuse reflectance spectra (DRS) for certain (oxy)nitrides containing transition-metal cations of Ti^{4+}, Nb^{5+} and Ta^{5+} with d^0 electronic configuration. It is clear that these (oxy)nitrides possess absorption bands at 500–750 nm, corresponding to band-gap energies of 1.7–2.5 eV that are estimated from the onset wavelengths of the absorption spectra. As expected from DFT calculations, these (oxy)nitrides have band-edge potentials suitable for overall water splitting, as revealed

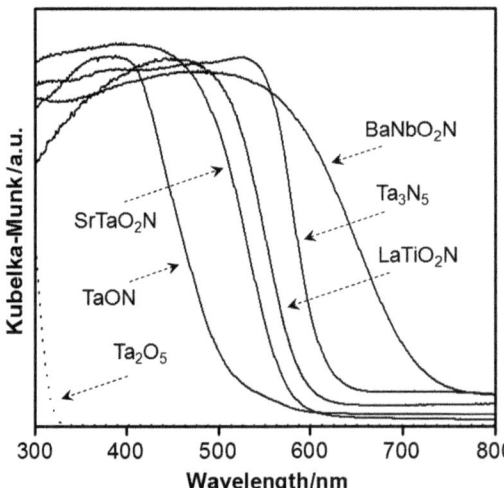

Figure 5.3   UV-visible diffuse reflectance spectra for (oxy)nitrides containing Ti^{4+}, Nb^{5+}, and Ta^{5+}.
Reprinted with permission from Reference 3. Copyright 2007 American Chemical Society.

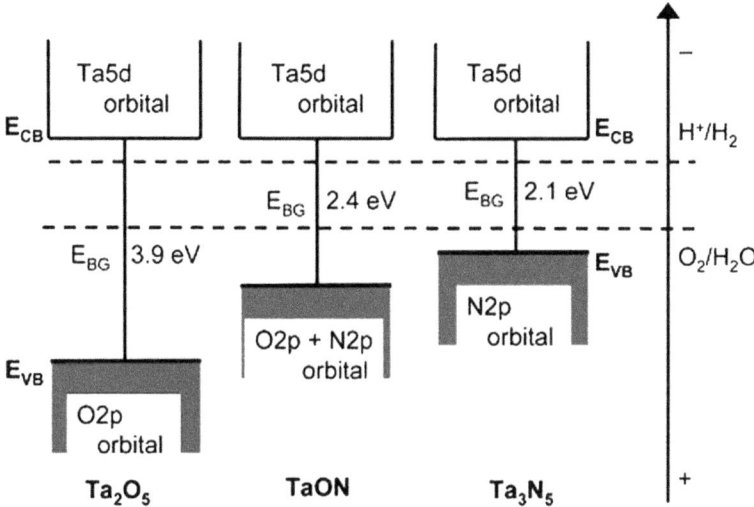

Figure 5.4 Schematic illustration of band structures of Ta_2O_5, TaON and Ta_3N_5. Reprinted with permission from Reference 16. Copyright 2003 American Chemical Society.

by UV photoelectron spectroscopy (UPS) and photoelectrochemical (PEC) analysis.[16] For example, it has been revealed that the tops of the valence bands are shifted to higher potential energies in the order $Ta_2O_5 < TaON < Ta_3N_5$, whereas the potentials of the bottoms of the conduction bands remain almost unchanged, as shown in Figure 5.4.

As overall water splitting is generally difficult to achieve due to the up-hill nature of the reaction, photocatalytic activities of a number of (oxy)nitrides for water reduction or oxidation were examined in the presence of methanol or silver nitrate as a sacrificial reagent. The reactions using sacrificial reagents are not "overall" water splitting reactions, but are often carried out as test reactions for overall water splitting.[3,4] Table 1 lists the photocatalytic activities of some d^0-(oxy)nitrides for H_2 or O_2 evolution in the presence of sacrificial reagents. These (oxy)nitrides exhibited relatively high photocatalytic activity for water oxidation to form molecular oxygen under visible irradiation ($\lambda > 420$ nm), with some exceptions, such as $ATaO_2N$ (A = Ca, Sr, Ba) and $LaTaON_2$. Among this group, TaON achieved the highest quantum efficiency of 34% (entry 5, Table 1).[12] It should be noted that water oxidation involving a 4-electron process can be achieved by (oxy)nitrides with high quantum efficiencies, although silver nitrate is used as an electron acceptor in such a system. In contrast, nitrogen-doped TiO_2 displayed negligible activity for this reaction under the same reaction conditions (entry 11, Table 1). (Oxy)nitrides also produced H_2 from an aqueous methanol solution upon visible irradiation when loaded with nanoparticulate Pt as a cocatalyst for H_2 evolution. In most cases, a low level of N_2 evolution accompanied the initial stage of photocatalytic reactions over these catalysts, indicating that the (oxy)nitride is partially

Table 5.1 Photocatalytic activities of (oxy)nitrides with d^0 electronic configuration for H$_2$ or O$_2$ evolution in the presence of sacrificial reagents under visible light ($\lambda > 420$ nm).^{a}

Entry	Oxynitride	Band gap energy^b/eV	Activity/$\mu mol \cdot h^{-1}$		Ref.
			H$_2$^{c}	O$_2$^{d}	
1	LaTiO$_2$N	2.0	30	41	15
2	Ca$_{0.25}$La$_{0.75}$TiO$_{2.25}$N$_{0.75}$	2.0	5.5	60	15
3	TiN$_x$O$_y$F$_z$	2.3	0.2	43	17, 19
4	CaNbO$_2$N	1.9	1.5	46	14
5	TaON	2.5	20	660	12
6	Ta$_3$N$_5$	2.1	10	420	13
7	CaTaO$_2$N	2.4	15	0	18
8	SrTaO$_2$N	2.1	20	0	18
9	BaTaO$_2$N	1.9	15	0	18
10	LaTaON$_2$	2.0	20	0	18
11	N-doped TiO$_2$	2.6^e	n.d.	1.6	17, 19

^{a}Reaction conditions: 0.2–0.4 g of catalyst, 200 mL aqueous solution containing sacrificial reagents, 300 W xenon lamp light source, Pyrex top irradiation-type reaction vessel with cutoff filter.
^{b}Estimated from onset wavelength of diffuse reflectance spectra.
^{c}Loaded with nanoparticulate Pt as a cocatalyst, reacted in the presence of methanol (10 vol%), sacrificial reagent.
^{d}Sacrificial reagent: silver nitrate (0.01 M).
^{e}Energy gap estimated from a shoulder peak (400–500 nm).

decomposed by the photogenerated holes (instead of water oxidation) according to the reaction:

$$2N^{3-} + 6h^+ \rightarrow N_2 \tag{4}$$

However, the production of N$_2$ was completely suppressed as the reaction progresses, and no change in the XRD pattern of the (oxy)nitrides as a result of the reaction has been detected. It thus appears that (oxy)nitride photocatalysts are essentially stable in the individual water reduction and oxidation reactions.

5.3 Improvement of Water Reduction Activity of d^0-(Oxy)nitrides

It should be noted that the activities for H$_2$ production are about an order of magnitude lower than for O$_2$ evolution, although the photocatalytic performance for H$_2$ evolution is stable. In this context, several new modifications to enhance the H$_2$ evolution rate of (oxy)nitrides have been pursued. Here we introduce such an attempt to reduce the particle size of Ta$_3$N$_5$. Nanoparticulate Ta$_3$N$_5$ can be readily prepared from nanoparticulate Ta$_2$O$_5$ precursor prepared through a precipitation method.[21] Compared to the conventional bulk-type Ta$_3$N$_5$ particles (300–500 nm), nanoparticulate Ta$_3$N$_5$ with a smaller particle size of 30–50 nm was shown to exhibit enhanced photocatalytic activity for H$_2$ evolution from aqueous solution containing methanol as an electron donor under visible light. Diffuse reflectance spectroscopy and PEC measurements suggested that the enhancement was due to the promotion of the water reduction process originating from the lower defect density of the material. It is also possible to prepare nanosized Ta_3N_5 particles through a template approach using a mesoporous carbon nitride (mpg-C_3N_4),[22] which has a potential to reduce or oxidize water with visible light as well.[27,28] This method allows one to prepare Ta_3N_5 nanoparticles with a primary particle size that in principle reflects the pore size of mpg-C_3N_4. Nanoparticles Ta_3N_5 having the smallest particle size, which was prepared from mpg-C_3N_4 with a pore size of 7 nm, showed approximately one order of magnitude higher activity than bulk Ta_3N_5, as shown in Figure 5.5. In addition to the low particle size and high surface area, the low density of defect sites would also result in an improvement of the activity, since fewer defect sites would be favorable for electron migration from the Ta_3N_5 bulk to the surface and/or electron transfer from the conduction band of Ta_3N_5 to the loaded Pt.[21]

Reducing the particle size of TaON, which exhibits higher activity than Ta_3N_5 (see Table 1), has been examined in a similar manner. However, it was difficult to suppress the generation of defects during the nitridation process, because TaON is an intermediate phase between Ta_2O_5 and Ta_3N_5.[23] As a result, the as-obtained TaON did not show an appreciable increase in photocatalytic activity, despite a relatively small particle size (50–80 nm). Although the calcination of solid photocatalysts after preparation has been demonstrated to be an effective approach for reducing the density of defects,[29,30] such post-calcination appears to be disadvantageous for TaON due to its low thermal

Figure 5.5 Time courses of 0.5 wt% Pt-loaded Ta_3N_5 samples for H_2 evolution from aqueous methanol solution under visible light ($\lambda > 400$ nm), along with a TEM image of Ta_3N_5 nanoparticles. Catalyst (0.2 g); an aqueous methanol solution (20 vol%, 400 mL); light source, high pressure mercury lamp (450 W); inner irradiation-type reaction vessel made of quartz or Pyrex with or without an aqueous $NaNO_2$ solution (2 M) filter.
Reprinted with permission from Reference 22. Copyright 2010 The Royal Society of Chemistry.

stability. In fact, post-calcination of TaON under O_2 or N_2 results in a decrease in activity.[31] The elimination of defects in TaON therefore remains a challenge.

To overcome this, a new method to reduce surface defects on TaON was applied using ZrO_2 as a "protector" through a surface modification technique.[24] Specifically, monoclinic ZrO_2 nanoparticles are dispersed on the surface of Ta_2O_5 precursor before nitridation, thereby protecting Ta^{5+} cations in the TaON surface from being reduced by H_2 (derived from disassociation of NH_3 at high temperatures) during nitridation, as schematically illustrated in Figure 5.6. As has been observed for other (oxy)nitrides such as $LaTiO_2N$[15] and $TiN_xO_yF_z$,[19] the reduction of Ta^{5+} cations in TaON during nitridation generates reduced tantalum species, which can act as recombination centers between photogenerated electrons and holes. On the other hand, when ZrO_2 is loaded on the surface of Ta_2O_5, tantalum cations at the interface between Ta_2O_5 (and/or TaON) and the loaded ZrO_2 are expected to interact with ZrO_2 and thereby become more cationic. As a result, the formation of reduced tantalum species on the TaON surface is effectively suppressed by prior loading with nanoparticulate ZrO_2. As expected, the as-prepared ZrO_2-modified TaON (represented as ZrO_2/TaON) exhibited enhanced water reduction behavior compared to unmodified TaON. This enhancement of activity is attributed to the reduced density of defects in ZrO_2/TaON, which contributes to a lower probability of undesirable electron-hole recombination in ZrO_2/TaON than in

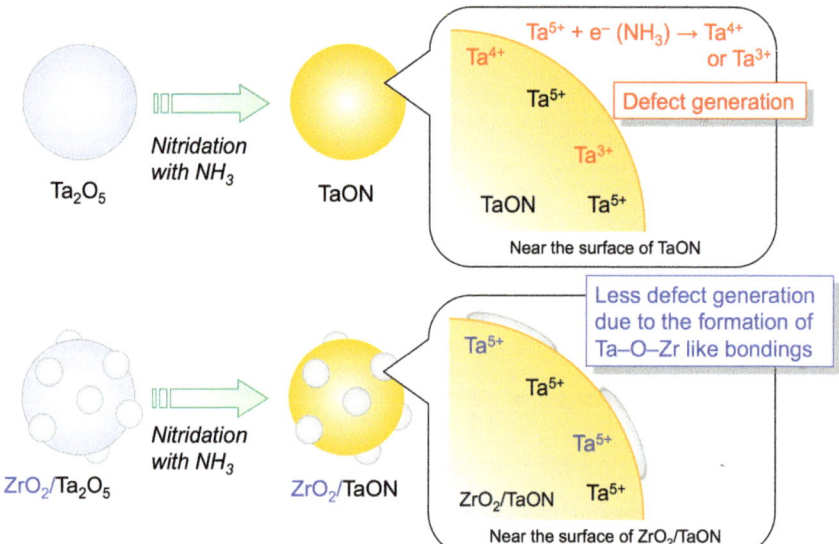

Figure 5.6   Nitridation of ZrO_2/Ta_2O_5 composite to produce ZrO_2/TaON while suppressing the production of reduced tantalum species (defect sites) near the surface of the material.
Reprinted with permission from Reference 24. Copyright 2008 The Chemical Society of Japan.

TaON. This material is also applicable to a building block for H_2 evolution in a two-step water splitting system under visible light, as will be presented later.

Improvement of water reduction was also attempted for a mixed metal oxynitride, $CaTaO_2N$.[26] $CaTaO_2N$ was successfully prepared by nitriding a mixture of layered Ca-Ta oxide ($RbCa_2Ta_3O_{10}$ or $HCa_2Ta_3O_{10}$) and $CaCO_3$ under NH_3 flow. The as-prepared $CaTaO_2N$ consisted of aggregated nano-particles with a primary particle size of 50–100 nm, and had a lower density of anionic defects. Compared to $CaTaO_2N$ prepared from a bulk-type Ca-Ta oxide in a conventional manner, $CaTaO_2N$ derived from the layered oxides exhibited an enhanced photocatalytic activity (about 2.5 times) for H_2 evolution under visible light.

5.4 Ge_3N_4, a Typical Metal Nitride with d^{10} Electronic Configuration

As described above, (oxy)nitride photocatalysts having the ability to reduce and/or oxidize water under visible light had been composed of transition-metal cations of Ti^{4+}, Nb^{5+}, or Ta^{5+} with d^0 electronic configuration. From the viewpoint of the electronic band structure, however, d^{10}-based semiconducting materials are advantageous over the d^0 configurations as a photocatalyst in that while the top of the valence band consists of O2p orbitals, the bottom of the conduction band is composed of hybridized s,p orbitals of typical metals.[5] The hybridized s,p orbitals possess large dispersion, leading to increased mobility of photogenerated electrons in the conduction band and thus high photocatalytic activity. This has stimulated study of (oxy)nitrides with the d^{10} electronic configuration as photocatalysts for overall water splitting.

(Oxy)nitrides with a d^{10} electronic configuration are therefore of interest as potentially efficient photocatalysts for overall water splitting. In evaluating such d^{10}-compounds according to this hypothesis, β-Ge_3N_4 was examined as a water-splitting photocatalyst.[32] This compound can be readily prepared by nitriding GeO_2 powder at 1123–1223 K for 15 h. HR-TEM images and an electron diffraction pattern of the as-prepared β-Ge_3N_4 indicated it to consist of primary well-crystallized rod-like particles with a hexagonal crystal system. The specific surface area of the as-prepared β-Ge_3N_4 was *ca.* 2.4 m^2/g.

It was found that β-Ge_3N_4 loaded with RuO_2 nanoparticles functions as a photocatalyst for overall water splitting. This was the first case involving a non-oxide photocatalyst for the cleavage of pure water. Since this discovery, the photocatalytic properties and effects of post-treatment to enhance the activity of Ge_3N_4 have been examined in detail.[33,34] The relationship between the structural characteristics and the photocatalytic performance of Ge_3N_4 has also been investigated.[35] Unfortunately, the band gap of β-Ge_3N_4 is about 3.8 eV, which is responsive only to ultraviolet light.[34] Figure 5.7 shows the dependence of the photocatalytic activity of RuO_2-loaded β-Ge_3N_4 for overall water splitting on the wavelength of incident light, along with a typical DRS of β-Ge_3N_4. The rate of H_2 and O_2 evolution in overall water splitting decreased

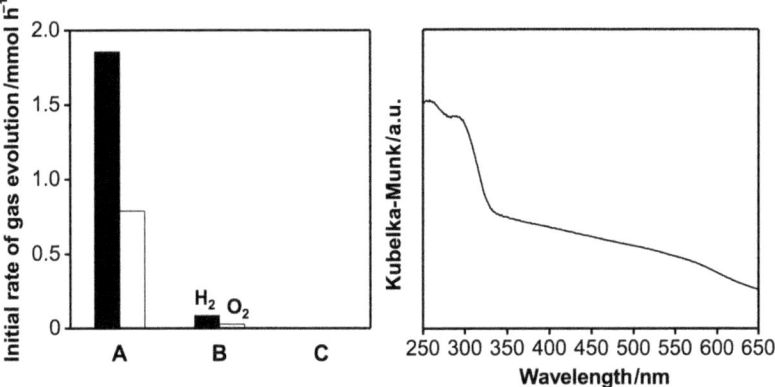

Figure 5.7   (Left) Dependence of photocatalytic activity of 1 wt% RuO$_2$-loaded β-Ge$_3$N$_4$ for overall water splitting on wavelength of incident light: λ > (A) 200, (B) 300, and (C) 400 nm. Catalyst (0.5 g); an aqueous H$_2$SO$_4$ solution (1 M, 350–400 mL); light source, high pressure mercury lamp (450 W); inner irradiation-type reaction vessel made of quartz or Pyrex with or without an aqueous NaNO$_2$ solution (2 M) filter. (Right) UV-visible diffuse reflectance spectrum of β-Ge$_3$N$_4$.
Reprinted with permission from Reference 34. Copyright 2007 American Chemical Society.

with increasing the wavelength of the incident light, due to a decrease in the number of incident photons from the light source. No H$_2$ and O$_2$ evolution was observed when the reaction was carried out at wavelengths longer than 400 nm or in the dark. These results indicate that the sharp absorption band at 350 nm, indicative of the band gap transition from the valence band formed by N2p orbitals to the conduction band formed by Ge4s,4p hybridized orbitals, contributes to promoting overall water splitting on RuO$_2$-loaded β-Ge$_3$N$_4$. In addition, it is clear that the broad absorption in the visible-light region assignable to impurities and/or defect sites is not available for the reaction. This situation forced us to search for a new d^{10} compound that can function under visible light.

5.5   GaN-ZnO and ZnGeN$_2$-ZnO Solid Solutions

To devise a new oxynitride with d^{10} electronic configuration that can decompose water under visible light, the solid solution of GaN and ZnO, $(Ga_{1-x}Zn_x)(N_{1-x}O_x)$, was examined.[36,37] The $(Ga_{1-x}Zn_x)(N_{1-x}O_x)$ solid solution is typically synthesized by nitriding a mixture of Ga$_2$O$_3$ and ZnO. Elemental analyses by inductively coupled plasma optical emission spectroscopy (ICP-OES) revealed that the ratios of Ga to N and Zn to O in the as-prepared material are close to unity, and N and O concentrations increase with the Ga and Zn concentrations, respectively. The atomic composition is controllable by changing the nitridation conditions.[37] X-ray diffraction and neutron powder diffraction analyses showed that the prepared samples are solid solutions of GaN and ZnO with no interstitial sites and no large disorder in the material.[38]

While GaN and ZnO both have band gap energies greater than 3 eV and thus do not absorb visible light, the $(Ga_{1-x}Zn_x)(N_{1-x}O_x)$ solid solution has absorption edges in the visible region. Figure 5.8 shows the UV-visible DRS for several samples. The absorption edge of $(Ga_{1-x}Zn_x)(N_{1-x}O_x)$ lies at a wavelength longer than those of GaN and ZnO and shifts to longer wavelengths with increasing Zn and O concentration (x) in $(Ga_{1-x}Zn_x)(N_{1-x}O_x)$. The band gap energies of the solid solutions are estimated to be 2.4–2.8 eV based on the DRS. It is thus clear that the visible-light-response of $(Ga_{1-x}Zn_x)(N_{1-x}O_x)$ originates from the presence of ZnO in the crystal. Initially, the origin of the visible-light absorption was thought to be band-gap narrowing of GaN due to p–d repulsion between Zn 3d and N 2p electrons in the upper part of the valence band.[36,37] Follow-up studies on the electronic structure of GaN-rich $(Ga_{1-x}Zn_x)(N_{1-x}O_x)$ using photoluminescence spectroscopy and DFT calculations suggested that this material absorbs visible light *via* electron transitions from the Zn acceptor level to the conduction band while maintaining the band-gap structure of the host GaN, as illustrated in Figure 5.9.[39,40] Because the concentration of Zn in $(Ga_{1-x}Zn_x)(N_{1-x}O_x)$ is relatively high, the acceptor level, which is filled with electrons derived from O donor levels (or thermal excitation), is likely to behave as an impurity band with a high density of states. Electrons can thus be transferred from the Zn acceptor level to the conduction band by visible-light absorption.

The as-prepared $(Ga_{1-x}Zn_x)(N_{1-x}O_x)$ exhibited little photocatalytic activity for water decomposition even under UV irradiation. However, modification by surface deposition of metal oxide nanoparticles as cocatalysts

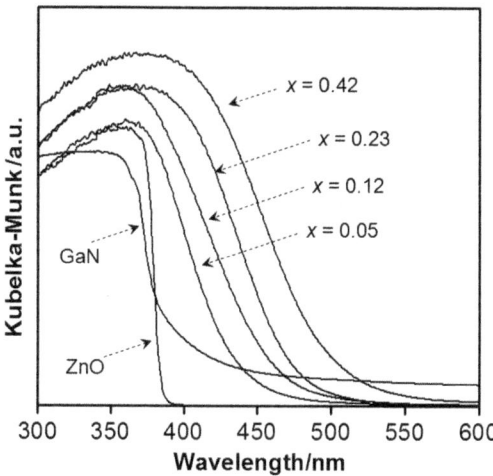

Figure 5.8 UV-visible diffuse reflectance spectra for various $(Ga_{1-x}Zn_x)(N_{1-x}O_x)$ solid solutions.
Reprinted with permission from Reference 3. Copyright 2007 American Chemical Society.

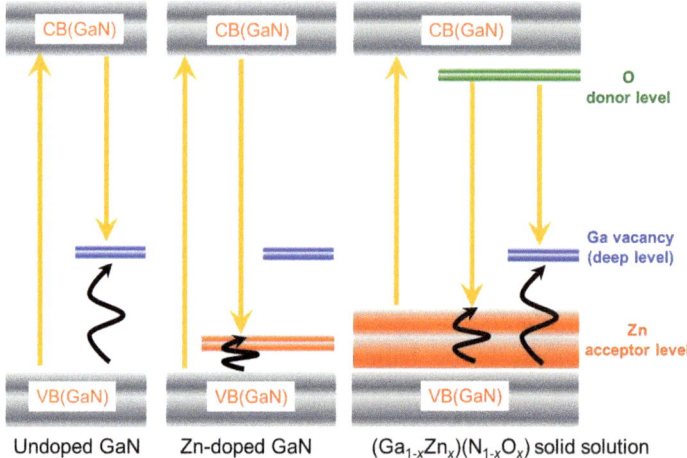

Figure 5.9   Expected energy level diagram for impurity levels in undoped GaN, Zn-doped GaN, and GaN-rich $(Ga_{1-x}Zn_x)(N_{1-x}O_x)$ solid solutions. Arrows denote photoabsorbtion, photoluminescence, and thermal relaxation.
Reprinted with permission from Reference 40. Copyright 2010 American Chemical Society.

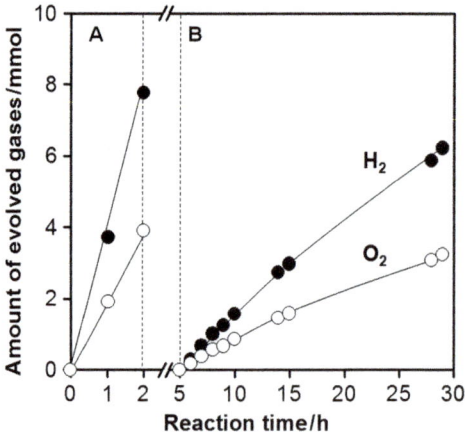

Figure 5.10 Time courses of overall water splitting on $(Ga_{1-x}Zn_x)(N_{1-x}O_x)$ loaded with $Rh_{2-y}Cr_yO_3$ (1 wt% Rh and 1.5 wt% Cr) under (A) UV ($\lambda>300$ nm) and (B) visible light irradiation ($\lambda>400$ nm). Reactions were performed in pure water (370 mL) with 0.3 g of catalyst powder under illumination from a high-pressure mercury lamp (450 W).
Reprinted with permission from Reference 42. Copyright 2011 Elsevier B. V.

resulted in clearly observable H_2 and O_2 evolution. The activity was strongly dependent on the loaded cocatalyst, and a mixed-oxide of Rh and Cr ($Rh_{2-y}Cr_yO_3$) is the most effective for $(Ga_{1-x}Zn_x)(N_{1-x}O_x)$ among the co-catalyst materials examined.[41,42] Figure 5.10 displays a typical time course of water splitting over $Rh_{2-y}Cr_yO_3$-loaded $(Ga_{1-x}Zn_x)(N_{1-x}O_x)$. As can be seen, this photocatalyst produced both H_2 and O_2 in the stoichiometric ratio ($H_2/O_2 = 2$) without noticeable degradation. The photocatalytic performance was also dependent on the pH of the reactant solution.[43] The photocatalyst exhibited stable and high photocatalytic activity in an aqueous solution at pH 4.5 for as long as three days. The photocatalytic performance at pH 3.0 and pH 6.2 was much lower, attributable to corrosion of the cocatalyst and hydrolysis of the catalyst. It has also been confirmed by X-ray diffraction, X-ray photoelectron spectroscopy, and X-ray absorption fine structure that the crystal structure of the catalyst and the valence state of both the surface and bulk do not change even after reaction for three days at optimal pH (4.5). These results offer useful guidelines for further research using other oxynitrides.

A solid solution of $ZnGeN_2$ and ZnO, $(Zn_{1+x}Ge)(N_2O_x)$, which is a similar material to $(Ga_{1-x}Zn_x)(N_{1-x}O_x)$ in terms of crystal structure and optical absorption, has also been found to be an active and stable photocatalyst for overall water splitting under visible light.[44,45] Both $ZnGeN_2$ and ZnO are wide-gap semiconductors with band gaps larger than 3 eV, although the band gap of $ZnGeN_2$ is dependent on the crystal structure and composition.[46] The $(Zn_{1+x}Ge)(N_2O_x)$ solid solutions can be prepared in a method similar to $(Ga_{1-x}Zn_x)(N_{1-x}O_x)$, but with a mixture of ZnO and GeO_2 as the starting material. The crystal structure of the synthesized sample was confirmed by both Rietveld analysis and neutron powder diffraction to be wurtzite with space group $P6_3mc$.[44] In the solid solution between $ZnGeN_2$ and ZnO, the oxygen atoms are replaced with nitrogen sites. The band gap of the material is estimated to be 2.5–2.7 eV based on the absorption edges, which is smaller than the band gaps of β-Ge_3N_4 (*ca.* 3.8 eV), $ZnGeN_2$ (*ca.* 3.3 eV), and ZnO (*ca.* 3.2 eV). Figure 5.11 shows DRS spectra of $ZnGeN_2$, ZnO, and a solid solution between the two, along with the proposed scheme of the origin of the visible-light response. The visible-light response of the material originates from the wide valence bands consisting of N2p, O2p, and Zn3d atomic orbitals and p–d repulsion between Zn3d and N2p and O2p electrons in the upper part of the valence bands. Therefore, the absorption edge of $(Zn_{1+x}Ge)(N_2O_x)$ tends to be located at longer wavelengths with increasing ZnO content.

Neither $ZnGeN_2$ nor ZnO alone exhibited photocatalytic activity for overall water splitting under UV irradiation. However, $(Zn_{1+x}Ge)(N_2O_x)$ became photocatalytically active under visible irradiation when loaded with a proper cocatalyst. Overall water splitting on a modified $(Zn_{1+x}Ge)(N_2O_x)$ proceeded by band gap photoexcitation from the valence band formed by N2p, O2p, and Zn3d atomic orbitals to the conduction band consisting of Ge4s,4p hybridized atomic orbitals. As demonstrated in the $(Ga_{1-x}Zn_x)(N_{1-x}O_x)$ section earlier, the photocatalytic activity of the transition metal-loaded catalysts was improved markedly by coloading Cr oxide.[45] The largest improvement in activity was obtained by loading the base catalysts with both 3 wt % Rh and 0.2 wt % Cr. It was confirmed that catalytic gas evolution of $Rh_{2-y}Cr_yO_3$-loaded $(Zn_{1+x}Ge)(N_2O_x)$ is stable for as long as 60 h.

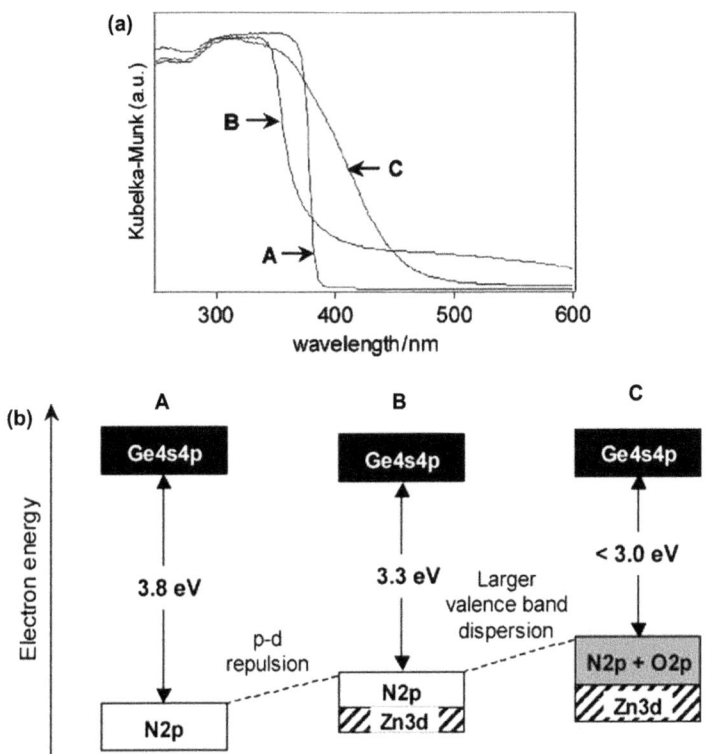

Figure 5.11 UV-visible DRS traces for (A) ZnO, (B) ZnGeN$_2$, and (C) a solid
solution between the two, (Zn$_{1.44}$Ge)(N$_{2.08}$O$_{0.38}$).
Reprinted with permission from References 44 and 45. Copyright 2007
American Chemical Society.

In addition to cocatalyst-loading, preparation conditions of (Zn$_{1+x}$Ge)(N$_2$O$_x$)
and the kind of precursors had a significant impact on the photocatalytic activity.
As mentioned earlier, (Zn$_{1+x}$Ge)(N$_2$O$_x$) are typically prepared by reaction of
GeO$_2$ and ZnO under a NH$_3$ flow at 1123 K. With increasing nitridation time, the
zinc and oxygen concentrations decreased due to reduction of ZnO and volatilization of zinc, and the crystallinity and band gap energy of the product increased.
The highest activity for overall water splitting was obtained for the sample after
nitridation for 15 h. Structural analyses revealed that the photocatalytic activity of
(Zn$_{1+x}$Ge)(N$_2$O$_x$) for overall water splitting depends heavily on the crystallinity
and composition of the material. The preparation route had a strong influence on
photocatalytic performance of (Zn$_{1+x}$Ge)(N$_2$O$_x$).[47] Improving the homogeneity
of the (Zn$_{1+x}$Ge)(N$_2$O$_x$) powder, reducing the number of superficial defects or
increasing the crystallinity, showed a direct positive impact on the photocatalytic
activity for overall water splitting. On the other hand, photoreduction of water, a
half reaction of overall water splitting, occurred with low Zn/Ge ratios and water
oxidation requires a high crystallinity. Moreover, overall water splitting was
achieved only if the crystal phase is active enough for photoreduction.

5.6   Two-Step Water Splitting Mimicking Natural Photosynthesis in Green Plants

Despite the efforts mentioned above, overall water splitting using these d^0-(oxy)nitrides has yet to be achieved. One plausible explanation for this is that they still have a high density of defect sites that can act as recombination centers for photogenerated electrons and holes. Nevertheless, Z-scheme overall water splitting under visible light has been achieved using several combination of (oxy)nitrides in the presence of an IO_3^-/I^- shuttle redox mediator.

TaON modified with Pt was first applied as a component for H_2 evolution in a two-step water splitting system.[48] When it was combined with Pt-loaded WO_3 as an O_2 evolution photocatalyst, overall water splitting was achieved under visible light in the presence of an IO_3^-/I^- shuttle redox. One reason for the effectiveness of TaON as a H_2 evolution component in an IO_3^-/I^--based Z-scheme system is the good ability of TaON to oxidize I^- into IO_3^-. However, this in turn implies that it is difficult to apply TaON into an O_2 evolution system; I^- ions are more susceptible to oxidation than water, thereby hindering water oxidation catalysis. Actually, neither Pt/TaON nor TaON are active for O_2 evolution half reaction from aqueous $NaIO_3$ solution. Interestingly, modification of TaON with nanoparticulate RuO_2 resulted in observable O_2 evolution.[49] Structural analyses and (photo)electrochemical measurements revealed that the activity of RuO_2/TaON is strongly dependent on the generation of optimally dispersed RuO_2 nanoparticles, which simultaneously promote both the reduction of IO_3^- and oxidation of water.[50] Thus, Z-scheme water splitting under visible light has been achieved using modified TaON catalysts.

ZrO_2/TaON, introduced earlier, exhibits higher performance for H_2 evolution in Z-scheme water splitting than TaON. In particular, combining Pt-loaded ZrO_2/TaON with Pt/WO_3 and an IO_3^-/I^- shuttle redox mediator achieved stoichiometric water splitting into H_2 and O_2 under visible light, yielding an AQY of 6.3% under irradiation by 420.5 nm monochromatic light under optimal conditions, six times greater than the yield achieved using a TaON analog.[51] To the best of the authors' knowledge, this is the highest reported value to date for a non-sacrificial visible-light-driven water-splitting system.

Using ZrO_2/TaON instead of TaON as a H_2 evolution photocatalyst also allowed one to apply a modified Ta_3N_5 as an O_2 evolution photocatalyst, extending the available wavelength for O_2 evolution up to 600 nm.[52] In this case, not only the use of ZrO_2/TaON but also modification of Ta_3N_5 is very important. Modification of Ta_3N_5 with nanoparticulate Ir and rutile titania (R-TiO_2) achieved functionality as an O_2 evolution photocatalyst in a two-step water-splitting system with an IO_3^-/I^- shuttle redox mediator under visible light ($\lambda > 420$ nm) in combination with a Pt/ZrO_2/TaON H_2 evolution photocatalyst. The loaded Ir nanoparticles acted as active sites to reduce IO_3^- to I^-, while the R-TiO_2 modifier suppressed the adsorption of I^- on Ta_3N_5, allowing Ta_3N_5 to evolve O_2 in the two-step water splitting system.

Although the absorption edge of TaON and ZrO_2/TaON (H_2 evolution photocatalysts) is limited to 520 nm in wavelength, replacing these

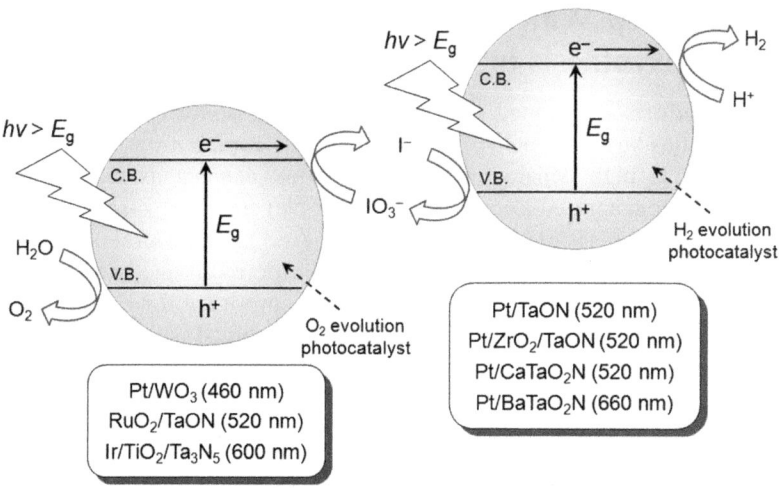

Figure 5.12   A schematic illustration of a two-step water-splitting system based on oxynitrides in the presence of an iodate/iodide shuttle redox mediator. The absorption edges of photocatalysts are also shown.

photocatalysts with $BaTaO_2N$ extended the wavelength available for H_2 evolution up to 660 nm.[53] The photocatalytic performance of $BaTaO_2N$ for H_2 evolution can be improved by 6–9 times as a result of forming a solid solution with $BaZrO_3$, due to an enhanced driving force for the redox reactions and the reduction of defect density.[54]

Thus, several combinations of (oxy)nitrides for Z-scheme water splitting in the presence of an IO_3^-/I^- shuttle redox mediator have been achieved, as schematically illustrated in Figure 5.12. The absorption wavelength available for the individual H_2 and O_2 evolution has been increased up to 660 nm and 600 nm, respectively. The next challenges remain in the promotion of electron transfer between two semiconductors and in the suppression of backward reactions involving shuttle redox mediators.

5.7   Photoelectrochemical Water Splitting Using d^0-Oxynitrides

(Oxy)nitrides are also useful as photoelectrode materials. While the separation of H_2 and O_2 produced in water splitting reaction is one of the biggest issues to be solved toward practical application, however, it is in principle possible to produce H_2 and O_2 separately in a PEC cell, as shown in Figure 5.13. Besides, in this approach, not only light energy but also electric energy can be put into the system for operation, resulting in a relatively high incident photon-to-current conversion efficiency (IPCE).

As mentioned earlier, most (oxy)nitrides have band edge potentials suitable for water splitting under visible light (<3 eV). This suggests that they are, in principle, capable of working as photoelectrodes to split water even without an

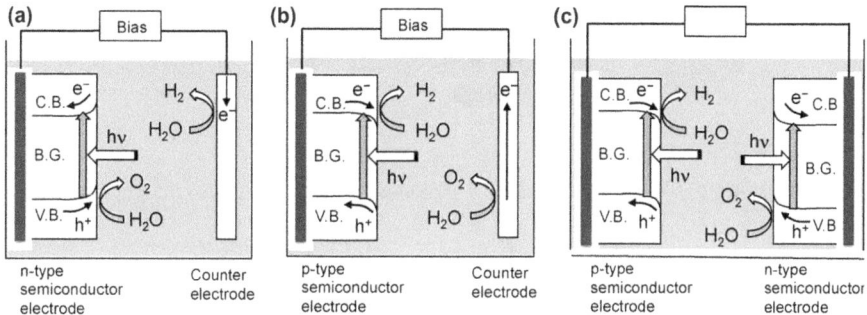

Figure 5.13   Photoelectrochemical water splitting systems using n-type semiconductor photoanode (a), p-type semiconductor photocathode (b), and tandem system (c).
Reprinted with permission from Reference 6. Copyright 2011 Elsevier B. V.

externally applied potential. This is distinctly different from the conventional metal-oxide based visible-light-responsive electrodes such as WO_3 and Fe_2O_3, which need an external bias for operation, due to their conduction band potential being located at more positive potential than water reduction potential.[55,56]

Several (oxy)nitrides such as TaON,[31,57] Ta_3N_5,[58–60] and $LaTiO_2N$,[61,62] have been studied as water-splitting photoelectrodes, and it was found that they function as photoanode materials due to their n-type semiconducting character. Abe *et al.* have developed a simple method for preparing efficient photoanodes of TaON and Ta_3N_5.[57,60] In this method, TaON or Ta_3N_5 particles are first deposited on a conductive glass (fluorine-doped tin oxide; FTO) substrate by electrophoretic deposition, and then a necking treatment is applied to form effective contacts between the deposited particles. Figure 5.14 displays the structure of electrodes after various treatments and the corresponding performance for PEC water splitting. The as-deposited TaON electrode, even after calcination at 673 K, gave little photoresponse (Figure 5.14(a)). However, $TaCl_5$ treatment followed by calcination at 723 K in air resulted in an appreciable increase in the photocurrent (Figure 5.14(b)). This increase in photocurrent was attributed to the presence of Ta_2O_5 bridges that promote electron transfer between TaON particles. Subsequent heating of the same electrode sample under NH_3 significantly increased the photocurrent (Figure 5.14(c)). XPS analysis indicated that most Ta_2O_5 bridges transformed into TaON, greatly facilitating the interparticle electron transport in the electrode structure. Modification of this electrode with colloidal IrO_2, which is known to be an effective catalyst for water oxidation,[63] further enhanced the photocurrent (Figure 5.14(d)). The onset potential of IrO_2-loaded post-necked TaON electrode was about −0.25 V *vs.* RHE, indicating that the electrode can potentially split water into H_2 and O_2 even without an applied bias. IrO_2 loading was also found to improve the stability of TaON electrodes during photoelectrolysis, with nearly stoichiometric H_2 and O_2 evolution in a two-electrode

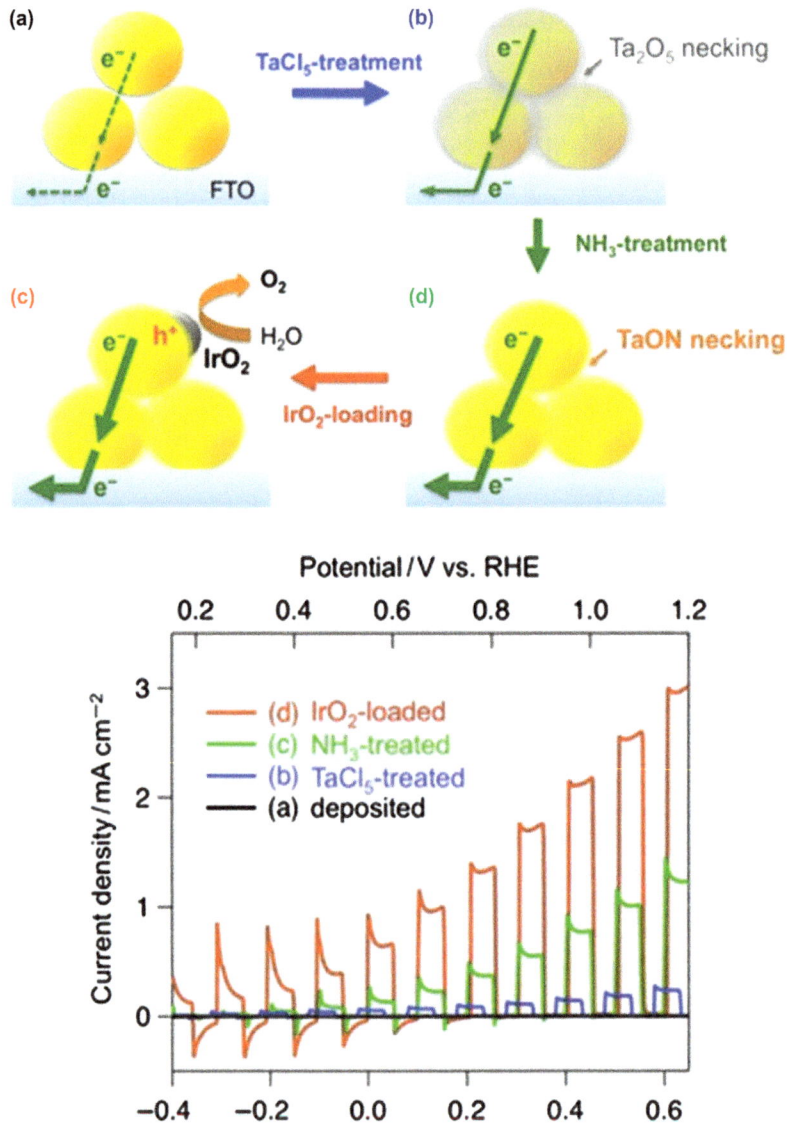

Figure 5.14   (Top) Schematic illustration of post-necking process and IrO_2 loading on TaON, and, (bottom) Current–potential curves in aqueous 0.1M Na_2SO_4 solution (pH 6) under chopped visible light irradiation ($\lambda > 400$ nm) for TaON electrodes as-prepared (a), treated by $TaCl_5$ (b), heated in NH_3 (c), and loaded with IrO_2.
Reprinted with permission from Reference 6. Copyright 2011 Elsevier B. V.

configuration in the presence of an applied bias greater than 0.6 V *vs.* Pt wire cathode. A similar result was observed when Ta_3N_5 was employed as an anode material. The IPCE of the optimized TaON and Ta_3N_5 electrode recorded at

1.15 V *vs.* RHE was *ca.* 76% at 400 nm and 31% at 500 nm, the former of which is the highest IPCE among oxynitride or nitride semiconductor photoelectrodes reported so far.

In terms of available wavelength, photons with wavelengths up to 600 nm are available for PEC water oxidation with Ta_3N_5[58–60] and $LaTiO_2N$.[61,62] However, no photoanode (including metal oxides) with a band gap smaller than 2.0 eV (corresponding to a 600 nm absorption band) that can oxidize water without application of an external potential (except through the use of a tandem cell configuration[64]) has been reported to date. As the band gap of a given material is decreased, the driving forces for water oxidation and reduction should inevitably decrease, making water splitting more difficult.

In such a situation, a niobium-based oxynitride, $SrNbO_2N$, was shown to achieve the functionality as a photoanode to oxidize water even without applying an external bias.[65] $SrNbO_2N$ is a perovskite-type oxynitride consisting of relatively earth-abundant metals that has a band gap of *ca.* 1.8 eV.[66] $SrNbO_2N$ powder also has the ability to photocatalytically reduce or oxidize water into H_2 or O_2 in the presence of proper electron donors or acceptors, respectively. When particulate $SrNbO_2N$ modified with colloidal IrO_2 is employed as a photoelectrode for visible-light-driven water splitting, it acts as an active photoanode to oxidize water, even without an externally applied potential. Nearly stoichiometric H_2 and O_2 evolution was observed during photoelectrolysis in a neutral electrolyte of Na_2SO_4 (pH $\approx$ 6) when a potential of +1.0-1.55 V *vs.* RHE was applied.

Figure 5.15 shows the dependence of photocurrent generated from the postnecked $SrNbO_2N$ electrode modified with colloidal IrO_2 at 0.95 V *vs.* RHE on the cutoff wavelength of the incident light, along with the DRS of $SrNbO_2N$.

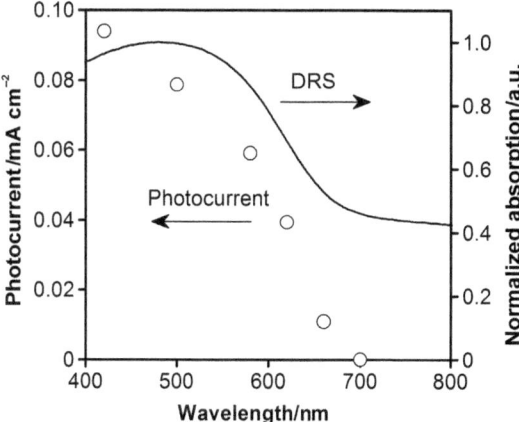

Figure 5.15 Dependence of photocurrent generated from the IrO_2-modified $SrNbO_2N$ electrode (6 cm²) at 0.95 V *vs.* RHE in aqueous 0.1 M Na_2O_4 solution upon the cutoff wavelength of the incident light. Diffuse reflectance spectrum for $SrNbO_2N$ is also shown.
Reprinted with permission from Reference 65. Copyright 2011 American Chemical Society.

The photocurrent was decreased with cutoff wavelength, reaching almost zero at 700 nm, which corresponds to the absorption edge of $SrNbO_2N$. This result indicates that PEC water oxidation occurs through light absorption by $SrNbO_2N$, and that an absorption band at longer wavelengths, assignable to reduced Nb species (*e.g.* Nb^{4+}) which are defects of the material, does not contribute to the reaction. The main problem of this electrode material is the low IPCE, which was 0.2% at 400 nm in the presence of a 1.23 V (*vs.* RHE) bias.

5.8 Conclusions

A range of (oxy)nitride materials for photocatalytic overall water splitting has been developed. (Oxy)nitrides have been found to function as stable photocatalysts for water reduction and oxidation under visible irradiation, and the $(Ga_{1-x}Zn_x)(N_{1-x}O_x)$ and $(Zn_{1+x}Ge)(N_2O_x)$ solid solutions with d^{10} electronic configuration have been shown to achieve overall water splitting under visible light without noticeable degradation. Some of d^0-type (oxy)nitrides are applicable to a two-step water splitting system that can harvest a wide range of visible photons. Carefully prepared photoelectrodes based on d^0-(oxy)nitrides such as TaON and Ta_3N_5 exhibit excellent IPCEs for splitting water under visible light in the presence of an external bias.

References

1. A. Fujishima and K. Honda, *Nature*, 1972, **238**, 37.
2. J. S. Lee, *Catal. Surv. Asia*, 2005, **9**, 217.
3. K. Maeda and K. Domen, *J. Phys. Chem. C*, 2007, **111**, 7851.
4. A. Kudo and Y. Miseki, *Chem. Soc. Rev.*, 2009, **38**, 253.
5. Y. Inoue, *Energy Environ. Sci.*, 2009, **2**, 364.
6. R. Abe, *J. Photochem. Photobiol. C: Photochem. Reviews*, 2010, **11**, 179.
7. R. Williams, *J. Chem. Phys.*, 1960, **32**, 1505.
8. A. B. Ellis, S. W. Kaiser, J. M. Bolts and M. S. Wrighton, *J. Am. Chem. Soc.*, 1977, **99**, 2839.
9. D. E. Scaife, *Solar Energy*, 1980, **25**, 41.
10. F. Tessier and R. Marchand, *J. Alloys and Compounds*, 1997, **262–263**, 410.
11. M. Jansen and H. P. Letschert, *Nature*, 2000, **404**, 980.
12. G. Hitoki, T. Takata, J. N. Kondo, M. Hara, H. Kobayashi and K. Domen, *Chem. Commun.*, 2002, 1698.
13. G. Hitoki, A. Ishikawa, T. Takata, J. N. Kondo, M. Hara and K. Domen, *Chem. Lett.*, 2002, **31**, 736.
14. G. Hitoki, T. Takata, J. N. Kondo, M. Hara, H. Kobayashi and K. Domen, *Electrochem.*, 2002, **70**, 463.
15. A. Kasahara, K. Nukumizu, G. Hitoki, T. Takata, J. N. Kondo, M. Hara, H. Kobayashi and K. Domen, *J. Phys. Chem. A*, 2002, **106**, 6750.

16. W.-A. Chun, A. Ishikawa, H. Fujisawa, T. Takata, J. N. Kondo, M. Hara, M. Kawai, Y. Matsumoto and K. Domen, *J. Phys. Chem. B*, 2003, **107**, 1798.
17. K. Nukumizu, J. Nunoshige, T. Takata, J. N. Kondo, M. Hara, H. Kobayashi and K. Domen, *Chem. Lett.*, 2003, **32**, 196.
18. D. Yamasita, T. Takata, M. Hara, J. N. Kondo and K. Domen, *Solid State Ionics*, 2004, **172**, 591.
19. K. Maeda, Y. Shimodaira, B. Lee, K. Teramura, D. Lu, H. Kobayashi and K. Domen, *J. Phys. Chem. C*, 2007, **111**, 18264.
20. K. Maeda, B. Lee, K. Teramura, D. Lu, H. Kobayashi and K. Domen, *J. Phys. Chem. C*, 2007, **111**, 18264.
21. K. Maeda, N. Nishimura and K. Domen, *Appl. Catal. A: Gen.*, 2009, **370**, 88.
22. L. Yuliati, J.-H. Yang, X. Wang, K. Maeda, T. Takata, M. Antonietti and K. Domen, *J. Mater. Chem.*, 2010, **20**, 4295.
23. K. Maeda, H. Terashima, K. Kase and K. Domen, *Appl. Catal. A: Gen.*, 2009, **357**, 206.
24. K. Maeda, H. Terashima, K. Kase, M. Higashi, M. Tabata and K. Domen, *Bull. Chem. Soc. Jpn.*, 2008, **81**, 927.
25. S. S. K. Ma, K. Maeda and K. Domen, *Catal. Sci. Technol.*, 2012, **2**, 818.
26. R. Sasaki, K. Maeda, Y. Kako and K. Domen, *Appl. Catal. B: Environ.*, 2012, **128**, 72.
27. X. Wang, K. Maeda, A. Thomas, K. Takanabe, G. Xin, J. M. Carlsson, K. Domen and M. Antonietti, *Nat. Mater.*, 2009, **8**, 76.
28. X. Wang, K. Maeda, X. Chen, K. Takanabe, K. Domen, Y. Hou, X. Fu and M. Antonietti, *J. Am. Chem. Soc.*, 2009, **131**, 1680.
29. K. Sayama, H. Arakawa and K. Domen, *Catal. Today*, 1996, **28**, 175.
30. H. Kominami, S. Murakami, Y. Kera and B. Ohtani, *Catal. Lett.*, 1998, **56**, 125.
31. R. Abe, T. Takata, H. Sugihara and K. Domen, *Chem. Lett.*, 2005, **34**, 1162.
32. J. Sato, N. Saito, Y. Yamada, K. Maeda, T. Takata, J. N. Kondo, M. Hara, H. Kobayashi, K. Domen and Y. Inoue, *J. Am. Chem. Soc.*, 2005, **127**, 4150.
33. Y. Lee, T. Watanabe, T. Takata, M. Hara, M. Yoshimura and K. Domen, *J. Phys. Chem. B*, 2006, **110**, 17563.
34. K. Maeda, N. Saito, D. Lu, Y. Inoue and K. Domen, *J. Phys. Chem. C*, 2007, **111**, 4749.
35. K. Maeda, N. Saito, Y. Inoue and K. Domen, *Chem. Mater.*, 2007, **19**, 4092.
36. K. Maeda, T. Takata, M. Hara, N. Saito, Y. Inoue, H. Kobayashi and K. Domen, *J. Am. Chem. Soc.*, 2005, **127**, 8286.
37. K. Maeda, K. Teramura, T. Takata, M. Hara, N. Saito, K. Toda, Y. Inoue, H. Kobayashi and K. Domen, *J. Phys. Chem. B*, 2005, **109**, 20504.

38. M. Yashima, K. Maeda, K. Teramura, T. Takata and K. Domen, *Chem. Phys. Lett.*, 2005, **416**, 225.
39. T. Hirai, K. Maeda, M. Yoshida, J. Kubota, S. Ikeda, M. Matsumura and K. Domen, *J. Phys. Chem. C*, 2007, **111**, 18853.
40. M. Yoshida, T. Hirai, K. Maeda, N. Saito, J. Kubota, H. Kobayashi, Y. Inoue and K. Domen, *J. Phys. Chem. C*, 2010, **114**, 15510.
41. K. Maeda, K. Teramura, D. Lu, T. Takata, N. Saito, Y. Inoue and K. Domen, *Nature*, 2006, **440**, 295.
42. K. Maeda, K. Teramura, N. Saito, Y. Inoue and K. Domen, *J. Catal.*, 2006, **243**, 303.
43. K. Maeda, K. Teramura, H. Masuda, T. Takata, N. Saito, Y. Inoue and K. Domen, *J. Phys. Chem. B*, 2006, **110**, 13107.
44. Y. Lee, H. Terashima, Y. Shimodaira, K. Teramura, M. Hara, H. Kobayashi, K. Domen and M. Yashima, *J. Phys. Chem. C*, 2007, **111**, 1042.
45. Y. Lee, K. Teramura, M. Hara and K. Domen, *Chem. Mater.*, 2007, **19**, 2120.
46. T. Misaki, X. Wu, A. Wakahara and A. Yoshida, *Proc. Int. Workshop Nitride Semiconductors IPAP Conf. Series 1*, 2000, 685.
47. F. Tessier, P. Maillard, Y. Lee, C. Bleugat and K. Domen, *J. Phys. Chem. C*, 2009, **113**, 8526.
48. R. Abe, T. Takata, H. Sugihara and K. Domen, *Chem. Commun.*, 2005, 3829.
49. M. Higashi, R. Abe, A. Ishikawa, T. Takata, B. Ohtani and K. Domen, *Chem. Lett.*, 2008, **37**, 138.
50. K. Maeda, R. Abe and K. Domen, *J. Phys. Chem. C*, 2011, **115**, 3057.
51. K. Maeda, M. Higashi, D. Lu, R. Abe and K. Domen, *J. Am. Chem. Soc.*, 2010, **132**, 5858.
52. M. Tabata, K. Maeda, M. Higashi, D. Lu, T. Takata, R. Abe and K. Domen, *Langmuir*, 2010, **26**, 9161.
53. M. Higashi, R. Abe, K. Teramura, T. Takata, B. Ohtani and K. Domen, *Chem. Phys. Lett.*, 2008, **452**, 120.
54. T. Matoba, K. Maeda and K. Domen, *Chem. Eur. J.*, 2011, **17**, 14731.
55. C. Santato, M. S. Ulmann and J. Augustynski, *J. Phys. Chem. B*, 2001, **105**, 936.
56. S. D. Tilley, M. Cornuz, K. Sivula and M. Grätzel, *Angew. Chem., Int. Ed.*, 2010, **49**, 6405.
57. R. Abe, M. Higashi and K. Domen, *J. Am. Chem. Soc.*, 2010, **132**, 11828.
58. A. Ishikawa, T. Takata, J. N. Kondo, M. Hara and K. Domen, *J. Phys. Chem. B*, 2004, **108**, 11049.
59. D. Yokoyama, H. Hashiguchi, K. Maeda, T. Minegishi, T. Takata, R. Abe, J. Kubota and K. Domen, *Thin Solid Films*, 2011, **519**, 2087.
60. M. Higashi, K. Domen and R. Abe, *Energy & Environ. Sci.*, 2011, **4**, 4138.
61. C. Le Paven-Thivet, A. Ishikawa, A. Ziani, L. L. Gendre, M. Yoshida, J. Kubota, F. Tessier and K. Domen, *J. Phys. Chem. C*, 2009, **113**, 6156.

62. N. Nishimura, B. Raphael, K. Maeda, L. Le Gendre, R. Abe, J. Kubota and K. Domen, *Thin Solid Films*, 2010, **518**, 5855.
63. A. Harriman, I. J. Pickering, J. M. Thomas and P. A. Christensen, *J. Chem. Soc., Faraday Trans. 1*, 1988, **84**, 2795.
64. O. Khaselev and J. A. Turner, *Science*, 1998, **280**, 425.
65. K. Maeda, M. Higashi, B. Siritanaratkul, R. Abe and K. Domen, *J. Am. Chem. Soc.*, 2011, **133**, 12334.
66. Y. Kim, P. M. Woodward, K. Z. Baba-Kishi and C. W. Tai, *Chem. Mater.*, 2004, **16**, 1267.

CHAPTER 6

Rapid Screening Methods in the Discovery and Investigation of New Photocatalyst Compositions

ALLEN BARD,* HEUNG CHAN LEE, KEVIN LEONARD, HYUN SEO PARK AND SHIJUN WANG

Chemistry and Biochemistry Department, University of Texas at Austin, Austin Texas TX 78712-1224, USA
*Email: ajbard@mail.utexas.edu

6.1 Introduction

The search for new photocatalysts is driven by the desire to find an efficient and stable, yet inexpensive material that will be useful for carrying out photoelectrochemical (PEC) reactions. Currently the generation of hydrogen or fuels by photoelectrolysis of water ("water splitting") is of interest. However, 40 years of experience in this field has shown that photocatalyst materials can be fabricated from many elements in the periodic table, and the efficiency for conversion of photons to electrons and holes can depend on the composition of the material, each involving 4 or more elements. Thus, the experimental exploration to discover and optimize photocatalysts involves millions of candidates. Rapid screening, involving the automated preparation of candidate arrays and their testing, is a promising approach. Moreover, quantitative comparisons of different elemental dopings of a given material can provide useful data that can guide the formulation of theories and models as well as indicate the compositional factors that are important in their behavior.

RSC Energy and Environment Series No. 9
Photoelectrochemical Water Splitting: Materials, Processes and Architectures
Edited by Hans-Joachim Lewerenz and Laurence Peter
© The Royal Society of Chemistry 2013
Published by the Royal Society of Chemistry, www.rsc.org

Rapid screening methods have been developed extensively for a variety of materials and films, *e.g.* catalysts, dielectrics, and phosphors.[1–3] In the examination of inorganic materials as PEC catalysts, a number of screening methods have been used, most of which center on creating electrodes consisting of arrays of spots of photocatalysts, with each spot having a different composition. There exists a variety of different methods of preparing arrays of photocatalysts in addition to different methods of screening the arrays. One example is inkjet printing, which is inexpensive, easy to use, and useful for the production of luminescent metal oxide libraries and multiple-metal oxides. When using inkjet printing to create photocatalyst arrays, the printers can either be inexpensive "off-the-shelf" consumer inkjet printers or research-grade ones. Consumer inkjet printers typically produce triangular gradient patterns created from three metal precursor solutions that replace the ink containers. From these patterns, qualitative information can be obtained about which combinations of metal precursors produce an increased photoresponse. Quantitative information can be obtained from research-grade inkjet printers, where known amounts of the precursor solution can be deposited. The Parkinson and Lewis groups utilized inkjet printing of overlapping patterns of soluble metal oxide precursors onto glass coated with fluorine-doped tin oxide (FTO).[4–8] Subsequent annealing in the air at temperatures below 550 °C yielded electrodes with patterns of mixed oxide compositions. The mixed metal oxide patterns were immersed in an electrolyte and the photocurrent responses were screened and mapped by scanning a green laser over the material. The McFarland group reported their work on the combinatorial production of materials by electrochemical synthesis and screening for water splitting through an automated electrochemical system.[9–11] This setup utilized a system with as many as 120 individual and isolated electrochemical cells (wells) into which different mixtures of precursors can be deposited, usually manually. A perforated Teflon block having an array of 10×12 holes of 6 mm diameter was used to create the individual and isolated electrochemical cells, and FTO glass placed under the Teflon block was used as the working electrode. A Pt wire counter electrode and Ag wire reference electrode were moved from one cell to the next to carry out electrodeposition at each location. In addition, the Maier group prepared a production array *via* sol-gel processes by controlled pipetting robots and screening strategies to search for novel hydrogen evolution photocatalysts.[12]

This chapter is mostly concerned with the techniques used in our laboratory, which involve robotic preparation of arrays from solution dispensed with a pL dispenser (described in Section 6.2) and screened by immersing the array in a given test solution and employing a scanning fiber optic in a commercial scanning electrochemical microscope (SECM) (described in Section 6.3). The photocurrent is measured as the fiber optic passes over a given spot in the array. It is also possible to detect photogenerated products electrochemically with a ring electrode around the fiber optic. Section 6.4 expands screening to electrocatalysts on the photocatalyst surface. Section 6.5 deals with establishing the composition and characteristics of the successful array spots ("hits") and their use in larger PEC cells. Section 6.6 briefly considers correlations of doping effects with theory.

6.2   Rapid Synthesis on Arrays

To screen photocatalysts rapidly, it is necessary to perform "synthesis on a chip" where one can have many photocatalysts of different compositions on a single compact electrode. These multi-component electrodes (referred to as spot array electrodes) are typically composed of an array of spots on a single support material, where each spot is a photocatalyst having a different composition. In most circumstances, one wants to screen photocatalysts rapidly to investigate doped photocatalysts, composites comprising two or more elements, or excesses and deficiencies of certain elements in multi-metal photocatalysts. Spot array electrodes provide a convenient method for making these comparisons because the photocatalytic behavior of each spot can be measured independently and used to determine the optimum photocatalyst composition.

An example of a spot array electrode in which the ratios of $Zn_xCd_{1-x}S_ySe_{1-y}$ were examined using SECM is shown in Figure 6.1.[13]

Here, the spot array electrode contains 50 spots, each having a different composition of $Zn_xCd_{1-x}S_ySe_{1-y}$. The diagram in Figure 6.1(a) shows the relative compositions of each element, where the numbers in each circle represent the Zn/Cd ratio and the numbers next to each two-row set represent the S/Se ratio. Figure 6.1(b) is a photograph of the finished spot array electrode.

As stated previously, spot array electrodes can be synthesized by different methods, depending on the technique to be used for rapid screening. Here, we focus mainly on fabricating spot array electrodes for rapid screening by SECM.[14] Generally, spot array electrodes for SECM are fabricated by dispensing drops of a photocatalyst precursor solution onto a support electrode

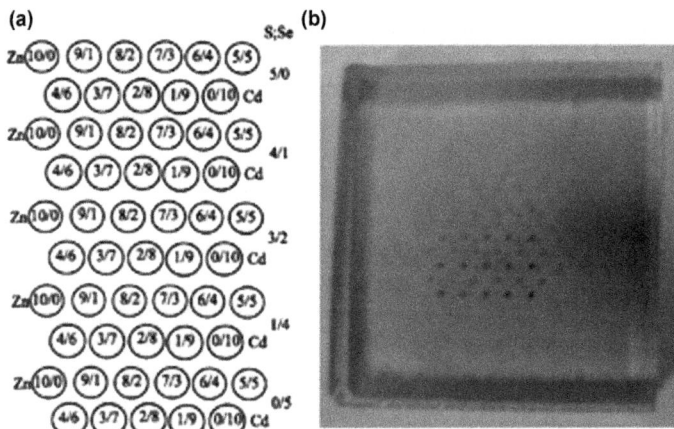

Figure 6.1   (a) Dispensed pattern of $Zn_xCd_{1-x}S_ySe_{1-y}$ array where each spot has different composition; number within a spot represent number of drops of Zn^{2+} and Cd^{2+} solution and listing on right side are relative number of drops of the S or Se precursor; (b) photograph of $Zn_xCd_{1-x}S_ySe_{1-y}$ array sample. Spots with high S content are difficult to see against the white background of the photo. Spot diameter about 230 μm.
Reprinted with permission from G. Liu, C. Liu and A. J. Bard, *J. Phys. Chem. C*, 2010, **114**, 20997. Copyright © 2010 American Chemical Society.

followed by annealing to form the desired photocatalysts. Here, each spot has a quantitatively different proportion of each photocatalyst component, and the ratio of components is determined by the number of drops of each precursor solution dispensed on each spot. The photocatalyst spots are typically created using a piezo dispenser that delivers pL-drops of the appropriate precursor solutions. These dispensers are computer-controlled to move and dispense drops at precise locations. Typically, one precursor solution is dispensed at a time. After one precursor solution is deposited, the solution is drained from the dispenser and a second (and perhaps a third or fourth) precursor solution is deposited at exactly the same position on the spot array electrode. The result is an array of drops containing different pre-programmed amounts of each metal. The distances between photocatalyst spots on the array are usually about 500–900 μm, with a spot diameter of approximately 200–500 μm as discussed in Section 6.3. Multidispenser systems are also available, and these speed up array preparation. With these, one must calibrate each dispenser drop size to establish each spot composition. Dispensers can have reproducibility problems because they are easily clogged with particles and damaged by inadvertent crashes with substrate. All solutions used in spot preparation should be filtered.

An example of dispensing equipment for creating spot array electrodes for SECM is a CH Instruments model 1550 Dispenser (Austin, TX) with a piezoelectric dispensing tip (MicroJet AB-01-60, MicroFab, Plano, TX) connected to an XYZ stage driven by a computer-controlled stepper-motor system (Newport). The applied potential and the pulse duration are controlled to change the amount dispensed from the piezo-dispensing tip. Typical drop sizes are ~ 100 pL. For example, to get ~ 100 pL/drop with the above equipment, a potential of 80 V is applied to the piezo-dispensing tip for 40 μs.[15] The total number of drops can vary depending on the resolution of concentrations needed. For example, a total of 10 drops will give a 10% change/drop in composition, while a total of 33 drops will give a 3% change/drop in composition.

When creating spot array electrodes, great care must be taken when producing the spots so that one can accurately compare their photocatalytic behavior. To make accurate comparisons, it is important that each spot has the same diameter and the same thickness. An example of an optical profile image showing the uniformity of a good spot array electrode created by the piezo-dispensing technique is shown in Figure 6.2.[13]

Here the optical profile image was measured by white light vertical scanning interferometry using a commercial optical surface profiler (NT9100, Veeco, New York).[16] The color in Figure 6.2 represents the surface height of the spot array electrode. In this example, each spot has a uniform thickness of 440 nm and uniform diameter of 230 μm.

To create high-quality spot array electrodes for which accurate comparisons between the spots can be made, one must consider several factors: the support electrode material, the photocatalyst precursors, the solvents for the photocatalyst precursors, pretreatment of the support electrode, the spot deposition method and conditions, spot size and spacing, and annealing of the electrode. Here we provide guidelines for each of these factors for creating spot array

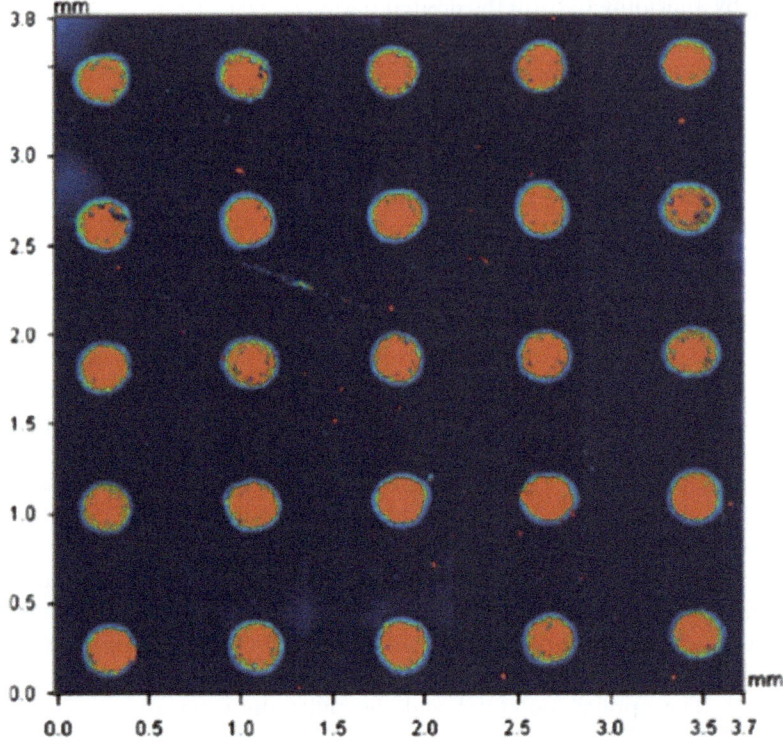

Figure 6.2 Optical surface profile image of a spot array electrode.
Reprinted with permission from G. Liu, C. Liu and A. J. Bard, *J. Phys. Chem. C*, 2010, **114**, 20997. Copyright © 2010 American Chemical Society.

electrodes to be screened by SECM. However, many of these guidelines translate well to other array preparation techniques.

The support electrode must be conductive so that the photocurrent from each individual spot can be measured from a single electrical contact. It should also show low photocatalytic and dark electrochemical activity, so that the background current density in the regions between the individual spots is significantly lower than the photocurrent from individual spots. For example, good photocatalysts, like doped $BiVO_4$, can achieve a 10-fold increase in spot current over the background current,[15,17] although a factor of 2 between the background current and spot current is also adequate.[14,18] Finally, the support electrode must be able to withstand the annealing temperatures required to form the photocatalysts. FTO coated glass, a widely used substrate, is conductive, shows low photocatalytic behavior, and can be safely heated to 550 °C without decomposition or significant changes in conductivity.

When depositing spots of photocatalysts, it is usual to deposit precursor solutions that contain metal salts and then to anneal these salts to obtain the desired oxide or other compound. For single metal oxides (*e.g.* Fe_2O_3), the precursor can be a simple nitrate or acetate salt (*e.g.* $Fe_3(NO_3)_3$),[18] where any component from the anion is removed by the annealing step. For multi-metal photocatalysts (*e.g.* BiVO$_4$ or CuInSe$_2$),[17,19] the individual metal salts are mixed to keep ratios constant, or they can be deposited one at a time if different ratios are to be studied. Typically, the precursor must be sufficiently soluble in the solvent to achieve at least 0.1 M concentrations,[14] which can create spots with enough mass to generate sufficient photocurrent, while keeping the volume of the precursor minimized to avoid intermixing and size variation between the spots on the array electrode. However, in some of the inkjet printing techniques, solution concentrations of 0.5 M are used.[6] There are some exceptions to using either nitrate or acetate salts; one example is tungsten precursors, which are typically $(NH_4)_{10}H_2(W_2O_7)\cdot xH_2O$ because of their high solubility.[15]

In addition to finding a solvent/precursor combination to achieve at least 0.1 M concentrations, the solvent should also have a volatility that ensures high spot uniformity and avoids "coffee rings". It is also important that the solvent has a high contact angle when deposited on the support electrode. This ensures adequate spacing between each spot and prevents individual spots from running together during dispensing. The most popular solvent for deposition of metal oxides onto FTO supports is ethylene glycol, which is able to dissolve many metal salt precursors, has an appropriate volatility, and has a high contact angle on well-cleaned FTO.

Preparation and cleaning of the support electrode is also important for creating high quality spot array electrodes. When FTO is used, cleaning is usually done by washing in water and ethanol and sonicating in ethanol for at least 30 minutes. However, freshly cleaned electrodes typically have improved wettability of the solvent on the substrate, which is not desirable for creating spot array electrodes. This can be remedied by allowing the FTO to dry in air for at least 12 hours before the dispensing step. This air drying process presumably provides an electrode that has adsorbed organics that cause a high contact angle between the substrate and the precursor solution. In addition, methanol[6] or a siliconizing solution[7] can also be used to increase the contact angle between the solvent and the support electrode.

After the precursor solutions are deposited, the photocatalysts are formed by an annealing process. For oxides, annealing in air is most convenient (most oxides can be formed by firing at 500 °C for 3 h). When non-oxides are studied, annealing can be performed in different environments. For example, in the case of $Zn_xCd_{1-x}S_ySe_{1-y}$, the metal precursors were 0.1 M $Zn(NO_3)_2$ and $Cd(NO_3)_2$ in a water/glycerol solvent. The solutions to produce the anions were 0.2 M thiourea and dimethyl selenourea (containing 0.05 M hydrazine). The arrays were held at 100 °C for 12 hours under an Ar atmosphere to produce the mixed sulfide–selenide photocatalysts.[13]

6.3 Rapid Screening with SECM

6.3.1 Method

An SECM typically involves an ultramicroelectrode (UME) that is scanned close to a substrate of interest while generating a species that interrogates the surface and reports back to the tip (called feedback).[20] In PEC SECM applications, sometimes called scanning photoelectrochemical microscopy (SPECM) (Figure 6.3[14,21]), the tip is replaced by a fiber optic.[14] This is coupled to a Xe lamp directly focused on one end of the fiber optic *via* a five-axis fiber aligner, the other end of the fiber optic irradiates spots on the substrate and the photoresponse of the spot is recorded. From the photoresponses of each spot, a photoactivity map of the total area of the substrate can be drawn.

An array of photocatalyst spots can be screened rapidly by this technique, typically in 10 to 45 min. By measuring the comparative photoresponse of spots with different compositions, the material compositions with the best

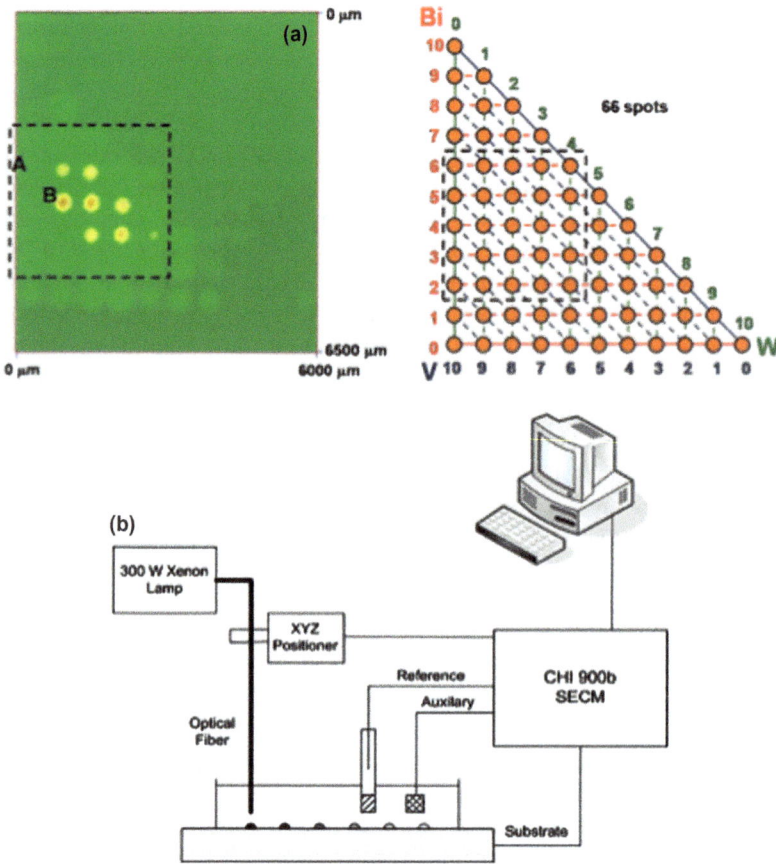

Figure 6.3   (a) Diagram of a spot array and resulting SECM image of the array involving W doping of $BiVO_4$. Brighter yellow spots indicate higher photocatalytic activity of the specific composition.[21] (b) Schematic diagram of scanning electrochemical microscope (SECM) with the fiber optic for photoelectrochemical screening of a photocatalytic material array.[14]
Reprinted with permission from J. Lee, H. Ye, S. Pan, A. J. Bard, *Anal. Chem.*, 2008, **80**, 7445; and H. Ye, J. Lee, J. S. Jang, A. J. Bard, *J. Phys. Chem. C*, 2010, **114**, 13322. Copyright 2008 and 2010 American Chemical Society.

photoactivity can be found[21] (Figure 6.4). The size of the fiber optic, the dimension of the spots on the array, and the scan rate are important to ensure high resolution of the SECM. Although a smaller irradiated area can provide higher geometric resolution, the intensity of the light through the smaller diameter optical fiber decreases the photocurrent, making fast scanning more difficult.

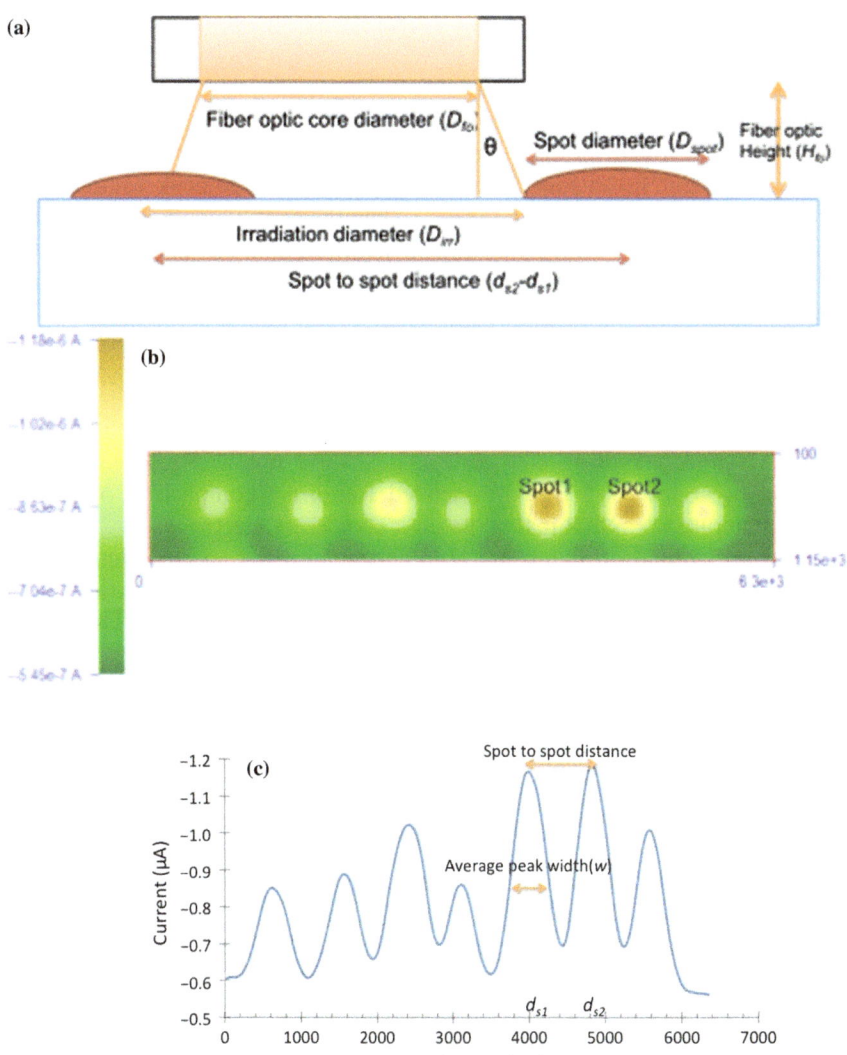

Figure 6.4   (a) Schematic view of the fiber optic and the array substrate in the SECM setup. (b) An image of spot array and (c) corresponding current *vs.* distance plot. Peaks in (c) represent photoactivity of each spot. The resolution of the peaks in (c) depends on dimensions in (a).

When designing the spacing of a spot array electrode and the positioning of the fiber optic cable, it is important to determine the signal resolution. The signal resolution (Rs) can be determined by considering the geometry of the SECM system assuming that (1) the photocurrent is sufficient to resolve the peak current from the background current and (2) the tip scan rate (fiber optic movement distance per unit time) is slow enough to a get steady state photo-current at all points on the scan. We will discuss the dependence of pixel resolution on scan rate below.

$$R_S = \frac{(d_{s2} - d_{s1})}{(w_{s1} + w_{s2})/2} \approx \frac{(d_{s2} - d_{s1})}{w_{s2}} \quad \text{when } w_{s1} \approx w_{s2} \quad (6.1)$$

In equation (6.1), d_{s1} and d_{s2} are the positions of two adjacent spots, spot 1 and spot 2, respectively along a line of spots, and w_{s1} and w_{s2} are the average peak width of spot 1 and spot 2, respectively. The average peak width can be approximated as the spot diameter (D_{spot}) when the spot diameter is larger than the irradiation diameter (D_{irr}). When the irradiation diameter is larger than the spot diameter, the average peak width can be approximated to the irradiation diameter.

$$w_{s2} = D_{irr}, \quad D_{irr} > D_{spot}$$
$$= D_{spot}, \quad D_{spot} > D_{irr} \quad (6.2)$$

When the signal resolution is 1 or less, peaks are not well resolved, and a resolution of at least 1.5 is recommended for direct comparison of photo-currents generated at individual spots on a spot array electrode.

The actual irradiation diameter (D_{irr}) depends on the specifications of the fiber optic (core diameter (D_{fo}) and numerical aperture (NA)) and the height (H_{fo}) (*i.e.* the distance between the fiber optic and the substrate). The irradi-ation diameter on the electrode surface can be calculated by,

$$D_{irr} = D_{fo} + 2\, H_{fo}\, \tan\theta \quad (6.3)$$

The angle θ can be determined from the definition of numerical aperture (NA),

$$NA = n \sin\theta \quad (6.4)$$

where n is the refractive index of the solution medium.

When a fiber optic (model FT-400-UMT with TECS 39) with a numerical aperture of 0.39 and a core diameter of 400 μm is set 200 μm above the sub-strate, the irradiation diameter is calculated to be 522 μm using the refractive index of water of 1.33. If the spot size on the array is 300 μm, the peak width can be estimated to be the irradiation diameter. Using a spot distance of 800 μm, the signal resolution is 1.5 and the spots will be sufficiently dis-tinguishable from each other (Figure 6.4). From our experience, a fiber optic with a 400 μm core diameter and a 750 μm coating diameter is suitable for SPECM screening; fiber optics with smaller dimensions have poorer structural strength and produce photocurrents that are too low to obtain useful results.

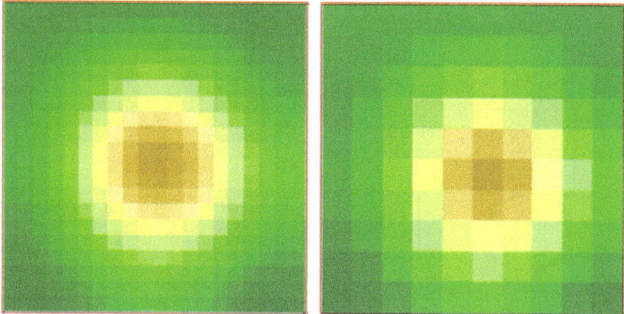

Figure 6.5   Images of an identical spot with two different geometric scan rates (500 μm/s and 1000 μm/s, increment distance = 50 μm and 100 μm for left and right image with increment times of 0.1 s for both). Pixel resolution on left-hand image is four times better.

In SECM, geometric scan rate is determined by two parameters – increment distance and increment time. Increment distance is the distance the tip moves in a given increment of time. Thus, the geometric scan rate is simply: Geometric Scan Rate (μm/sec) = Increment Distance (μm) / Increment Time (s). The increment distance determines the pixel size in an image, while the increment time determines how long the current will be collected for one pixel. Because the photocurrent of a semiconductor/electrolyte junction is limited by the photon flux, the response should be immediate, and the increment time can be small (~ 0.1 s) so as to reduce the overall scanning time. Figure 6.5 shows an example of two pixel resolutions of an identical spot.

One can also detect the products of the photoreaction (*e.g.* oxygen) using an optical fiber with a ring electrode as the SECM tip (Figure 6.6(b)).[14] This tip provides typical SECM-type measurements and collection of substrate generated products, during simultaneous irradiation of the substrate[17,22] (Figure 6.6(a)). Using the tip-collection/substrate-generation (TC/SG) mode of SECM, the photogenerated products can be collected at the ring electrode when it is held at an appropriate potential. These types of measurements are especially useful when the activities of electrocatalysts for specific reactions on semiconductor electrodes are of interest.

6.3.2   Guidelines

The SECM cell for photocatalyst rapid screening consists of a Teflon cell body, an O-ring, the spot array electrode, and a plastic piece that holds the array against the O-ring and in contact with electrolyte in the cell (Figure 6.7). All components are tightly assembled with screws. The cell body has one small chamber for the counter electrode and a larger chamber for the reference electrode. The O-ring limits the size of the array and is consistent with the moving range of the stepper motors in the SECM. The size of the array should be designed to this limitation.

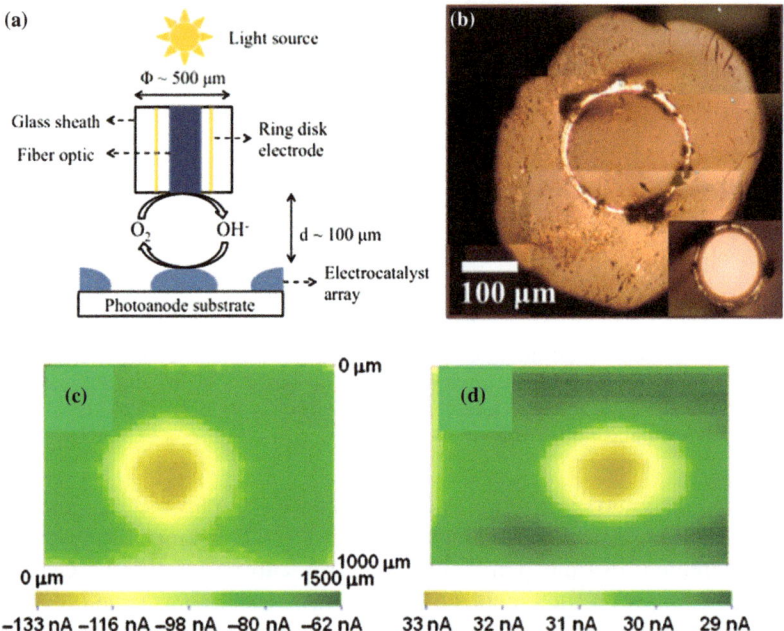

Figure 6.6 (a) Schematic diagram of tip-collection/substrate-generation mode of SECM on the electrocatalyst array prepared on photoanode. (b) Photographic images of the ring-disk optical fiber electrode.[22] Inset shows the bright optical fiber when the light is on. SECM images of (c) photocurrent at Co_3O_4 spot on W-doped $BiVO_4$ with the substrate at 0.3 V (*vs* Ag/AgCl) and (d) oxygen reduction current at the ring at -0.2 V (*vs* Ag/AgCl) in Ar-saturated 0.2 M sodium phosphate buffer (pH 6.8) solution under UV-visible light irradiation.[17] Indicated negative current in (c) is anodic current and positive current in (d) is cathodic current.
Reprinted with permission from H. Ye, H. S. Park, A. J. Bard, *J. Phys. Chem. C*, 2011, **115**, 12464 and Y. Cong, H. S. Park, S. Wang, H. X. Dang, F.-R. F. Fan, C. B. Mullins, A. J. Bard, *J. Phys. Chem. C*, 2012, **116**, 14541. Copyright 2011 and 2012 American Chemical Society.

The electrolyte for the screening is determined by the nature of the desired photoreaction. If the goal of the experiment is to screen for the maximum photoactivity of an n-type material, one can use a sacrificial reagent (*e.g.* SO_3^{2-}) as an electron donor that undergoes an irreversible reaction. If one is interested in a specific reaction (for example water oxidation with an n-type semiconductor), the corresponding solution would be water with buffer electrolyte at a particular pH. In this case, recombination of photogenerated carriers with reaction intermediates will often produce a smaller photocurrent than that seen with a sacrificial donor. If the photoreaction at the semiconductor involves the production or consumption of protons, the electrolyte should be adequately buffered to prevent significant pH changes at the electrode surface.

The adhesive copper tape attached FTO of the array electrode substrate is placed between the cell body and the back of the cell. The holes for working and

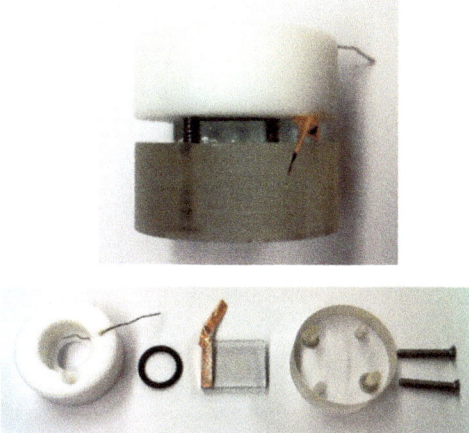

Figure 6.7 Photos of an SECM cell for PEC array screening. From the top, assembled cell, cell body, O-ring, spot array electrode substrate with adhesive copper tape attached, bottom side of the cell and screws.

reference electrodes prevent them from interfering with the movement of the SECM tip, *i.e.* the fiber optic and its holder. The optical fiber is positioned 100 to 200 μm above the substrate to prevent it from touching the substrate while screening. This is accomplished by first touching the fiber optic to the FTO of the substrate, tightening the fiber in the SECM tip holder, and then withdrawing the fiber away from the substrate to the desired distance using the piezoelectric or stepping motor actuator. The array can be irradiated with the full light of the Xe lamp or at a given wavelength by inserting a bandpass filter between the fiber optic aligner and the Xe lamp.

Many factors affect the image quality and current magnitude of the screened array. Light intensity must be constant during the scanning. Thus, one needs to wait until the Xe lamp has stabilized to prevent drifting of the light intensity with time. Higher light intensities generally provide better signal to noise ratios.

Generally, screening is carried out under chronoamperometric conditions. The potential of the substrate should be selected at a position where the dark current is near zero while the photocurrent reaches its maximum level. For example, for the current-potential profile of the photocatalyst shown in Figure 6.8, holding the potential at +0.2 to +0.3 V *vs.* Ag/AgCl for screening produces a high photocurrent while the dark current is small. The current scale for the color-coded image needs to be adjusted for a good signal-to-noise ratio and contrast between the different spots.

Even when the potential is set where the dark current is small, the background current from the substrate should be minimized. This often requires adjusting the potential and allowing the background current to stabilize by holding it in the dark for a significant time before scanning. For spots of the size described here, a typical background current is on the order of 10^{-8} to 10^{-7} A. If the background current is significantly larger, leakage of solution to the

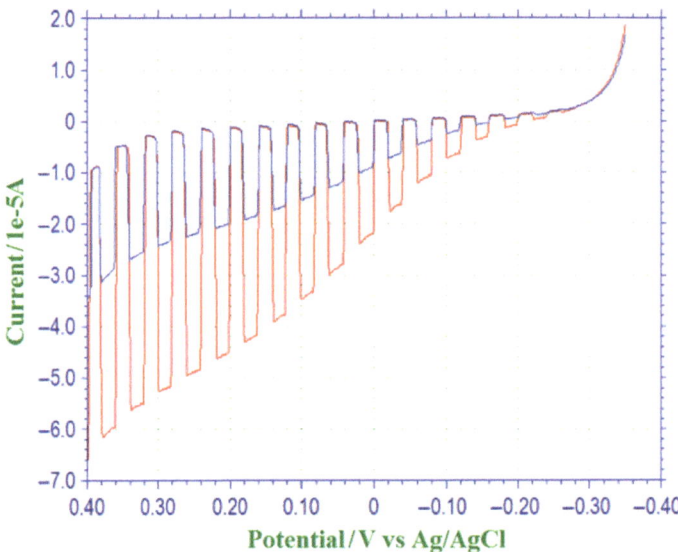

Figure 6.8   An example of a linear sweep voltammogram under chopped light. The holding potential of the array electrode for the rapid screening can be determined by the potential where the dark current is zero while photo-current is maximized (here, 0.2 to 0.3 V).

contact might be the cause. An unusually small background current, *e.g.* 10^{-9} A, might indicate bad contact to the FTO.

6.3.3   Data Analysis

In assessing the relative photocurrents of different spots and in searching for "hits", one should not set the current scale over too small or too large a range. Typically the maximum should be at least 4 times the photocurrent of known compositions, and thus the current scale has to be adjusted over a large enough range that observed differences among spots are meaningful and that current densities represent reasonable photo responses at the given light intensity. Also, having two lines on the array (at top and bottom) with identical composition to serve as internal controls is useful in avoiding misinterpretations, since these two lines should be identical. Figure 6.9[13] shows examples of well resolved SECM images taken from the spot array in Figure 6.1(a) in a 0.1 M Na_2SO_4/ Na_2S solution with the potential held at -0.4 V *vs* Ag/AgCl with (a) UV-vis and (b) visible light irradiation.

Figure 6.10 shows 5 spots with the same composition as a control line, a spot array with the same component and the same concentration. Although the first spot shows less photocurrent, the actual photocurrent difference is only about 40 nA, which is not significant. In this case, the smaller photocurrent was the result of non-uniform spot thickness. Non-uniformity of spot thickness can cause errors in the relative spot evaluations and, as described above (*e.g.*

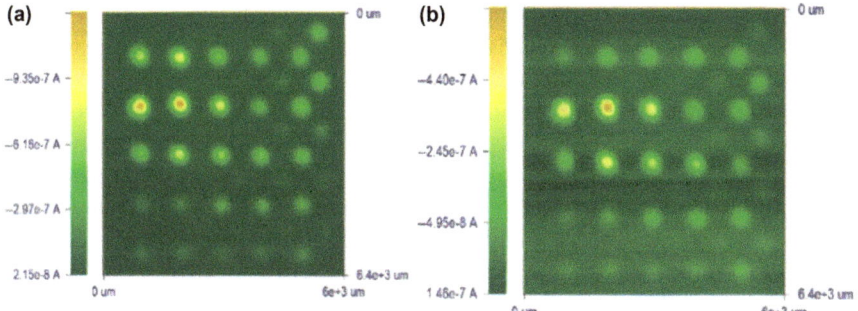

Figure 6.9 SECM images of the spot array in Figure 7.1(a) in a 0.1 M Na_2SO_4/Na_2S solution with the potential held at -0.4 V *vs* Ag/AgCl with (a) UV-vis and (b) visible light irradiation.
Reprinted with permission from G. Liu, C. Liu and A. J. Bard, *J. Phys. Chem. C*, 2010, **114**, 20997. Copyright © 2010 American Chemical Society.

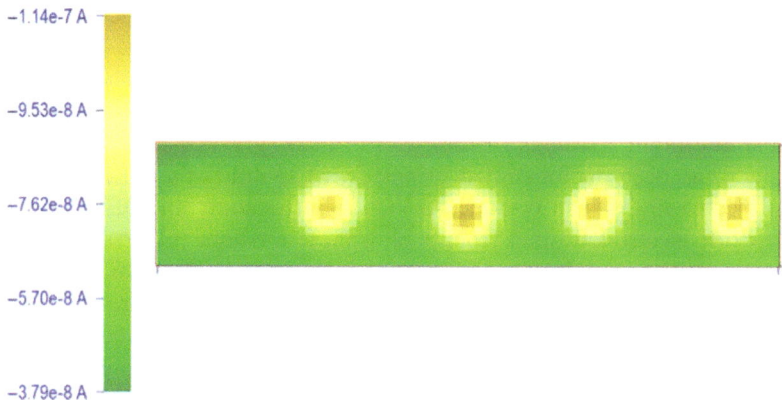

Figure 6.10 Spot array with the same component and the same concentration. Although the first spot shows less photocurrent due to slight mismatch of thickness, the actual photocurrent difference is only about 40 nA, which is not significant.

Figure 6.2), one can evaluate spot thickness by white light interferometry or atomic force microscopy. It is also possible to obtain absorbance spectra of spots in an array with a fiber optic light source and a detector.[13]

6.4 Rapid Screening with Electrocatalysts

For many PEC reactions of interest, using electrocatalysts at the photocatalyst/solution interface can promote the reaction, decrease surface recombination, and improve stability.[23,24] For example, for an n-type semiconductor one might want to promote photogenerated holes to oxidize water molecules rather than recombine with electrons or react with lattice anions leading to subsequent decomposition. Two reactions of interest are those involved in water photolysis, *i.e.* the hydrogen evolution and oxygen evolution reactions (HER and OER). Both are heterogeneous, multi-electron inner-sphere reactions, which can limit the overall water photolysis rate because of slow kinetics.[25] Consequently, electrocatalysts are often required to achieve reasonable reaction rates. Extensive studies have been done to find good electrocatalysts for the HER and OER on both conductor and semiconductor electrodes. For example, Pt, Pd, Ru, Ni, Ni-Mo, MoS$_3$, and many other materials have been studied as HER catalysts[26–29] and have been deposited on Si, Cu$_2$O, SrTiO$_3$, and other photocathodes.[30–32] Similarly, Co$_3$O$_4$, RuO$_2$, MnO$_2$, IrO$_x$, Pt and many other metal (hydr)oxides and complexes have been reported as good OER electrocatalysts[33–35] and have been applied to Fe$_2$O$_3$, BiVO$_4$, TaON, WO$_3$, TiO$_2$, and many other photoanodes.[36–41] Thus, it is of interest to screen different electrocatalysts on semiconductor photocatalysts. The existence of catalyst-support interactions as well as molecular interactions between electrocatalysts and adsorbates (*e.g.* hydroxides and protons) can affect the activity of the electrocatalysts. This implies that the activity of a given electrocatalyst in a PEC system may be different than that of the same electrocatalyst on a conductive support in the dark.[34,42,43]

Rapid screening of photocatalysts with electrocatalysts is a valuable approach to finding a good combination of electrocatalyst and photocatalyst. For example, the rapid screening of the electrocatalyst arrays prepared on photoelectrodes has been demonstrated using SECM to study the OER activity of several electrocatalysts on W-doped BiVO$_4$.[17] In this study, the OER activity of cobalt oxide electrodeposited from a phosphate medium (Co-Pi), Co$_3$O$_4$, Pt, and IrO$_x$ on W-doped BiVO$_4$ arrays were investigated (Figure 6.11).[17] These included Ir/Co oxide electrocatalyst arrays with different atomic ratios of Ir/Co deposited using the piezoelectric dispenser on the photoanode substrate (Figure 6.11(a)). The deposition and heat treatments were similar to those described previously for the photocatalyst arrays, but a photoanode was used in place of the conducting substrate. Moreover, photochemical deposition was used to deposit Pt electrocatalyst arrays on the W-doped BiVO$_4$ (Figure 6.11(b)). This was accomplished by placing an optical fiber over the photocatalyst substrate to generate electron-hole pairs, where the metal precursor, H$_2$PtCl$_6$, was reduced to form the electrocatalyst while the generated hole was consumed by a sacrificial electron donor, MeOH. From the rapid screening, Co$_3$O$_4$ and Pt were found to be good electrocatalysts with W-doped BiVO$_4$. In contrast, IrO$_x$ formed from nanoparticles showed weaker OER activity on W-doped BiVO$_4$, although it is a good OER electrocatalyst on conducting substrates in the dark. Again, the results indicate strong interactions between the electrocatalysts and the photocatalysts, which may significantly affect the kinetics of desired chemical reactions on the electrocatalysts. Using the rapid screening technique, optimum compositions of the complex electrocatalyst pairs, as well as the deposition conditions, can be studied on photoelectrodes.

There are several advantages to using a photoelectrode as the substrate for rapid screening of electrocatalyst arrays including: (i) the activities of the

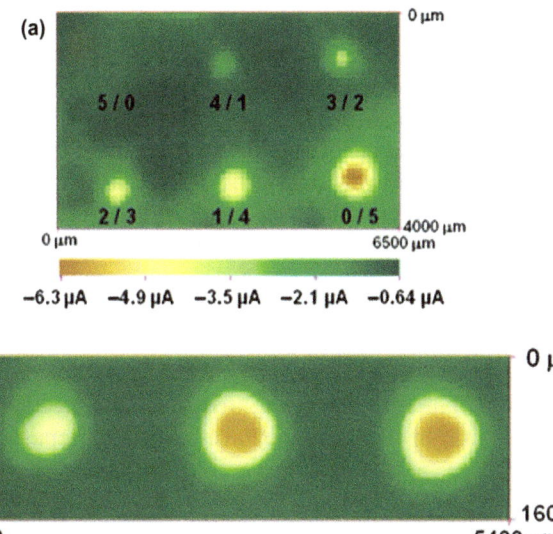

Figure 6.11   (a) SECM images of a Ir/Co oxide array on W-doped BiVO$_4$ film and (b) the images of Pt arrays on W-doped BiVO$_4$ film at 0.3 V *vs.* Ag/AgCl in 0.2 M sodium phosphate buffer (pH 6.8) under UV-visible irradiation. Numbers under each spot represent the number of drops of dispensed Ir and Co solutions, respectively in (a). Pt spots were fabricated by photodeposition of Pt precursor under UV-visible light irradiation through an optical fiber for 20, 30, and 40 min (left to right) in (b).[14]
Reprinted with permission from J. Lee, H. Ye, S. Pan, A. J. Bard, *Anal. Chem.*, 2008, **80**, 7445. Copyright 2008 American Chemical Society.

photoelectrodes alone and in combination with electrocatalysts are obtained quantitatively at the same time, (ii) the effect of material density and compositional variations of electrocatalysts on the photocatalytic activities can be screened rapidly, (iii) the products of the catalytic reactions, such as oxygen from OER and hydrogen from HER, can be titrated quantitatively by the tip-collection/substrate-generation (TC/SG) mode of the SECM with the optical fiber/ring configuration, and (iv) effects of undesirable sample-to-sample deviation of photoelectrode activities are avoided because various electrocatalyst arrays are prepared and studied on the same substrate under the same conditions. The TC/SG study of SECM is invaluable for identifying the origin of photoreactions on the electrocatalyst/photocatalyst electrodes and for measuring the faradaic efficiency of the desired reaction from the electrocatalysts. For example, irradiation of an n-type photocatalyst often leads to both the OER and semiconductor decomposition; collection of O$_2$ allows one to determine the ratio of these two reactions.[17,22]

The fact that the preparation methods for electrocatalysts vary with the nature of the electrocatalyst (*e.g.* direct application, photodeposition, electrodeposition) may limit the application of rapid screening. For example, if different electrocatalysts cannot be prepared simultaneously on the same photoelectrode using identical preparation methods or if the photoelectrodes are degraded during the preparation of the electrocatalysts, rapid screening cannot be used. The interactions between the catalyst precursor solution and the photoelectrode substrates can also limit array preparation with a piezoelectric dispenser. For example, if a drop of precursor solution is not confined to a certain photocatalyst spot and spreads over the substrates, the thickness or amount of electrocatalyst on the spots cannot be controlled. Consequently, quantitative comparisons between the spots in an array may not be possible. Thus, in screening arrays of the electrocatalysts on a photoelectrode, it is important that they are properly prepared as described in 6.2.

6.5   Follow-Up Studies with Large Electrodes

After finding photocatalyst compositions that show good photocurrents ("hits") using rapid screening, it is necessary to synthesize and characterize bulk electrodes of the optimum compositions to verify the rapid screening results. One of the goals of creating bulk film electrodes is to mimic the composition of the spot on the array electrode with amounts that are easier to characterize and study in cell configurations. Thus, it is often convenient to use the same support electrode as was used with the spot array electrode (typically FTO). While there are several methods for creating bulk film electrodes (*e.g.* electrodeposition, spin-coating, spray-coating), one simple method is to drop-cast the same precursor solutions used in the spot array electrode onto the FTO substrates.[15] For example, on an FTO substrate with dimensions of $1.5\,\text{cm} \times 1.5\,\text{cm}$, drop-casting $100\text{--}200\,\mu\text{L}$ of a $0.02\,\text{M}$ precursor solution in ethylene glycol will create a uniform bulk film after annealing. To create bulk films of mixed compositions, it is better to premix the precursor solutions as opposed to depositing one at a time as is done for the spot array preparation. After drop-casting, the bulk film electrodes are annealed using the same conditions as the spot array electrode.

Bulk films can then be characterized by standard electrochemical techniques such as chopped-light linear sweep voltammetry and Mott-Schottky analysis. In addition, other materials characterization techniques, such as XRD, XPS, and UV-Visible absorbance measurements, can be used on bulk electrodes. Figure 6.12 is an example showing the chopped-light linear sweep voltammograms of undoped $BiVO_4$, $BiVO_4$ doped with W, and $BiVO_4$ doped with W and Mo, which were compositions selected *via* rapid screening.[15] If the bulk films bear out the screening results, alternative methods of film preparation and the effect of structure on performance can be probed to optimize the material.[44]

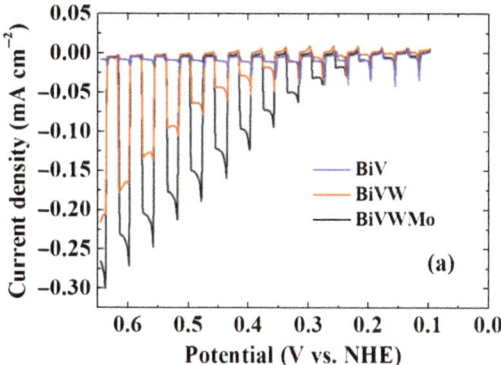

Figure 6.12   Linear sweep voltammograms of undoped $BiVO_4$ (blue), W-doped $BiVO_4$ (red), and W/Mo-doped $BiVO_4$ (black) with chopped light under UV-visible irradiation in a 0.1 M Na_2SO_4 aqueous solution (pH 7, 0.2 M sodium phosphate buffered). Beam intensity was $\sim 120\,mW\,cm^{-2}$ from a full Xe lamp and the scan rate was $20\,mVs^{-1}$.
Reprinted with permission from H. S. Park, K. E. Kweon, H. Ye, E. Paek, G. S. Hwang and A. J. Bard, *J. Phys. Chem. C*, 2011, **115**, 17870. Copyright © 2010 American Chemical Society.

6.6   Correlation of Doping Effects and Theory

Many factors that affect the activity of photocatalysts have to be considered when selecting materials for screening, *e.g.* the band gap energy, the position of band edges, light absorptivity, carrier mobility, recombination rate of the photogenerated carriers in bulk and at the interface, and the stability of the semiconductor material under irradiation.[45,46] However, many semiconductor materials do not have the appropriate characteristics to be used as photoelectrodes for OER and HER. Thus, metal doping into the photocatalyst has been tried to modify the semiconductor properties. For example, H, C, N, Nb, and many other elements have been doped into TiO_2 to decrease its large band-gap so as to utilize more radiant energy.[47–51] Modification of band-gap size has also been demonstrated for chalcogenide semiconductors, like CdSe using the SECM screening technique (Figure 6.1(b)).[13] Ni, Zn, Pt, and Ru were doped into WO_3 to improve its chemical stability and to enhance the carrier mobility.[52,53] Also, W, Mo, and P were doped into $BiVO_4$ and Mo, Cr, and Sn were doped into Fe_2O_3 to improve the conductivity of the metal oxides.[15,21,54–58]

One example of how doping affects photocatalysts is an increase in the carrier density and improvement in the conductivity of the photoelectrodes. However more subtle effects of dopants are important as well. While it is beyond the scope of this chapter to consider in any depth the application of theory, *e.g.* density functional theory (DFT), to predict semiconductor properties, we briefly consider the case of $BiVO_4$ and the effect of W and Mo doping (Figure 6.13[15]). In Figure 6.13(a) and (b), the donor densities can be estimated by capacitance measurements in Mott-Schottky plots where the slope of the plot is inversely

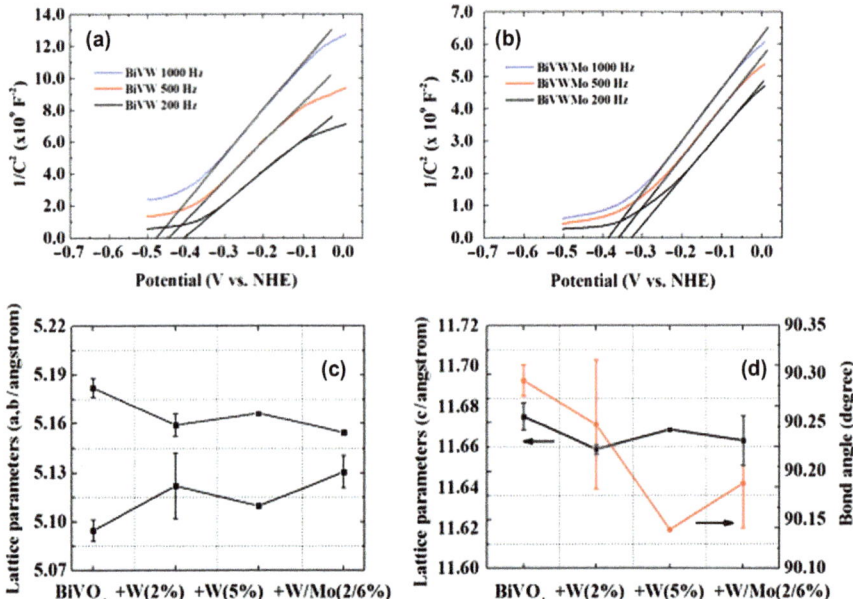

Figure 6.13   (a) Mott-Schottky plots of (a) W-doped $BiVO_4$ and (b) W/Mo-doped $BiVO_4$ obtained from the capacitance measurements. For the capacitance measurements, AC amplitude of 10 mV was applied for each potential and three different AC frequencies were used for the measurements: 1000 Hz (blue), 500 Hz (red), and 200 Hz (black). Tangent lines of the M-S plots are drawn to obtain the flat band potential. (c), (d) Lattice parameters estimated from Rietveld refinement of the XRD patterns for undoped $BiVO_4$, 2 at % W-doped $BiVO_4$, 5 at % W-doped $BiVO_4$, and 2 at % W and 6 at % Mo-doped $BiVO_4$.[15]
Reprinted with permission from H. S. Park, K. E. Kweon, H. Ye, E. Paek, G. S. Hwang, A. J. Bard, *J. Phys. Chem. C*, 2011, **115**, 17870. Copyright 2011 American Chemical Society.

proportional to the donor density. The dopants can also modify the electronic states of the host semiconductor, *e.g.* the band-gap energy of photocatalysts as reported for TiO_2.[59] In addition to the electronic modification of the semiconductor, heavily doped metal oxides show a deformation of the crystal structure as shown in Figure 6.13(c) and (d) for $BiVO_4$.[15] In Figure 6.13(c) and (d), the lattice parameters of monoclinic scheelite-like structures were changed as the doping concentration of W and Mo increased in the $BiVO_4$ with an improvement in performance. The relationship of doping with the crystal and electronic structures and the activity of photocatalysts is discussed in depth in reviews specifically dealing with the water splitting reaction.[15,60–63]

References

1. E. W. McFarland and W. H. Weinberg, *Trends Biotechnol.*, 1999, **17**, 107.
2. R. Potyrailo, K. Rajan, K. Stoewe, I. Takeuchi, B. Chisholm and H. Lam, *ACS Comb. Sci.*, 2011, **13**, 579.

3. H. Koinuma and I. Takeuchi, *Nature Mater.*, 2004, **3**, 429.
4. M. Woodhouse and B. A. Parkinson, *Chem. Soc. Rev.*, 2009, **38**, 197.
5. M. Woodhouse, G. S. Herman and B. A. Parkinson, *Chem. Mater.*, 2005, **17**, 4318.
6. M. Woodhouse and B. A. Parkinson, B. A, *Chem. Mater.*, 2008, **20**, 2495.
7. J. E. Katz, T. R. Gingrich, E. A. Santori and N. S. Lewis, *Energy Environ. Sci.*, 2009, **2**, 103.
8. J. He and B. A. Parkinson, *ACS Comb. Sci.*, 2011, **13**, 399.
9. A. Kleiman-Shwarsctein, P. Zhang, Y. Hu and E. W. McFarland, *On Sol. Hydrogen Nanotechnol.*, 2009, 401.
10. T. F. Jaramillo, S.-H. Baeck, A. Kleiman-Shwarsctein and E. W. McFarland, *Macromol. Rapid Commun.*, 2004, **25**, 297.
11. T. F. Jaramillo, S.-H. Baeck, A. Kleiman-Shwarsctein, K.-S. Choi, G. D. Stucky and E. W. McFarland, *J. Comb. Chem.*, 2005, **7**, 264.
12. M. Seyler, K. Stoewe and W. F. Maier, *Appl. Catal. B*, 2007, **76**, 146.
13. G. Liu, C. Liu and A. J. Bard, *J. Phys. Chem. C*, 2010, **114**, 20997.
14. J. Lee, H. Ye, S. Pan and A. J. Bard, *Anal. Chem.*, 2008, **80**, 7445.
15. H. S. Park, K. E. Kweon, H. Ye, E. Paek, G. S. Hwang and A. J. Bard, *J. Phys. Chem. C*, 2011, **115**, 17870.
16. J. Chang, K. C. Leonard, S. K. Cho and A. J. Bard, *Anal. Chem.*, 2012, **84**, 5159.
17. H. Ye, H. S. Park and A. J. Bard, *J. Phys. Chem. C*, 2011, **115**, 12464.
18. J. S. Jang, J. Lee, H. Ye, F.-R. F. Fan and A. J. Bard, *J. Phys. Chem. C*, 2009, **113**, 6719.
19. G. Liu and A. J. Bard, *J. Phys. Chem. C*, 2010, **114**, 17509.
20. A. J. Bard, M. V. Mirkin, *Scanning Electrochemical Microscopy*, Taylor and Francis, 2nd Ed., 2012.
21. H. Ye, J. Lee, J. S. Jang and A. J. Bard, *J. Phys. Chem. C*, 2010, **114**, 13322.
22. Y. Cong, H. S. Park, S. Wang, H. X. Dang, F.-R. F. Fan, C. B. Mullins and A. J. Bard, *J. Phys. Chem. C*, 2012, **116**, 14541.
23. D. K. Zhong, S. Choi and D. R. Gamelin, *J. Am. Chem. Soc.*, 2011, **133**, 18370.
24. M. Higashi, K. Domen and R. Abe, *J. Am. Chem. Soc.*, 2012, **134**, 6968.
25. A. J. Bard, *J. Am. Chem. Soc.*, 2010, **132**, 7559.
26. B. D. Stubbert, J. C. Peters and H. B. Gray, *J. Am. Chem. Soc.*, 2011, **133**, 18070.
27. J. R. McKone, E. L. Warren, M. J. Bierman, S. W. Boettcher, B. S. Brunschwig, N. S. Lewis and H. B. Gray, *Energy Environ. Sci.*, 2011, **4**, 3573.
28. A. Sobczynski, A. Yildiz, A. J. Bard, A. Campion, M. A. Fox, T. Mallouk, S. E. Webber and J. M. White, *J. Phys. Chem.*, 1988, **92**, 2311.
29. H. I. Karunadasa, C. J. Chang and J. R. Long, *Nature*, 2010, **464**, 1329.
30. J. R. McKone, E. L. Warren, M. J. Bierman, S. W. Boettcher, B. S. Brunschwig, N. S. Lewis and H. B. Gray, *Energy Envrion. Sci.*, 2011, **4**, 3573.

31. A. Paracchino, V. Laporte, K. Sivula, M. Gratzel and E. Thimsen, *Nature Mater.*, 2011, **10**, 456.
32. K. Maeda and K. Domen, *J. Phys. Chem. Lett.*, 2010, **1**, 2655.
33. Y. Surendranath, M. Dinca and D. G. Nocera, *J. Am. Chem. Soc.*, 2009, **131**, 2615.
34. R. Subbaraman, D. Tripkovic, K.-C. Chang, D. Strmcnik, A. P. Paulikas, P. Hirunsit, M. Chan, J. Greeley, V. Stamenkovic and N. M. Markovic, *Nature Mater.*, 2012, **11**, 550.
35. A. Minguzzi, M. A. Alpuche-Aviles, J. R. Lopez, S. Rondinini and A. J. Bard, *Anal. Chem.*, 2008, **80**, 4055.
36. A. Kay, I. Cesar and M. Gratzel, *J. Am. Chem. Soc.*, 2006, **128**, 15714.
37. D. K. Zhong, M. Cornuz, K. Sivula, M. Gratzel and D. R. Gamelin, *Energy Environ. Sci.*, 2011, **4**, 1759.
38. D. Wang, R. Li, J. Zhu, J. Shi, J. Han, X. Zong and C. Li, *J. Phys. Chem. C*, 2012, **116**, 5082.
39. S. K. Pilli, T. G. Deutsch, T. E. Furtak, J. A. Turner, L. D. Brown and A. M. Herring, *Phys. Chem. Chem. Phys.*, 2012, **14**, 7032.
40. K. Maeda and K. Domen, *J. Phys. Chem. C*, 2011, **115**, 3057.
41. Z. Zou, J. Ye, K. Sayama and H. Arakawa, *Nature*, 2001, **414**, 625.
42. Z. Chen, J. J. Concepcion, X. Hu, W. Yang, P. G. Hoertz and T. J. Meyer, *Proc. Natl. Acad. Sci. USA*, 2010, **107**, 7225.
43. T. F. Jaramillo, K. P. Jorgensen, J. Bonde, J. H. Nielsen, S. Horch and I. Chorkendorff, *Science*, 2007, **317**, 100.
44. S. P. Berglund, D. W. Flaherty, N. T. Hahn, A. J. Bard and C. B. Mullins, *J. Phys. Chem. C*, 2011, **115**, 3794.
45. A. J. Nozik, *Ann. Rev. Phys. Chem.*, 1978, **29**, 189.
46. A. J. Bard, *J. Photochem.*, 1979, **10**, 59.
47. X. Chen, L. Liu, P. Y. Yu and S. S. Mao, *Science*, 2011, **331**, 746.
48. Z. Xiong and X. S. Zhao, *J. Am. Chem. Soc.*, 2012, **134**, 5754.
49. S. U. M. Khan, M. Al-Shahry and W. B. Ingler Jr., *Science*, 2002, **297**, 2243.
50. J. Tang, A. J. Cowan, J. R. Durrant and D. R. Klug, *J. Phys. Chem. C*, 2011, **115**, 3143.
51. X. Ma, Y. Wu, Y. Lu, J. Xu, Y. Wang and Y. Zhu, *J. Phys. Chem. C*, 2011, **115**, 16963.
52. S. H. Baeck, T. F. Jaramillo, C. Brandli and E. W. McFarland, *J. Comb. Chem.*, 2002, **4**, 563.
53. W. Liu, H. Ye and A. J. Bard, *J. Phys. Chem. C*, 2009, **114**, 1201.
54. A. Kleiman-Shwarsctein, Y.-S. Hu, A. J. Forman, G. D. Stucky and E. W. McFarland, *J. Phys. Chem. C*, 2008, **112**, 15900.
55. Y. Ling, G. Wang, D. A. Wheeler, J. Z. Zhang and Y. Li, *Nano Lett.*, 2011, **11**, 2119.
56. W. Luo, Z. Yang, Z. Li, J. Zhang, J. Liu, Z. Zhao, Z. Wang, S. Yan, T. Yu and Z. Zou, *Energy Environ. Sci.*, 2011, **4**, 4046.
57. S. P. Berglund, A. J. E. Rettie, S. Hoang and C. B. Mullins, *Phys. Chem. Chem. Phys.*, 2012, **14**, 7065.

58. W. J. Jo, J.-W. Jang, K. J. Kong, H. J. Kang, J. Y. Kim, H. Jun, K. P. S. Parmar and J. S. Lee, *Angew. Chem. Int. Ed.*, 2012, **51**, 3147.
59. W.-J. Yin, H. Tang, S.-H. Wei, M. M. Al-Jassim, J. Turner and Y. Yan, *Phys. Rev. B*, 2010, **82**, 045106–1.
60. A. Kubacka, M. Fernandez-Garcia and G. Colon, *Chem. Rev.*, 2012, **112**, 1555.
61. K. Maeda, *J. Photochem. Photobiol. C*, 2011, **12**, 237.
62. R. Abe, *J. Photochem. Photobiol. C*, 2010, **11**, 179.
63. A. Kudo and Y. Miseki, *Chem. Soc. Rev.*, 2009, **38**, 253.

CHAPTER 7

Oxygen Evolution and Reduction Catalysts: Structural and Electronic Aspects of Transition Metal Based Compounds and Composites

SEBASTIAN FIECHTER* AND PETER BOGDANOFF

Institute for Solar Fuels, Helmholtz-Zentrum für Materialien und Energie, Hahn-Meitner-Platz 1, 14109 Berlin, Germany
*Email: Fiechter@helmholtz-berlin.de

7.1 Introduction

The direct generation of fuels from sunlight, water and CO_2 to establish a sustainable energy supply for mankind by chemical means can be considered as a paramount challenge. Such energy converting devices could solve the inherent problem of energy storage of sunlight, having also the discontinuous availability of sunlight and regenerative energies in general in mind. Fuels, which are in demand to guarantee mobility especially for air transportation, can be obtained by converting harvested hydrogen into hydrocarbons.

Among biomimetic approaches using inorganic systems, one possible solution is the conversion of sunlight into chemical energy via photonic excitation of a thin film PV structure which is directly combined with corrosion-stable layers at the front and back contact of the PV system, catalyzing the process of

RSC Energy and Environment Series No. 9
Photoelectrochemical Water Splitting: Materials, Processes and Architectures
Edited by Hans-Joachim Lewerenz and Laurence Peter
© The Royal Society of Chemistry 2013
Published by the Royal Society of Chemistry, www.rsc.org

water electrolysis at the electrode-electrolyte-interfaces.[1] Generated hydrogen can be stored as compressed gas, liquid-H_2, metal hydride or methanol. Alternatively, CO_2 reduction and hydrocarbon production can be achieved on the cathode side of the device. In both cases, noble metal-free catalysts are needed to develop electrodes for a mass market with respect to oxidation/ reduction of water (light-induced hydrogen and oxygen evolution) and reduction of CO_2, respectively. In this context, CO_2 reduction to organic fuels at gas diffusion electrodes (GDE) based on noble and non-noble catalysts in combination with a photoanode also appears a promising route.

7.1.1 Chemical Energy Storage

It is well known that solar cells, thermal concentrators and wind generators are able to convert sunlight directly and indirectly into electrical energy. As solar energy in the form of irradiation and wind energy is not permanently available, a part of this energy has to be stored. There are only two possibilities to do this on a large scale (*i.e.* GW range), namely by hydrogen or hydrocarbon generation and hydropower. In case of hydrogen generation, water photolysis can be used. Compared with battery systems and hydrocarbons such as benzene or diesel, hydrogen has an extraordinarily high energy density and can be stored in pressurised bottles or as metal hydride. The low weight of hydrogen and its high energy content also make it attractive as fuel for mobile applications. By way of comparison: 3 kg H_2 corresponds to about 100 kWh of chemical energy, whereas the weight of lithium ion battery needed to deliver 100 kWh of electrical energy is presently 540 kg. Subsequent conversion of this chemical energy into electricity by a fuel cell (typical system efficiency of 50%) means that the overall energy conversion efficiencies are comparable. Since cheap metal hydrides are not yet available for technical application, compressed hydrogen has to be stored in dedicated containers. Based on present technology, the system weight of 125 kg (H_2) still compares favourably with a battery system weighing 830 kg.Thus to store chemical energy in the form of hydrogen efficiently, novel storage systems (*e.g.* metal hydrides, methanol, NH_3) as well as efficient energy converting systems have to be developed in order to use sunlight as a virtually inexhaustible, renewable energy source. Therefore to realize a future hydrogen economy, it is necessary to develop efficient catalysts and efficient energy converting devices.

In contrast to the process of photosynthesis, in which non-noble metal catalysts such as Mn- and Fe-Ni-containing complexes convert CO_2 and water into O_2 and carbohydrates (or, under certain conditions, hydrogen), existing systems, which connect solar cells or wind generators to electrolysing systems, employ platinum and ruthenium oxide as catalysts for water splitting in the so called Polymer Electrolyte Membrane (PEM) electrolyzer.[2] Non-noble catalysts such as Ni (hydrogen evolution) and Mn-Co-Ni oxides/hydroxides (oxygen evolution) are only used in alkaline electrolyzers, which are best suited for stationary applications but show disadvantages under energy fluctuating conditions. One of the intriguing research goals is therefore the mimicking of the thylakoid membrane in plants. In this context, development of an artificial

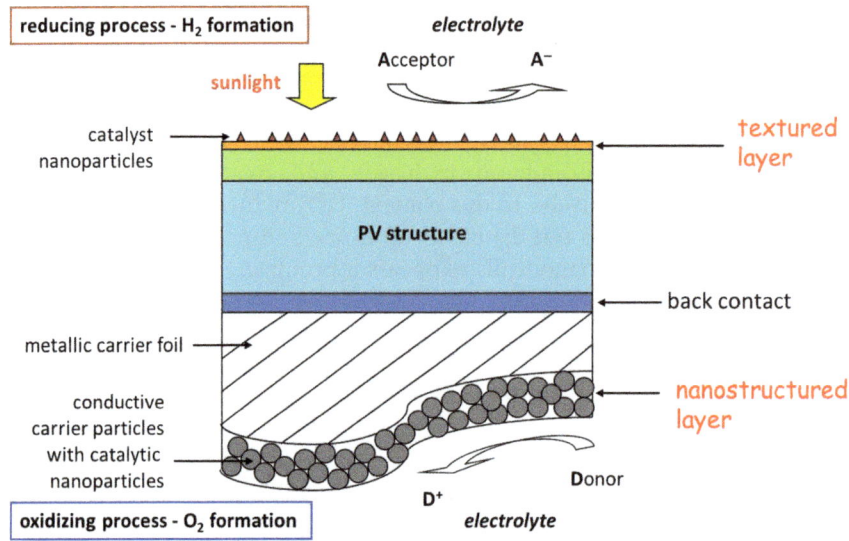

Figure 7.1   Schematic cross section of a water splitting membrane combining a thin film
               tandem solar cell structure with nano-scaled catalysts to generate hydrogen
               at the cathode and oxygen at the anode side under irradiation of the device
               with sunlight after immersion in water (adapted from reference 3).

leaf, *i.e.* a "water splitting membrane", allowing the direct conversion of
sunlight into hydrogen and oxygen, is particularly important. Approaches that
include organic and inorganic systems to achieve this goal have been proposed.
The basic challenges relate to the stability of organic artificial systems, and
an alternative research target pursues a design that hitherto uses inorganic
components: a thin film photovoltaic monolithic so called back-to-back
tandem structure with integrated catalysts at photocathode (hydrogen generation) and the photoanode side (oxygen generation) to facilitate water splitting
by sunlight as shown in Figures 7.1 and 7.28.[1,3]

7.1.2   Oxidation and Reduction Catalysts in Nature

Catalytic centers, which enable photolysis of water in plants, algae and
cyanobacteria, are associated with photosystem I and II, which are part of the
thylakoid membrane of the chloroplasts in plant cells (Figure 7.2).

While the Oxygen Evolving Complex in photosystem II is responsible for
oxidation evolution of water molecules according to equation (7.1)

$$2\,H_2O + 4\,h^+ \rightarrow 4\,H^+ + O_2 \qquad (7.1)$$

photosystem I can produce in algae and cyanobacteria hydrogen according to
equation (7.2)

$$4\,H^+ + 4\,e^- \rightarrow 2\,H_2 \qquad (7.2)$$

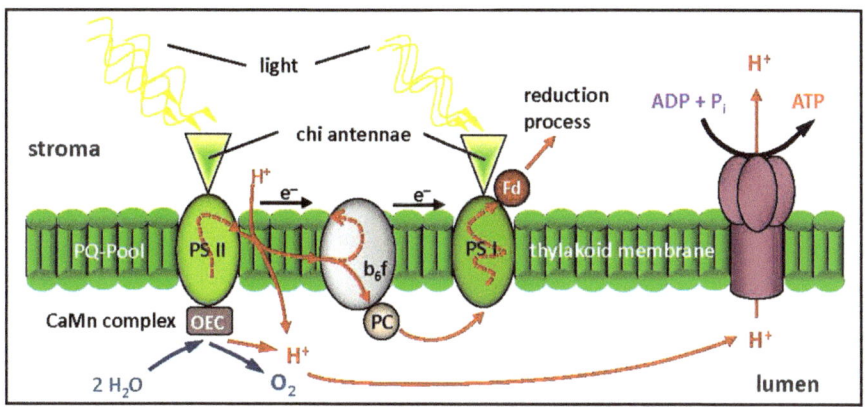

Figure 7.2 In a simplified picture a section of a thylakoid membrane is shown. Integrated in the membrane are the complexes photosystem I (PSI) and photosystem II (PSII) as well as the b_6-f complex and the enzyme ADP synthease. While after light excitation electrons (e^-) are transported along the thylakoid membrane protons (H^+) migrate from the stroma to the lumen side and form together with protons produced by water splitting a pH gradient between stroma (pH 8.5) and lumen (pH 5.5), which is the driving force to form at the one hand the reduction equivalent NADPH + H^+ (nicotinamide adenine dinucleotide phosphate) from $NADP^+$ or *via* hydrogenase hydrogen and the energy equivalent ATP (adenosine triphosphate) from ADP (adenosine diphosphate) *via* the ADP synthease complex. Ferredoxin (Fd) is responsible for the transfer of protons to other enzymes, b_6f is a cytochrome complex transporting together with plastocyanin (PC) and plastoquinone (PQ) protons. In the Oxygen Evolving Complex (OEC) oxygen will be generated.

The structure of the oxygen evolving complex of composition Mn_3CaO_3MnO in PS II is shown in Figure 7.3.[4] On the basis of a recent structure determination of PS II with a resolution of 1.9 Å by Umena *et al.*,[5] the OEC can be described as a distorted cubane-type cluster consisting of 3 Mn, 1 Ca and 4 O atoms. A further MnO unit is attached at one edge of the cubane-type unit (Figure 7.3). The five oxygen atoms serve as oxo-bridges linking four Mn and one Ca atom. In addition, Umena *et al.* found that four water molecules were associated with the cluster: two associated with the terminal Mn (Mn4) atom and two with the Ca atom. Some of these water molecules may therefore serve as the substrate for water splitting.[5]

Dau and Zaharieva describe the energetics of electrons and protons after excitation of pigment 680* ($\Delta E_m = 1.83\,eV$) in PS II (Figures 7.4a and 7.4b).[6] The excited electron acts on the one hand to reduce a bonded quinone (Q_A) after oxidation of tyrosine Tyr_z ($Tyr_{160/161}$) within 1 µs. This radical pair is stable for 1 ms against recombination losses. These authors assume that in maximum 50% of the excited states are stored via this chemical form. They suggest a maximum efficiency of $\eta = 10.5\%$ for hydrogen generation by interaction of PS II and PS I, taking into account losses in the chain of electron transfer. The overvoltage in the process of water oxidation at the Mn_4Ca

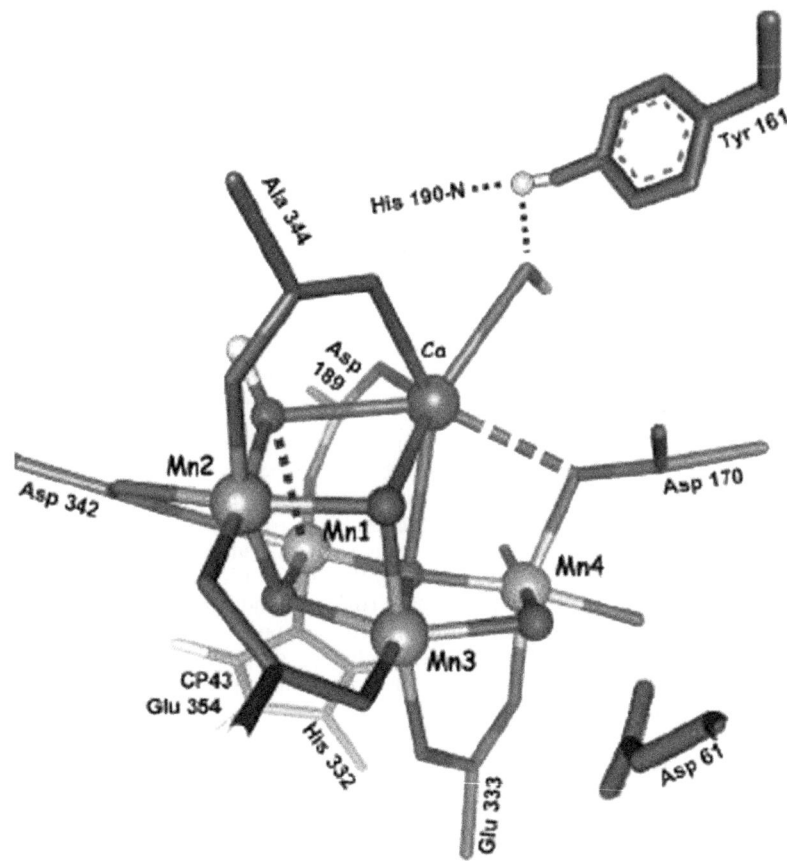

Figure 7.3 Oxygen Evolving Complex (OEC) in Photosystem II (PS II). The complex can be understood as a distorted cubane cluster embedded in a protein matrix consisting of 3 Mn (Mn1-Mn3), 1 Ca and 4 O atoms. An additional Mn-O pair (Mn4-O) is attached at one side of the cubane cluster (pentanuclear complex).[4]

complex is expected to be at least 0.3 V. The tyrosine radical oxidizes the Mn_4Ca complex as a function of already accumulated oxidizing equivalents. Driven by four photons, one oxygen molecule will be released.

The fractional energy yield (FEY) of the overall reaction of H_2 production from water is given by

$$FEY(H_2) = \frac{q\left(E_{O_2-} - E_{mH_2}\right)}{E_{680} + E_{700}} \tag{7.3}$$

or

$$FEY(H_2) = \frac{1.23\,eV + 0.059\,eV\,pH_{O_2} - qE_{mH_2}}{3.60\,eV} \tag{7.4}$$

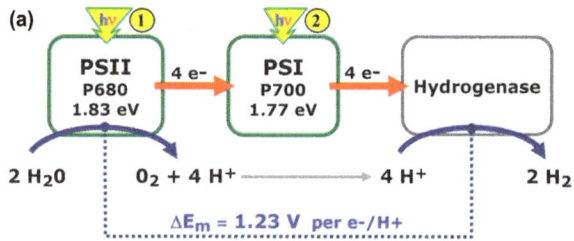

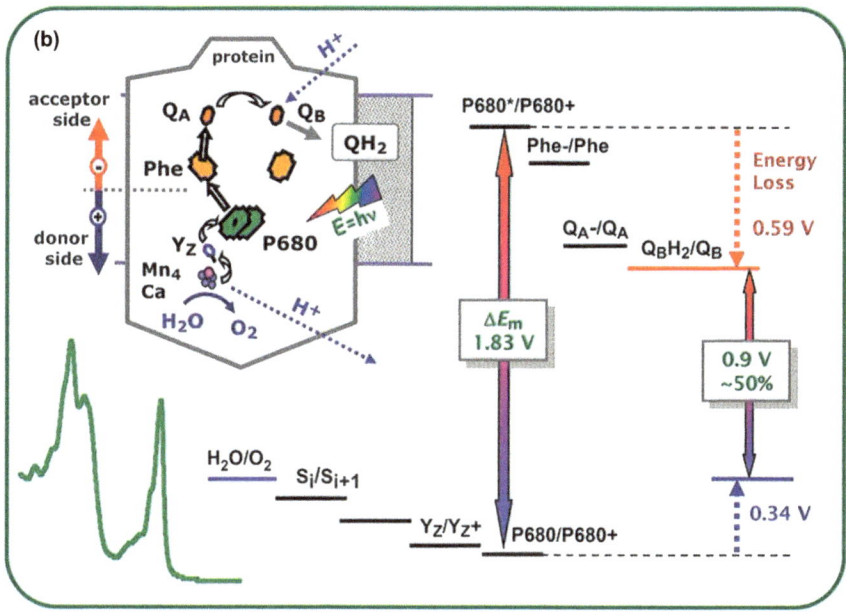

Figure 7.4 (a) Scheme on light-driven H_2 formation by PSII and PSI. The corresponding maximal η_{SOLAR} is estimated to be 10.5%. The reducing equivalents provided by PSI normally are used for CO_2 reduction and eventually carbohydrate formation, but they can also be used for driving proton reduction and H_2 formation by enzymes denoted as hydrogenases.[6] (b) Arrangement of redox factors in PSII. The protein subunits harbor all essential redox factors. The aqueous phases at the acceptor side (stroma) and donor side (lumen) are separated by a lipid bilayer (thylakoid membrane). Formation of the excited state of P680 is followed by electron transfer steps, which at the donor side resembles movement of a positively charged "hole".[6]

where pH_{O2} and pH_{H2} denote the pH at the lumen (water oxidation) and at the stroma side (proton reduction), respectively (see Figure 7.2). The pH dependence is formally given, but it is of minor interest with respect to the "harvestable" dihydrogen. Neglecting the pH difference, we arrive at a FEY value of 34%. This figure represents a maximal value assuming that all electrons removed from water are used for hydrogen reduction.[6]

In oxygenic photosynthesis, the reduced plastoquinone molecules (QH_2) feed electrons in an electron transfer chain (Figure 7.4b), thereby (i) increasing the proton motive force by intricate coupling of redox chemistry with proton transport and (ii) providing reducing equivalents to PSI. The PSI threshold energy corresponds to 700 nm (P700 instead of P680, $E_{700} = 1.77$ eV). Its design is similar with respect to the fast energy transfer and electron transfer processes that are mediated by pigments and redox factors in the low dielectric, membrane-intrinsic regions of the cofactor-protein complex but necessarily different for the slower redox reactions at the donor and acceptor side.[4,6]

The energy loss is clearly lower at the PSII donor side than at the acceptor side (see Figure 7.4b). This relates to the low "overpotential" of PSII water oxidation compared to other energy losses. In electrochemical water oxidation, the overpotential (η) is here defined as the difference between the H_2O/O_2 equilibrium potential and the anodic potential V^+ at room temperature, *i.e.* $\eta(V) \approx V^+ - (1.23 - 0.06\,\mathrm{pH})$ V at a current density of 1 mA cm^{-2}.

Siegbahn has performed density functional theory (DFT) calculations to describe the process of water splitting.[7] The best O-O bond formation mechanism found can be explained with a reaction between an oxygen radical and a μ-oxo ligand. The reconstruction found in the S_2 to S_3 transition provides energetically improved binding the water molecule to a manganese atom in the Mn_3Ca cubane-type unit (Figure 7.5).

The different steps of water oxidation assigned as S-transitions, are shown schematically in Figure 7.6. The starting point is a state, S_0', where O_2 was released in the previous cycle. After O_2 removal, the cubane-type cluster is incomplete, exhibiting an opening in the bond. In the transition S_0' to S_0, a water molecule has reacted with the cluster. An OH group is now bonded, completing the cubane-type unit which is connected with a simultaneous loss of a proton to the bulk. An arrow indicates the position of the substrate OH. In the S_0 to S_1 transition, an electron is removed from the Mn4 atom in the Mn-O unit which is attached to the cube cluster (an * marks the oxidation in S1), and a proton from the cube is released to the bulk. In the transition S_1 to S_2, a first Mn atom of the cubane-type unit changes its oxidation state from Mn^{3+} to Mn^{4+} (Mn2 in Figure 7.3) releasing an electron. In S_2 to S_3 transition, the next substrate water binds. In this process, a proton is removed simultaneously from the substrate water, exactly as in the formation of S_0, and an electron is taken from a further Mn atom (Mn1). In S_3 to S4 transition, all manganese are now in the valence state Mn(IV) (oxidation of Mn3). In the final transition S_3 to S_4, a proton is taken from the substrate OH and an electron from the substrate oxygen. Finally, the O–O bond is formed between the oxygen radical and the oxo-group remaining from the first substrate water, O_2 is released, and the cycle starts all over again.

As described above, protons and electrons are removed in an alternating fashion from the $CaMn_4$ cluster, reacting with water molecules in two steps. The model with alternating removal of charges has been used experimentally to analyze water oxidation in PSII and can essentially explain the experimental results.[4,7]

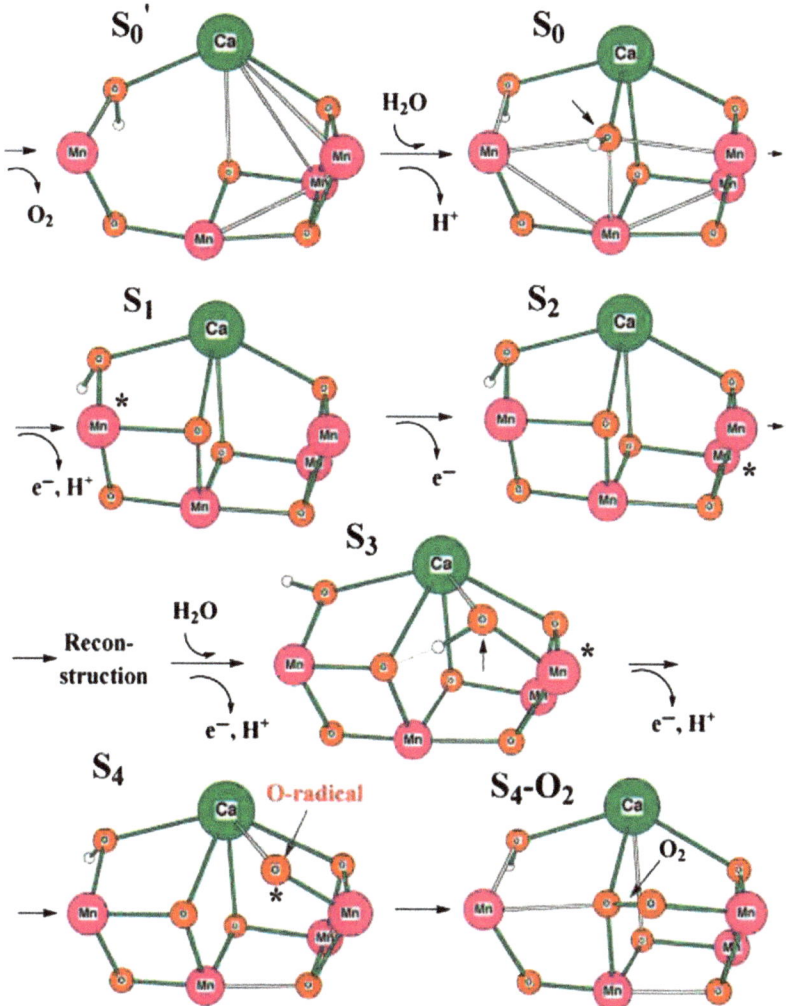

Figure 7.5 Schematic picture of the different S-transitions. The structures have been optimized, but only the most important atoms are shown. The * marks the atom that has been oxidized in that transition.[7]

7.2 Oxygen Evolving Catalysts – an Inorganic Approach

In the last 15 years, water splitting membranes and (photo)electrodes that mimick photosynthesis have been designed and developed.[1,8,9] Presently, PV hybrid electrolyzing systems integrating non-noble catalysts in the front and back contact of a photovoltaic thin film device are most promising.[1,8] The advantage of this strategy is that the PV system is separated from the electrochemical processes, *i.e.* the thin film PV device only acts as a light-driven power supply to enable water splitting because the catalytically active layers are

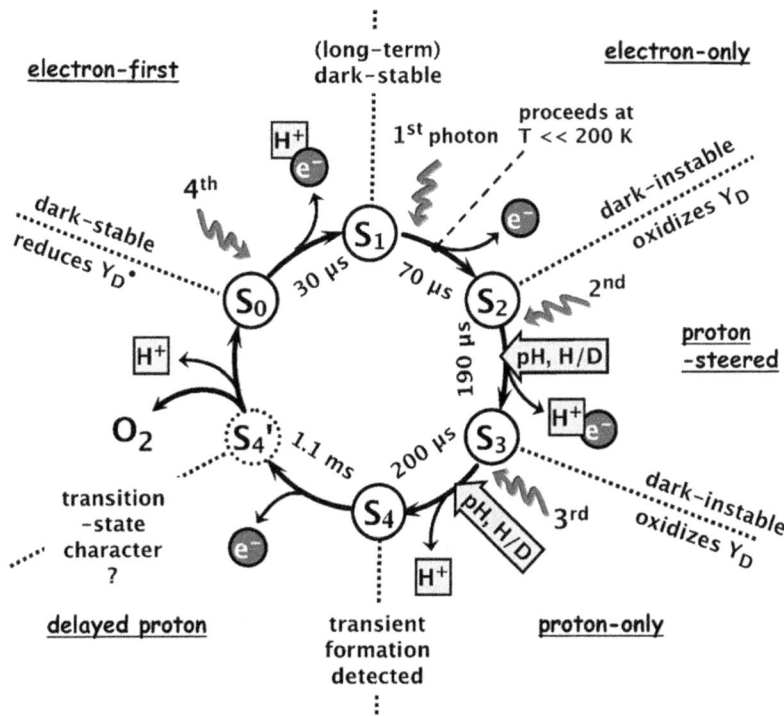

Figure 7.6 Extended S-state cycle and its relation to electron and proton removal in the individual S-state transitions (see reference 4 and references therein). A complete cycle requires sequential absorption of four photons. For each transition it is indicated whether an electron is transferred from the Mn complex to the oxidized tyrosine Tyr_z, whether a proton is released from the Mn complex or its ligand environment, and whether or not the rate constant is sensitive to pH and H_2O/D_2O exchange. The transition times given are indicating that the processes between the single transitions are relatively slow ($30 \mu s$–1.1 ms).[4]

deposited on the back contact or on the transparent conductive film protecting the PV device from contact with the electrolyte (Figures 7.1 and 7.7). To mimic the catalytic behavior of the $CaMn_4O_x$ cluster in photosystem II, different calcium manganates have been synthesized and their catalytic behavior with respect to the oxygen evolution reaction (OER) investigated.[10–13] Catalyst layers consisting of nano-sized particles showed the best catalytic properties and lowest overvoltages in the oxygen evolution reaction.[13] Special interest has been focussed on birnessite-type phases, which crystallize in a layer structure where alkali and earth alkaline metals as well as water molecules are inter-calated between manganese oxide sheets.[12]

Another approach involves using pure cobalt and manganese oxides. CoO_x nanoparticles deposited electrochemically from a cobalt nitrate solution in presence of a phosphate buffer were described by Reece *et al.* as an OER catalyst.[1] With this material, a water splitting device employing a

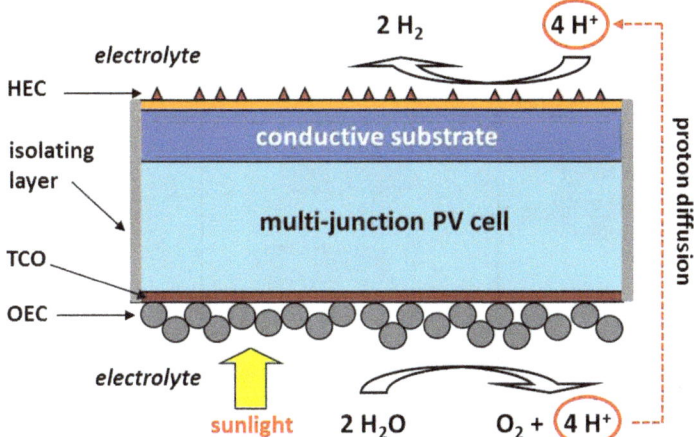

Figure 7.7 "Artificial leaf" consisting of a triple-junction a-Si PV structure (3jn-a-Si) deposited on a stainless steel foil. A NiMoZn layer was used as cathode catalyst, and the anode catalyst consisted of CoO_x clusters interlinked by phosphate groups.[1]

triple-junction amorphous silicon solar cell (3jn-a-Si) was realized showing a Solar To Hydrogen (STH) efficiency of 2.5% (Figure 7.7). The wired counterpart where the dark cathode, consisting of a metal mesh with NiMoZn catalyst, was located in front of the PV hybrid anode yielded a STH of 4.7%. The difference in efficiency can be explained by the low migration velocity and mobility of protons in water (3.27×10^{-5} cm^2 V^{-1} s^{-1}) at pH 7 according to the Grotthus mechanism.

Recently Gorlin and Jaramillo described $\alpha\text{-}Mn_2O_3$ as a catalyst that can be employed for the OER as well as for Oxygen Reduction Reaction (ORR). The bifunctional catalyst was electrodeposited potentiostatically. At a current density of 5 mA cm^{-2}, this Oxygen Evolving Catalyst (OEC) showed an overvoltage of about 450 mV.[14]

7.2.1 Structural Features

The crystal structures of mineral birnessite[15] ($NaMn_7O_{14}$ x nH_2O), $Ca_2Mn_3O_8$ and other layer compounds consisting of manganese and oxygen can be derived from the structure of manganese dioxide, $\delta\text{-}MnO_2$, by intercalation of alkaline and alkali earth elements as well as water molecules between the manganese oxide sheets. Birnessite, buserite and related phases are amongst the strongest known natural oxidants.[16] In nature, $Mn^{III}Mn^{IV}(O,OH)_x$ phases are produced by the bacterial oxidation of soluble Mn^{II}. In these phases, the manganese atoms are octahedrally coordinated by oxygen, and layers are formed by edge-sharing of the octahedra. All these manganates belong to the structure family of phyllomanganates. Alternatively to $\delta\text{-}MnO_2$-type phases, in

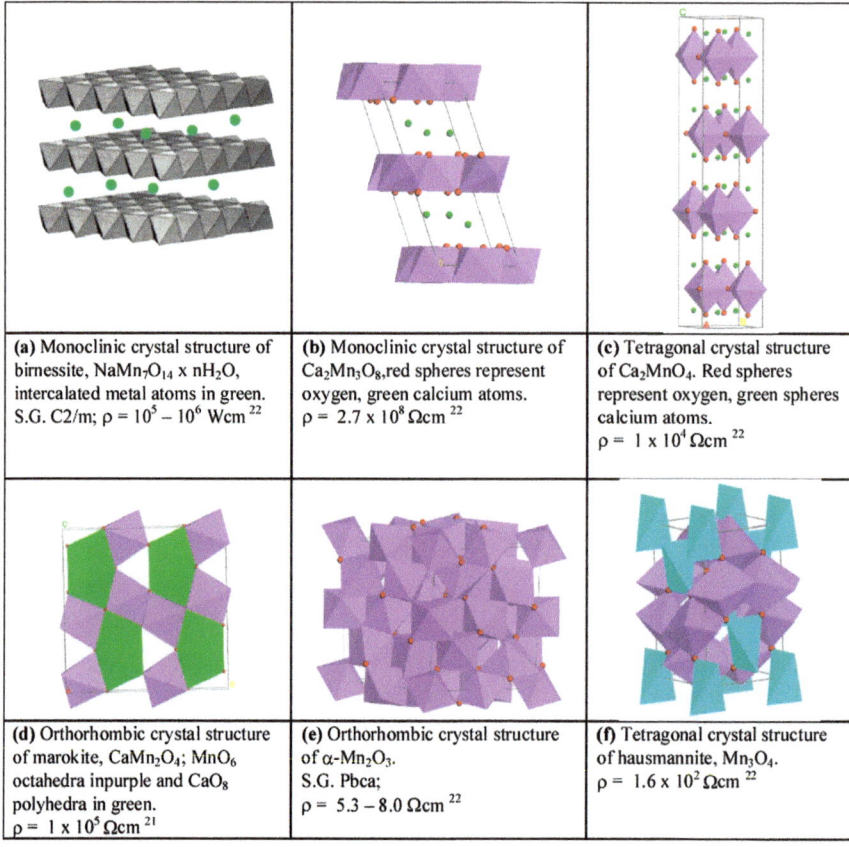

(a) Monoclinic crystal structure of birnessite, $NaMn_7O_{14} \times nH_2O$, intercalated metal atoms in green. S.G. C2/m; $\rho = 10^5 - 10^6$ Wcm [22]	**(b)** Monoclinic crystal structure of $Ca_2Mn_3O_8$, red spheres represent oxygen, green calcium atoms. $\rho = 2.7 \times 10^8$ Ωcm [22]	**(c)** Tetragonal crystal structure of Ca_2MnO_4. Red spheres represent oxygen, green spheres calcium atoms. $\rho = 1 \times 10^4$ Ωcm [22]
(d) Orthorhombic crystal structure of marokite, $CaMn_2O_4$; MnO_6 octahedra in purple and CaO_8 polyhedra in green. $\rho = 1 \times 10^5$ Ωcm [21]	**(e)** Orthorhombic crystal structure of α-Mn_2O_3. S.G. Pbca; $\rho = 5.3 - 8.0$ Ωcm [22]	**(f)** Tetragonal crystal structure of hausmannite, Mn_3O_4. $\rho = 1.6 \times 10^2$ Ωcm [22]

Figure 7.8   Structural features of Ca-Mn-O and Mn-O phases.

Ca_2MnO_4 layer units are formed by corner-sharing of the coordination octahedra (see Figure 7.8a–c). Recently, Najafpour *et al.*[11] discussed the mineral marokite ($CaMn_2O_4$) as a possible oxygen evolving catalyst. Its structure can be interpreted as a post-spinel structure consisting of two distorted edge-sharing MnO_6 octahedra and calcium atoms that are eight-fold coordinated by oxygen.[17] The material has a bifunctional character because it can also be used as oxygen reduction reaction (ORR) catalyst.[18] In all these compounds, structural motifs can be found that show similarities with the cubane-type $CaMn_3O_4MnO$ cluster in PSII. Alkaline earth and alkali-metal-free manganese compounds also exhibit remarkable catalytic effects in the OER. Most prominent examples are α-Mn_2O_3, which crystallizes in a distorted bixbyite structure, and Mn_3O_4, known as mineral hausmannite. The structure of the latter can be understood as a tetragonally distorted spinel structure where the Mn^{2+} ions are coordinated tetrahedrally by oxygen, while the Mn^{3+} ions are located in an octahedral ligand field of oxygen atoms. As described above, the oxidation state of manganese atoms in the $CaMn_4O_x$

cluster in PSII varies from $+3$ to $+4$ during the OER. In birnessite of ideal composition, manganese occurs in a valence state of $+3.86$. As known from chemical analysis of different birnessites, up to 20% of the sodium atoms can be replaced by calcium in the lattice. Under these conditions, the valency of manganese approaches $+3.8$, a value equal to the mean oxidation state of manganese in the $CaMn_3O_4MnO$ cluster in PSII.[11] Therefore, it is thought that manganese oxide cluster units showing a mixed Mn^{3+}/Mn^{4+} valence state in a MnO_x electrode in contact with an aqueous electrolyte should fulfill the requirements of good oxygen evolving catalysts. However, as will be shown below, the best catalytic effects were found in porous α-Mn_2O_3 layers. As a sesquioxide of orthorhombic symmetry, α-Mn_2O_3 crystallizes in a distorted bixbyite structure.[17] Characteristic features are MnO_6 octahedra sharing six edges with neighboring coordination units. In comparison to the other structures discussed above, α-Mn_2O_3 is characterized by two different types of coordination octahedra. The type I octahedra show three pairs of Mn-O distances varying from 1.955 to 2.067 Å. In the type II octahedra, all distances in the octahedral are different: four shorter Mn-O bonds are located in a common plane with values ranging from 1.875–2.011 Å, and two longer bonds are located each side of this plane, forming the tops of an elongated octahedron with distances varying from 2.192 to 2.306 Å. The observed distortion can be explained by the Jahn-Teller effect due to the d^4 state of the Mn^{3+} atoms in high spin configuration.[17] This variety in bond lengths is thought to be advantageous for an electrocatalyst, offering a wide range of different Mn-O distances in the multistep process of water cleavage and oxygen formation.

α-Mn_2O_3 and Mn_3O_4 have a further advantage compared to many other phyllomanganates: they show low resistivities compared to the mentioned compounds. Resistivities are given in the figure captions of Figure 7.8. Here α-Mn_2O_3 and Mn_3O_4 have highest conductivity, while all other compounds are highly insulating.[19–24]

7.2.2 Preparation Routes and Phase Formation

Films of the above mentioned phyllomanganates and binary manganese oxides have been deposited on conductive $F:SnO_2$/glass substrate by sol-gel processing, reactive magnetron sputtering and potentiostatic as well as galvanostatic deposition.[12,24,25,29]

7.2.2.1 Phyllomanganates

Phyllomanganates are characterized by arrays of MnO_6 octahedra ordered in planes of repeating distances along one of the crystallographic axes of the related crystal lattice. Formed at low temperatures they show a high degree of disorder such as manganese and oxygen point defects.

(i) Phyllomanganates with birnessite-type structure were prepared by Wiechen *et al.*[10] by a conproportionation reaction of Mn^{2+} and MnO_4^- ions in aqueous solution in presence of different alkali and alkaline earth cations at high pH. It is thought that the first product obtained after three days ripening belongs to buserite structure, showing a higher number of water molecules intercalated between the sheets consisting of MnO_6 octahedra in comparison to birnessite. After drying the precipitate at 65 °C in air, birnessite was formed.

(ii) Another method to prepare thin films of calcium phyllomanganates by sol-gel and screen printing techniques was reported by Ramirez et al.[13] The Pechini sol-gel method was employed following a procedure described by Fawcett et al. to prepare a thin film of $Ca_2Mn_3O_8$ by dipping into a solution.[23] For this purpose, stoichiometric amounts of $Ca(NO_3)_2 \times 4H_2O$ and $Mn(NO_3)_2$ were dissolved in 25 mL of distilled water. Three moles of citric acid monohydrate and ethylene glycol were added to the mixture and stirred for fifteen minutes. To obtain a powder from the $Ca_2Mn_3O_8$ sol-gel solution, the yellowish solution was heated slowly (6 K/min) up to 773 K and held at this temperature for two hours to eliminate water, nitrates and carbonaceous material. The resulting voluminous brown powder was pressed into pellets and heated at 1023 K for two hours in air. To deposit thin films, $F : SnO_2$/quartz substrates were submerged at slow speed (1 mm s^{-1}) into a Ca_2MnO_4 sol-gel solution. The films obtained were then heated at a rate of 6 K min^{-1} up to 773 K and held at this temperature for two hours. A subsequent heat treatment at 1073 K for one hour was performed to assure the formation of the crystalline phase. Screen-printed films were prepared from a $Ca_2Mn_3O_8$ paste, which consisted of 1 g of $Ca_2Mn_3O_8$ powder, 1.5 g of polyethylene glycol 20 000, 1.6 mL ethanol and 0.016 mL acetic acid. The paste was homogenized in a silicon carbide ball mill at 400 rpm for five hours. After screen-printing with a 90 T mesh, the film was dried for five minutes at room temperature before heating at 393 K for two minutes and annealing at 723 K for thirty minutes.[13]

(iii) Hocking et al. described a further method to prepare a potassium birnessite.[16] After impregnating a synthetic tetranuclear-manganese cluster ($[Mn_4O_4L_6]^+$ (L = diaryl-phosphinate)) into a Nafion matrix, formation of a potassium birnessite nanoparticles in the membrane was observed by coalescence of the metal organic precursor after electrochemical treatment and catalytic cycling.

7.2.2.2 Binary Oxides

Films of high electrocatalytic activity towards OER were found using the following material deposition protocols.

(i) Films were prepared by galvanostatic deposition onto $F : SnO_2$/glass substrates at a current density of $0.25 mA/cm^2$, as reported by Wu et al.[22] Electrochemical deposition was carried out in a continuously stirred solution of 0.5M $MnSO_4$ and 0.5M Na_2SO_4 (1 : 1 ratio) for 25 minutes. After removal from the solution, the films were rinsed

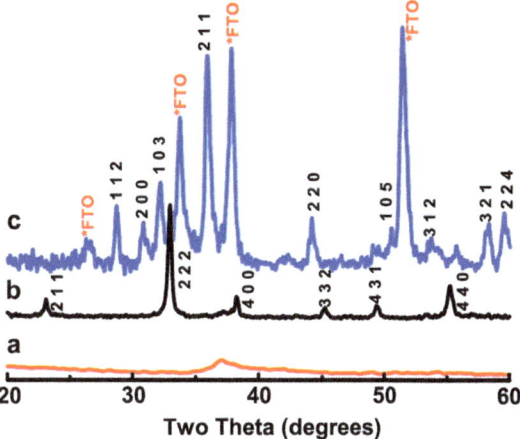

Figure 7.9 X-ray diffraction patterns of galvanostatically deposited films before (pattern (a)) and after heat treatment (patterns (b) and (c)). Pattern (a) is related to the amorphous phase $Mn(O,OH)_x$, pattern (b) to α-Mn_2O_3 and pattern (c) to Mn_3O_4. The as grown layers and those treated at 773 K could be removed from the substrate and measured independently while the layer treated at 873 K was measured on the TCO layer ($F:SnO_2$ on glass).[24]

thoroughly with demineralized water and dried at 373 K in air. The as-deposited material appeared mainly amorphous in X-ray powder diffractograms, which showed only a single broad peak of small intensity identified as $Mn(O,OH)_x$. After heating of the film at 773 K for one hour in air, α-Mn_2O_3 was formed. Further annealing at 873 K for three hours under nitrogen atmosphere resulted in the formation of Mn_3O_4 (Figure 7.9).[24]

Raman spectra of the films showed a strong phonon band in the 648–659 cm^{-1} region and weak phonon bands in the range from 200–500 cm^{-1}. The vibrations found in these spectra were related to the motion of the oxygen atoms within the MnO_6 octahedral units in $Mn(O,OH)_x$, α-Mn_2O_3 and Mn_3O_4.[24]

Cross section TEM pictures of the deposited films showed that $Mn(O,OH)_x$, α-Mn_2O_3 and Mn_3O_4 films had thicknesses of 900 nm, 580 nm and 360 nm, respectively. The mean crystallite size of the layers amounted to 26 nm, 25 nm and 48 nm, respectively. The layers of $Mn(O,OH)_x$ and α-Mn_2O_3 were highly porous, those of Mn_3O_4 rather compact. Scanning electron micrographs shown in Figure 7.10 revealed a surface morphology of bended and intergrown nanosheets in case of $Mn(O,OH)_x$, of intergrown nanoplatelets in case of α-Mn_2O_3 and compact nanograins in layers of Mn_3O_4. A crack in the Mn_3O_4 film is visible in Figure 7.10c. The micrograph with this untypical feature was chosen to demonstrate that the film is also uniform in the cross section.[24]

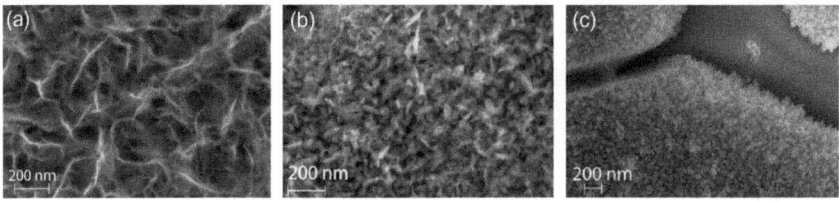

Figure 7.10   (a) $Mn(O,OH)_x$ film after galvanostatic deposition from $MgSO_4$ solution annealed at 373 K in air. (b) $Mn(O,OH)_x$ film annealed at 773 K in air. Formation of α-Mn_2O_3 could be confirmed by XRD (see Figure 7.9). (c) α-Mn_2O_3 film annealed at 873 K under N_2 atmosphere. α-Mn_2O_3 was transformed into Mn_3O_4 by release of oxygen.[24]

(ii) Zaharieva *et al.* described Mn oxide layers electrodeposited on transparent conductive substrates by cycling a voltage 25 times in a range from 2.15 to -0.75 V at a sweep rate of $100\,mVs^{-1}$ (*i.e.* 58 s per cycle) in an aqueous Mn^{2+} solution (0.5mM $Mg(CH_3COO)$). Oxide films deposited by this technique showed high activity in OER while layers deposited at a constant potential were catalytically inactive.[12] Gorlin and Jaramillo used potentiostatic deposition at two potentials (5 min at 1 V followed by deposition of 25 mC at 1.2 V) and a calcination of the films at 480 °C for 10 h.[12,26]

(iii) MnO_x layers were deposited by reactive magnetron sputtering using a manganese target. The temperature of the heated $F:SnO_2$/glass substrates was varied in the range from 300 K to 873 K. Layers of about 50 nm thickness could be deposited within 20 minutes. The layers exhibiting best OER activity crystallized in the α-Mn_2O_3 structure when deposited at 630 °K. Scanning and transmission electron micrographs of these layers as shown in Figure 7.11 show a uniform coverage of the substrate with grain sizes in the film of several ten nanometers.

7.2.3   Electrochemical Behavior and Structure-function Analysis

In this section, the electrochemical behaviour of the materials (films and powders) prepared by different methods will be discussed with respect to oxygen evolution reaction as a function of morphology, crystal structure and surface properties.

7.2.3.1   *Phyllomanganates*

(i) *Powder precipitated from a Mn^{2+}, MnO_4^- solution*
The highest catalytic activity with respect to oxygen evolution was obtained using a slurry of nearly amorphous Ca-birnessite particles and Ce^{IV} as oxidant in an aqueous solution.[10] The presence of a layer oxide structure composed of MnO_6 octahedra was confirmed by X-ray

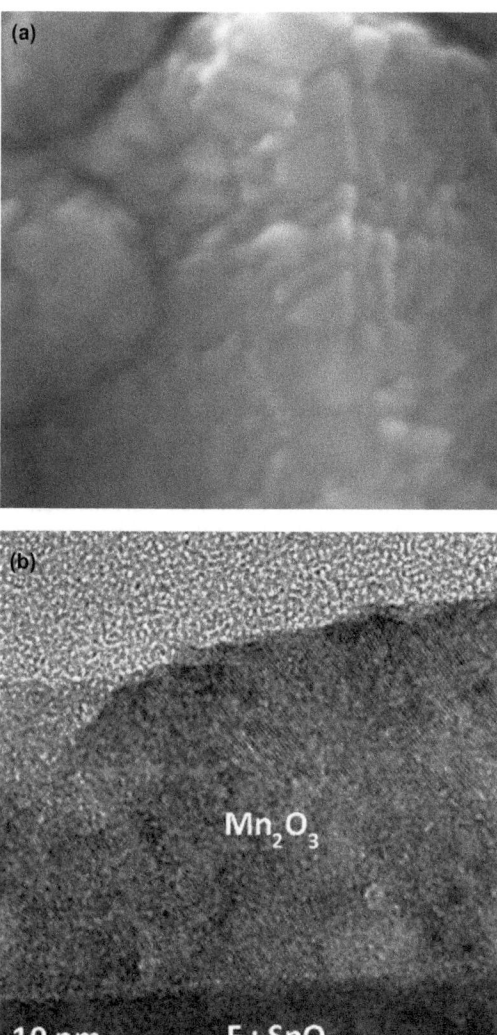

Figure 7.11 (a) Scanning electron micrograph of a α-Mn_2O_3 layer deposited at 630 K. The edge length of the stacked crystallites is ≤ 50 nm. (b) Cross-section transmission electron micrograph of a Mn_2O_3 film deposited on a F:SnO_2 coated glass substrate at 630 K. The thickness of the layer amounts to ~ 45 nm. Crystal lattice plains belonging to individual grains of 10–25 nm size are visible.[29]

absorption spectroscopy. From a careful evaluation of the spectra, mean manganese oxidation states ranging from $+3.5$ to $+3.8$ were found depending on the intercalated metals M (M = K, Mg, Ca, Sr). The structural motif of $CaMn_3(\mu\text{-}O)_4$ cubes was deduced from the analysis of the Fourier-transformed Extended X-ray Absorption Fine Structure (EXAFS) spectra of the materials.

(ii) *Thin films of* $Ca_2Mn_3O_8$ *prepared by the sol-gel method and by screen printing*

Cyclic voltammetry measurements were performed using a scan rate of $50\,mV\,s^{-1}$ at pH 7 in a phosphate buffer. All films showed an onset of anodic current density starting at 1.8 V (RHE) (see Figure 7.14). The current density for the screen-printed $Ca_2Mn_3O_8$ film was six times higher at 2.3 V than for dip-coated films. Micrographs revealed that screen-printed $Ca_2Mn_3O_8$ layers had a pronounced higher surface area than dip-coated films because of their microporosity at a thickness of 6 µm compared two 100 nm film thickness of a sol-gel film. The Tafel slopes of screen-printed and dip-coated $Ca_2Mn_3O_8$ layers were 223 mV/dec and 370 mV/dec, respectively. It is assumed that the kinetics at the electrode/ electrolyte interface were not limited by electron transfer but by previous adsorption processes of the reactants. The differences in adsorption processes for both films could also be due to variations in the surface compositions. An exchange current density j_o of less than $10^{-7}\,A\,cm^{-2}$ was obtained for both films.[13,24]

(iii) *Nafion films with integrated birnessite nanoparticles*

The films were cycled in unbuffered and buffered solutions at pH 7. After electrooxidation at a potential of 1.0 V (versus Ag/AgCl) in aqueous electrolyte (0.1M Na_2SO_4), an oxidation state of the manganese ions in formed oxide nanoparticles between +3.75 and +3.85 was derived from the analysis of XANES spectra. EXAFS spectra allowed the conclusion that a disordered birnessite type phase had been formed. Illumination of the film with visible light for 40 minutes in an electrolyte resulted in dissolution of the nanoparticles, which could be correlated with a reduction of manganese ions from Mn^{III}/Mn^{IV} to Mn^{II}. It could be demonstrated that this process is reversible (see Figure 7.12).

The experiments of Hocking *et al.* demonstrated that, under oxidative conditions of an electrode in contact with an aqueous electrolyte, Mn^{II}-containing oxide precursors are spontaneously transformed in a Mn^{III}/Mn^{IV} layer compound such as birnessite-type. As shown in the next section, oxidation of a Mn_3O_4, hausmannite, thin film electrode leads to an oxidation of manganese from Mn^{II} and Mn^{III} to a mean oxidation state higher than 3 + (Table 7.1). It can be expected that this change in oxidation state is also correlated with a structural change at the electrode/electrolyte interface.[16]

7.2.3.2 Binary Oxides

(i) Cyclic voltammograms of galvanostatically deposited films were recorded using N_2-purged electrolytes at a scan rate of 5 mV s^{-1} at pH 7 (phosphate buffer) and at pH 13, respectively. At potentials > 1.8 V vs. RHE, all films exhibited significant anodic currents which could be either due to an oxidation of the electrodes or due to an electrooxidation of water (OER). In order to distinguish between these two possible

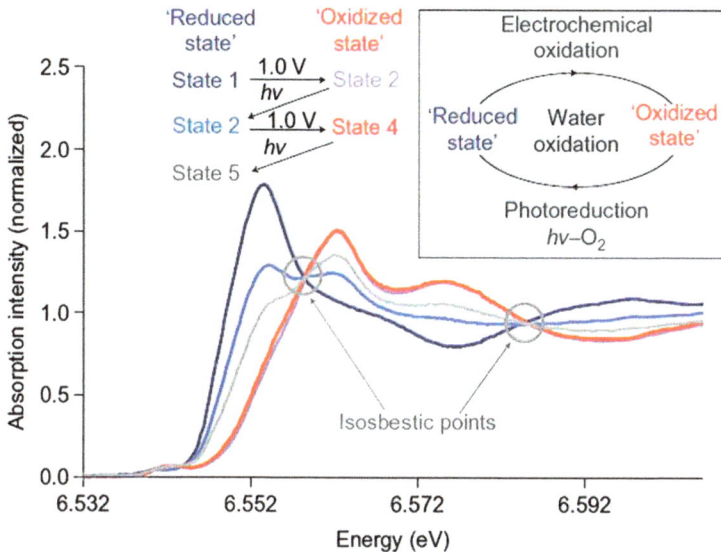

Figure 7.12 Comparison of XANES spectra during various states of catalytic cycling. Mn K-edge XANES of a Nafion-coated $[Mn_4O_4L_6]^+$-loaded glassy carbon electrode measured in different "states" of photochemical cycling. State 1 = initial load; State 2 = State 1 + 1.0 V (versus Ag/AgCl) applied potential in electrolyte; State 3 = State 2 + 40 minutes of light excitation in electrolyte; State 4 = State 3 + 1.0 V applied potential in electrolyte; State 5 = State 4 + 20 minutes of light excitation in electrolyte. Two clean isosbestic points can be observed indicating that repeated cycling between an oxidized birnessite-like state and a reduced Mn(II) state can be achieved.[16]

processes, differential electrochemical mass spectroscopy (DEMS) measurements were performed. Experimental details of this technique are given in references 27 and 28. The inlet system between the electrochemical cell and the differential pumped vacuum system of the mass spectrometer consists of a porous hydrophobic membrane. The working electrode covered by the manganese oxide layer was directly attached to the membrane of the inlet system. The remaining thin layer of electrolyte between the membrane and electrode is sufficient to perform electrochemical experiments (Figure 7.13). Some of the oxygen which is formed at the working electrode surface diffuses into the mass spectrometer where it is detected simultaneous to the electrochemical data (Figure 7.14).

DEMS measurements performed during the first CV sweep at pH 7 revealed that the onset of positive currents of manganese oxide films were not correlated with the detected oxygen evolution. This effect is explained by an oxidation of the electrode surfaces in the beginning and to some extent also by capacitive currents. After several potential scans, the current densities for the water oxidation process of amorphous

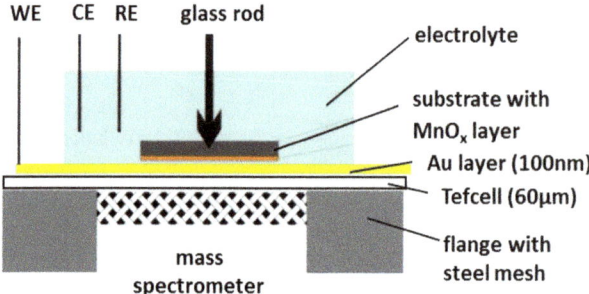

Figure 7.13 Set-up of a differential electrochemical mass spectrometer (DEMS) used to characterize manganese oxides.[27,28]

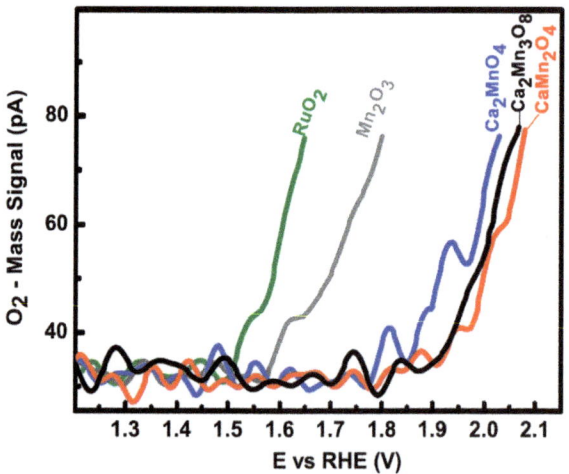

Figure 7.14 DEMS measurements to detect oxygen evolution studying RuO_2, Mn_2O_3 and different calcium manganate electrodes as oxygen evolving catalysts.[13,24]

$Mn(O,OH)_x$ were found to be improved visible by a negative shift of their O_2 evolution onset potentials from 1.8 V and 1.7 V vs. RHE, respectively (compare solid and dotted line in Figure 7.15). In the case of α-Mn_2O_3 and Mn_3O_4, no significant improvement was observed even after several scans.

X-ray Photoelectron Spectroscopy (XPS) spectra of the Mn 2p region were measured to get insight of the manganese oxidation state close to the surface before and after cycling of the electrodes in the voltage range from 1.2 to 2.05 V. The observed binding energies were assigned to Mn^{IV}, Mn^{III} and Mn^{II}, respectively. The Mn $2p_{2/3}$ double peak structure of Mn_3O_4 (curve c in Figure 7.16) is thought to be related to Mn^{III} (641.4 eV) and Mn^{II} (640.8 eV). After OER, the dotted red curve of this

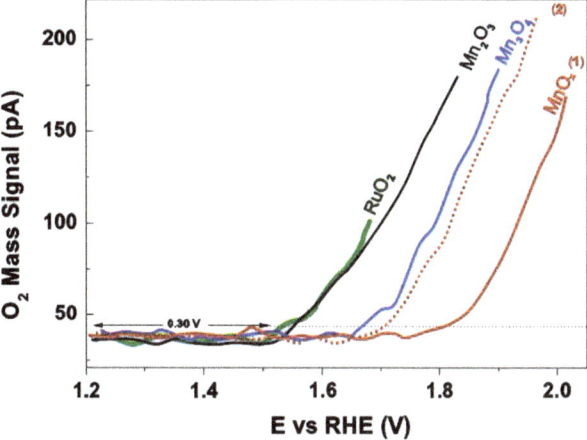

Figure 7.15   Oxygen mass signal as a function of applied voltage investigating a RuO_2 reference electrode (green), a $Mn(O,OH)_x$ electrode (red), a α-Mn_2O_3 electrode (black) and a Mn_3O_4 electrode (blue). All films were deposited on F:SnO_2/glass substrates. The figure shows the DEMS curves after the first scan (solid lines) and after repeated cycling (dotted lines) in phosphate buffer (pH 7).[23]

material becomes similar to the curve of $Mn(O,OH)_x$ (curve a in Figure 7.16), which remains unchanged before and after the electrochemical treatment. The $Mn(O,OH)_x$ spectrum shows two shoulders at 641.4 eV and 643.3 eV and a peak position at 642.4 eV assigned as Mn^{IV}. XPS spectrum of α-Mn_2O_3 exhibits a main peak at 641.4 eV addressed as Mn^{III} and a shoulder at 642.4 eV. After electrochemical oxygen evolution, the XPS spectrum indicates a slight shift towards higher binding energies which is tentatively interpreted as a partial oxidation of Mn^{III} to Mn^{IV} at the surface near region of the electrode. Since the absolute binding energies are difficult to compare with measurements performed by other groups the relative position of the Mn $2p_{1/2}$ satellite structure (ΔE $2p_{1/2}$) are used to characterize the oxidation state of manganese oxide. Matsumoto and Sato[26] as well as Gorlin and Jaramillo[14] reported for MnO_2, Mn_2O_3 and Mn_3O_4 values of 11.9 eV (11.8 eV) and 10.5 eV (10.0 eV) and 11.3 eV (10.0 eV) using powders as references. From the spectra of the as deposited films of own work, shown in Figure 7.16, values of 11.2 eV, 10.0 and 10.2 eV as well as 11.4 eV, 10.1 eV and 11.2 eV after electrochemical treatment were determined.[24]

Cross section TEM pictures of the deposited films showed that $Mn(O,OH)_x$, α-Mn_2O_3 and Mn_3O_4 film had thicknesses of 900 nm, 580 nm and 360 nm, respectively. Mean crystallite sizes of the layers amounted to 26 nm, 25 nm and 48 nm. The layers of $Mn(O,OH)_x$ and α-Mn_2O_3 were highly porous, those of Mn_3O_4 compact.[24]

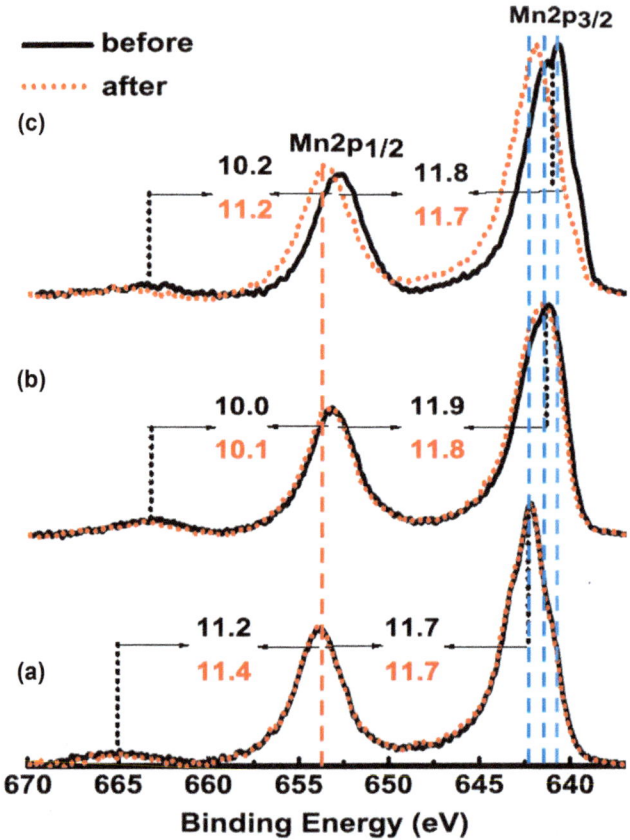

Figure 7.16 XPS- Mn 2p transitions of Mn(O,OH)$_x$ (a), α-Mn$_2$O$_3$ (b) and Mn$_3$O$_4$ (c) films on F:SnO$_2$/glass before (black) and after (red) electrochemical measurements.[23]

Even in initial CV scans, Mn$_2$O$_3$ exhibited highest current densities at 1.8 V vs. RHE compared to Mn$_3$O$_4$ and Mn(O,OH)$_x$. After further cycling, Mn(O,OH)$_x$ electrodes improved their performance drastically while α- Mn$_2$O$_3$ and Mn$_3$O$_4$ remained mainly unaffected (Figure 7.15). These results led us to conclude that although MnO$_x$ and Mn$_2$O$_3$ films possess similar surface areas, particle sizes and porosities, the difference in the onset potential values is connected with the Mn^{3+}/Mn^{4+} ratio at the surface of the films. From Table 7.1 it can be inferred that all three films experience oxidation during the electrochemical process. This effect is most pronounced in the case of Mn$_3$O$_4$, which is obviously corroborated with a degradation of the film. Oxidation of Mn^{III} to Mn^{IV} is clearly visible in Mn(O,OH)$_x$, but smallest in α-Mn$_2$O$_3$. The high catalytic effect of α-Mn$_2$O$_3$ is tentatively attributed to the presence of few Mn^{IV} centres on the surface of the grains, leaving the Mn^{III} ions in the bulk unaffected.

Table 7.1 Mn $\Delta E2p_{1/2}$ splitting evaluated from X-ray photoelectron spectra of powders and films of different manganese oxides (powders and films). The powder values were taken from reference 14 (values in brackets) and 26.

Sample	Mn $\Delta E2p_{1/2}$ Powder/eV	Mn $\Delta E2p_{1/2}$ Film as grown/eV	Mn $\Delta E2p_{1/2}$ Film after EC
MnO_2	11.9 (11.8)		
$Mn(O,OH)_x$	–	11.2	11.4
Mn_2O_3	10.5 (10.0)	10.0 (10.0)	10.1
Mn_3O_4	11.3 (10.0)	10.2	11.2

At pH 13, only Mn_2O_3 and Mn_3O_4 films showed a stable electrochemical behavior. According to the pH-shift of the water oxidation reaction, the onset potentials of these currents and the oxygen signals are shifted by about 380mV to more negative potentials. Compared to the measurements at pH 7, significantly enhanced current densities and oxygen signals in the mass spectrometer were obtained at pH 13. This effect could be explained by a higher reaction rate of OER due to the high concentration of OH^- ions present at pH 13. Furthermore the lower conductivity of the phosphate electrolyte and chemisorption of the phosphate ions on the surface of the film have to be taken into account.[24]

(ii) When comparing layers deposited by cycling in CV and those deposited at constant potential, inactive behavior of the latter was found, but increasing activity towards OER at potentials > 1.66V (RHE) measured in a 0.1 M phosphate buffer in the counterparts. Scanning electron micrographs of the oxides revealed that the electrodeposited material did not form closed films, but was characterized by upright standing platelets not fully covering the substrate surface. The inactive material consisted of bent and intergrown nanosheets of 100 to 200 nm length and few nm thickness. The film morphology is comparable to the samples shown in Figure 7.10a. The active films showed an appearance comparable to Figure 7.10b, but the shape of the individual upright standing nanosheets was more uniform and the individual sheets remained unbent, but were also intergrown (Figure 7.17).[12]

By performing in-line X-ray absorption spectroscopy using synchrotron radiation, it has been proven that the active film when immersed in an electrolyte and after application of a potential of 1.77 V (RHE) exhibits a mixed manganese valence state of +3.8 visible in Near Edge X-ray Absorption Fine Structure (NEXAFS) while the inactive films exhibited a value of +4.0. Figure 7.18 shows the Fourier transform of the Extended X-ray Absorption Fine Structure (EXAFS). The first peak of these spectra represents the nearest Mn-O distance related to an octahedral coordination of the manganese atoms. The second peak represents di-μ-oxo bridging of Mn atoms at 2.86 Å, which is at a

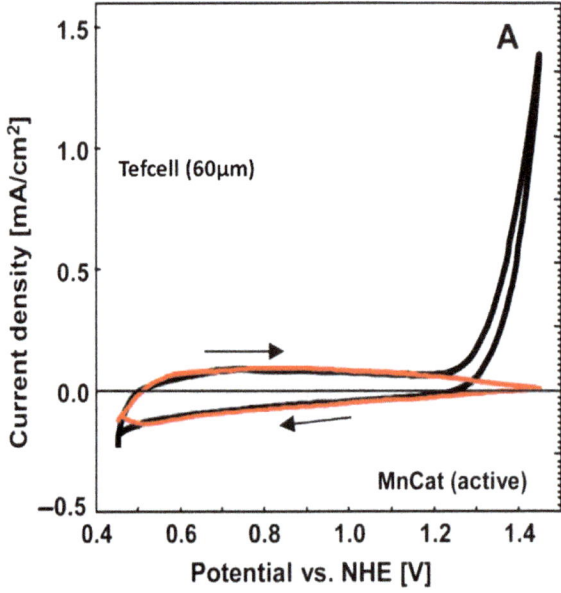

Figure 7.17   Cyclic voltammograms of electro-deposited Mn films, which are active
(A) or inactive (B) in water oxidation (sweep rate $20\,mVs^{-1}$). The
electrolyte was $0.1\,M$ phosphate buffer. At low electrode potentials,
partial dissolution of the oxide film may occur.[12]

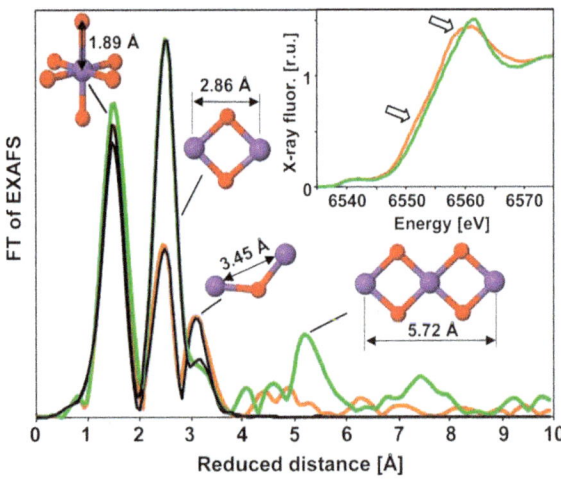

Figure 7.18   X-ray absorption spectra of the MnCat (orange) and the inactive oxide
(green). The edge region of the spectrum (XANES) is shown in the inset;
the arrows mark shoulders in the MnCat spectrum. Each peak in the
Fourier-transformed EXAFS spectra relates to a specific structural motif
that is schematically depicted (O in red, Mn in purple). The spectra
obtained by EXAFS simulations are shown as thin black lines.[12]

coordination number of six in the second shell also typical for birnessite type materials due to six edge-sharing octahedra. While the inactive layer also shows long range order represented by the EXAFS peak at 5.72 Å again fitting to a MnO_x layer compound (see Figure 7.8a), the active material was characterized by an increased intensity of mono-μ-oxo bridges and corner sharing coordination octahedra. Unfortunately the electrodes with active MnO_x nano-sheets began to dissolve after few CV cycles.

(iii) Cyclic voltammograms (CVs) of sputtered MnO_x films deposited at 420 K and 620 K, respectively are shown in Figure 7.19. Layers deposited at temperatures below 573 K are belonging to MnO_2. It is assumed that films deposited in the temperature range from 600 to 650 K crystallize in an α-Mn_2O_3 structure. Figure 7.19 displays CV-diagrams of MnO_2 and α-Mn_2O_3 electrodes at voltages above 1.5 V and 1.6 V, respectively. The related films were produced in two different sputter systems to screen reproducibility of the electrode performances. Their behavior fits well into the trend for OECs shown in Figure 7.15. Figure 7.20 shows the current density as a function of deposition temperatures measured at a potential of 1.785 V/RhE. The red line depicts the behavior during the first cycle of CV where the surface of the electrodes in contact with the electrolyte are established by a forming process, the blue line represents the performance after the

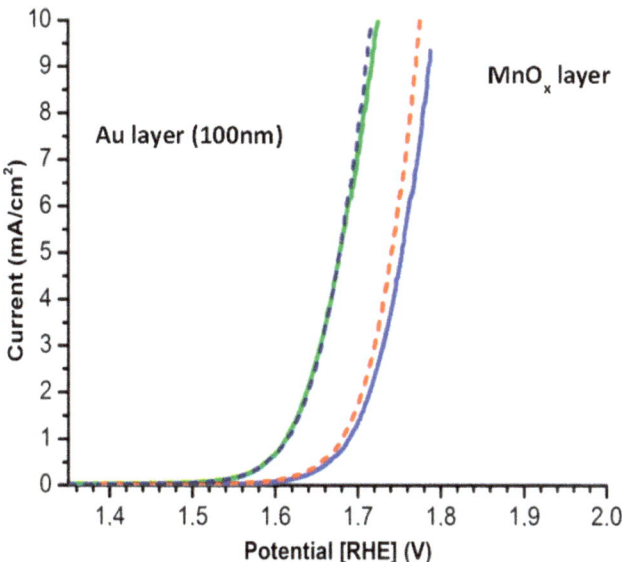

Figure 7.19   CV curves of 50 nm thick films of α-Mn_2O_3 and MnO_2. Dashed and solid lines represent the performance of electrode films deposited in two different sputter systems.[29]

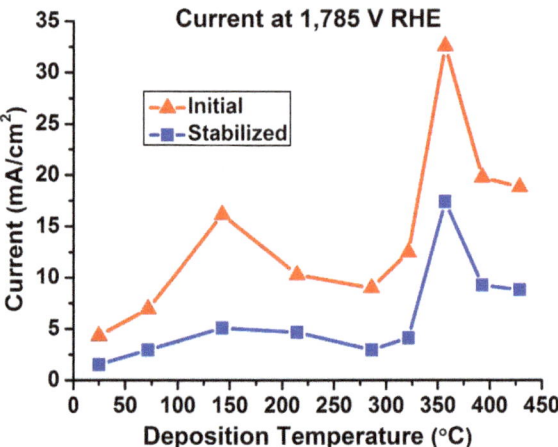

Figure 7.20   Current density of MnO$_x$ phases deposited by reactive sputtering as a function of substrate temperature. The red curve shows the behavior of layers during the 1st CV cycle, the blue one that of the electrodes after the fourth cycle when the CVs show no further change in electrochemical behavior. The peak position belongs to α-Mn$_2$O$_3$ electrode material.[29]

4th cycle when CVs are not subjected by further changes in the waves. In the temperature range belonging to the peak at 357 °C (630 K; Figure 7.20), the α-Mn$_2$O$_3$ phase is stable while at lower temperatures MnO$_2$ and at higher temperatures Mn$_3$O$_4$ are stable. All CVs were performed in 1M KOH (pH 14) with a scan rate of 20 mV/s in deaerated solutions.[29]

7.3   Functionalization of Electrode Surfaces and *In-Operando* studies

As described above, the process of forming an electrocatalyst depends on the structure of the material in the bulk and on the reconstruction and conformation of the surface in contact with an electrolyte as a function of potential, pH value and constitution of the electrolyte. Here the spontaneous formation of a birnessite-type catalyst by coalescence of nanoparticles in a Nafion film might be a convincing example.[16] The metal organic manganese complexes in the film obviously react with each other under formation of a layer compound, intercalating K$^+$ ions from the electrolyte. In addition, it is known from layer compounds that their charge carrier transport is highly anisotropic and therefore dependent on the orientation of the grains in a polycrystalline film with respect to the substrate and the electrolyte. It is also known that electron transport within the layers is typically several orders of magnitude higher than perpendicular to them. Furthermore, defect chemistry and structural disorder play an important role. In TiO$_2$, for example, it is thought that titanium point defects at the surface play an important role in water photolysis. Disorder within the layers of layer-type compounds could enhance and facilitate reactions on the surface.

From Figure 7.20 it is obvious that manganese oxides have stability ranges which are defined by temperature and oxygen activity. The peak position in this figure can be explained by a beneficial defect chemistry in the oxide with respect to OER. In addition, the morphology of the films contributes to its catalytic activity because a porous film of high surface area offers a higher number of catalytically active sites in contact with the electrolyte than a rather bulky layer. Film morphology, nanoparticle orientation, phase formation and electronic properties are accessible by electron microscopy, X-ray diffraction, X-ray absorption spectroscopy and conductivity measurements. Elucidation of surface formation and point defect interaction with water molecules on an atomic scale in order to unravel the reaction steps of water oxidation still remains challenging. Two methods will be presented which can contribute to these open questions.

7.3.1 In-Line Synchrotron Radiation X-ray Photoelectron Spectroscopy

Figure 7.21 shows a picture and the corresponding scheme of an electrochemical cell where electrode surfaces in contact with a droplet of an

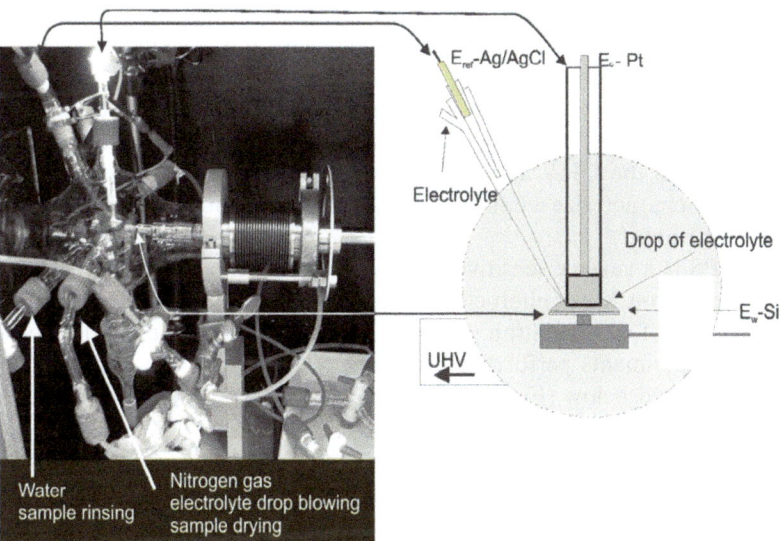

Figure 7.21 Picture and corresponding scheme of an electrochemical cell (three-electrode arrangement) for surface preparation of an electrode in nitrogen atmosphere. The electrode is fixed to a transporting rod which allows the transfer of the electrode from the glass vessel to the UHV chamber for surface characterization.[30,31]

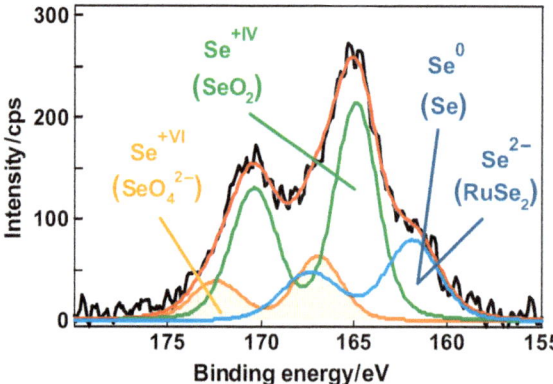

Figure 7.22 X-ray photoelectron spectrum of a $RuSe_{0.3}$ catalyst in the Se3p region after oxidation of the sample. As found by *in situ* NEXAFS (see Figure 7.25) selenium shows the oxidation states -2, $+4$ and $+6$.

electrolyte can be prepared under defined gas atmosphere. The cell is specifically designed for the combined in-system analysis of surfaces using electrochemical and surface analytical (XPS) tools. The figure also shows the glass vessel of this setup connected to an ultrahigh vacuum (UHV) chamber. The capillary with the Ag/AgCl reference electrode works also as an electrolyte inlet system. Using N_2, the pressure created in the capillary forms an electrolyte drop below the Pt counter electrode and the working electrode. The electrical contact between the three electrodes and the drop of electrolyte is then formed by moving the sample holder with the working electrode upwards. After the electrochemical treatment, the sample is rinsed with water and dried in N_2. The sample is then transferred into a buffer chamber for outgassing and, after the pressure reached values in the 10^{-9} mbar range, it is transferred into the UHV analysis chamber.[30,31] This setup was successfully employed to characterize a conductive indium oxide phosphate film on top of an InP electrode.[32,33]

Since XPS is a surface sensitive method it can be expected that in-line synchrotron radiation X-ray photoelectron spectroscopy (SR-XPS) can contribute to clarification of the oxidation states present on the electrode as a function of potential. Experiments performed using this technique on the Ru-Se-S-O system are described below (Figure 7.22).

7.3.2 *In-Situ* X-ray Absorption Spectroscopy

Another technique which should contribute to elucidation of the surface structure of the investigated electrodes after electrochemical treatment is *in-situ* X-ray absorption spectroscopy (XAS). The setup shown in Figure 7.23 can be used to perform *in-situ* X-ray Absorption Near Edge Spectroscopy (XANES) and Extended X-ray Absorption Fine Structure (EXAFS) spectroscopy. XAS measurements can be carried out under *in-situ* conditions using an

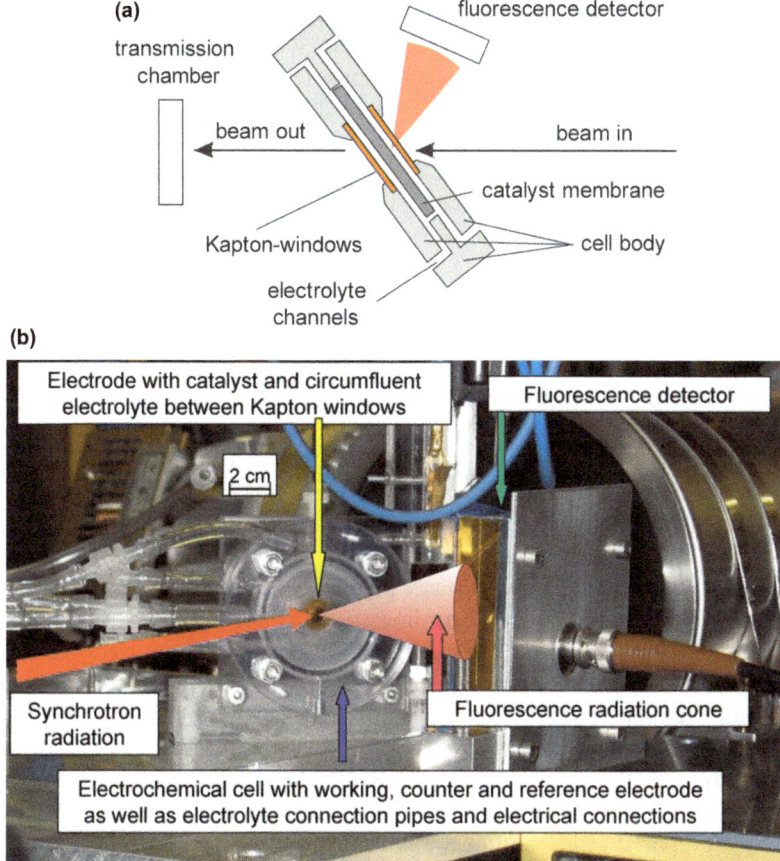

Figure 7.23 (a) Scheme and picture of the electrochemical setup of half-cell for *in-situ* XAS measurements of electro-catalysts in transmission and fluorescence mode. The working electrode consists of carbon paper coated with the catalyst. Reference and counter electrodes are not shown for the sake of simplification. (b) During the measurement, the surface normal of the cell is adjusted with an angle of 45° towards the incoming beam. The cell is flushed with the requested electrolyte. A potentiostatic regime is maintained during the measurement.

electrochemical cell consisting of a plastic body where a graphite paper with a thin catalyst layer on top was fixed between acrylic glass rings (Figure 7.23). Kapton-windows fixed at the edges of these rings and located closely to the coated carbon paper allow XAS measurements in the transmission as well as in the fluorescence mode. Simultaneous electrochemical experiments to study the related electrochemical processes are performed under the flow of a thin electrolyte film driven by a peristaltic pump.

As an example to characterize an electrocatalyst by this technique, the system Ru-S-Se-O is used to demonstrate how a surface modified catalyst can change its properties from an oxygen reduction catalyst to an oxygen evolving material. Carbon-supported Ru_xS_y nanoparticles were prepared by annealing a commercial ruthenium catalyst (carbon-supported Ru catalyst: Ru 40 wt%) in H_2S gas flow at temperatures ranging from 300 K to 1073 K. The Se counterpart, Ru_xSe_y, was prepared by impregnation of a specifically pre-treated commercial carbon black with $RuCl_3$ followed by reduction and subsequent selenization using a reductive annealing technique at elevated temperatures.[33,34] Phase analysis and morphology of the catalysts were carried out using TEM and ex-situ as well as *in-situ* XRD. It could be demonstrated that the carbon-supported Ru_xS_y catalyst exclusively exhibited ruthenium as a crystalline phase up to 673 K. At higher temperatures the pyrite-type phase RuS_2 (laurite) becomes dominant whose particles continuously grow from 9 to 14 nm with increasing annealing temperature in the range from 773 to 1073 K (Figure 7.24).

A totally different behavior was found in the Ru-Se-system. With increasing temperature, an optimized Ru-Se-ratio adjusts itself to a ratio of $Ru : Se = 1 : 0.3$, while in the case of the system Ru-S the amount of sulphur dissolved in the metallic nanoparticles is a function of the annealing temperature (Figure 7.24).[34,35]

Figure 7.24 depicts the continuously probed X-ray diffractogram taken as a function of temperature. Analyzing line scans at 300 °C, 700 °C, 900 °C and 925 °C, it was found that coalescence of the carbon supported Ru nanoparticles occurred with increasing temperature, leading to a decrease of full width half maxima (FWHM) of the Ru peaks. Furthermore, new diffraction lines at $2\theta \approx 53.8°$ and 45.5° appeared at temperatures above 530 °C, pointing to the formation of the pyrite-type phase RuS_2. At temperatures above 900 °C, these lines disappeared again, indicating decomposition of the RuS_2. A $Ru_xS_y/$ C catalyst prepared at RT and heated did not show any RuS_2 diffraction lines, proving that the amount of sulfur attached to the catalyst can be controlled by the annealing temperature.

The structure of the most active $RuSe_x/C$ ORR catalysts (Ru : Se ratio $\approx 1 : 0.3$) at a mean Ru particle size of 2.5 nm cannot be consistently described by an idealized core-shell model assuming a closed Se coverage around the metallic Ru nano-cores. More realistically, as elucidated from Anomalous Small Angle X-ray Scattering (ASAXS), selenium tends to form clusters with sizes of about 0.5 nm firmly bonded to the Ru cores of approx. 2.5 nm in diameter.[35] It was therefore concluded that a significant part of the surface of the Ru cores has to be accessible to reactants participating in the electrochemical process.[34]

This perception has been confirmed by evaluating *in-situ* XAS spectra in the XANES range varying the potential of the electrode (Figure 7.25a). A gradual shift of the absorption edge with increasing potential was observed in the case of the $RuSe_x$ catalyst, which can be correlated with an increasing oxidation of the surface of the ruthenium nanoparticle from Ru^0 and Ru^{2-} to Ru^{4+} and presumably even higher oxidation states (see Figure 7.22). The principal course and shape of the curve change at potentials > 850 mV. Above this value, the surface of the ruthenium nano-particles oxidizes cumulatively, forming a

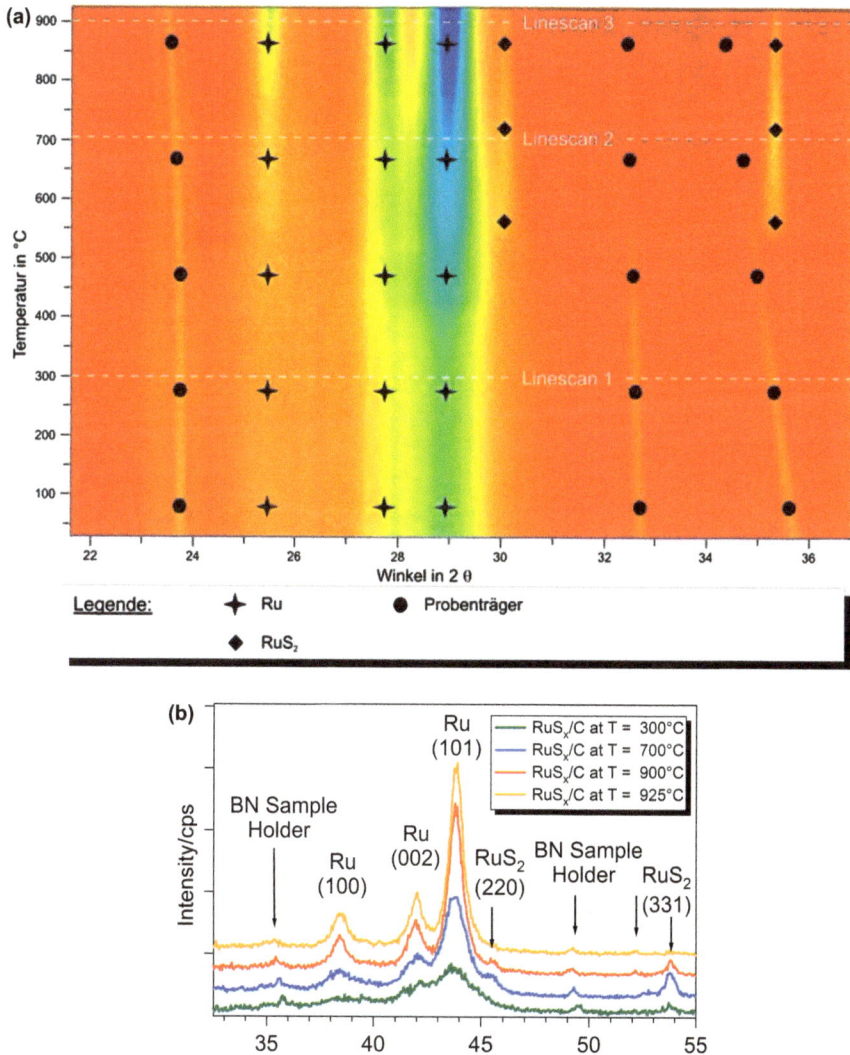

Figure 7.24 (a) 2D representation of diffractograms as a function of temperature heating RuS$_x$/C nano-particles under vacuum from RT to 925 °C. Red color represents low intensity of the diffraction signal, yellow and blue colors represent medium and high intensity, respectively. The blue stripe is caused by coalescence of Ru nano-particles. Additional lines appearing at $2\theta = 29.9°$ and $35.2°$ above 550 °C indicate the formation of RuS$_2$. (b) X-ray diffractograms of a RuS$_x$/C catalysts heated at 300 °C, 700 °C, 900 °C and 925 °C. To allow comparison of the *in-situ* measurements with those performed under Cu kα radiation conditions, the 2θ scale was transformed appropriately.[36]

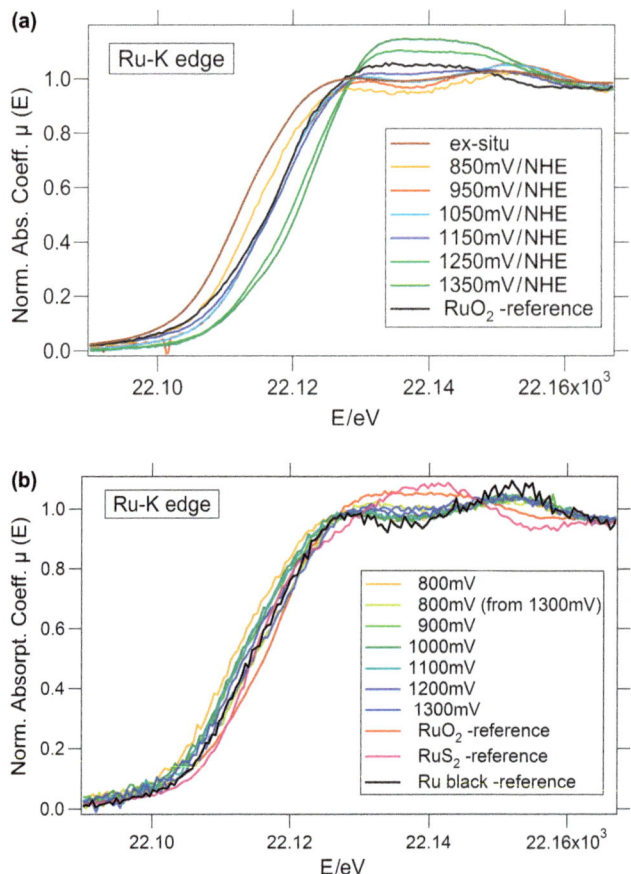

Figure 7.25   (a) *In situ* NEXAFS measurements of a RuSe$_x$ catalyst (prepared at 800 °C) performed at the Ru-K edge in an electrochemical half cell at different potentials. The minimum of the NEXAFS curves at 22.137 eV disappears at voltages higher than 950 mV indicating oxidation of the catalyst particles. (b) *In situ* NEXAFS measurements of a RuS$_y$ catalyst (prepared at 70 °C) performed at the Ru-K edge in an electrochemical half-cell at different potentials. Even at voltages of 1300 mV no oxidation is visible.[36]

ruthenium oxide layer of increasing thickness (Figure 7.25b). Looking at the RuS$_y$ counterpart, an opposite behaviour was found. The particles show neither a shift of the absorption edge with increasing potential of the electrochemical half-cell nor a change of the shape of the curves, even under anodizing conditions. The absorption edges of the curves are located at lower energies than that of the Ru-black reference, but in principle follow the shape of its curve. Obviously the surfaces of the sulfur-modified Ru nano-particles are stabilized against oxidation by an ultrathin layer of sulfur since no RuS$_2$ was formed even heating the sample up to 800 °C. The reference NEXAFS curves of both, RuS_2 and RuO_2, differ essentially from the curves of the catalyst measured at different potentials (Figure 7.25b).[36,37]

To further prove this concept, *in-situ* XANES measurements were performed at the Se-K-edge (Figure 7.26). It can be inferred from the spectra shown in Figure 7.26 that selenium bonded to the surface of the ruthenium particles can be switched between an oxidized (*e.g.* Se^{+IV}) and a reduced (Se^{-II}) oxidation state. Under open circuit conditions, the original material possesses selenium in both oxidation states (black line in Figure 7.26).

The amount of chalcogen on the surface of the Ru nanoparticles is of significant importance for the electrocatalytic activity of the final electrocatalyst. The corresponding catalytic activities for the oxygen reduction reaction were analyzed by cyclic voltammetry (CV) and rotating disk electrode (RDE) measurements studying sulfur modified Ru/C catalyst. The highest activity in ORR is realized as long as the ruthenium nanoparticles are not completely covered by chemisorbed sulfur. With increasing amounts of sulfur (resulting from treatment of the carbon-supported ruthenium nanoparticles in an H_2S gas flow at higher temperatures) the kinetic current related to the ORR drops rapidly. As soon as RuS_2 has been formed, the catalyst becomes active towards H_2O_2 production (Figure 7.27). H_2O_2 formation increases with increasing size of the RuS_2 particles.[36]

RuS_2 is a compound semiconductor with a direct gap of 2.2 eV. Thin films of this sulfide were deposited by reactive sputtering to investigate the material as a photoactive electrode with catalytically-active centres formed in contact with

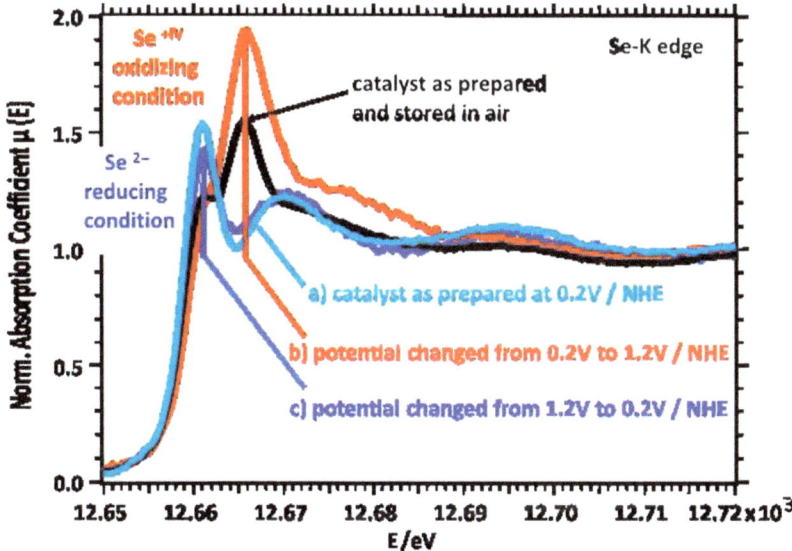

Figure 7.26 XANES spectra of a $RuSe_x$ catalyst measured at different potentials of the half cell. The oxidation state of selenium bonded on the surface of the Ru nanoparticles can be switched between an oxidized and a reduced oxidation state of selenium.[36]

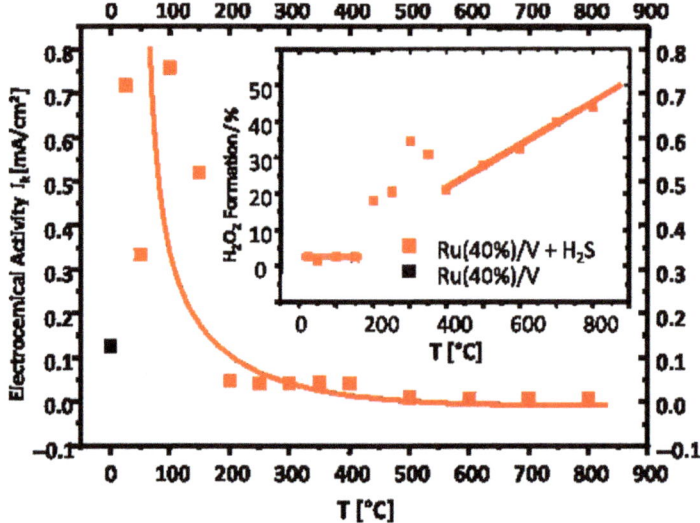

Figure 7.27 Kinetic current densities i_k measured at 0.7 V (NHE) as a function of annealing temperature of RuS_x/C catalysts. The inlet shows the temperature dependence on the formation of H_2O_2 in rotating ring disc electrode (RRDE) measurements at a rotation speed of 400 rpm at 0.4 V (NHE).[36]

the electrolyte.[37] It could be demonstrated that OER activity increases with increasing size of the RuS_2 particles deposited on Ti substrates. However, in contrast to RuS_2 single crystals, no photoeffect could be observed on the sputtered layers under illumination with a tungsten lamp. Time-resolved microwave conductivity analysis indicated the presence of mobile charge carriers after illumination, but apparently these carriers cannot participate in the electrooxidation of water. After an activation step involving cycling the electrode in 0.5 M H_2SO_4 of a deaerated electrolyte, the formation of an oxide/hydroxide layer was confirmed by analysis of the O1s-XPS spectrum, which shows a significant increase of the oxygen peak (531.1 eV) measured at the electrochemical treated surface. The layer thickness can be roughly estimated from the electrical charge passed to be ≤ 0.5 nm.[37]

7.4 Integrated Concepts and Challenges in Catalysts Characterization

Section 7.3 discusses characterization techniques that can contribute to analysis of the changes at the surface of the catalyst in contact with an electrolyte. Since XPS is in principle a surface sensitive method, electrodes consisting of the transition metal chalcogenides mentioned above can be investigated. However to learn more about the coordination and bonding distances in the near-surface region, the investigated catalysts have to be prepared in an appropriate manner. As described in section 7.3.2, it is possible to modify and functionalize a catalyst by chemisorption of an element that enhances the catalytic effect remarkably. In the case of Ru nanoparticles, this was possible by modifying their surface by chalcogen atoms. This concept cannot be transferred directly to transition metal oxides, as discussed in section 7.2. It is necessary to enlarge the surface of the catalytically-active material to perform XAS by impregnation of a conductive support with nanoparticles of the material to be investigated. In case of ruthenium catalyst, carbon black was used as support. However under anodic conditions, this support is chemically unstable and cannot be employed. Recently Liu *et al.*[38] have reported that oxides such as TiO_x and ZrO_x can be used as support for the electrocatalyst to be investigated. The conductivity of the carrier particles can be adjusted by control of their defect chemistry (oxygen point defect or doping concentration). The carrier particles should have a typical size of several 10 nm and can be loaded by the transition metal oxide particles to be investigated by XAS. If the particle size is fixed at 2 nm, about 50% of the atoms are located on the surface. Under these conditions, the sensitivity of *in-situ* XAS measurement can be increased significantly with respect to changes at the surface.

The complex structure of the light-driven water splitting device as shown in Figure 7.1 could in principle be simplified by depositing at the back and front sides of a conductive support highly light-absorbing semiconductors in form of thin films that are not only able to separate electron-hole pairs under illumination in the bulk of the films, but also to catalyse water oxidation and proton reduction at the related electrode/electrolyte interphases. So far, no system is known where photoactivity and catalytic activity are realized in one electrode material. Presently the following approaches for a water splitting device are under development.

(i) A photoelectrochemical setup where a single photoactive electrode is used as photoanode or photocathode, the band gap of which is adapted to the energetic needs of water splitting. Prominent examples are n-type α-Fe_2O_3, n-type $BiVO_4$ (W alloyed) and p-type Cu_2O.[39,40]

(ii) A p-type photoactive electrode combined with a n-type electrode. Illumination of both electrodes needs to achieve a photovoltage that is high enough to split water. Typical potentials needed are 1.6–1.8 V at current densities of 5–10 mA cm^{-2}. Best efficiencies might be obtained using a transparent conductive substrate for one of the electrodes (back side illumination) while the counter electrode is illuminated via the electrolyte.[41–43]

(iii) Using a PV hybrid system where the front or back contacts of a thin film solar cell device are in direct contact with an aqueous electrolyte. In such a geometry, the PV system is decoupled from the electrochemical system (see Figure 7.28). Transition metal oxides such as MnO_x, NiO_x and CoO_x can be employed as efficient dark catalyst in order to lower charge carrier barriers at the electrode/electrolyte interface.

(iv) State of the art triple junction silicon thin film solar cells (3j a-Si/μc-Si solar cells) possess the capability to generate a photovoltage of 1.8V at the maximum power point at a current density of 5 mA cm^{-2}. However a more economic version can be realized by combining a photoactive oxide semiconductor with an easier to fabricate tandem-junction Si thin film solar cell (2j a-Si/μc-Si solar cell). The special geometry displayed in Figure 7.27 uses a 2j Si thin film cell in superstrate configuration. Both sides of a TCO glass substrate are used to deposit on the one side a POS and on the other side a 2j Si solar cells structure. Since the photoactive oxide semiconductor is acting as anode and should have a higher band gap than the photo-cathode α-Fe$_2$O$_3$, n-type BiVO$_4$ and n-type TaON can be used.[39,40,43]

Transition metal oxides are in demand for all of the systems described above. In case of the photoactive oxide semiconductors (POS) shown in Figure 7.28, a

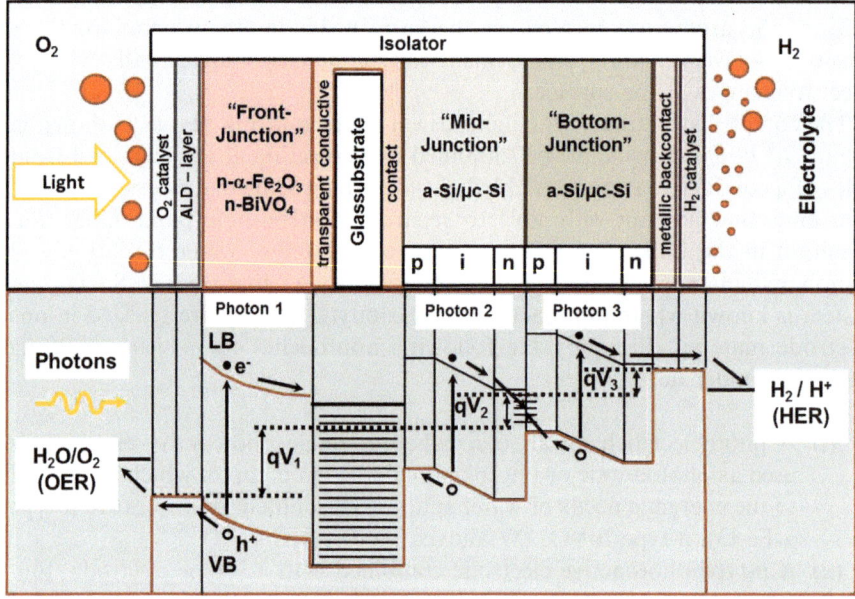

Figure 7.28   Schematic cross section of a water splitting light-driven device combining a tandem Si solar cell and a photoactive oxide semiconductor. In the upper part of the scheme a tandem junction thin film amorphous (a-Si) - microcrystalline (μc-Si) silicon solar cell deposited on the right side of aTCO glass substrate is shown, on the left side a photoactive oxide semiconductor layer is displayed. The water splitting device is illuminated *via* the photoactive oxide semiconductor layer, the surface of which is functionalized by an oxygen-evolving catalyst. While the oxygen evolution reaction is taking place at the anode side, hydrogen evolution happens at the metallic back contact equipped with a related catalyst. The lower part of the figure is displaying the band structure of the device. Starting from the highest in case of the POS and the lowest close to the back contact to absorb a high fraction of incident solar light (adapted from reference 43).

current density of up to $10\,mA\,cm^2$ is required. Presently no POS can be prepared that gives such high current densities. Remarkable success has been achieved during the last decade by doping and defect control of the POSs α-Fe_2O_3 and $BiVO_4$.[39,40] The latter material presently shows the highest current density (up to $4\,mA\,cm^{-2}$) measured at pH 7 after deposition of a CoO_x layer as co-catalyst. The generically small carrier mobilities and charge carrier life times could be influenced by tailoring the defect chemistry of the POS. Further improvement can be expected by designing a convenient nano-architecture of the light absorbing films. In addition, doping gradients and modification of the surface defect chemistry should lead to an improved POS performance.

For all these water splitting devices, transition metal oxide catalyst can be used to lower overvoltages at the anode side. In the case of the system Mn-O, the best results with respect to current density at 1.8 V and the highest electrochemical stability were obtained with a galvanostatically deposited thin film of α-Mn_2O_3 using a TCO-coated glass substrate.

7.5 Summary and Outlook

Coming from the $CaMn_4O_x$ cluster in photosystem II, which acts as oxygen evolving catalyst in photosynthesis, different inorganic materials were considered focusing on manganese oxides and alkali metal and earth alkaline metal manganates that have in common bonding distances similar to those known the $CaMn_4O_x$ complex in PSII. Phyllomanganates are considered as model substances assuming that the catalytically active centres are located on the surface of layers composed of edge-sharing MnO_6 octahedra coordinated to intercalated calcium and water molecules. Among binary manganese oxides, α-Mn_2O_3 is characterized by remarkable properties towards oxygen evolution. The electrochemical stability of thin films of this oxide can be controlled by the preparation conditions. The most stable layers are those deposited galvanostatically and treated in air at 673 K. X-ray absorption techniques using synchrotron radiation were used to distinguish between different manganese oxides. This technique is without any alternative, because also amorphous or highly disordered materials can be characterized. Using the example of ruthenium catalysts functionalized by chalcogen atoms which are efficient catalysts for the oxygen reduction reaction (ORR) as well as the oxygen evolution reaction (OER), it was demonstrated how sensitive X-ray absorption near edge structure (XANES) and extended X-ray absorption fine structure (EXAFS) analyses can be to characterize changes at the surface of nano-sized catalyst particles during electrochemical treatment. Depending on the preparation conditions, the investigated dark catalysts have to be integrated in a light-driven water splitting device. One option is to separate the photovoltaic part from the part where the electrochemical processes occur. As a mid-term solution, triple junction a-Si/μc-Si thin film solar cells in superstrate geometry can be used, where the back contact can be used as cathode to evolve hydrogen and the dark anode for oxygen evolution. An advantage of this configuration is that protons formed at the anode surface have a short distance to migrate to the cathode surface, where they are reduced under hydrogen formation. A further advantage is that the PV part is not shadowed by the catalysts and that incident light cannot be scattered by evolving gas bubbles in this geometry.

The following questions have to be taken into account in the development of an electro-catalytically active material.

- Are the components available, cheap and nontoxic?
- How far can the defect chemistry be used to tailor the conductivity in the bulk?
- Can dopants improve the electrocatalytic properties?
- Can an electrochemical stable surface be prepared by interaction of the electrolyte with the as grown material cycling the electrode in a certain potential range?
- How far is the electrocatalytic performance of the catalyst influenced by the morphology of the individual nanoparticles and their texture in the deposited film?
- Which role plays disorder and amorphous regions in the electrocatalysts?

In common with materials development in general, the design of suitable electrocatalysts for water splitting is a challenge where a matrix of different properties has to be considered.

Acknowledgements

The assistance of I. Zizak during the high temperature XRD measurements at the BESSY synchrotron beam-line KMC-2 is gratefully acknowledged. Furthermore, we thank A. Webb and M. Herrmann for their comprehensive support performing the EXAFS measurements at Hasylab beam-line X1 and U. Bloeck for TEM work. Financial support by the Federal Ministry of Research and Technology BMBF under the contract number 03SF0353A is gratefully acknowledged.

References

1. S. Y. Reece, J. A. Hamel, K. Sung, T. D. Jarvi, A. J. Esswein, J. J. H. Pijpers and D. G. Nocera, *Science*, 2011, **334**, 645.
2. T. Smolinka, S. Rau, C. Hebling in *Hydrogen and Fuel Cells: Fundamentals, Technologies and Applications* ed. D. Stolten, Wiley-VCH, 2010, ISBN: 978-3-527-32711-9, 271–290.
3. B. Neumann, P. Bogdanoff and H. Tributsch, *J. Phys. Chem. C*, 2009, **113**, 20980.
4. H. Dau and M. Haumann, *Coord. Chem. Rev.*, 2008, **252**, 273.
5. Y. Umena, K. Kawakami, J. R. Shen and N. Kamiya, *Nature*, 2011, **473**, 55.
6. H. Dau and I. Zaharieva, *Acc. Chem. Res.*, 2009, **42**, 1861.
7. P. E. M. Siegbahn, *Acc. Chem. Res.*, 2009, **42**, 1871.

8. R. E. Rocheleau, E. L. Miller and A. Misra, *Energy Fuels*, 1998, **12**, 3.
9. O. Khaselev and J. A. Turner, *Science*, 1998, **280**, 425.
10. M. Wiechen, I. Zaharieva, H. Dau and P. Kurz, *Chem. Sci.*, 2012, **3**, 2330.
11. M. M. Najafpour, T. Ehrenberg, M. Wiechen and P. Kurz, *Angew. Chem. Int. Ed.*, 2010, **49**, 2233.
12. I. Zaharieva, P. Chernev, M. Risch, K. Klingan, M. Kohlhoff, A. Fischer and H. Dau, *Energy Environ. Sci.*, 2012, **5**, 7081.
13. A. Ramírez, P. Bogdanoff, D. Friedrich and S. Fiechter, *Nano Energy*, 2012, **1**, 144.
14. Y. Gorlin and T. F. Jaramillo, *J. Am. Chem. Soc.*, 2010, **132**, 13612.
15. P. Ramdohr, H. Strunz, Klockmanns Lehrbuch der Mineralogie, 1975, Ferdinand Enke Verlag Stuttgart.
16. R. K. Hocking, R. Brimblecombe, L.-Y. Chang, A. Singh, M. H. Cheah, C. Glover, W. H. Casey and L. Spiccia, *Nature Chemistry*, 2011, **3**461 and references therein.
17. R. Norrestam, *Acta Chem. Scand.*, 1967, **21**, 2871 and references therein.
18. J. Du, Y. Pan, T. Zhang, X. Han, F. Cheng and J. Chen, *J. Mater. Chem.*, 2012, **22**, 15812.
19. M. Morita, C. Iwakura and H. Tamura, *Electrochim. Acta*, 1979, **22**, 325.
20. D. P. Dubal, D. S. Dhawale, R. R. Salunkhe, V. J. Fulari and C. D. Lokhande, *J. Alloys Compounds*, 2010, **497**, 166.
21. R. N. De Guzman, A. Awaluddin, Y.-F. Shen, Z. R. Tian, S. L. Suib, S. Ching and C.-L. O'Young, *Chem. Mater.*, 1995, **7**, 1286.
22. P. H. Klose, *J. Electrochem. Soc.*, 1970, **117**, 854; V. G. Bhide and R. H. Dani, *Physica*, 1961, **27**, 821; M.-S. Wu, P.-C. Chiang, J.-T. Lee and J.-C. Lin, *J. Phys. Chem.B*, 2005, **109**, 23279.
23. I. D. Fawcett, J. E. Sunstrom, M. Greenblatt, M. Croft and K. V. Ramanujachary, *Chem. Mater.*, 1998, **10**, 3643.
24. A. Ramírez, M. May, D. Stellmach, P. Bogdanoff, S. Fiechter (submitted).
25. B. A. Pinaud, Z. Chen, D. N. Abram and T. F. Jaramillo, *J. Phys. Chem. C*, 2011, **115**, 11830.
26. Y. Matsumoto and E. Sato, *Mater. Chem. Phys.*, 1986, **14**, 397.
27. P. Bogdanoff and N. Alonso-Vante, *J. Electroanal. Chem.*, 1994, **379**, 415.
28. P. Bogdanoff, N. Alonso-Vante and Ber. Bunsenges, *Phys. Chem.*, 1993, **97**, 940.
29. B. de Raadt, A. Kratzig, P. Bogdanoff, K. Ellmer, S. Fiechter (submitted).
30. K. Skorupska, *J. Solid Stat. Electrochem.*, 2009, **13**, 205.
31. K. Jacobi, M. Gruyters, P. Geng, T. Bitzer, M. Aggour, S. Rauscher and H. J. Lewerenz, *Phys. Rev. B*, 1995, **51**, 5437.
32. H. J. Lewerenz, C. Heine, K. Skorupska, N. Szabo, T. Hannappel, T. Vo-Dinh, S. A. Campbell, H. W. Klemm and A. G. Muñoz, *Energy Environ. Sci.*, 2010, **3**, 748.
33. A. G. Muñoz, C. Heine, H. W. Klemm, T. Hannappel, N. Szabo and H. J. Lewerenz, *ECS Trans.*, 2011, **35**, 141.

34. S. Fiechter, I. Dorbandt, P. Bogdanoff, H. Tributsch, M. Bron, G. Zehl, J. Radnik, M. Fieber-Erdmann and H. Schulenburg, *J. Phys. Chem. C*, 2007, **111**, 477.
35. G. Zehl, G. Schmithals, A. Hoell, S. Haas, C. Hartnig, I. Dorbandt, P. Bogdanoff and S. Fiechter, *Angewandte Chemie, Int. Ed.*, 2007, **46**, 7311.
36. S. Fiechter, K. Mette, P. Bogdanoff, G. Zehl, I. Dorbandt, U. Kramm (submitted).
37. P. Bogdanoff, C. Zachäus, S. Brunken, A. Kratzig, K. Ellmer and S. Fiechter, *Phys. Chem. Chem. Phys.*, 2013, **15**, 1452.
38. Y. Liu, A. Ishihara, S. Mitsushima, N. Kamiya and K. Ota, *Electrochem. Solid-State Lett.*, 2005, **8**, A400.
39. K. Sivula, F. Le Formal and M. Grätzel, *ChemSusChem*, 2011, **4**, 432.
40. F. F. Abdi, N. Firet and R. van de Krol, *ChemCatChem*, 2013, **5**, 490.
41. B. Kaiser, W. Jaegermann, S Fiechter and H. J. Lewerenz, *Bunsen-Magazin*, 2011, **13**, 104.
42. S. Fiechter, P. Bogdanoff, T. Bak and J. Nowotny, *Adv. Appl. Ceramics*, 2011, 39.
43. E. L. Miller, B. Marsen, D. Paluselli and R. Rocheleau, *Electrochem. Solid-State Letters*, 2005, **8**, A247.

CHAPTER 8

The Group III-Nitride Material Class: from Preparation to Perspectives in Photoelectrocatalysis

RAMÓN COLLAZO*[a] AND NIKOLAUS DIETZ[b]

[a] Department of Material Science & Engineering, NC State University, Raleigh, NC, 27606, US; [b] Department of Physics & Astronomy, Georgia State University, Atlanta, Georgia, 30303, US
*Email: rcollaz@unity.ncsu.edu

8.1 Introduction

Renewable solar fuel generation either *via* photovoltaic (PV) cells or *via* photo-electrocatalytic solar fuel cells is expected to play a central role in the way energy is produced in the coming century to mitigate environmental problems associated with continued consumption of fossil fuels. Fuels generated by photoelectrocatalytic conversion may be hydrogen (H_2), methanol (CH_3OH), or other alkanes – molecules that can store tremendous amounts of energy per mass unit. These solar fuels can release energy in usable fashion upon reaction with oxygen (O_2), without producing environmentally harmful byproducts. The formation of these envisioned solar fuels can be readily addressed through the development of compound semiconductor-based electrolytic cells that have the advantage of producing the products at separate locations. Group III-nitrides compound semiconductors are ideally suited to drive such endergonic reactions

RSC Energy and Environment Series No. 9
Photoelectrochemical Water Splitting: Materials, Processes and Architectures
Edited by Hans-Joachim Lewerenz and Laurence Peter
Published by the Royal Society of Chemistry, www.rsc.org

with light, acting as the light absorbing units in hybrid photoelectrochemical (PEC) cells. The surfaces of these semiconductors have to be stabilized and modified with molecular or inorganic catalysts in order to catalyze specific desired reactions pathways. Research in this area will require materials of high quality with well-defined tailored properties and reproducibility. In the following sections, we provide a brief review of present knowledge of the properties the group III-nitride compound semiconductors and of efforts to improve their structural and physical materials properties as well as the integration of dissimilar materials for the envisioned device functionalities.

8.1.1 Properties of the Binaries InN, GaN, AlN and their Ternaries

Binary group III-nitrides – *e.g.* AlN, GaN, InN, TlN, and their ternary and quaternary alloys – possess a number of attractive physical, optical, and electronic properties that allow the fabrication of novel materials and device structures, which have been reviewed over the last two decades.[1-11]

As depicted in Figure 8.1, the group III-nitrides can crystallize in a hexagonal wurtzite (wz) or in a cubic zinc blende (zb) crystal structure and span a wide range of band gap energies, depending on composition (cf. Table 8.1). The III-nitrides are partially ionic solids due to large differences in the electronegativity of the group-III metal cations and nitrogen anions,[7] and the most stable crystalline structure of is the hexagonal wurtzite crystal structure. Due to the strong ionic III-N bonds, the unit cell of the III-nitrides is distorted from the ideal wurtzite unit cell, resulting in a large spontaneous polarization along the c-axis. The large differences in the ionic radii and bonding energies of the

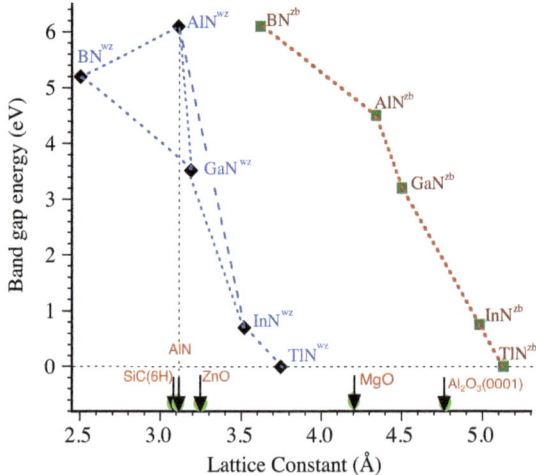

Figure 8.1   Lattice constants and band gap energies for wurtzite (wz) and cubic zinc blende (zb) group III-nitride compound semiconductors, together with utilized substrates used in the growth of epilayers.

Table 8.1 Lattice parameter and band gap values for group III-nitride alloys.

Binary alloy	a (Å)	c (Å)	c/a ratio	u	Literature
wz-InN	3.517	5.685	1.616	1.616	Ref. 13
wz-GaN	3.18940	5.18614	1.62606	0.3789	Ref. 17
wz-AlN	3.1120	4.9808	1.60054	0.3869	Ref. 17

Ternary band gap values: $E_g(A_x C_{1-x} N) = E_g(A) \cdot x + E_g(c)(1-x) - b \cdot X \cdot (1-x)$

Ternary alloy	Bandgap $E_g(A)$ (eV)	Bandgap $E_g(C)$ (eV)	Bowing parameter b	Literature/Remarks
$In_x Ga_{1-x} N$	0.70 ± 0.05	3.52 ± 0.1	1.6 ± 0.2	b = 1.65 ± 0.07 Ref. 18
				b = 1.36 Ref. 19
$Al_x Ga_{1-x} N$	6.10 ± 0.1	3.52 ± 0.1	0.7 ± 0.1	
$In_x Al_{1-x} N$	0.70 ± 0.05	6.10 ± 0.1	3.4 x + 1.2	Ref. 19

group III metal cations give rise to different lattice constants, band gap energies and electron affinities, which challenges the epitaxial deposition of the III-nitrides and its alloys. At present, the lattice mismatch between the different III-nitride materials and commonly used substrates result in high dislocation densities, which impacts the optoelectronic and mechanical properties in the epilayers. Such dislocations also critically influence the surface morphology of heteroepitaxial III-nitride epilayers. Recently, due to advances in III-nitride native substrates, homoepitaxial and pseudomorphic heteroepitaxial films with lower dislocation densities have been achieved. The lower dislocation densities lead to an improved surface morphology, where the single crystalline surfaces are determined by surface energy minimization.

The band gaps of three binaries InN, GaN, and AlN span from $0.7 \pm 0.05 \, eV$[12,6,13] to $3.5 \pm 0.1 \, eV$[7] through $6.1 \pm 0.1 \, eV$[7], where the low band gap value of InN is still undergoing revisions (cf. Table 8.1).[6] Recent assessments of the ternaries InGaN and InAlN alloy formation within a cluster expansion approach[14] substantiate the contribution of compositional instabilities to the wide-spread band gap bowing of the indium-containing ternaries InGaN and InAlN,[15,16] affecting the short-range and long-range ordering of ternary alloys.

One unique lattice-matched system comprises the $Al_{1-x}In_x N - GaN$ (x = 0.174) heterostructures, which provide for a large band-offset as well as a refractive index contrast $\Delta = (n_{2_{GaN}} - n_{2_{AlInN}}) / n_{2_{GaN}}$ of 7% with respect to GaN. This lattice matched $Al_{0.83}In_{0.17}N/GaN$ system is being explored, for instance, for resonant cavity light emitting diode (RCLED) vertical cavity surface emitting laser (VCSEL)[20,21] or field effect transistor (FET)[22] devices. However, the band gap energies and electron affinities of this system are not suited for photoelectrocatalytic applications.

8.1.2 Band Gap Alignments for InN-GaN-AlN-InN Alloys and Heterostructures

Ternary Ill-nitride alloys are promising for photonic absorber structures in photoelectrocatalytic architectures for fuel generation from sunlight. The design and fabrication of photoelectrocatalytic structures based on group III-nitride heterostructures or electrolyte-semiconductor interfaces requires reliable values for the valence band positions and band gap values of the three binaries InN, GaN, AlN, as well as the valence band and band gap value shifts for their ternary alloys. A wide spread of values is reported and summarized in several reviews.[8,10,11] The most recent review, by Moses *et al.*[11] summarizes the present status on the band gap and band alignment values and highlights the uncertainty in the absolute values of the valence band positions, which are affected by interfacial and surface termination effects. The evolution of the band gap values and the band alignments for the binaries InN, GaN, AlN and their ternaries are depicted in Figure 8.2, which shows that the band gap energy values range from 6.1 eV for AlN to approximate 0.7 eV for InN. In the ternary alloy system, the valence and conduction bands can be engineered over a wide energy range, allowing the energetic alignment of the semiconductor bands with the HOMO and LUMO levels of catalysts for efficient charge transfer. The system thus provides the possibility to investigate the photoanode behavior of low-gap compounds. For InGaN alloys with a composition of 25% to 30% indium, the energy gap of 2.2 eV as well as the band edge positions are such that single junction water splitting with light could also be viable.

The ternary band gap values for the InN-GaN-AlN alloys, together with bowing parameters, *b*, are given in Table 8.2. Pelá *et al.*[23] has reviewed the band gap values for the binary and ternary alloys and utilized the local-density

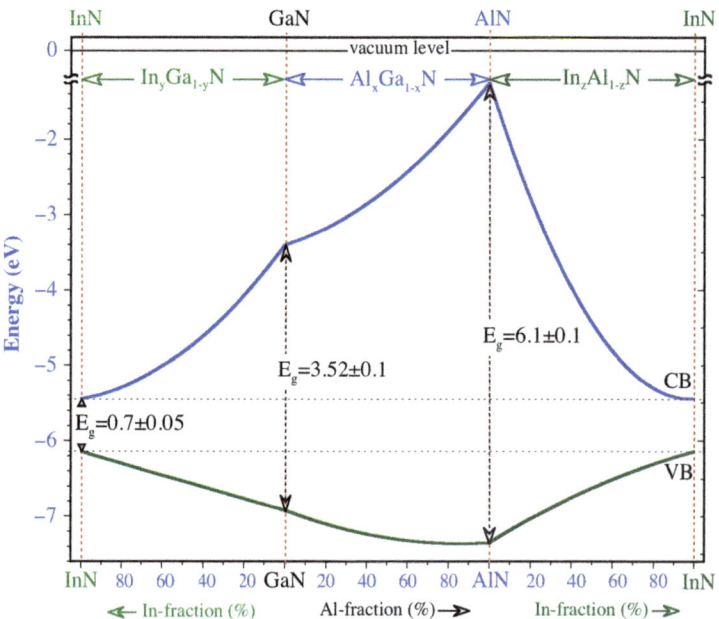

Figure 8.2   Valence band position and offsets for group III-nitrides alloys. Also marked are the hydrogen redox (H_{ad}^{+}/H_2 – located at about 4.6 eV) and oxygen redox (O_2/H_2O – about 1.23 eV lower) potentials for water.

Table 8.2 Valence band position and offsets for group III-nitrides alloys.

Valence band position absolute to vacuum level		
Alloy	$E_{\text{vac}} - E_v(XN)$	*Literature*
c-plane '*wz*-InN' (In-terminated)	6.03 eV	Ref. 13
c-plane '*wz*-InN' (N-terminated)	6.41 eV	Ref. 13
m-plane '*wz*-InN'	6.52 eV	Ref. 13
m-plane '*wz*-InN' relaxed	5.81 eV	Ref. 11
m-plane '*wz*-GaN' relaxed	6.42 eV	Ref. 11
m-plane '*wz*-AlN' relaxed	6.88 eV	Ref. 11

Valence band offsets between binaries: $E_v(AN) - E_v(CN) = E_{\Delta V}(A_X C_{1-X} N) \cdot X - b$
$\cdot X \cdot (1 - x)$

Ternary alloy	*Valence band offset between binaries (eV)*	*Bowing parameter b*	*Literature*
$In_x Ga_{1-x}N$	− 0.71	0 (linear with x)	Ref. 19,11
	− 0.85 ± 0.15	linear with x	exp. Ref. 26
	−0.58 ± 0.08	N/A	exp. Ref. 27
	− 0.78	PE on InN/GaN HJ	exp. Ref. 28
$Al_x Ga_{1-x}N$	+ 0.7xxx	0.6	
	+ 0.43	–	Ref. 11
$In_x Al_{1-x}N$	−1.52 ± 0.17	–	exp. Ref. 25
	−1.32	–	exp. Ref. 29
	−0.96	–	Ref. 11

approximation LDA-1/2 method[24] to establish more accurate band gap parameters, taking experimental data into account. The band gap value for InN was estimated by LDA-1/2 to be 0.95 eV,[23] which is inconsistent with experimental values that converge to a band gap value of approximate 0.70 eV. The valence-band positions relative to the vacuum level, $E_{\text{Vac-v}}$ vary strongly in literature and seem to be highly influenced by surface orientation and surface polarity.[11,10] The valence band offset between binaries InN and AlN was determined by x-ray photoelectron spectroscopy to be 1.52 eV[25] with an associated conduction band offset of 4.0 ± 0.2 eV, forming a type-I heterojunction between the binaries. The valence band offset between binaries InN-GaN positions was estimated to be 0.62 eV with an band gap bowing parameter, b, of 1.36 eV.[19] The relative valence band is assumed to shift linearly with composition x[19]; the observed band gap bowing is therefore assigned to a nonlinear conduction band shift.

8.2 Fabrication of Epitaxial Alloys and Heterostructures

8.2.1 Epitaxial Growth Techniques

Several methods have been developed for the growth of InN, GaN, AlN, and their alloys. The choice of which method is used is based on the desired properties of the crystals. In general, two parameters must be considered: thickness and quality. The major growth methods are discussed briefly here.

(a)   Low-pressure MOCVD

Metal-organic chemical vapor deposition (MOCVD or MOVPE) is a well-established growth method for the growth of AlGaN and InGaN.[30] Using nitrogen and/or hydrogen as carrier gas, metal-organic (MO) precursors are used as the metal sources (for example, TMG or TEG as a gallium precursor), and ammonia (NH_3) is used as nitrogen source. The metal-organic precursor and NH_3 are both transported to the reaction zone above a substrate to form the desired group-III-nitride compound. Typically, MOCVD is used in the mass transport limited regime, in which the growth rate solely depends on the amount of metal-organic supplied.[31] Growth temperatures are in the range from 600 °C for InN up to 1500 °C for AlN (with growth temperatures for GaN lying around 1050 °C). Reactor pressures may range from 100 Pa to close to 100 KPa (ambient pressure). Growth rates as high as 5 μm/hr can be achieved by MOCVD. Major impurities found are oxygen, hydrogen, carbon, and silicon.

(b)   Hydride Vapor Phase Epitaxy (HVPE)

This method uses the reaction between gallium trichloride ($GaCl_3$) gas, indium trichloride ($InCl_3$) gas, aluminum trichloride ($AlCl_3$) gas, and ammonia (NH_3) gas to form GaN,[32–34] InN,[35,36] AlN, and ternary III-nitride alloys. The reaction between the source gases takes place in a 2-zone temperature-controlled furnace, with growth temperatures in the raction zone comparable to those of MOCVD. Similar to MOCVD, the growth with HVPE takes place close to thermal equilibrium. Due to the very high gas flows that can be employed in HVPE, high growth rates of 10–50 μm/hours can be achieved, enabling the growth of thick bulk-like GaN and AlN layers.

(c)   MBE and Plasma-assisted MBE

In contrast to MOCVD and HVPE, molecular beam epitaxy (MBE) is kinetically driven rather than typical thermally based chemical reactions.[37,38] For deposition *via* MBE, the source materials (metallic gallium, aluminum, or indium) are heated in a Knudsen effusion cell under ultra-high vacuum. A nitrogen beam from a plasma source is employed typically as the nitrogen source. Metal and nitrogen both impinge and react on the substrate to form the group-III-nitride compound. MBE is a non-equilibrium growth method with very low deposition rate in the range of some hundreds of nanometers per hour. A major improvement in the growth of III- nitrides by MBE came when plasma-sources where added to growth systems to assist the activation of nitrogen precursors.

(d) Remote-plasma-enhanced or Reactive Rf-sputtering CVD

Due to the low desorption pressure of nitrogen in InN, growth temperatures for this material are comparably low (around 500 °C). At these temperatures, the dissociation of NH_3 to provide nitrogen for the growth of InN is reduced, leading to InN films with low nitrogen concentration. RF-sputtering CVD uses an RF plasma to enhance the dissociation of NH_3 and increase the incorporation of nitrogen.[39,40] Thus, the growth of III-nitrides can be accomplished at much lower growth temperature, while the growth rates are comparable to MOCVD. Defect levels comparable to that of MBE can be achieved.

(e) Superatmospheric MOCVD (also Denoted as High-pressure CVD)

This technique is an extension of MOCVD towards the superatmospheric reactor pressure regime. Promising results were achieved in the growth of InN and In-rich InGaN as this technique enables higher temperatures for these materials.[41–43] However, controlled precursor injection schemes become critical as gas phase reactions become imminent with increasing gas phase densities. An additional challenge is the reduction of the surface boundary layer with increased reactor pressure, which leads to a decreased growth rate for higher pressures. The importance of this approach lies in the reduction of the growth temperature gaps between the III-nitride binaries, which is critical for the stabilization of embedded ternary III-nitride alloys and heterostructures.

8.2.2 Substrate Issues

Different substrates are used for the growth of group-III-nitrides. These include sapphire, SiC and silicon. Sapphire, the most widespread substrate, is used since it is relatively inexpensive and can be bought commercially with diameters up to 6 inches. In contrast, SiC is comparatively expensive but offers more desirable properties such as better heat transport and decreased lattice mismatch.[31] The use of silicon as a substrate is motivated mainly by the possibility to integrate group-III-nitrides into the well-established silicon technology and the ease of availability of large area substrates.[44] However, especially for AlN and InN, growth on silicon is not well understood so far. All these substrates have a thermal and lattice mismatch to AlN, GaN, and InN, leading to strain in the layers. Thus, native substrates, *i.e.* bulk AlN, GaN, and InN are widely desired. Lately, progress in the growth of bulk GaN and AlN has been tremendous, leading to the commercial availability of these substrates.[45] Unfortunately, no bulk InN crystals are available, inspiring the exploration of different concepts such as strain relaxing interlayers.[46] In the future, strongly improved quality is to be expected from layers and devices grown on these bulk crystals, especially for In-rich InGaN and Al-rich AlGaN.[47–50]

8.3 The Al$_x$Ga$_{1-x}$N System: from the Binaries to Ternary Alloys and Heterostructures

The Al$_x$Ga$_{1-x}$N system represents the wider band gap part of the III-nitride family. Certainly, GaN has been the most researched member of this family, allowing for successful device applications and future directions. Recently, interest in UV optoelectronics and high-power electronics has led to focused research on the Al$_x$Ga$_{1-x}$N alloys, as they provide a suitable material system for the realization of these applications. As the band gap range of this system covers the UV part of the spectrum, they are not particularly useful for the development of photovoltaic (PV) cells or photoelectrocatalytic solar fuels. Nevertheless, they present a proper test system for the development of the general concepts for nitrides and in general the universality of point defect control in such wide band gap materials. The following section summarizes the basic properties of the GaN and AlN binaries, focusing on their polarity applications and surface morphology. In addition, it discusses the important thermodynamic implications for describing the full miscibility of the Al$_x$Ga$_{1-x}$N alloy system and its demonstration in term of MOCVD growth of these alloys. Finally, it finishes with a discussion of current understanding of point defect with the ultimate goal of controlling their electronic properties. This is currently one of the most researched areas within this material system.

8.3.1 Properties of Epitaxial GaN-AlN Alloys

AlN and GaN possess a number of interesting physical, optical, and electronic properties that allow the fabrication of novel devices. III-nitrides typically form a hexagonal wurtzite unit cell and are partially ionic solids due to large differences in the electronegative of the III-metal cations and nitrogen anions. Due to the largely ionic behavior of the III-N bonds, the unit cell of the III-nitrides is distorted from the ideal wurtzite unit cell, and a large spontaneous polarization exists along the c-axis. As the III-metal cations have different ionic radii and bonding energies with nitrogen, the III-nitrides have different lattice constants, spontaneous polarizations, band gap energies, and electron affinities. The non-centrosymmetric crystalline structure and the lack of inversion symmetry along the c-axis gives rise to two distinct polarities of the {0001}-surfaces. The cation and anion atoms are arranged in separate layers along the basal planes to form hexagonal close packed planes, and the orientation of the bonds between the cation and anion layers defines the polar orientations: III-polar ($+c$ orientation) if the single bond points from III-metal cation to nitrogen anion along the <0001> direction or N-polarity ($-c$ orientation) if the single bond points from nitrogen anion to III-metal cation (Figure 8.3). Polar orientation should not be confused with surface termination, as each orientation may be terminated with either species. The direction of the spontaneous polarization vector and the type of charge induced at a surface or interface is also determined by the polar orientation (Figure 8.3). As the spontaneous polarization

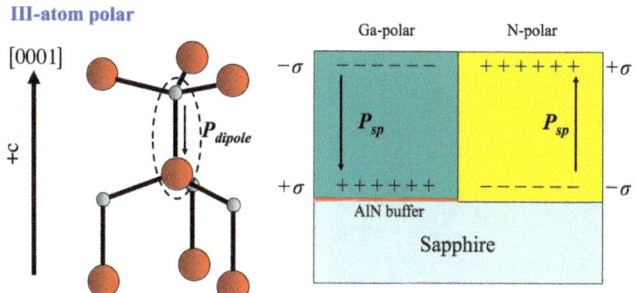

Figure 8.3 The vector direction of spontaneous polarization, Psp, in III-nitride crystals relative to the <0001> direction. The signs of the interface/ surface charge, σ, are also indicated for Ga-polar and N-polar GaN grown using a patterned AlN buffer to control film polarity.

constitutes a non-zero dipole moment per unit volume in the crystal, there exists an internal electric field in the order of a few MV/cm.[51] These internal electrical fields can enhance electron or hole accumulation at surfaces or interfaces, and surface charge densities on the order of $10^{13}\,cm^{-2}$ can be generated. In addition to a spontaneous polarization, large piezoelectric polarization is present in strained crystals such as thin films deposited by heteroepitaxy. The combined effects of spontaneous and piezoelectric polarization in III-nitrides can strongly influence the electrical behavior of a device and makes the control of polarity a necessity for device fabrication.

As most III-nitride growth is preformed on {0001} surfaces, the two polarities present different challenges to crystal growth due to differences in surface configurations, compositions, and chemistries.[2] These dissimilarities lead to different levels of impurity incorporation and the formation of difference microstructures. The principle technique for polarity control during heteroepitaxial deposition is to produce a substrate surface that favors the nucleation of grains with the preferred polarity through the use of polar substrates (Si-polar or C-polar SiC) or combinations of surface treatments for non-polar substrates (sapphire).[52,53] For GaN deposition on c-sapphire wafers, a surface nitridation with ammonia at high temperature yields N-polar GaN while the deposition of a low-temperature AlN buffer layer yields Ga-polar GaN.[53] Using conventional lithography techniques to pattern and selectively etch a low-temperature AlN buffer layer of ~ 30 nm thickness deposited on a sapphire wafer allows Ga-polar and N-polar domains to be grown simultaneously (see Figure 8.4).[53] In the case of homoepitaxial deposition, the polarity of the deposited film is inherited from the substrate.

In addition to requiring polarity control techniques, epitaxial deposition of the III-nitrides presents other challenges. The large lattice mismatch between the different III-nitride materials and other common substrate materials result in dislocation densities of 10^9–$10^{10}\,cm^{-2}$ for heteroepitaxially deposited thin films. Such large dislocation densities can have a substantial impact on electrical and mechanical properties. The surface morphology of

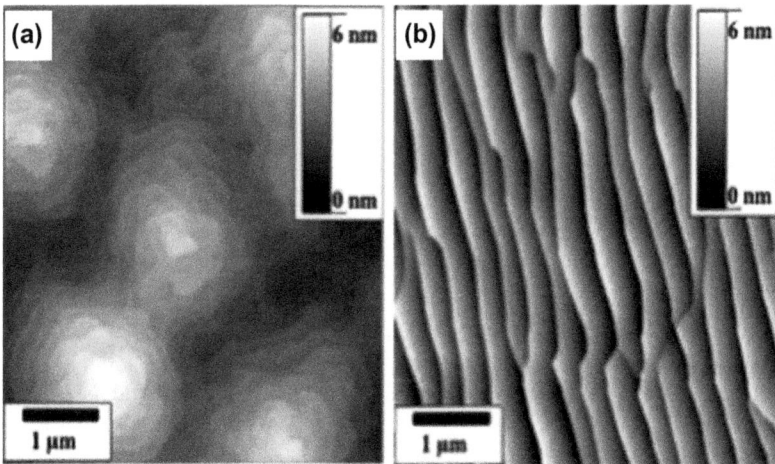

Figure 8.4 5×5 µm² AFM images of (a) Ga-polar GaN film deposited on c-sapphire and (b) Al-polar AlN deposited on bulk AlN with a miscut of about 1 deg. The bi-layer step spacing is approximately 150 nm for both samples. The GaN bi-layer steps are arranged in spirals emanating from screw dislocations. The AlN surface show step-bunches of 3 nm height with a spacing of 350 nm.

heteroepitaxially-grown III-nitride films is also controlled by dislocations, and III-nitride crystals grown in supersaturated environments develop bi-layer step spirals emanating from screw dislocations.[54] Figure 8.4a shows an atomic force microscopy (AFM) image of a GaN thin film deposited by MOCVD on c-sapphire. The spacing between successive bi-layer steps in a growth spiral can be controlled through process supersaturation during deposition and can be varied from ~ 100 nm to ~ 200 nm.[55] In contrast, homoepitaxial and pseudomorphic heteroepitaxial films inherit a much lower dislocation density from their substrate material, typically 10^6 cm^{-2} for GaN and 10^4 cm^{-2} for AlN substrates.[56,57] The surface morphology of films with reduced dislocation densities is more alike to single crystalline surfaces and is determined by surface energy minimization.

Figure 8.4b shows an AFM image of an AlN thin film deposited by MOCVD on bulk (0001)-AlN with a miscut of about 1° where a combination of crystalline faceting and bi-layer steps is evident. This morphology is the result of step-flow growth by individual bi-layers and step-bunching of these bi-layers to form alternating (0001)-AlN facets and high Miller-index or unreconstructed facets to conserve the macroscopic misorientation of the AlN substrate.[58] As with heteroepitaxial films, the spacing of the bi-layer steps of homoepitaxially deposited AlN can be controlled by process supersaturation during deposition and can be varied from 90 nm to 150 nm.[59] The spacing of the step-bunches can be varied from 80 nm to 500 nm by varying the miscut of the AlN substrates. The capability of using AlN single crystalline wafers for homoepitaxial growth of AlN and pseudomorphic Al-rich AlGaN films was developed recently. This capability arises from the development of the technology for growth and wafering of bulk single crystal AlN grown by physical vapor transport.

The III-nitrides show promise for use in photochemical processes due to the large range of electron affinities. It is generally accepted that the electron affinity of GaN is 4.1 eV. Reported values of the electron affinity of AlN and InN are ∼2 eV and 5.8 eV but are topic of ongoing research.[60,61] As discussed in section 8.1, the electron affinities of ternary alloys have a high uncertainty, but they are expected to exhibit electron affinities intermediate to the binaries.

8.3.2 Ternary AlGaN Alloys and Heterostructures

The enthalpy of mixing for III/V (including III-N), IV, and II/VI semi-conductor solid solutions is always below a certain critical temperature.[62] A positive enthalpy of mixing in terms of atomic interactions indicates a "higher bond energy" between two distinct species due to "chemical" energies (related to partial charge transfer due to differences in electronegativity), and "strain" energies related to distortions in the lattice due to differences in the sizes of the constituent species. This can result in a single, homogeneous solid phase having a higher free energy than a mixture with two solid phases.

The regular solution model satisfies the condition of a non-zero enthalpy of mixing. A regular solution, as defined by Hildebrand, is a solution with a non-zero enthalpy of mixing and no excess entropy of mixing, suggesting neither clustering or ordering.[63] This model is used to predict the stability of the different AlGaN solutions within a pseudobinary phase diagram. Following the regular solution model, the molar free energy of mixing for $Al_xGa_{1-x}N$ is given by:

$$\Delta G_M = \Omega x(1 - x) + RT[x\ln x + (1 - x)\ln(1 - x)] \qquad (8.1)$$

where R is the molar gas constant, x is the AlN molar fraction within AlGaN, and Ω is the interaction parameter. Using a modified valence-force-field (VFF) model, Ho *et al.* calculated the interaction parameter to be 3.6 kJ/mol, with the main contribution coming from the microscopic strain energy associated with the bond distortion in the alloy.[64] Figure 8.5 shows how the molar free energy of mixing depends on the molar fraction of AlN within the solid solution of $Al_xGa_{1-x}N$ for three different temperatures characteristic of the processing of these alloys. For these temperatures, all composition ranges within the free energy curve have lower free energy of mixing than the extremes of the ranges. There are no binodal points that would define a binode (where a homogeneous phase will be metastable or unstable). The binodal critical points are defined by the extreme or critical points of the molar free energy of mixing, that is, $d\Delta G_M/dx = 0$, defining the binodal curve. The critical temperature, the temperature at which at any temperature above the solid solution is stable, is $-56\,^\circ$C, well below the processing temperatures of these alloys. A binode region exists within the pseudobinary phase diagram of the AlGaN alloy below this temperature.

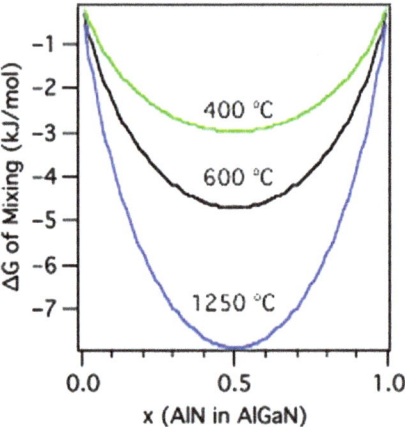

Figure 8.5   Molar free energy of mixing for AlGaN as a function of the molar fraction of AlN within AlGaN for three different temperatures.

All alloy compositions will be stable against phase separation and segregation above this temperature (for temperatures used during processing).

The stability of these alloys is demonstrated by growing layers of different Al composition by metalorganic vapor phase epitaxy (MOVPE). As mentioned above, all alloy compositions are expected to be stable against phase segregation under common growth conditions. The concentration of Al in the alloy is simply given by $x_{Al} = f'_{Al}/(f'_{Ga} + f'_{Al})$, where f'_{Al} and f'_{Ga} are the actual fluxes arriving to the growing surface of the corresponding Al and Ga carrying species. These actual flows can be related to the intended flows as $f'_{Ga} = \alpha f_{TEG}$ and $f'_{Al} = 2\beta f_{TMA}$, where f_{TEG} and f_{TMA} are the corresponding triethylgallium (TEG) and trimethylaluminum (TMA) intended metalorganic flows. In this definition, the factors α and β are introduced as correction factors, and the factor of 2 in the Al flow is due to TMA being a dimer in the gas phase at room temperature. From this, concentration of the Al alloy can be written as:

$$x_{Al} = \frac{2\beta f_{TMA}}{\alpha f_{TEG} + 2\beta f_{TMA}} = \frac{2\frac{\beta}{\alpha} f_{TMA}}{f_{TEG} + 2\frac{\beta}{\alpha} f_{TMA}} \tag{8.2}$$

where a correction factor ratio can be defined as $\gamma = \beta/\alpha$. According to this definition, the aluminum concentration in the alloy should be expressed as:

$$x_{Al} = \frac{2\gamma f_{TMA}}{f_{TEG} + 2\gamma f_{TMA}} \tag{8.3}$$

taking the correction factors into account.

Pre-reactions or parasitic reactions in the gas phase are expected for aluminum and ammonia within the MOVPE process, since it is a spontaneous and strongly exothermic reaction. This depletes the source flow of the aluminum component, increasing the gallium fraction within the alloy.[65] In order for Equation (8.3) to be satisfied if the pre-reaction occurs, the reduction in the aluminum flow or the reaction rate needs to be linearly proportional to the aluminum concentration in the source gas flow. This implies that the pre-reaction needs to be first order with respect to aluminum in its kinetic rate law. Any deviation from the trend described by Equation (8.3) will represent either instability in the AlGaN alloy or a different order kinetic rate law. The two effects cannot be distinguished from just a deviation from Equation (8.3).

The γ correction factor depends on particularities of the process, specifically on: temperature, total pressure, ammonia partial pressure, and gas velocity through the total flow. These parameters influence the pre-reaction rate that depletes the aluminum flow.[66] In addition, these parameters determine the gallium and aluminum supersaturation, which ultimately determines the incorporation rate. Certain conditions, like high temperatures (above 1100 °C) and low ammonia partial pressures may lead to low gallium supersaturation, inducing a deviation from linearity.

Figure 8.6 shows the AlN molar fraction within the AlGaN alloy as a function of the ratio of the molar flow rates as given by Equation (8.3). Only one composition was observed per intended film. The Al compositions graphically represented this way can be fitted with a line of slope 1, if the γ factor is equal to 1.4, throughout the whole composition range, from 0 to 1. As mentioned above, this implies that the alloy compositions are stable against phase separation and segregation. In addition a γ factor greater than 1 implies that no significant amounts of aluminum (if any) are being lost to pre-reactions. This is dependent on the particular process conditions and reactor geometry used to grow the films.

The film relaxation is defined as $R = (a - a_s)/(a_R - a_s)$, where a is the measured in-plane film lattice constant, a_R is the relaxed in-plane film lattice constant, and a_S is the in-plane substrate lattice constant, that in this case is a for GaN, is easily determined. A relaxation of 0 indicates a fully coherent interface between film and substrate, while 100% relaxation implies a fully relaxed film.

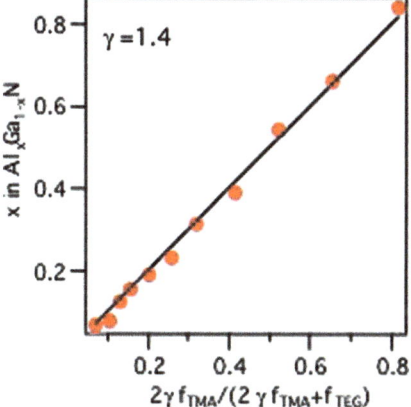

Figure 8.6 AlN molar fraction within the AlGaN alloy as a function of the ratio of the molar flow rates as given by Equation (8.3).

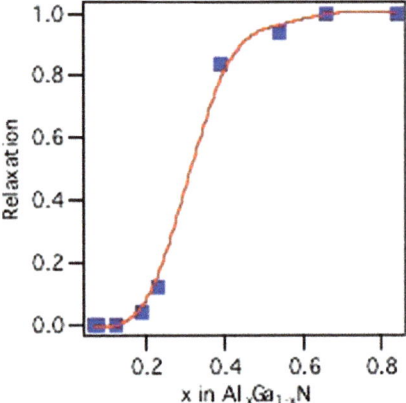

Figure 8.7   Film relaxation as a function of the AlN molar fraction within the AlGaN alloy for a film thickness of 200 nm.

Figure 8.7 shows the film relaxation as a function of the AlN molar fraction within the AlGaN alloy. For compositions below 20%, the relaxation is 0, implying a fully coherent interface, while for a composition around 20% there is a slight degree of relaxation, reaching 12%. Nevertheless, there is an abrupt transition at about 25 to 30% in Al content, where the film achieves nearly full relaxation (for compositions above 30% to below 60%) and full relaxation for compositions above 60%. This abrupt change in the relaxation as a function of composition is indicative of cracking as the relaxation mechanism. AlGaN films on GaN are in tension since the in-plane lattice constant of AlGaN is smaller than that of GaN. Extensive cracking is observed on the higher Al-content AlGaN films.

Figure 8.8 shows a $2\theta - \omega$ triple crystal scan on the direction normal to the substrate for the films represented above. The reflections from the (0002) planes from GaN and AlGaN can be observed along with Pendellösung thickness fringes around the AlGaN peak for AlGaN films with compositions below 25% Al. The occurrence of these fringes has been qualitatively related to the high quality of these films.

Although the previous discussion relates to AlGaN films on GaN templates, recent work describes the progress of AlGaN on AlN bulk crystals substrates. Dalmau *et al.* describe this state-of-the-art system, AlGaN films with compositions above 60% Al remain pseudomorphic for thickness below 1 μm.[47] Partial relaxation is observed, but not *via* the cracking mechanism as these films are expected to be in compression, contrary to those grown on GaN that are expected to be in tension.

8.3.3   Point Defects in AlGaN Alloys: Electrical Properties

Three types of point defects have been identified as determining the conductivity properties of n-type AlGaN: Si, O, and V_{III}. Silicon is commonly used as

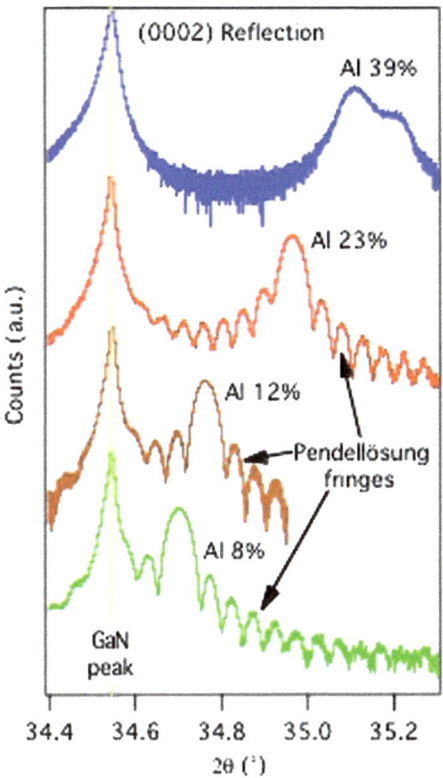

Figure 8.8 $2\theta - \omega$ triple crystal scan of the AlGaN/GaN films grown on sapphire.

the impurity of choice for intentional doping of n-type AlGaN. It is readily incorporated as a substitutional impurity with a charge state (q) of $+1$ in a III-site, acting as a donor, covering the whole alloy range, from GaN to AlN. The intrinsic energy of formation of this impurity has been estimated as -0.8 eV and less than -1.5 eV for GaN and AlN, respectively.[67] Activation energy of 17 meV at a donor concentration of 3×10^{17} cm^{-3} has been observed in GaN. For AlN, values higher than predicted by a hydrogenic model (75 meV) have been reported, but these measurements do not take into account the high degree of compensation observed. After taking into account the high degree of compensation, the activation energy of Si for Al$_{0.7}$Ga$_{0.3}$N was found to be 55 meV, compared with a value of 60 meV predicted by the hydrogenic model.[68] Thus, it is reasonable to assume that the actual value for the activation energy of Si in AlN is around the predicted value (from 75 meV to 95 meV). These activation energies correspond to activation ratios of 0.5 to 0.06 in GaN and AlN, respectively. For the alloy compositions of technological interest, the activation ratio is expected to be around 0.1. At high Si incorporation, it has been suggested that Si may occupy a N-site, self-compensating the free carriers. The intrinsic energy of formation of Si in a N-site (Si$_N$) is much higher than that of the Si$_{III}$ due to the size mismatch with N, making its formation highly unlikely and thus it cannot be considered as an amphoteric dopant.[69] Highly conductive n-type $Al_{0.7}Ga_{0.3}N$ has been achieved by heavy doping with Si (6×10^{19} cm^{-3}) yielding a carrier concentration of $\sim1\times10^{19}$ cm^{-3} and mobilities of ~25 cm^2 V^{-1}s^{-1}.[2] Lower activation energies are observed by band gap renormalization due to the heavy doping.

For the next point defect, oxygen is a common impurity in III-nitrides that is unintentionally incorporated during growth. Similarly to Si, it is readily incorporated as a substitutional impurity with a charge state (q) of +1 in a N-site, acting as a donor. The intrinsic energy of formation of this impurity has been estimated as 0.3 eV and –0.1 eV for GaN and AlN, respectively. The n-type background carrier concentration observed in GaN has been attributed to this impurity. On the other hand, O is expected to undergo a *DX* transition at $x>0.3$ in $Al_xGa_{1-x}N$ that can act as a deep acceptor, effectively compensating intentional donors in the semiconductor.[70,71] Experimental observations suggest that this transition may not occur up to x ~0.6 from observations of unintentionally doped n-type $Al_{0.67}Ga_{0.23}N$ with resistivity of 85 Ωcm at room temperature.[72] This indicates that the intrinsic energy of formation of this defect is slightly higher than expected, thus complete compensation from O in the alloy range of interest is not expected, although some compensation is still present.

III-vacancies are the last important point defects that need to be taken into account. Three charged states (-1, -2, -3) and one neutral state could be stable in the system, along with their corresponding complexes with oxygen.[73–75] In both cases, the three charged states act as ionized acceptors, strongly compensating intentional donors. As discussed above, native compensation as provided by these vacancies and their complexes are the main source of compensation for n-type wide band gap semiconductors. The formation energy of these defects becomes lower as the band gap increases. It is important to note that the intrinsic energy of formation of vacancies or their complexes is lowered near a dislocation due to the presence of the associated strain field: in other words, the equilibrium concentration of these defects increases near a dislocation. Therefore, reduction of dislocations is desirable to further enhance the electrical conductivity in n-type material.

Magnesium is the p-type dopant of choice for AlGaN alloys. It is readily incorporated in a III-site as an acceptor, whereas its incorporation on an interstitial site or a N-site is energetically unfavorable. A high ionization energy (around 150 meV for GaN as presented in previous reports) yields low hole concentrations. In AlN, this energy is expected to be around 400 meV, thus for the AlGaN compositions of technological interest yields an activation ratio of 1×10^{-5}.[76] In this case, high levels of Mg incorporation are needed to achieve technical conductivities in p-type AlGaN.

Similar to the scenario presented above for n-type AlGaN, native compensation with N vacancies as a donor of different ionization states is a fundamental problem in p-type doping. In the case of a N vacancy, the intrinsic energy of formation is around 1.5 eV to 2 eV for GaN and AlN, respectively. These energies are lower than that for a group-III vacancy, making their formation highly favorable. The lowering of the energy of formation as the Fermi level moves to the top of the conduction band should make the existence of p-type material impossible. It has been established that the mechanism responsible for p-type conduction in a standard process is the possibility of co-doping with hydrogen. As hydrogen is present in a typical growth environment, it readily forms a complex with Mg, which is incorporated as a neutral species. Since this does not alter the position of the Fermi level, lowering the energy of formation of the compensating native point defect, its formation is practically inhibited. Further post-processing of the layer is used to remove the hydrogen and activate the Mg as a p-type dopant. This processing is performed at temperatures high enough for possible complex dissolution and diffusion but low enough to kinetically inhibit the otherwise favorable formation of the N vacancies. In other words, we can think of p-type conductivity as a kinetically stabilized property.

Other species have been explored theoretically and experimentally as alternatives to Mg doping, but still Mg remains as the best alternative. Other IIA and B family species suffer from low solubility due to higher size mismatch and higher ionization energies. Be has been explored as a realistic alternative since it has similar ionization energies as Mg and higher solubility due to its reduced size. However, its smaller size facilitates incorporation as interstitial species, where it acts as a donor, thus causing self-compensation.[77] Since no better alternative to Mg seems possible, current research focuses on establishing process routes to inhibit the formation of the native compensating defects without using H as a co-dopant, maximizing in this way the p-type carrier concentration possible from the incorporated acceptor.

8.4 The $In_xGa_{1-x}N$ System

Even though group III-nitride alloys have been investigated for several decades, the breakthrough for the material system came with improvements in GaN thin film growth technology and the successful demonstration of the first InGaN based blue LED[78] and laser diode,[79] which delivered on the promise of a blue solid state light source with the potential for an all spectral tunable LED based on InGaN alloys. The vision became even more prominent with revelation of the low band gap value of InN ($Eg = 0.7\,eV$), which made the InGaN alloys system also of interest for optical communication applications and multi-junction photovoltaic (PV) solar cells. The following section provides a short review of the present state of materials growth and materials structures formation, leading towards emerging materials and device structures.

8.4.1 Gallium-rich Ternary InGaN Alloys and Heterostructures

The binary GaN discussed in section 8.3, is one of the most studied group III-nitride binary. Tuning the optical band gap between the binaries GaN and InN from the UV to the mid IR wavelength region depends on the ability of InGaN lattice to incorporate indium, maintaining a common processing window during the growth process. Therefore, one of the most important objectives during recent years was to understand the influence of nanoscale and microscopic fluctuations in the indium content on the structural, electrical and optoelectronic properties of ternary epilayers and MQW structures.

The incorporation of indium into the InGaN alloys and the growth of InGaN/GaN MQW's are affected by several factors:

(a) The presence of spontaneous and piezoelectric polarization-related electric fields that occur parallel to the c-axis of wurtzite InGaN growth surface. Heterostructures grown along the c-axis may therefore exhibit large internal electric fields that affect the radiative recombination rates as well as the carrier transport of the heterostructure barriers.[80]

(b) The polar nature of wurtzite III-nitride materials leads to two growth surface polarities, which challenges growth process due to differences in growth surface chemistries, impurity incorporation, and surface morphology evolution. The control of the surface polarity is therefore essential for the structural and physical properties of layers.

(c) The large lattice mismatch between the different III-nitride materials and other common substrate materials results in dislocation densities of 10^9–10^{10} cm^{-2} for heteroepitaxial layers. Such large dislocation densities impact on the structural, mechanical and optoelectronic properties of layers.

(d) The vast difference in partial pressures between the binaries leads to significant different growth temperatures for InN, GaN, and AlN, challenging the stabilization of their ternaries and quaternaries under optimum processing conditions.[81] Lateral spinodal decomposition within the growth surface is expected to contribute to the phase separation due to the miscibility gap in InGaN alloys grown under low-pressure MOCVD and MBE growth conditions.[82]

To address issues related to piezoelectric field at interfaces, the recent review by Farrell *et al.*[80] summarized growth efforts on nonpolar and semipolar GaN templates and the effect of the substrate misorientation on growth surface morphology and device performance. Recent advances in understanding and controlling the extending defects in such structures led to improved high-performance LDs on freestanding (20$\bar{2}$1) GaN substrates. However, issues related to compositional induced lattice-strain as well as the stabilization of the different partial pressures between the binaries InN and GaN have to be solved to utilize the full compositional range.

8.4.2 InN and Indium-Rich InGaN Epilayers and Heterostructures

The growth of InN and the integration of higher indium concentrations of InGaN into ternary III-nitrides is a major challenge for presently employed low-pressure deposition techniques such as MBE and low-pressure metal-organic chemical vapor deposition (LP-MOCVD) due to thermodynamic limitations of stabilizing materials with vastly different partial pressures.[81,83] For instance, in order to integrate InN or indium-rich InGaN epilayers into GaN or gallium-rich InGaN heterostructures, a common growth processing window has to be established in which different compositions of InGaN can be stabilized. However, the presently established LP-MOCVD processing conditions revealed a temperature gap of almost 400 °C between InN and GaN, which has to be bridged by any kind of off-equilibrium means of growth surface stabilization. If the growth process itself does not provide sufficient growth surface stabilization, the growth temperature difference between the two binaries provides a significant driving force for structural degradation of the layer quality and the predicted miscibility gap in the ternary InGaN alloys.[84,82] Consequences associated with this problem are discussed in the context of spinodal decomposition, compositional fluctuations and phase segregations in the ternary InGaN[85–92,10] system – an added problem to compositional induced lattice strain, interfacial piezoelectric polarization effects, and extended defect related effects that have to be addressed.

To address the stabilization of InN and indium-rich III-nitrides, off-equilibrium growth techniques such as plasma-assisted MBE[37,38] or remote-plasma-enhanced CVD[39,40] concepts are being explored. A further expansion of the processing space is assessed by superatmospheric or high-pressure CVD.[41–43] High-pressure or superatmospheric MOCVD extends the growth parameters space by enabling reactor pressures of up to 100 bar.[42,93,94] Growing indium-rich InGaN at elevated pressures allows an increase of the growth temperature of an compound with highly-volatile constituents, closing the growth temperature gap between the binaries. Therefore, utilizing the reactor pressure to stabilize the growth surface is a viable approach to overcome problems of off-equilibrium techniques arising from different partial pressures and low growth temperatures. Recent research on HPCVD growth of InGaN has demonstrated that this approach reduces the gap in the growth temperature processing window of group III-nitride alloys.

As depicted in Figure 8.9, the growth temperatures of InN can be increased by more than 120 °C in the pressure regime between 1 and 20 bar, which is a significant advantage over conventional low-pressure MOCVD, where the growth temperatures are around 650 °C. For InGaN, the increase of growth temperature as function of reactor pressure is lowered due to the reduced temperature gap between InN and GaN. Therefore the MOCVD reactor pressure (high pressure *vs.* low pressure) represents a critical balance between the partial pressures of the alloy compositions that have to be integrated/stabilized at a specific growth temperature.

A major drawback of the high-pressure CVD approach lies in the significant reduction in growth rates with increasing reactor pressure as depicted in Figure 8.10. In the investigated pressure range of 20 bar, the growth rate falls almost by one order of magnitude, suggesting that even if this approach is feasible to stabilize and integrate MQW with vastly different partial pressures,

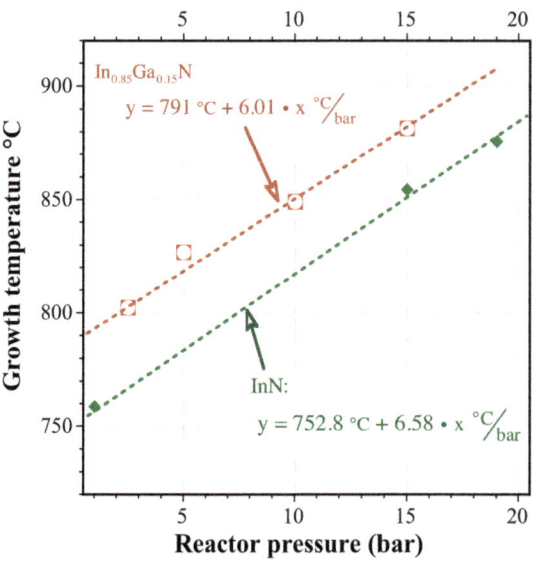

Figure 8.9 Growth temperature versus reactor pressure for the growth of InGaN epilayers under HPCVD growth conditions.

it might not be viable for growth of thick material layers. As demonstrated, the stabilization of indium-rich InGaN can be addressed by high-pressure CVD,[42,94–97] using a pulsed injection of the precursor scheme, which minimizes gas phase reactions and has the significant advantage to prevent phase segregations. This scheme facilitates the control of the growth process on a sub-monolayer level and enables a thorough real-time optical analysis of surface chemistry processes during growth. Current research is exploring the phase stability of InGaN and digital InGaN alloy formation, which may not only prevent phase segregations but also enable the fabrication of III-nitride superlattices.[98–100]

Grandal *et al.*[100] showed that multiple quantum wells (MQWs) of $In_{0.83}Ga_{0.17}N$-InN-$In_{0.83}Ga_{0.17}N$ grown on GaN can be fabricated by plasma-assisted molecular beam epitaxy (PA-MBE) with light emission at 1.5 µm. The PA-MBE approach uses the kinetic energy of the incoming atoms/ions to stabilize the growth surface, using relative low growth temperatures (below 500 °C). Both PA-MBE and HPCVD have the same goal of controlling the partial pressure at the surface to integrate dissimilar materials – but different process control parameters. A critical issue is the type of growth mode: 2-dimensional (2-D) versus 3-dimensional (3-D) film growth. In 2-D growth mode, material is deposited layer-by-layer. Conversely, 3-D growth consists of formation of islands and their subsequent coalescence. The latter results in grain boundaries that detrimentally influence the topographical and electrical properties of the deposited epilayers, *e.g.* carrier mobility and free carrier concentration.[101]

Control of the topographic properties of InGaN layers (*i.e.* a smooth surface) is essential for the fabrication of heterostructures and the integration of

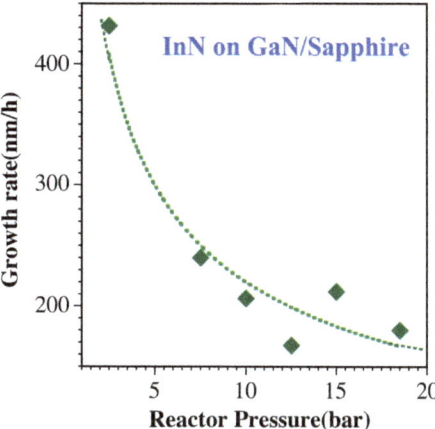

Figure 8.10 Decrease of growth rate with increasing reactor pressure for InN growth.

multiple quantum wells MQWs and engineered assembled nanocomposites. The unique temporal precursor injecting system utilized in the high-pressure CVD approach is promising for controlling the growth surface chemistry processes to stabilize indium-rich InGaN alloys that are difficult to achieve otherwise. However, the alloys may exhibit new defects related to ordering processes on a microscopic scale, which requires careful analysis as they affect the overall quality of the gain media. One of the present challenges in the epitaxial growth of ternary III-nitrides is the potential segregation of one of the constituent column III elements (Ga or In) at the growth front and at the interfaces with the binary material. The ejection of indium from an underlying InGaN layer with Ga deposition thus results in a lower free energy for the surface. The transport of indium to the surface is mediated by the surface exchange of Ga for In. The lower free energy of the GaN layer accounts for the asymmetry in the In diffusion profile with growth order in compositional modulated structures. This preferential segregation has to be carefully controlled and suppressed for the growth of InGaN/InN/InGaN quantum wells (QW) and multiple quantum wells (MQWs). A further aspect arises from the strong polarity of Group III-nitride crystals. A higher concentration of indium in InGaN/InN/InGaN QW/MQWs results in more strain and more polarization.[102] The quantum confined Stark effect (QCSE) is caused by spontaneous polarization and by a strain-induced piezoelectric field. Increasing the indium composition increases the piezoelectric field.[103] The resulting QCSE will cause a blue shift at high current densities moving further away from the desired wavelength, and at lower current densities efficiency will be low due to charge separation.[103–105]

Recent results show that high-pressure MOCVD is a viable method to address the challenges in the growth and stabilization of indium-rich InGaN alloys – even though much more work needs to be done to improve the structural and physical properties of these alloys. At this point, the assessed

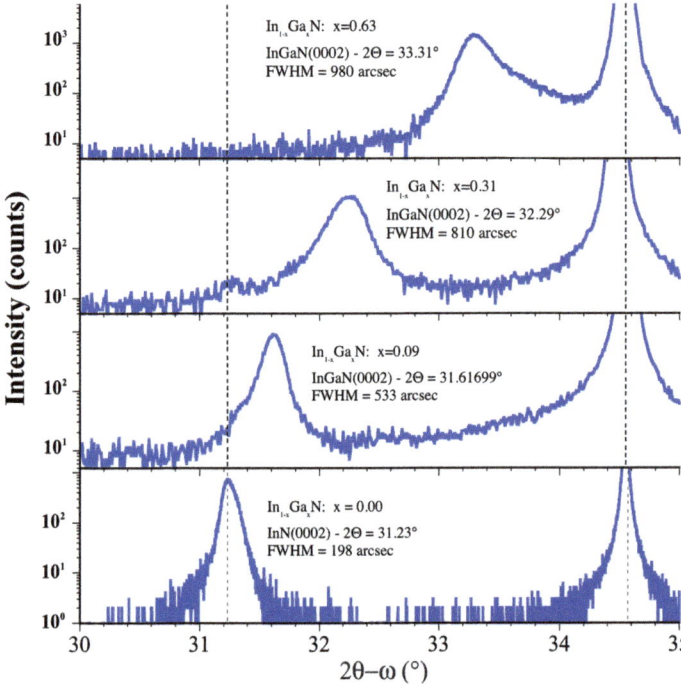

Figure 8.11 Structural quality of InN and indium-rich InGaN alloys grown by HPCVD.

thermal stability window of indium-rich InGaN alloys is focused on growth temperatures in the range of 850–950 °C, reactor pressures from 1 to 18 bar, and simultaneous group-III precursor injection. The XRD analysis for selected indium-rich InGaN summarized in Figure 8.11 shows that a macroscopic single-phase material can be obtained. The broadening of the InGaN(0002) Bragg reflexes (see Figure 8.11) and the rocking curve with FWHM's around 3000 arcsec for $x = 0.31$ indicates a high density of point and extended defects in these layers. The structural degrading with increasing gallium incorporation is presently addressed by exploring a sequential group-III precursor injection approach, adjusting the surface chemistry to each group III element (Ga and In) separately.

8.5   Present Challenges in Materials Improvement and Materials Integration

The development of native substrates, either GaN or AlN, has led to the realization of high crystalline quality AlGaN for the eventual characterization and classification of the main factors that determine the alloy properties. For AlGaN films with Al content below 50%, a GaN native substrate is preferred, while for high Al content, AlN is the substrate of choice. Besides these recent developments, work based on heteroepitaxy on foreign substrates (*i.e.* Si and

SiC) is still ongoing due to other factors, such as availability and other market pressures. Nevertheless, research in these native substrates actually frames the work on the other foreign substrates, as the observed reduction in dislocations bring about what otherwise may be more important limitations in the technology; without the films' high crystalline quality, these limitations cannot be properly distinguished and identified.

As discussed in section 8.3, due to the magnitude of their band gap, the direct applicability of AlGaN to photovoltaic (PV) cells or photoelectrocatalytic solar fuels is limited. On the other hand, integration within the InGaN or AlInN system as part of more complex device structures, such as cladding layers or current injection layers, make them useful materials as a whole for the complete III-nitride family. In light of this, integration with the complete systems becomes rather difficult as typically the three ternary alloy systems are grown on completely different process regimes. Even though the AlGaN alloy system is stable over the whole composition range, the InGaN and specially the AlInN system have miscibility gaps within their phase diagrams limiting the process possibilities and integration capabilities. Several new process alternatives are being explored, some of which where discussed in the previous sections. These alternatives, especially those exploiting kinetic limitations, will open the process space for realizing these novel structures based on the nitride material system.

The main problem that AlGaN technology is currently facing is the difficulty in achieving efficient doping for n- and p-type behavior. This is a fundamental problem with wide band gap materials that arises from two aspects related to the magnitude of the band gap: dopant activation energy, and compensating defects. It is important to realize that the larger the energy gap, the higher the concentration of compensating point defects that is at equilibrium with the system, effectively transforming a semiconductor that was intentionally doped into an insulator. This realization is leading research efforts to develop novel approaches for point defect control that lead to efficient doping, after the achievement of high crystalline quality films. Otherwise, this technology may not realize its intended functionality and applicability.

As outlined in section 8.4, the growth and integration of high-quality ternary InGaN epilayers over a wide range of composition has many challenges to overcome. The successful integration of InGaN alloys with almost 30% indium leads to not only important light emitting device structures and promising photovoltaic solar cell structures, but also emerging new optoelectronics ranging from structures for optical communications in the 1.3–1.5 μm range to structures that cover the whole visible spectral range within one material system. The huge potential of InGaN-based materials structures for envisioned energy saving and energy-generating devices provides sufficient incentives to expand research efforts towards new avenues and approaches to stabilize InGaN alloys over the entire composition range. Various concepts – from superatmospheric growth to off-equilibrium techniques such as RF plasma-assisted MOCVD – have shown that alloys over the entire InGaN can be formed. The improvement of the material quality and integration of highly dissimilar group III-nitride alloys will remain an ongoing research effort.

8.6   InGaN-based Photoelectrochemical Cells for Hydrogen Generation

PEC solar fuel cell structures use solar energy for driving chemical processes, in particular the reduction of water for hydrogen generation. Theoretically a semiconductor with band gap greater than 1.23 eV and conduction and valence band edges that straddle the electrochemical potentials for $H + /H_2$ and H_2O/O_2, respectively, is required, but due to losses and over-potential, 1.6 eV or greater is required experimentally.[106] For thermodynamic stability against photodecomposition, the conduction band edge must be more negative than the reduction decomposition potential (stable against cathodic decomposition) and the valence more positive than oxidation decomposition potential (stable against anodic decomposition) in water.[107] Very few semiconductors meet all these requirements. One of the most studied semiconductors, TiO_2 meets the hydrogen generation requirements and is stable against cathodic decomposition and shows good corrosion resistance, but its large band gap results in very poor efficiency. A dual band gap configuration with two different semiconductors results in higher efficiencies. However, stability and corrosion resistance when immersed remain as issues.

InGaN (up to 50% indium: Moses *et al.*[19]) meets the requirements for hydrogen generation as both photoanode and photocathode, and it has a highly tunable band gap (0.7 to 3.4 eV) that may be optimized to match the solar spectrum for improved efficiency. Figure 8.12 shows a band diagram for of the n-InGaN/electrolyte interface illustrating the alignment of the semiconductor bands with respect to the reduction/oxidation potentials for water. High corrosion resistance and stability is expected due to a small lattice constant and strong bonding, and InGaN has been shown to be resistant against wet etchants and stable in aqueous solutions.[108] Fujii *et al.*[110] have fabricated Si-doped n-$In_xGa_{1-x}N$ (x = 0.02 and x = 0.09) PEC cells and have found increased photocurrents for $In_xGa_{1-x}N$ with an applied bias compared to GaN but lower currents at zero bias (attributed to a reduction in conduction band edge potential). However Li *et al.*[109] have reported PECs based on Si-doped n-$In_xGa_{1-x}N$ epilayers on GaN that gave higher photocurrent with x = 0.4 than x = 0.2 for all biases. Photogenerated holes are considered to be strong oxidizers and could oxidize the semiconductor itself,[110] and Theuwis *et al.*[111] have reported that photo-anodic etching of n-type InGaN occurs under illumination with forward bias. It has been reported that p-type semiconductors have characteristics required for hydrogen production at the surface[110] and exhibit resistance against oxidation due to electron accumulation under illumination.[112] Hence p-type InGaN is a candidate for high efficiency hydrogen generation. Aryal *et al.*[113] have observed hydrogen generation with p-type Mg-doped $In_xGa_{1-x}N$ (0 < x < 0.22) electrodes with additional bias with higher conversion efficiencies compared to p-GaN. The stability of this system in HBr solution was verified. Hwang *et al.*[114] have fabricated $In_{0.1}Ga_{0.9}N$ nanowires on Si wires for improved photocatalytic activity through increased surface area and improved charge separation.

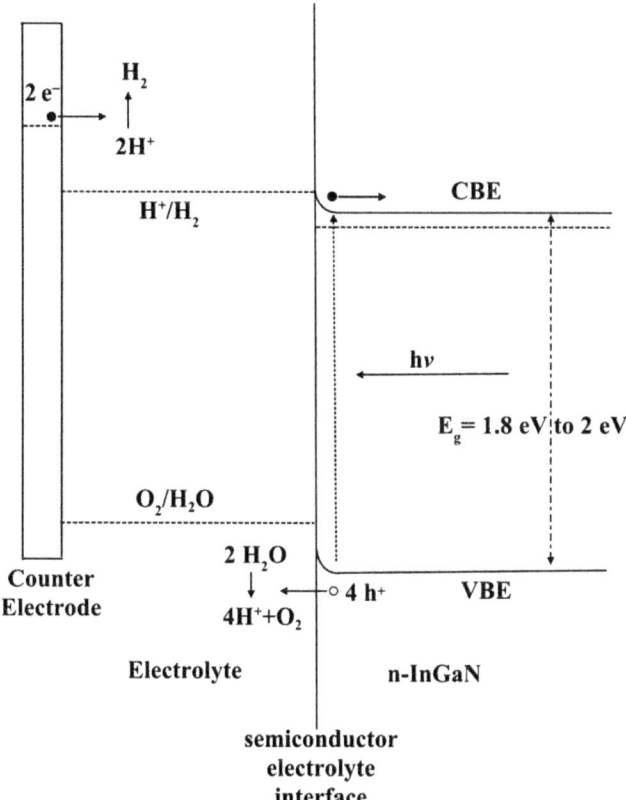

Figure 8.12 Band diagram for n-InGaN/electrolyte interface. The In-fraction is expected to be less than 50%.

8.7 Conclusions

We provided in this contribution a review of the physical properties of group III-nitrides – *e.g.* AlN, GaN, InN, and their ternary and quaternary alloys and discussed the present state and challenges in the materials growth, materials integration, and the physical materials properties improvements. As shown, group III-nitride compounds have a significant potential as semiconducting photoelectrode materials in PEC solar fuel cell structures. The band alignments in the ternary InGaN and InAlN alloy systems can be engineered over a wide energy range, providing for suitable energetic alignments of the semiconductor bands with the HOMO and LUMO levels of catalysts and for efficient charge transfer. Recent advances in group III-nitride materials stabilization and materials quality are sufficiently encouraging to proceed with their integration in PEC cells and the evaluation of such device structures. Further advances are needed in the stabilization of ternary, indium-rich InGaN and InAlN alloys and their integration into dissimilar group III-nitride alloys in order to take advantage of the full potential of the group III-nitride system.

Acknowledgement

ND acknowledges the support by AFOSR under award # FA9550-10-1-0097.

References

1. S. Strite and H. Morkoc, *J. Vac. Sci. Technol. B*, 1992, **10**, 1237–1266.
2. O. Ambacher, *J. Phys. D: Appl. Phys.*, 1998, **31**, 2653–2710.
3. C. Wood, *Naval Research Reviews*, 1999, **51**(1), 4–76.
4. I. Vurgaftman, J. R. Meyer and L. R. Ram-Mohan, *J. Appl. Phys.*, 2001, **89**, 5815–5875.
5. I. Vurgaftman and J. R. Meyer, *J. Appl. Phys.*, 2003, **94**, 3675–3696.
6. V. Yu., Davydov and A. A. Klochikhin, *Semiconductors*, 2004, **38**(8), 861–898.
7. H. Morkoc, *Handbook of Nitride Semiconductors and Devices*, Vol. 3, ISBN-13:978-3527408399, 2008.
8. F. K. Yam and Z. Hassan, *Superlattices Microstruct*, 2008, **43**(1), 1–23.
9. S. Chattopadhyay, A. Ganguly, K. Chen. and L. Chen, *Crit. Rev. Sol. State Mat. Sci*, 2009, **34**, 224.
10. J. Wu, *J. Appl. Phys.*, 2009, **106**(1), 011101–28.
11. P. G. Moses, M. Miao, Q. Yan and C. G. Van de Walle, *J. Chem. Phys.*, 2011, **134**, 084703.
12. K. M. Yu, Z. Liliental-Weber, W. Walukiewicz, W. Shan, J. W. Ager III, S. X. Li, R. E. Jones, E. E. Haller, Hai Lu and William J. Schaff, *Appl. Phys. Lett.*, **86**(7), 071910–2.
13. A. Belabbes, J. Furthmüller and F. Bechstedt, *Phys. Rev. B*, 2011, **84**(20), 205304.
14. L. C. de Carvalho, A. Schleife, J. Furthmüller and F. Bechstedt, *Phys. Rev. B*, 2012, **85**, 115121.
15. M. Łopuszyński and J. A. Majewski, *Phys. Rev. B*, 2012, **85**(3), 035211–25.
16. I. Gorczyca, S. P. Łepkowski, T. Suski, N. E. Christensen and A. Svane, *Phys. Rev. B*, 2009, **80**, 075202.
17. W. Paszkowicz, S. Podsiadło and R. Minikayev, *J. Alloys Comp.*, 2004, **382**, 100–106.
18. E. Sakalauskas, Ö. Tuna, A. Kraus, H. Bremers, U. Rossow, C. Giesen, M. Heuken, A. Hangleiter, G. Gobsch and R. Goldhahn, *Phys. Stat. Solidi (b)*, 2012, **249**, 485–488.
19. P. G. Moses and C. G. Van de Walle, *Appl. Phys. Lett.*, 2010, **96**, 021908–3.
20. G. Cosendey, J.-F. Carlin, N. A. K. Kaufmann, R. Butte and N. Grandjean, *Appl. Phys. Lett.*, 2011, **98**, 181111–3.
21. C. Kruse, H. Dartsch, T. Aschenbrenner, S. Figge and D. Hommel, *Phys. Stat. Solidi (b)*, 2011, **248**, 1748–1755.
22. J. Xie, X. Ni, M. Wu, J. H. Leach, U. Ozgur and H. Morkoc, *Appl. Phys. Lett.*, 2007, **91**, 132116.
23. R. R. Pela, C. Caetano, M. Marques, L. G. Ferreira, J. Furthmuller and L. K. Teles, *Appl. Phys. Lett.*, 2011, **98**, 151907–151903.

24. L. G. Ferreira, M. Marques and L. K. Teles, *Phys. Rev. B*, 2008, **78**, 125116.
25. P. D. C. King, T. D. Veal, P. H. Jefferson, C. F. McConville, T. Wang, P. J. Parbrook, H. Lu and W. J. Schaff, *Appl. Phys. Let.*, 2007, **90**, 132105–3.
26. T. Makimoto, K. Kumarkura, T. Nishida and N. Kobayashi, *J. Electron. Mater.*, 2002, **31**(4), 313–315.
27. P. D. C. King, T. D. Veal, C. E. Kendrick, L. R. Bailey, S. M. Durbin and C. F. McConville, *Phys. Rev. B*, 2008, **78**, 033308.
28. C.-L. Wu, H.-M. Lee, C.-T. Kuo, C.-H. Chen and S. Gwo, *Appl. Phys. Lett.*, 2008, **92**, 162106.
29. M. Akazawa, B. Gao, T. Hashizume, M. Hiroki, S. Yamahata and N. Shigekawa, *J. Appl. Phys.*, 2011, **109**(1), 013703–8.
30. W. G. Breiland, M. E. Coltrin, J. R. Creighton, H. Q. Hou, H. K. Moffat and J. Y. Tsao, *Mat. Sci. Eng. R*, 1999, **24**, 241–274.
31. R. F. Davis, S. M. Bishop, S. Mita, R. Collazo, Z. J. Reitmeier and Z. Sitar, *AIP Conf. Proc.*, 2007, **916**, 520–540.
32. K. Motoki, T. Okahisa, S. Nakahata, N. Matsumoto, H. Kimura, H. Kasai, K. Takemoto, K. Uematsu, M. Ueno, Y. Kumagai, A. K. Tu and H. Seki, *J. Cryst. Growth*, 2002, **237–239**, 912–921.
33. P. Kempisty, B. Łucznik, B. Pastuszka, I. Grzegory, M. Boćkowski, S. Krukowski and S. Porowski, *J. Cryst. Growth*, 2006, **296**, 31–42.
34. K. Hanaoka, H. Murakami, Y. Kumagai and A. Koukitu, *J. Cryst. Growth*, 2011, **318**, 441–445.
35. Y. Kumagai, K. Takemoto, A. Koukitu and H. Seki, *J. Crys. Growth*, 2001, **222**, 118–124.
36. Y. Kumagai, J. Kikuchi, Y. Nishizawa, H. Murakami and A. Koukitu, *J. Cryst. Growth*, 2007, **300**, 57.
37. H. Lu, W. J. Schaff, J. Hwang, H. Wu, W. Yeo, A. Pharkya and L. F. Eastman, *Appl. Phys. Let*, 2000 , 2548–2550.
38. M. Higashiwaki and T. Matsui, *J. Cryst. Growth*, 2003, **251**, 494–498.
39. K. S. A. Butcher, A. J. Fernandes, P. P.-T. Chen, M. Wintrebert-Fouquet, H. Timmers, S. K. Shrestha, H. Hirshy, R. M. Perks and B. F. Usher, *J. Appl. Phys.*, 2007, **101**, 123702–13.
40. S. Valdueza-Felip, J. Ibáñez, E. Monroy, M. González-Herráez, L. Artús and F. B. Naranjo, *Thin Solid Films*, 2012, **520**, 2805–2809.
41. K. J. Bachmann, H. T. Banks, C. Höpfner, G. M. Kepler, S. LeSure, S. D. McCall and J. S. Scroggs, *Mathematical and Computer Modelling*, 1999, **29**(8), 65–80.
42. N. Dietz, Indium-nitride growth by HPCVD: Real-time and ex-situ characterization, in *III-Nitrides Semiconductor Materials*, ed. Z. C. Feng, 2006, Imperial College Press, pp. 203–235.
43. M. Buegler, S. Gamage, R. Atalay, J. Wang, M. K. I. Senevirathna, R. Kirste, T. Xu, M. Jamil, I. Ferguson, J. Tweedie, R. Collazo, A. Hoffmann, Z. Sitar and N. Dietz, *Phys. Stat. Sol.*, 2011, **8**, 2059.

44. A. Dadgar, P. Veit, F. Schulze, J. Bläsing, A. Krtschil, H. Witte, A. Diez, T. Hempel, J. Christen, R. Clos and A. Krost, *Thin Solid Films*, 2007, **515**, 4356–4361.

45. D. Ehrentraut and Z. Sitar, *MRS Bulletin*, 2009, **34**, 259–265.

46. Y. Kobayashi, K. Kumakura, T. Akasaka and T. Makimoto, *Nature*, 2012, **484**, 223–227.

47. R. Dalmau, B. Moody, R. Schlesser, S. Mita, J. Xie, M. Feneberg, B. Neuschl, K. Thonke, R. Collazo, A. Rice, J. Tweedie and Z. Sitar, *J. Electrochem. Soc.*, 2011, **158**, H530.

48. J. R. Grandusky, S. R. Gibb, M. C. Mendrick, C. Moe, M. Wraback and L. J. Schowalter, *Appl. Phys. Express*, 2011, **4**, 082101–3.

49. M. Kneissl, T. Kolbe, C. Chua, V. Kueller, N. Lobo, J. Stellmach, A. Knauer, H. Rodriguez, S. Einfeldt, Z. Yang, N. M. Johnson and M. Weyers, *Sem. Sci. Technol.*, 2011, **26**, 014036.

50. T. Wunderer, C. L. Chua, Z. Yang, J. E. Northrup, N. M. Johnson, G. A. Garrett, H. Hen and M. Wraback, *Appl. Phys. Express*, 2011, **4**, 092101–3.

51. F. Bernardini, V. Fiorentini and D. Vanderbilt, *Phys. Rev. B*, 1997, **56**, 10024–3.

52. C. J. Sun, P. Kung, A. Saxler, H. Ohsato, E. Bigan, M. Razeghi and D. K. Gaskill, *J. Appl. Phys.*, 1994, **76**, 236.

53. S. Mita, R. Collazo and Z. Sitar, *J. Cryst. Growth*, 2009, **311**, 3044–3048.

54. W. K. Burton, N. Cabrera and F. C. Frank, *Philos. Tr. R. Soc. S-A.*, 1951, **243**, 299–358.

55. S. Mita, R. Collazo, A. Rice, R. F. Dalmau and Z. Sitar, *J. Appl. Phys.*, 2008, **104**, 013521.

56. T. Paskova and K. R. Evans, *IEEE J. Sel. Top. Quant*, 2009, **15**(4), 1041–1052.

57. P. Lu, R. Collazo, R. F. Dalmau, G. Durkaya, N. Dietz, B. Raghothamachar, M. Dudley and Z. Sitar, *J. Cryst. Growth*, 2009, **312**, 58–63.

58. C. Herring, *Phys. Rev.*, 1951, **82**, 87–93.

59. A. Rice, R. Collazo, J. Tweedie, R. Dalmau, S. Mita, J. Xie and Z. Sitar, *J. Appl. Phys.*, 2010, **108**, 43510.

60. C. I. Wu, A. Kahn, E. S. Hellman and D. N. E. Buchanan, *Appl. Phys. Lett.*, 1998, **73**, 1346.

61. J. W. Ager, N. Miller, R. E. Jones, K. M. Yu, J. Wu, W. J. Schaff and W. Walukiewicz, *Phys. Stat. Sol. B*, 2008, **245**, 873–877.

62. G. B. Stringfellow, *Organometallic Vapor-Phase Epitaxy: Theory and Practice*, 1998, Academic Press.

63. J. H. Hildebrand, *J. Am. Chem. Soc.*, 1929, **51**, 66–80.

64. I. H. Ho and G. B. Stringfellow, *Appl. Phys. Lett.*, 1996, **69**, 2701.

65. Y. A. Xi, K. X. Chen, F. W. Mont, J. K. Kim, W. Lee, E. F. Schubert, W. Liu, X. Li and J. A. Smart, *Appl. Phys. Lett.*, 2007, **90**, 051104.

66. A. Rice, R. Collazo, J. Tweedie, J. Xie, S. Mita and Z. Sitar, *J. Crysy. Growth*, 2010, **312**, 1321.

67. C. G. Van de Walle and J. Neugebauer, *J. Appl. Phys.*, 2004, **95**, 3851–3879.
68. K. Zhu, M. L. Nakarmi, K. H. Kim, J. Y. Lin and H. X. Jiang, *Appl. Phys. Let.*, 2004, **85**, 4669.
69. C. Stampfl and C. G. Van de Walle, *Phys. Rev. B*, 2002, **65**, 155212–10.
70. C. G. Van de Walle, *Phys. Rev. B*, 1998, **57**, R2033–R2036.
71. R. Zeisel, M. W. Bayerl, S. T. B. Goennenwein, R. Dimitrov, O. Ambacher, M. S. Brandt and M. Stutzmann, *Phys. Rev. B*, 2000, **61**, R16283–R16286.
72. M. L. Nakarmi, N. Nepal, J. Y. Lin and H. X. Jiang, *Appl. Phys. Lett.*, 2005, **86**, 261902.
73. J. Neugebauer and C. G. Vandewalle, *Phys. Rev. B*, 1994, **50**, 8067–3.
74. N. Nepal, M. L. Nakarmi, J. Y. Lin and H. X. Jiang, *Appl. Phys. Lett.*, 2006, **89**, 092107–3.
75. J. Slotte, F. Tuomisto, K. Saarinen, C. G. Moe, S. Keller and S. P. DenBaars, *Appl. Phys. Lett.*, 2007, **90**, 151908–3.
76. T. Tanaka, A. Watanabe, H. Amano, Y. Kobayashi, I. Akasaki, S. Yamazaki and M. Koike, *Appl. Phys. Let.*, 1994, **65**, 593–594.
77. C. G. Van de Walle, C. Stampfl, J. Neugebauer, M. D. McCluskey and N. M. Johnson, *Doping of AlGaN alloys*, Mrs Internet JNSR, 1999, **4**, pp. G10.14.
78. S. Nakamura, T. Mukai and M. Senoh, *Appl. Phys. Lett.*, 1994, **64**, 1687–3.
79. S. Nakamura, M. Senoh, S. Nagahama, N. Iwasa, T. amada, T. Matsushita, H. Kiyoku and Y. S. ugimoto, *Jpn. J. Appl. Phys.*, 1996, **35**, L74–L76.
80. R. M. Farrell, E. C. Young, F. Wu, S. P. DenBaars and J. S. Speck, *Semicond. Sci. Technol.*, 2012, **27**, 024001–15.
81. J. MacChesney, P. M. Bridenbaugh and P. B. O'Connor, *Mater. Res. Bull.*, 1970, **5**, 783–792.
82. G. B. Stringfellow, *J. Cryst. Growth*, 2010, **312**(6), 735–749.
83. V. Vasil'ev and J. Gachon, *Inorganic Materials*, 2006, **42**(11), 1176–1187.
84. Md. Tanvir Hasan, Ashraful G. Bhuiyan and Akio Yamamoto, *Solid-State Elect.*, 2008, **52**, 134.
85. T. P. Bartel and C. Kisielowski, *Ultramicroscopy*, 2008, **108**, 1420.
86. M. G. Ganchenkova, V. A. Borodin, K. Laaksonen and R. Nieminen, *Phys. Rev. B*, 2008, **77**, 075207.
87. J. Zheng and J. Kang, *Mat. Sci. in Semicond. Proc*, 2006, **9**, 341.
88. M.-K. Chen, Y.-C. Cheng, J.-Y. Chen, C.-M. Wu, C. C. Yang, K.-J. Ma, J.-R. Yang and A. Rosenauer, *J. Cryst. Growth*, 2005, **279**(1–2), 55–64.
89. S. Yu. Karpov, *MRS Internet J. Nitride Semicond. Res*, 1998, **3**, 16.
90. N. A. El-Masry, E. L. Piner, S. X. Liu and S. M. Bedair, *Appl. Phys. Let.*, 1998, **72**, 40–42.
91. I. Ho and G. B. Stringfellow, *Appl. Phys. Lett.*, 1996, **69**, 2701–03.
92. G. Stringfellow, *J. Elect. Mat*, 1982, **11**(5), 903–918.
93. M. Alevli, G. Durkaya, A. Weerasekara, A. G. U. Perera, N. Dietz, W. Fenwick, V. Woods and I. Ferguson, *Appl. Phys. Lett.*, 2006, **89**, 112119.

94. N. Dietz, M. Alevli, R. Atalay, G. Durkaya, R. Collazo, J. Tweedie, S. Mita and Z. Sitar, *Appl. Phys. Lett.*, 2008, **92**(4), 041911–3.

95. G. Durkaya, M. Alevli, M. Buegler, R. Atalay, S. Gamage, M. Kaiser, R. Kirste, A. Hoffmann, M. Jamil, I. Ferguson and N. Dietz, Growth temperature - phase stability relation in In1-xGaxN epilayers grown by high-pressure CVD, *Mater. Res. Soc. Symp. Proc.*, 2001, **1202**, I5.21, 1–6.

96. R. Kirste, M. R. Wagner, J. H. Schulze, A. Strittmatter, R. Collazo, Z. Sitar, M. Alevli, N. Dietz and A. Hoffmann, *Phys. Stat. Solidi A*, 2010, **207**, 2351–2354.

97. M. Buegler, S. Gamage, R. Atalay, J. Wang, I. Senevirathna, R. Kirste, T. Xu, M. Jamil, I. Ferguson, J. Tweedie, R. Collazo, A. Hoffmann, Z. Sitar and N. Dietz, *Proc. of SPIE*, 2010.

98. D. J. As, *Journal Microelectronics*, 2009, **40**(2), 204–209.

99. S. Choi, H. J. Kim, J.-H. Ryou and R. D. Dupuis, *J. Cryst. Growth*, 2009, **311**(12), 3252–3256.

100. J. Grandal, J. Pereiro, A. Bengoechea-Encabo, S. Fernandez-Garrido, M. A. Sanchez-Garcia, E. Munoz, E. Calleja, E. Luna and A. Trampert, *Appl. Phys. Lett.*, 2011, **98**, 061901.

101. V. Lebedev, V. Cimalla, F. M. Morales, J. G. Lozano, D. González, Ch. Mauder and O. Ambacher, *J. Cryst. Growth*, 2007, **300**, 50–56.

102. A. Yoshikawa, S. Che, Y. Ishitani and X. Wang, *J. Cryst. Growth*, 2009, **311**(7), 2073–2079.

103. T. Takeuchi, S. Sota, M. Katsuragawa, M. Komori, H. Takeuchi, H. Amano and I. Akasaki, *Jpn. J. Appl. Phys.*, 1997, **36**, L382–L385.

104. S. Chichibu, T. Azuhata, T. Sota and S. Nakamura, *Appl. Phys. Lett.*, 1996, **69**(27), 4188–4190.

105. S. F. Chichibu, A. Uedono, T. Onuma, B. A. Haskell, A. Chakraborty, T. Koyama, P. T. Fini, S. Keller, S. P. DenBaars, J. S. Speck, U. K. Mishra, S. Nakamura, S. Yamaguchi, S. Kamiyama, H. Amano, I. kasaki, J. Han and T. Sota, *Nature Materials*, 2006, **5**, 810–816.

106. M. G. Walter, E. L. Warren, J. R. McKone, S. W. Boettcher, Q. Mi, E. A. Santori and N. S. Lewis, *Chem. Rev.*, 2010, **110**(11), 6446–6473.

107. H. Gerischer, *J. Electroanal. Chem.*, 1977, **82**(1–2), 133–143.

108. C. B. Vartuli, S. J. Pearton, C. R. Abernathy, J. D. MacKenzie, F. Ren, J. C. Zolper and R. J. Shul, *Solid-State Elect*, 1997, **41**(12), 1947–1951.

109. J. Li, J. Y. Lin and H. X. Jiang, *Appl. Phys. Lett.*, 2008, **93**(16), 162107–162103.

110. K. Fujii, K. Kusakabe and K. Ohkawa, *Jpn. J. Appl. Phys.*, 2005, **44**, 7433–7435.

111. A. Theuwis, K. Strubbe, L. M. Depestel and W. P. Gomes, *J. Electrochem. Soc.*, 2002, **149**(5), E173–E178.

112. K. Fujii and K. Ohkawa, *Jpn. J. Appl. Phys.*, 2005, **44**, L909.

113. K. Aryal, B. N. Pantha, J. Li, J. Y. Lin and H. X. Jiang, *Appl. Phys. Lett.*, 2010, **96**, 052110.

114. Y. J. Hwang, C. H. Wu, C. Hahn, H. E. Jeong and P. Yang, *Nano Letters*, 2012, **12**(3), 1678–1682.

Epitaxial III-V Thin Film Absorbers: Preparation, Efficient InP Photocathodes and Routes to High Efficiency Tandem Structures

THOMAS HANNAPPEL,*[a,b] MATTHIAS M MAY[b] AND
HANS-JOACHIM LEWERENZ[b,c]

[a] Ilmenau University of Technology, Institute of Physics, Department
Photovoltaics, Gustav-Kirchhoff-Straße 5, 98693 Ilmenau, Germany;
[b] Helmholtz-Zentrum Berlin für Materialien und Energie, Institute for Solar
Fuels, Hahn-Meitner-Platz 1, 14109 Berlin, Germany; [c] California Institute of
Technology, Joint Center for Artificial Photosynthesis, 1200 East California
Boulevard, CA 91125, USA
*Email: thomas.hannappel@tu-ilmenau.de

9.1 General Introduction

Photovoltaic solar energy conversion and photoelectrochemical water splitting differ due to the variability of the output power in photovoltaics, whereas light-induced water dissociation can only be carried out and optimized for voltages greater than the thermodynamic threshold of 1.23 V at room temperature and ambient pressure.[1,2] This severely restricts the acceptable range of variability of photoelectrochemical systems. Since only the product of the current and

RSC Energy and Environment Series No. 9
Photoelectrochemical Water Splitting: Materials, Processes and Architectures
Edited by Hans-Joachim Lewerenz and Laurence Peter
© The Royal Society of Chemistry 2013
Published by the Royal Society of Chemistry, www.rsc.org

voltage values at the maximum power point of a photovoltaic (PV) solar cell defines its efficiency, one can trade photocurrent *vs.* photovoltage in the design, yielding a substantially larger choice in materials selection. By contrast, the selection criteria for photoelectrochemical water splitting are more restrictive.

In monolithically integrated water splitting systems, absorber and catalyst surfaces are exposed to the electrolyte, and stable operation must be achieved at the reactive solution interface under non-equilibrium conditions.[3–5] Separate PV/electrolyzer systems are also characterized by materials issues due to operation at high currents, resistivity resulting from excessive bubble formation and the limited output for each stand-alone type system. Also the costs of the noble metal catalysts used in electrolyzer technology makes integrated structures with earth-abundant absorbers and catalysts considerably more promising.[6–8] Realization of the integrated approach needs a multidisciplinary procedure that encompasses materials development (absorbers, catalysts, linker groups), customized membrane preparation and architectural aspects that relate also to chemical engineering issues such as bubble suppression, hydrogen scavenging, gas separation and electrolyte flow.

The stability of light-driven water splitting systems can be addressed by using transition metal oxide absorbers that are known to be stable against oxidative processes.[9–11] Alternatively, one can use passivating coatings that are either produced *in situ* by surface transformation processes[12–14] or by employing recent advances in coating technology such as atomic layer deposition.[15,16]

In the design of a water splitting photosystem one can chose to use a single absorber structure, as has been historically introduced by Honda and Fujishima,[9,17] or to follow the Z-scheme of natural photosynthesis where two visible light photons are absorbed.[18–20] For a single junction structure, the energy gap must be large enough ($E_g \sim 2.2\,eV$) for water splitting including the thermodynamic minimum voltage together with the overpotentials on anode and cathode that provide the driving force for the reaction.[21] The achievable efficiency for a single junction structure is limited to about 15% for AM 1.5 conditions.[22] Inorganic Z-scheme analogues are based on semiconductor tandem structures[23,24] that can easily provide the voltage necessary to split water and can achieve significantly higher efficiencies.

In this chapter, we discuss tandem approaches that use epitaxial III-V thin film semiconductor compounds as absorber materials. We show that high-quality device layer structures can be grown either on re-usable III-V substrates (if lift-off processes are employed) or on IV-valent semiconductor substrates, *e.g.* Ge or Si. We present work on the development of highly photoactive III-V epitaxial thin films, emphasizing preparative aspects, surface analyses and control for photocathode half-cells in photoelectrochemical water splitting systems. We also discuss potential future device developments.

9.2 Introduction to III-V Materials

The material class of III-V semiconductor compounds allows flexibility in the design of both, electronic structure and device architecture. Opportunities for tailoring the optoelectronic properties of III-V semiconductors to specific applications are reflected in highly efficient photovoltaic solar cells that show conversion efficiencies above 40% in multi-junction concentrator designs.[25,26]

High-performance optoelectronic semiconductor devices such as lasers or solar cells are typically based on compound semiconductors of the third and the fifth group of the periodic table, the so-called III-V semiconductors. Their properties, such as band gap, lattice constant or refractive index, can be fine-tuned to the specific needs of the particular application by a variation of the compounds' stoichiometry. III-V semiconductors hold the current world record efficiencies for photovoltaic concentrator systems (currently 44%)[27] as well as for water splitting photoelectrochemical (PEC) devices.[28–30] The cell with the hitherto highest direct solar-to-hydrogen efficiency of 12.4% was realized with a monolithic GaInP$_2$/GaAs absorber stack more than a decade ago,[20] but was limited by corrosion problems.

The history of III-V compounds in photoelectrochemistry dates back to the early days of solar water splitting. Already, in 1975, a tandem approach for light-induced hydrogen production was realized using GaP as photocathode.[31] The closely related absorber material InP was the basis of an efficient and stable half-cell that was developed and analyzed a few years later.[28,32] This system showed excellent stability due to the formation of an interfacial oxide film by *in situ* surface transformation in a vanadium (II)/(III) electrolyte.[33] The recent advances of growth techniques for III-V semiconductors and heteroepitaxial approaches[34,35] allow the use of the III-V material class in thin film efficient water splitting structures and devices.

9.3 Epitaxial III-V Systems for Solar Energy Conversion

9.3.1 Efficiency Considerations and Multi-Junction Absorbers

The theoretical limit of the conversion efficiency for a multiple solar cell configuration consisting of an infinite number of p-n junctions with different band gaps reaches $\sim 86\%$ at maximum solar concentration.[36] Current high-efficiency solar cells use a combination of multiple absorbers with different energy gaps that are connected *via* tunnel junctions in a monolithic stack[37] by which photovoltaic efficiencies beyond 40% can be achieved under concentrated sunlight.[38] Next-generation multi-junction cells with four or more junctions and optimized band gaps are expected to exceed the present record efficiency, surpassing the 50% mark.[39] Present triple-junction cells are based on a III-V/IV material combination such as Ga$_{0.35}$In$_{0.65}$P/Ga$_{0.83}$In$_{0.17}$As/Ge. Energy gaps of binary and ternary III-V semiconductors and their lattice constants are shown in Figure 9.1. Direct band gaps (conduction band (CB) minimum at the Brillouin zone center Γ) are advantageous because of their high oscillator strength resulting in a small absorption length. Conduction band minima at other high-symmetry points (X, Δ or L) imply the presence of an indirect band gap and,

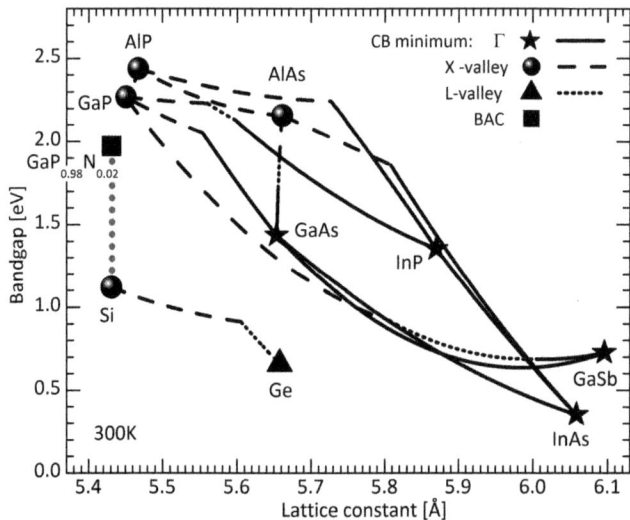

Figure 9.1 Band gap energies over lattice constants of typical III-V compound semiconductors, Si and Ge. Stars indicate direct band gaps, circles and triangles indirect gaps. Lines show the properties of ternary (binary) compounds. The black square gives the gap of GaPN lattice-matched to Si according to the BAC model (see text), and the gray dotted line exemplifies lattice match.
Adapted from reference 35.

consequently, a larger absorption length. The challenges of developing an efficient device for solar water splitting encompass (i) adjustment of the energy gaps for optimized current matching under solar irradiation, (ii) adaption of the lattice constants required to achieve a defect-free, monolithic stacking and (iii) adjustment of the absolute band edge positions with regard to the water splitting energies. Considering the size of the energy gap and matching of lattice constants, one observes in Figure 9.1 that $Ga_xIn_{1-x}P$ and $Ga_yIn_{1-y}As$ can be lattice-matched to a Ge substrate with appropriate stoichiometry. In addition, the direct nature of their band gaps allows application in thin film devices.

Before discussing the suitability of the considered materials in terms of their absolute band edge positions relative to the redox potentials for water splitting, an efficiency plot for photovoltaic devices (calculated with EtaOpt)[40] is presented in Figure 9.2. The contour plot gives the theoretical conversion efficiency for a tandem PV cell under standard terrestrial AM 1.5 g (global) solar irradiation as a function of the energy gaps of the bottom and top cell. Empirical results lead to the assumption that the achievable photovoltages of semiconductors are approximately 0.3 to 0.5 V lower than their energy gaps. Therefore, the sum of the energy gaps of tandem cells designed for water splitting should lie in the range of 2.6 to 3 eV, as indicated by the area between the two red lines in Figure 9.2. One sees that a tandem structure with Si (band gap 1.12 eV) as bottom cell should ideally employ a top cell with a band gap in the range of 1.7 to 1.8 eV.

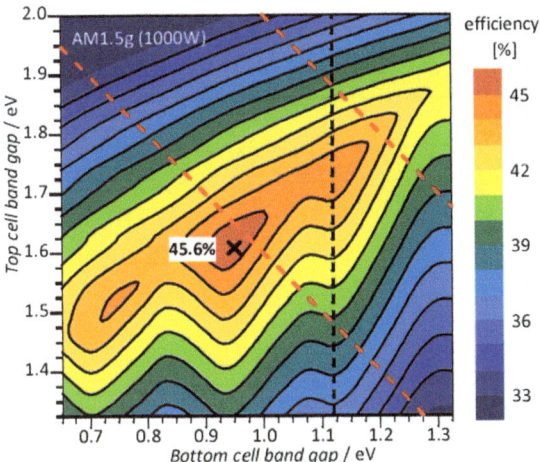

Figure 9.2   Theoretical efficiencies (AM1.5 g irradiation) of a tandem photovoltaic device, generated with EtaOpt,[40] with the absolute maximum and the band gap of Si (black dashed line). Red dashed lines indicate the area, where the sum of the band gaps lies between 2.6 and 3.0 eV.

9.3.2   Band Alignment

The absolute energetic positions of valence and conduction bands play a crucial role in the design of tandem heterojunctions and for light-induced water splitting. Band alignments of binary and ternary III-V semiconductors are given in Figure 9.3 (tabulated data of valence band offsets with respect to InSb and bowing parameters can be found in reference 41). The diagram plots maxima and minima of valence (VB) and conduction bands, respectively, versus lattice constants. The band offset of the dilute nitride $GaP_{0.98}N_{0.02}$, lattice-matched to Si, shows according to the BAC model a reduced "direct-like" energy gap compared to pristine GaP with a constant valence band maximum position.[42] The data demonstrate the large parameter space that is available for the design of III-V heterojunction structures which can be further extended by the use of quaternary compounds.

The grey horizontal bars in Figure 9.3 indicate the H_2/H_2O potential, whose work function against the vacuum scale has been taken to be 4.6 to 4.7 eV[43,44] as well as the O_2/H_2O potential at +1.23 V. Among the binary semiconductors, GaP is a candidate for unbiased water splitting: its conduction band minimum lies above the redox potential for hydrogen evolution, and the valence band maximum is positioned below the redox potential for oxygen evolution. The (indirect) energy gap of ~ 2.26 eV is still in the range for efficient use of the solar spectrum, allowing maximum photocurrents of about $10 \, mAcm^{-2}$, which is considered an acceptable for solar water splitting.[45] Its indirect energy gap complicates applications but, recently, GaP nanowire structures with MoS_2 catalysts have been prepared that show surprisingly good performance.[46]

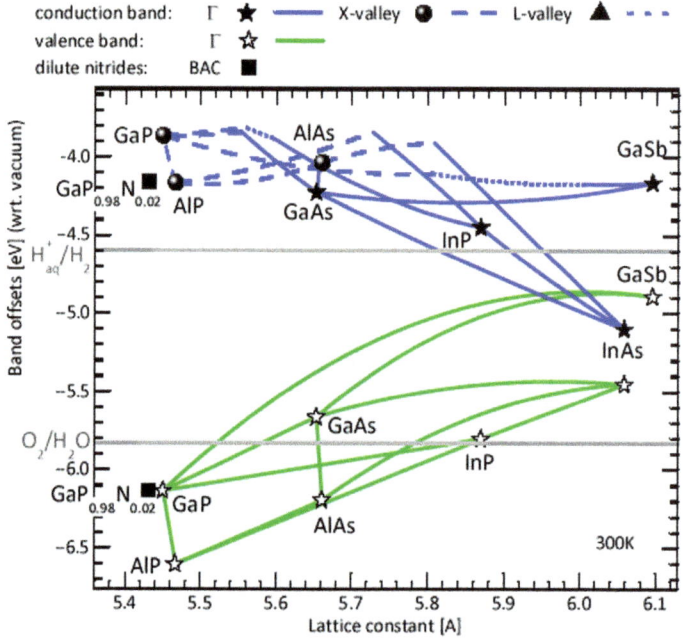

Figure 9.3   Band alignments over lattice constants for III-V semiconductors.[41] Blue lines indicate CB minima, green lines VB maxima. The gray horizontal lines represent the redox potentials of hydrogen and oxygen evolution, respectively. For convenience of illustration, the scale relative to InSb was replaced by the vacuum scale calibrated to the electron affinity of InP (4.4 eV).
Courtesy of O. Supplie.

InP, on the other hand, would allow use of the solar spectrum close to the theoretical maximum for a single absorber. Apart from the limitation that the maximum attainable photovoltage is approximately given by $E_g - 2(E_F - E_V)$, which corresponds to $1.35\,\text{eV} - 0.4\,\text{eV} \sim 0.95\,\text{eV}$ and thus much smaller than even the thermodynamic value for water splitting, its valence band maximum just matches the redox potential for oxygen evolution. Therefore, InP has been used to demonstrate efficient hydrogen evolution in half cells.[47]

According to the Anderson model,[48] an ideal heterojunction is determined by the respective electron affinities, work functions and energy gaps of the involved semiconductors. Figure 9.4 shows such heterojunctions for two isolated semiconductors, neglecting charge transfer and the resulting band bending at the interface for simplicity reasons. In a type I ("straddling gap") heterojunction (Figure 9.4a), the energy gap of the first semiconductor straddles the smaller gap of the second semiconductor grown on top. This results in a barrier, ΔE_c, for electrons moving from the right to the left in Figure 9.4(a) and simultaneously a barrier, ΔE_v, exists for hole movement in the same direction. If semiconductor 1 is grown on top (dashed lines in Figure 9.4a) and the layer of semiconductor 2 is thin enough (*i.e.* in the order of several nm), a quantum well

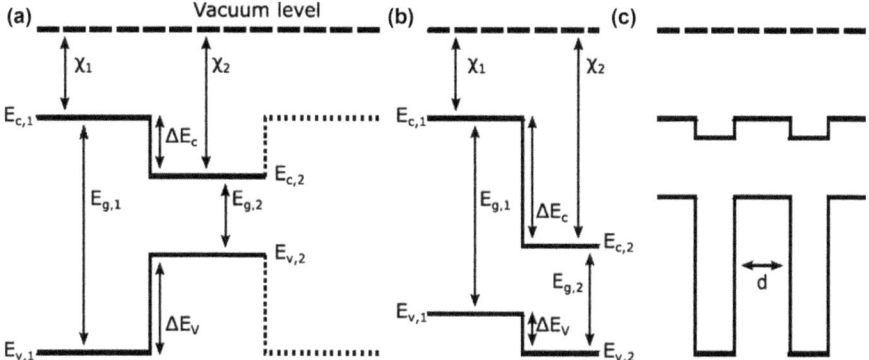

Figure 9.4 Idealized hetero-junction of two semiconductors. (a) Illustrates a type I and (b) a type II hetero-junction. χ denotes the electron affinity, E_g the band gap, E_c and E_v positions of conduction and valence band, respectively, and ΔE their offsets. The second hetero-junction in (a) (dotted lines) completes a quantum well structure. (c) illustrates the principle of a resonant tunneling diode with distance d between the two barriers.

is created. Charge carriers within this well experience energy quantization, which is used in quantum well lasers[49] or in advanced third-generation concepts for solar absorbers, as outlined in section 9.6.3. Type I heterojunctions can be created utilizing compounds such as $GaAs/Al_xGa_{1-x}As$ (almost unstrained), or the semiconductor pair InP/InAs (highly strained). Type II heterojunctions, ("staggered lineup") are characterized by band offsets where ΔE_C and ΔE_V have the same sign (compare Figure 9.4b). This results in a barrier for one type of charge carrier and a cliff for the other. For carriers moving from right to left in Figure 9.4(b) a barrier exists for electrons and a cliff for holes. Type II heterojunctions are used in resonant tunneling diodes,[50] where the height of two barriers and their distance d define the energies for which charge carriers experience resonant tunneling (Figure 9.4c). Charge carriers at the resonant energy can tunnel with a theoretical transmission probability of unity. Type II heterojunctions have been realized with InP/GaSb (highly strained) or, lattice-matched, with $InP/In_{0.53}Ga_{0.47}As$.

9.3.3 Water Splitting with III-V Semiconductors

The earliest tandem approach for water splitting used TiO_2 as photoanode and p-GaP as photocathode.[31] Having energy gaps of $\sim 3\,eV$ and $2.26\,eV$, this tandem configuration was far from using the solar spectrum efficiently. Recently, GaP-based layers incorporating nitrogen were proposed to enhance limited charge transfer efficiency to the electrolyte.[51] Besides the influence of physical and chemical surface treatment procedures on surface morphology and electronic structure,[52] a pronounced influence of surface defects on corrosion was noted.[53]

In contrast to GaP, InP is characterized by a high electron mobility and a direct energy gap (Figure 9.1). Recently, the material has been used as custom-made thin film homoepitaxial absorbers and, combined with stabilization layers and electrocatalysts, high efficiency in light-induced hydrogen evolution has been reported.[54] The results show that the design of III-V absorbers must address stability under operation. In addition, a promising approach would be to combine the III-V thin films with an earth abundant substrate such as Si.[35]

9.3.4   III-V Heteroepitaxy

Early realizations of solid state multi-junction solar cells applied metal-organic chemical vapor deposition (MOCVD) to grow a top cell on GaAs substrates.[55] Metal interconnects were fabricated in a post-processing step for serial connection, in the absence of appropriate tunnel junctions.[56] Later, monolithic stacks of record multi-junction solar cells including appropriate tunnel junctions were developed, also utilizing MOCVD.[57] Present approaches that result in the highest efficiencies typically employ Ge as substrate and bottom sub-cell with subsequent heteroepitaxial growth of the upper cells by MOCVD, interconnected with several functional layers such as tunnel diodes, grading and buffer layers.[26] While the tunnel junctions ensure an efficient serial connection of the subcells, the need for grading and buffer layer structures arises from metamorphic growth, where lattice mismatch leads to relaxation and dislocation defects. To minimize Shockley-Read-Hall recombination at these growth defects, sophisticated grading and buffer layers with a thickness in the order of several 100 nm are introduced.[58,59] These layer structures are designed to allow relaxation in a confined range of the monolithic stack to enable defect-free growth of the subsequent sensitive material, upon changing the lattice constant. The highest epitaxial quality can be achieved with pseudomorphic, lattice-matched growth avoiding relaxation as a source of crystal defects and non-radiative recombination.

Si as growth substrate has the advantage of being available in abundance and of high quality in the earth's crust. Potential ternary candidates for lattice-matched heteroepitaxy on Si substrates are shown in Figure 9.1. Dilute nitride GaPN is of special interest here, as its energy gap of 1.95 eV is located relatively close to the optimum of 1.73 eV for a tandem structure using a Si bottom cell.[35] The incorporation of As into the quaternary compound GaPNAs further reduces the energy gap towards the maximum PV efficiency with a theoretical value of 44.8%.[35,60] However, the III-V/Si interface is known to introduce defects,[61] and this has limited the efficiencies of III-V on Si structures.[62] These aspects are discussed in more detail in section 9.6.

The method of choice for the preparation of high-quality III-V semiconductors in an industrially scalable manner is MOCVD. It enables the preparation of abrupt interfaces, which is essential for well-defined and efficient tunnel junctions as well as for other device components. The control of the surface/interface properties is mandatory for fabrication of high-quality heterojunctions. The method of choice here is optical *in situ* growth characterization by reflection anisotropy spectroscopy (RAS).[63–65] The method is outlined in section 9.4.

9.4 Preparation of Epitaxial III-V Thin Film Absorbers Using Optical *In Situ* Control

MOCVD is employed under near-ambient pressure. The preparation method generally uses hydrogen gas and toxic precursor material such as arsine (alternatively tert-butylarsine), phosphine (tert-butylphosphine) or silane (di-tert-butylsilane). However, typical surface analysis tools require ultra-high vacuum (UHV). As an optical method, RAS is applicable in both, UHV and under MOCVD conditions. It features high surface sensitivity on an atomic level and thus enables *in situ* control of epitaxial III-V MOCVD preparation.

9.4.1 Optical *In Situ* Control: Reflectance Anisotropy Spectroscopy

RAS is an ellipsometric reflectance technique operating at near-normal incidence, probing the azimuthal anisotropy of surfaces.[66–68] It measures the difference in the normal-incidence reflectance, r_x and r_y, utilizing linearly polarized light directed onto the sample surface as shown in Figure 9.5. The signal is normalized with respect to the total reflection r.

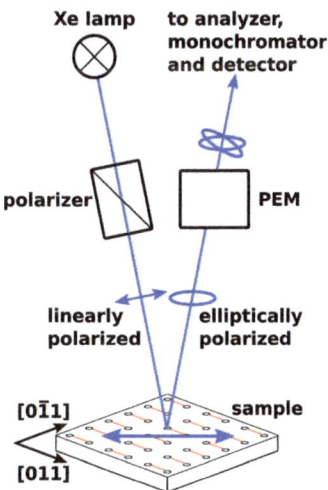

Figure 9.5 Principle of RAS: White light is linearly polarized and focused on a sample. If anisotropy exists, the reflected beam is elliptically polarized. A photoelastic modulator (PEM) creates a phase modulation of the light exhibiting polarization perpendicular to the axis of the PEM. The phase modulation is then converted to an amplitude modulation in a polarization analyzer, monochromated and finally detected by a photodiode. For details, see ref. 80.

As the figure shows, the basic principle of RAS is the measurement of the complex reflectance anisotropy $r_x - r_y/r$ that results from the reduced symmetries of the (reconstructed) surfaces of, for instance, a cubic crystal.[69,70] Regarding the anisotropy along the crystal axes $[0-11]$ and $[011]$ of a (100) surface, the normalized reflectance anisotropy is given by

$$\frac{\Delta r}{r} = 2\frac{r'_{[0\bar{1}1]} - r'_{[011]}}{r'_{[0\bar{1}1]} + r'_{[011]}} \tag{9.1}$$

Accurate RAS measurements require a strain-free optical window. In case of residual strain, one can correct by calibrating the measured reflection anisotropy signal with that of a calibration standard (*e.g.* a Si(110)-wafer) or by measuring an isotropic surface (such as an oxidized Si(100) wafer) as baseline and/or by measuring the sample upon rotation by $90°$.

RAS signals can arise from both the bulk as well as the surface structure. The latter allows for correlating atomic surface structure and optical anisotropy. Regarding a sample consisting of a cubic crystal, a change of the reflectance anisotropy may have various origins such as:

1. mechanical strain;[71]
2. bulk ordering of sub-lattices occurring in ternary or quaternary compositions (*e.g.* CuPt-ordering in InGaP on GaAs);[72]
3. Fabry-Pérot-like interferences of heteroepitaxial layers;[73]
4. atomic steps and terraces on the surface;[71]
5. surface electric fields inducing the so-called linear electro-optical effect (LEO) *via* doping;[74]
6. surface roughness and 3D structures;[75,76] and
7. atomic surface reconstruction: surface-state signatures and bulk-related features.[63]

9.4.2 Metal-Organic Chemical Vapor Deposition

The MOCVD deposition process (also called metal-organic vapor phase epitaxy) operates at or near ambient pressure conditions of the order of 10 to 1000 mbar, but initial findings of the *in situ* RAS analysis need to be correlated with additional surface science techniques. For undistorted measurements of the prepared surfaces with UHV-based surface analysis tools, a transfer system (Figure 9.6) has been developed to transfer samples from the MOCVD apparatus to UHV within a very short time (in the order of seconds) without contaminating the interfaces under study.[77] The transfer routine is outlined in the following for the example of atomically-ordered, epitaxial InP and GaP(100)-films prepared with only non-gaseous precursors tert-butylphosphine (TBP), trimethylindium (TMIn), and triethylgallium (TEGa) in a commercial MOCVD apparatus equipped with an optical window for *in situ* spectroscopy. RAS signals were benchmarked with surface science data from low-energy

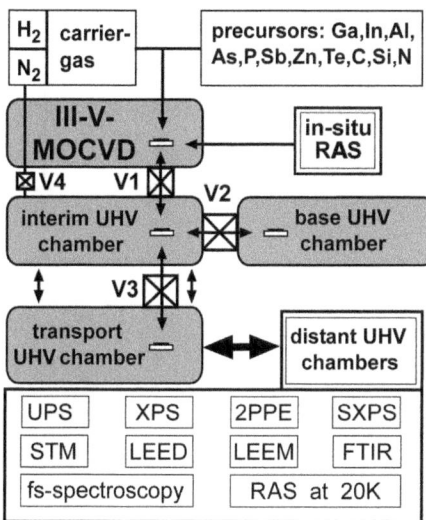

Figure 9.6 Scheme of MOCVD reactor and UHV transfer system. For details, see text.

electron diffraction/microscopy (LEED/LEEM), scanning tunneling microscopy (STM) and soft X-ray photoemission (UPS, SXPS, XPS). Fourier transform infrared spectroscopy (FTIR) confirmed hydrogen termination of the reconstructed surfaces of InP and GaP. Femtosecond-resolved 2-photon photoemission (fs-2PPE) enabled the study of unoccupied states as well as charge carrier dynamics.[78] All surface science data, collected in UHV, could then be correlated with the particular RA spectrum measured already in the MOCVD reactor and thus, the electronic and atomic structure can be determined *in situ* during preparation.

The transfer system attached to the commercial MOCVD reactor (AIX200) consists of an interim chamber, where a total pressure $< 10^{-8}$ mbar is reached in less than 20 seconds, a basic UHV chamber and a mobile UHV transport chamber (compare Figure 9.6). The latter is equipped with a battery-driven ion getter pump supplying a base pressure of $p < 3 \times 10^{-10}$ mbar. All chambers and the MOCVD reactor are separated by UHV-valves.

After standard wet-chemical substrate preparation, samples were dried in a stream of N_2 and attached by clamping to a molybdenum sample holder, placed in the center of a modified graphite susceptor inside the MOCVD reactor. After thermal deoxidation in H_2 atmosphere, buffer layers were grown under supply of TMIn (TEGa for GaP) and TBP at temperatures in the order of 850 K and then cooled down under TBP stabilization. This stabilization with the phosphorous-precursor prevents desorption of P-clusters from the surface, which would eventually lead to a P-deficient bulk.[79] To obtain a specific, well-defined surface reconstruction, *e.g.* P-rich (Figure 9.7b) InP(100), the flow of TBP was switched off at around 600 K, the purge gas was

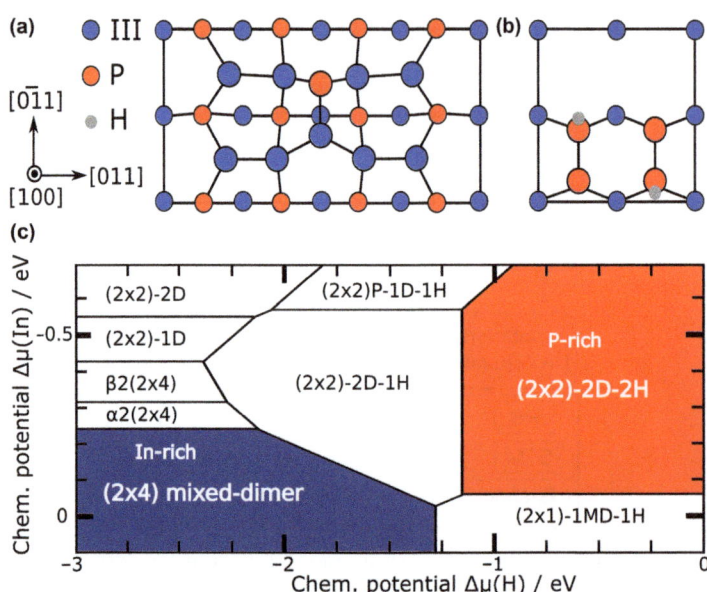

Figure 9.7   Ball and stick model (top view) of the III-rich (2×4) mixed-dimer
(a) and P-rich (2×2)-2D-2 H (b) surface reconstruction (for details see
section 9.5). Large (small) symbols indicate positions of the first and
second (third and fourth) atomic layers. The calculated phase diagram[81]
for InP in (c) relates the growth parameter space with possible surface
reconstructions.

changed to N_2 and the reactor pressure was reduced to about 1 mbar.
Next, the UHV-valve V1 separating the MOCVD-reactor from the interim
chamber was opened and the hot sample was transferred to the interim
chamber which was purged with the carrier gas. Subsequently, valve V1 was
closed and valve V2 to the UHV basis chamber was opened. After 20 s, a
pressure in the 10^{-9} mbar range was reached. At $p < 3 \times 10^{-10}$, the sample was
transferred into the UHV transport-chamber and, maintaining UHV conditions, to any other UHV system that is equipped with an appropriate UHV
load-lock system. XPS was performed with a hemispherical analyzer (Specs
Phoibos 100) and a monochromatized Al K_α X-ray source. The equipment for
the RAS (LayTec EpiRAS 200) measurements has been documented in the
literature.[68,80]

P-rich surfaces, which are relevant for preparing InP- and GaP-based III-V
devices, and which could previously not be investigated in the state they were
originally prepared, were analyzed with the transfer method developed in our
laboratory. For example, it was found that MOCVD-prepared P-rich InP and
GaP (100) surface reconstructions differ distinctively from those prepared by
MBE: for MOCVD, H atoms are attached to buckled P dimers due to the
supply of H_2 in the MOCVD environment (see Figure 9.7b and calculated
phase diagram[81] in Figure 9.7c).

9.5 InP(100) Absorbers: Heterostructures and Photocathodes

Surface electronic properties, such as band bending are known to depend critically on the surface condition; defects can act as recombination centers or as points of attack for (photo)corrosion. As will be shown in section 9.6.1 below, different surface reconstructions also impact the interaction of the semiconductor surface with H_2O. A microscopic understanding of the surfaces is therefore essential for a knowledge-based surface and interface preparation. The presence of hydrogen in an MOVCD growth environment can promote surface reconstructions that differ significantly from those of UHV-prepared material. In contrast to the In-rich surface, which appears to be generally accepted as the hetero-dimer unit cell,[87] the P-rich MOCVD prepared surface reconstruction had not yet been observed before. Both surface reconstructions have been studied in detail regarding morphology,[82] band structure[83] and carrier dynamics.[78]

The In-rich, (2×4)-reconstructed InP(100) surface is characterized by mixed In-P dimers oriented along the [0-11] direction terminating the surface (Figure 9.7a). In-In-bonds along [011] form the second layer. The reconstruction spans 2 bulk unit cells in the [0-11] direction and 4 in the [011] direction, hence the notation (2×4). The hydrogen-stabilized, P-rich surface is illustrated in Figure 9.7(b). Its (2×2) unit cell contains two oppositely buckled P-dimers, where in each case one of the dangling bonds is saturated by an H-P bonding. These dimers can conduct a flipping motion, as recently observed by STM investigations.[84] The notation (2×2)-2D-2 H symbolizes 2 dimers and 2 hydrogen atoms per unit cell of the reconstruction. Surface configurations accessible in MOCVD-typical hydrogen ambient[81] are illustrated in a phase diagram in Figure 9.7(c).

The transition from the P-rich to the In-rich surface reconstruction (red to blue area in Figure 9.7c) can be clearly distinguished *in situ* with RAS, as displayed in Figure 9.8. Schmidt *et al.*[63,81] relate the minima around 1.6 eV, present in both reconstructions, to electronic transitions at the surface Brillouin zone. A prominent maximum around the interband transition E_1 is characteristic for the (2×2) reconstructed P-rich surface (red curve), while the (2×4) reconstructed In-rich surface shows a minimum.

9.5.1 InP-based Heterostructures

The practical significance of a well-defined surface preparation was demonstrated in the development of a low band gap tandem cell, which was designed on the lattice constant of InP (see Figure 9.1) to facilitate lattice-matched growth on InP substrates.[39] State-of-the-art triple-junction photovoltaic solar cells presently employ three different semiconductors as light absorbers. Ge typically serves simultaneously as substrate and bottom cell. Next-generation quadruple-junction solar cells utilizing four different absorbers should replace the Ge substrate by a low band gap tandem bottom cell to further exploit the

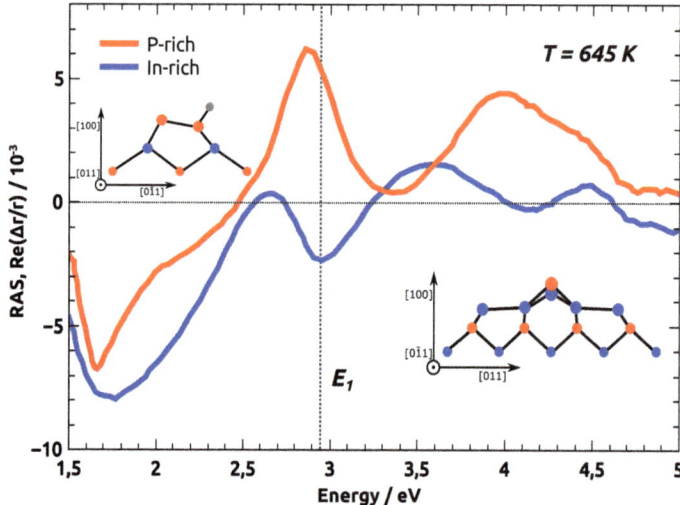

Figure 9.8   RA spectra of the P-rich and In-rich surfaces of InP(100) at 645 K. Insets
show the side-view of the surface reconstructions.

solar irradiation. Calculations show that the optimal energy gap relation for the
bottom tandem is in the order of 0.7 and 1.0 eV enabling theoretical efficiencies
of 61%.[39,85] A realization would comprise lattice-matched InGaAsP/InGaAs
(0.73 and 1.03 eV) in combination with an InGaP/GaAs top tandem (1.86 and
1.42 eV) with a tunnel junction between InGaAs and InGaAsP (on InP substrates) using highly-doped n^{++}-InGaAs and p^{++}-GaAsSb (Figure 9.9). To
enable efficient tunneling through the barrier between InGaAs and GaAsSb,
the junction has to be as abrupt as possible. It has been found that III-rich
prepared InGaAs surfaces exhibit more abrupt interfaces to GaAsSb due to a
reduced diffusion of Sb resulting in higher efficiencies.[39,86]

   In contrast, resonant tunneling diodes (RTD) based on GaAsSb/InP showed
a higher performance with V-rich (As) prepared interfaces.[50] The RTD (see
Figure 9.4c) consists of type II heterojunctions and requires thin (7 nm) abrupt
InP barriers sandwiching a 7 nm thick GaAsSb quantum well leading to the
development of a resonant tunneling bipolar transistor with high power efficiency.[50] The role of surface reconstructions on interface electronic properties
has been clearly demonstrated by these actual applications.

9.5.2   Surface-Functionalized InP(100) for Efficient Hydrogen Evolution

Epitaxial thin-film InP was prepared by de-oxidation of a p^+-doped InP(100)
wafer and subsequent growth of a 3 μm thick p-doped buffer. Surfaces were
prepared In-rich from a 2×4 reconstructed surface termination where only one
P atom exists in the topmost surface layer (compare Figure 9.7a). Ohmic
contacts were made using Zn/Au alloying as reported earlier.[29] The doping

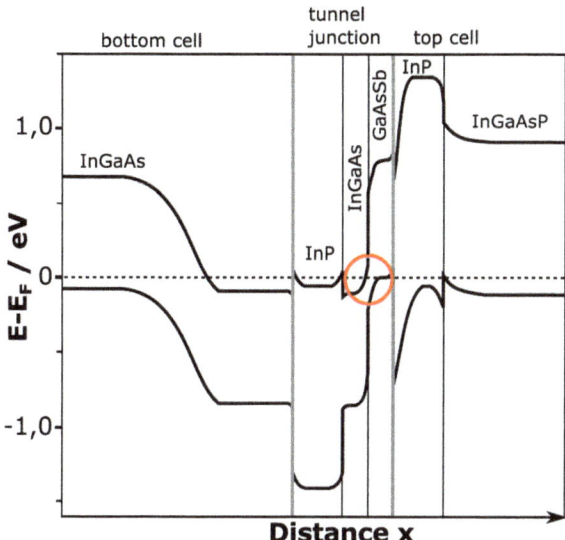

Figure 9.9 Tunnel junction of a low band gap tandem cell InGaAs/InGaAsP on the lattice constant of InP.[39] The red circle marks the actual tunneling region.

level of these films was in the low 10^{17} cm^{-3} range. Surfaces obtained that way show atomic terraces even in ambient air AFM experiments. It is found that the electrodes are rather passive in acidic solutions such as 1 M HCl when illuminated. This can be attributed to the overpotential for proton reduction and the slow In atom dominated surface kinetics or surface oxidation upon exposure to ambient air. Only after a specific surface conditioning, performed *in situ*, an increase in the cathodic photocurrent is observed, and after deposition of noble metal catalysts, highly efficient hydrogen evolution.[30] Here, the focus is on the analysis of the surface condition after the respective chemical and (photo)electrochemical surface treatments.

9.5.2.1 *In Situ Surface Conditioning*

Prior to photoelectrochemical treatments, a bromine-methanol etch was applied. The surface topography is shown in Figure 9.10. The $2 \times 2\,\mu$m image shows atomic terraces and growth defects corresponding to a defect density of about 10^7 cm^{-2}. The observed depressions were already visible on as-grown samples, as were the atomic terraces. In cross-sectional AFM images, the height variations could not clearly be related to the lattice parameters of InP in the <100> direction, probably due to oxidation of the surface with different oxide coverages along step edges and on terraces which could contribute to the passive behavior at lower cathodic potentials (see below).

The analysis of the surface conditions was performed using a photoelectrochemical cell that can be directly attached to UHV surface analysis apparatus, developed by Bitzer *et al.*[87] and later improved by Rauscher *et al.*[88] The

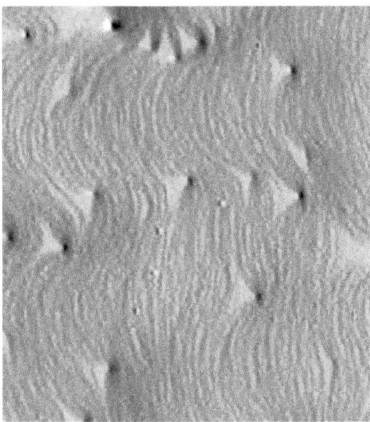

Figure 9.10   Tapping mode AFM amplitude images after etching of InP(100): 2×4 in
a solution of 0.05 weight% Br_2 in methanol for 30 s; the image size is
2×2 μm.

system is characterized by a very low surface contamination due to exclusion of
ambient atmosphere contamination, with about 0.2 ML of hydrocarbons on
the surface after electrochemical currents have been passed.[87] The system
allows application of ultraviolet photoelectron spectroscopy, LEED (low energy electron diffraction), SXPS (soft x-ray spectroscopy), HREELS (high
resolution electron energy loss spectroscopy) and provides a wealth of information on (photo)electrochemical processes. Figure 9.11 (left) shows a
photograph of the vessel and a schematic drawing. The sample treatment
procedure is given in the figure caption.

Figure 9.12 shows in a combinatory manner the electrochemical treatments
that comprise the photoactivation step, the electrodeposition of Rh and the
photocurrent-voltage characteristics of the completed device. These experiments,
originally performed in a three-electrode standard potentiostatic arrangement,
have been repeated in an identical way in the glass vessel attached to the Solid-
Liquid Analysis System of the collaborating research group "Electronic Material
Interfaces" at the undulator U49/2 at Bessy in Berlin, Germany.

It can be clearly seen that the photoactivity of the as-prepared InP thin-film
in HCl is very low and only after repeated cyclic polarization within judiciously
chosen potential limits, an increase of the cathodic photocurrent is seen. Rh
photoelectrodeposition was performed at a potential where the hydrogen
evolution reaction was in its early stage (see Figure 9.12). The output power
characteristic of the completed half-cell is presented, too, showing high solar to
hydrogen conversion efficiencies. The photoelectrochemical behavior of the
samples can be understood by a combination of surface chemical and topographical analyses, which follows below.

Figure 9.13 shows an energy band schematic that relates the potential scans
for the surface conditioning with the electron energy of the semiconductor. The
anodic and cathodic decomposition potentials are indicated and the scan region

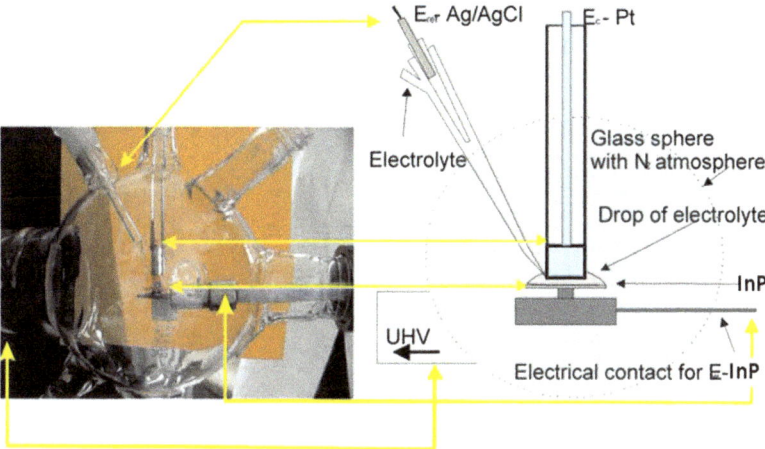

Figure 9.11 Photoelectrochemical cell for in-system surface analysis after electrochemical conditioning; the three-electrode potentiostatic arrangement consists of the working electrode (here: InP), mounted on a vacuum stub, a cylindrical counter electrode (Pt) at the end of a glass slab that also serves as light guide and of a Luggin capillary as Ag/AgCl reference electrode. The photoelectrochemical process is interrupted by jet-blowing of the electrolyte, then, the sample is rinsed with N_2-saturated deionized milli-Q water and subsequently dried in a nitrogen stream. The sample is then transferred into a buffer chamber where outgasing of the volatile residual species occurs. Once the pressure reaches the low 10^{-8} mbar range, the sample is transferred into the analysis chamber of the respective UHV system for measurements.

between $+0.3$ V and -0.1 V (SCE) (see top of Figure 9.12) has been included. Photoelectrodeposition at -0.2 V is seen to lie in a potential range where the InP surface is inverted but it is not in strong inversion which would screen out static electrical fields. Since the scan in HCl only shows total currents, at potentials positive from 0.2 V, anodic and cathodic partial currents obviously compensate each other.

The anodic (dark) partial current is related to the corrosion reactions indicated in Figure 9.13. They include the formation of indium oxide but also that of indium phosphates. Since In metal is a cathodic corrosion product, its oxidation to an indium oxide is likely. The anodic dissolution reactions include the formation of indium oxide and of two types of indium phosphates according to:

$$2In + 6h^+(VB, SS) + 3H_2O \rightarrow In_2O_3 + 6H_{aq}^+ \qquad V^0 = -0.45 V(SCE)$$

$$3InP + 15H_2O + 24h^+(VB, SS)$$

$$\rightarrow In(PO_3)_3 + 2In(OH)_3 + 24H_{aq}^+ \qquad V^0 = -0.51 V$$

$$InP + 4H_2O + 8h^+(VB, SS) \rightarrow InPO_4 + 8H_{aq}^+ \qquad V^0 = -0.72 V$$

$$(9.2)$$

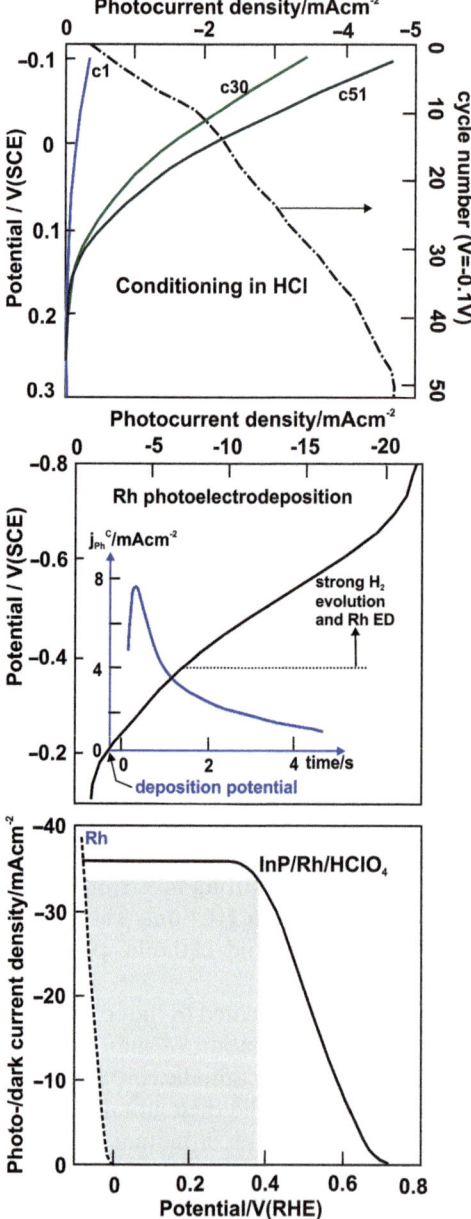

Figure 9.12 (Photo)electrochemical conditioning of InP(100) for photoelectrocatalytic device preparation; top: cyclic polarization in 1 M HCl; c1 – c50
indicate cycle numbers; dashed dotted line: cathodic photocurrent at the
cathodic scan limit – 0.1 V(SCE); middle: Rhodium photoelectrodeposition protocol; the insert shows the time profile of the cathodic photocurrent at the deposition potential of –0.2 V(SCE); bottom:
photocurrent-voltage characteristic for the device in comparison to the
dark current of a Rh wire (dotted line).

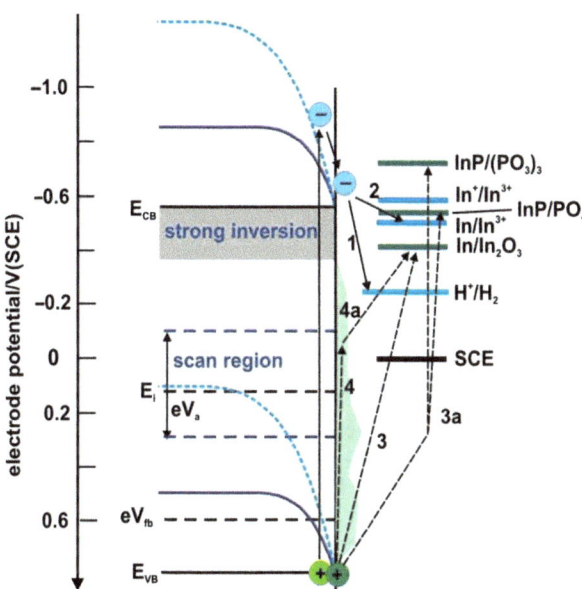

Figure 9.13   Energy *vs.* electrode potential schematic including the respective de-composition potentials; the scan region in HCl is indicated by horizontal blue dashed lines; process 1: hydronium reduction by photo-induced excess electrons; process 2: cathodic corrosion reaction resulting in In formation; processes 3, 3a: anodic corrosion by holes from the valence band top; process 4, 4a: anodic oxidation by holes from surface states that are filled or empty depending on the respective scan potential; eV_a: applied voltage, eV_{fb}: flatband energy; E_i: intrinsic Fermi level.

The cathodic corrosion reaction, energetically located close to that of In phosphate formation, is:

$$\text{p-InP} + 3e^-_{CB}(h\nu) + 3H^+_{aq} \rightarrow In^0 + PH_3 \qquad (9.3)$$

The complexity of the surface chemistry of InP in HCl and, also following photoelectrodeposition of Rh metal, suggests the use of advanced surface analysis methods to analyze the induced surface transformations. Accordingly, the surface chemical and electronic conditions of the samples after the three main steps of the conditioning procedure, *i.e.* (i) etching, (ii) photoelectrochemical conditioning and (iii) photoelectrodeposition of Rh electrocatalyst have been analyzed using UPS and XPS employing synchrotron radiation for enhanced surface sensitivity. Figure 9.14 shows a schematic that combines the inelastic photoelectron escape length of specific core levels of the InP system with the photon energy for highest surface sensitivity. Since highest surface sensitivity is obtained for a kinetic energy of approximately 50 eV, one tunes the photon energy for each core level such that the kinetic energy of the emitted photoelectrons lies in that range. This results in a similar surface sensitivity for

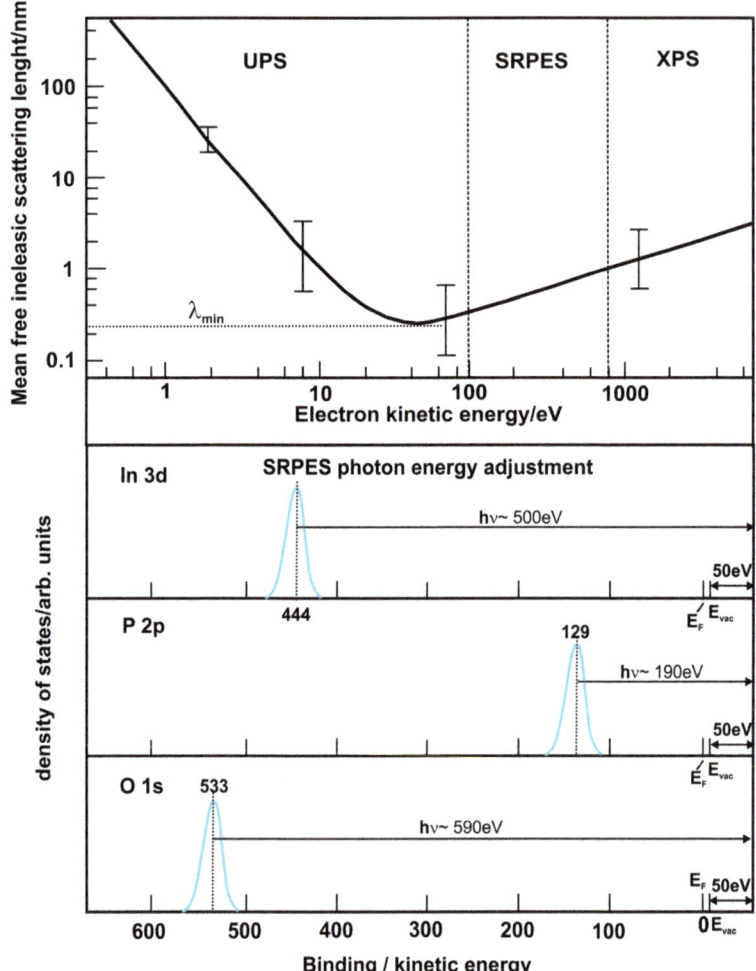

Figure 9.14 Mean inelastic scattering length for photoelectrons (top, note error bars) and the adjusted photon energy for the core level lines of the investigated elements for high surface sensitivity; a kinetic energy of 50 eV has been assumed for highest surface sensitivity, e.g. smallest λ. The energetic distance between Fermi level and vacuum level is taken to be roughly 5 eV and the position of the Fermi level has been taken as zero binding energy.

the analyzed elements, which is not the case for laboratory XPS, where the photon energy (Mg K$_\alpha$ or Al K$_\alpha$) is fixed thereby providing spectra with different depth information for different elements. In the following, we present the data for each treatment separately and summarize them to derive a complete as possible picture of the surface chemical processes that induce the surface transformations and of the energy relationship of the structure.

9.5.2.2 In-system Surface Analysis: Synchrotron Radiation Photoelectron Spectroscopy

Figure 9.15 shows SRPES data, recorded after the photoelectrochemical cycling procedure that is shown in Figure 9.12 and after the Rh photoelectrodeposition following the protocol given in the same figure. Whereas the lines for In, InP and In_2O_3 are well-known from the literature, the distinction between the phosphates and phosphites of indium is not as clear-cut. One can, however, calculate the electronegativity of In in these complexes using Sanderson electronegativities.[89] An important aspect of Sanderson's electronegativity is the so-called electronegativity equilibration which states that if two or more atoms (in a compound, for example) with different electronegativities form a compound, they will adjust to assume the same intermediate electronegativity which is given by the geometric mean of the individual atoms that form the compound. In other words, compound electronegativity underlies the concept that electrons distribute themselves around a molecule to minimize or to equalize the Mulliken electronegativity. The geometric mean of the electronegativity is calculated according to:

$$\chi_{mol} = \left(\prod_{k=1}^{p} \chi_k \right)^{1/p} \tag{9.4}$$

Here, the molecule has been assumed to consist of p atoms of electron affinities χ_k. The method has also been applied to describe charge relations at junctions between solids.[90]

Considering the electron affinities for P, O and In, the values for $InPO_4$ and $In(PO_3)_3$ can be calculated using the respective Sanderson element electronegativities of 2.52 (P), 3.65 (O) and 2.14 (In). It suffices to calculate the ligand electronegativities of PO_4 and $(PO_3)_3$ as their values indicate the partial charge shift that is related to the binding energy shift in an XPS experiment. The calculated differences between $\chi(PO_4) = 3.39$ and $\chi((PO_3)_3) = 3.29$ are small, but indicate that In has a larger positive partial charge with the phosphate ligand as indicated in Figure 9.15. In addition, this assignment is in accordance with literature data.[91] Figure 9.15 shows that In metal as well as InP signals originate from the top monolayer ($\lambda_{esc} \sim 0.3$–0.4 nm, see Figure 9.14) after cycling in HCl as well as after Rh deposition. Both signals decrease after the electrodeposition step, but, nevertheless, the data show that contrary to the experiments performed in a laboratory cell, the oxide film and the electrodeposited Rh leave parts of the InP surface exposed. The origin of this difference could be attributed to the limited volume of the droplet, the altered coupling of the light and scattering due to hydrogen bubble formation. Formation of indium oxide is attributed to the photoreduction of InP to In and, upon anodic potential scan, In oxidation. In addition, anodic decomposition of InP towards P containing compounds is noted (see eqn. 9.2). Upon Rh deposition, the relative amount of indium oxide increases. Since at the Rh photoelectrodeposition potential, the reduction of In_2O_3 and that of In^{3+} to In can

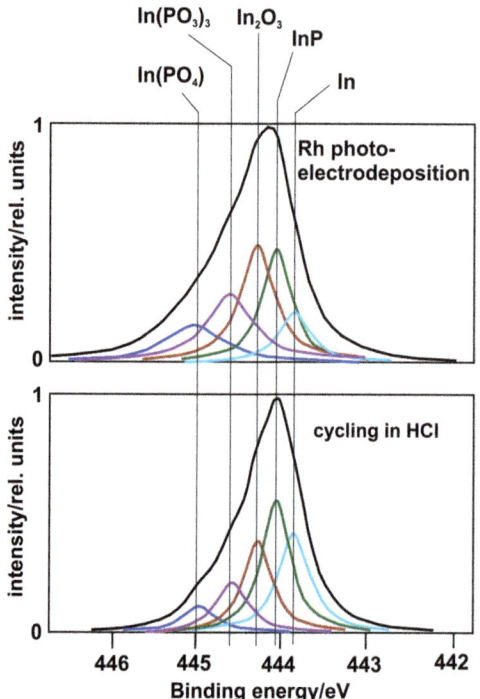

Figure 9.15   Deconvoluted SRPE spectra of the $In_{3d5/2}$ core level after cycling in 1 M
HCl (bottom) and after photoelectrodeposition of Rh (top); photon
energy: 600 eV; the identified species are indicated in the figure along with
their binding energy (see text).

occur as side reaction (compare Figure 9.13), the formation of oxides under
these conditions is attributed to hole injection from the Rh redox couple into
occupied surface states, similar to the behavior found upon Pt electrodeposition
onto Si.[14] This observation is corroborated by the increased relative amount of
the P-O groups at higher binding energy, pointing to a chemical route during
photoelectrodeposition of Rh.

The core level analysis of the P 2p line allows for assessing relative changes of
the P-O group contributions upon surface conditioning. Figure 9.16 shows P 2p
spectra recorded at hv = 200 eV ($\lambda_{esc} \sim 0.3$ nm). The relative contributions from
the phosphorus groups increase substantially after the Rh deposition. Since the
areas related to the InP surface decreases, the data show continuing chemical
oxidation of the uncovered InP surface as described above. Another aspect
concerns the relative amounts of the PO_4 *vs.* the $(PO_3)_3$ groups. After Rh deposition, the former is considerably decreased compared to the situation after
cycling in HCl.

The depth dependence of the contributions in the film can be investigated by
changing the photon energy. The evaluation of such an experiment is shown in
Figure 9.17 together with a schematic that indicates the elastic escape depth of
photoelectrons.

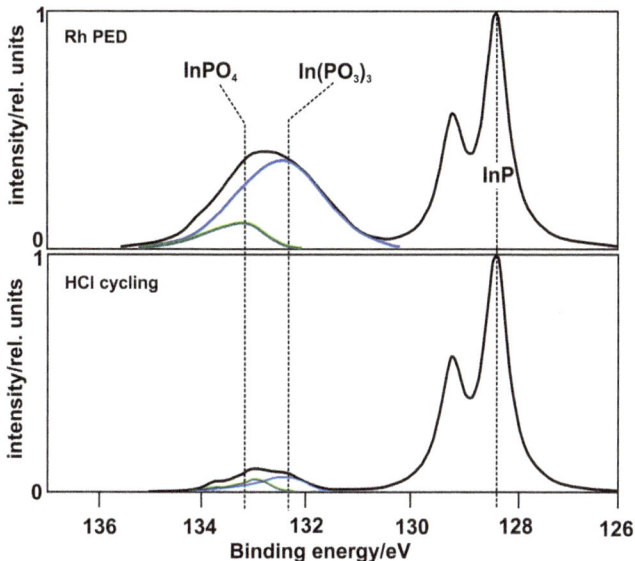

Figure 9.16 SRPES data for the P 2p line at high surface sensitivity for cycling in HCl (bottom) and after Rh photoelectrodeposition (PED) (top); photon energy 200 eV; (see text).

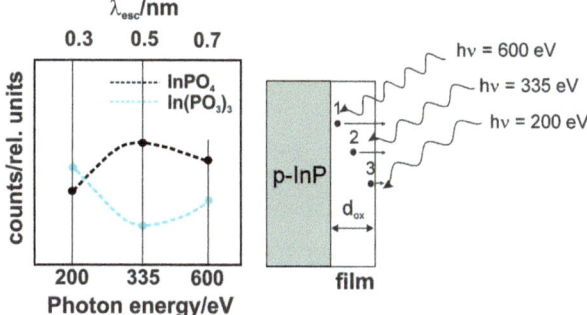

Figure 9.17 Photon energy dependence and mean inelastic photoelectron scattering length of the P-O related signals in the oxidic film on InP (left); right hand side: schematic on the surface sensitivity of the signals (compare also Fig. 9.14) (see text).

The sum of the signals from P-O compounds appears to be rather constant with depth. Therefore, the contribution from indium oxide is assumed to be also rather constant within the probed surface layer, as indicated by the non-graded color scheme in Figure 9.18(a), where the findings regarding the film composition are summarized. In Figure 9.18(b), the partial coverage is indicated in a schematic way for the case of cycling in HCl and, in Figure 9.18(c), after Rh photoelectrodeposition. Due to the time dependence of the photon flux of measurements at the synchrotron, a calibration of the signals is difficult.

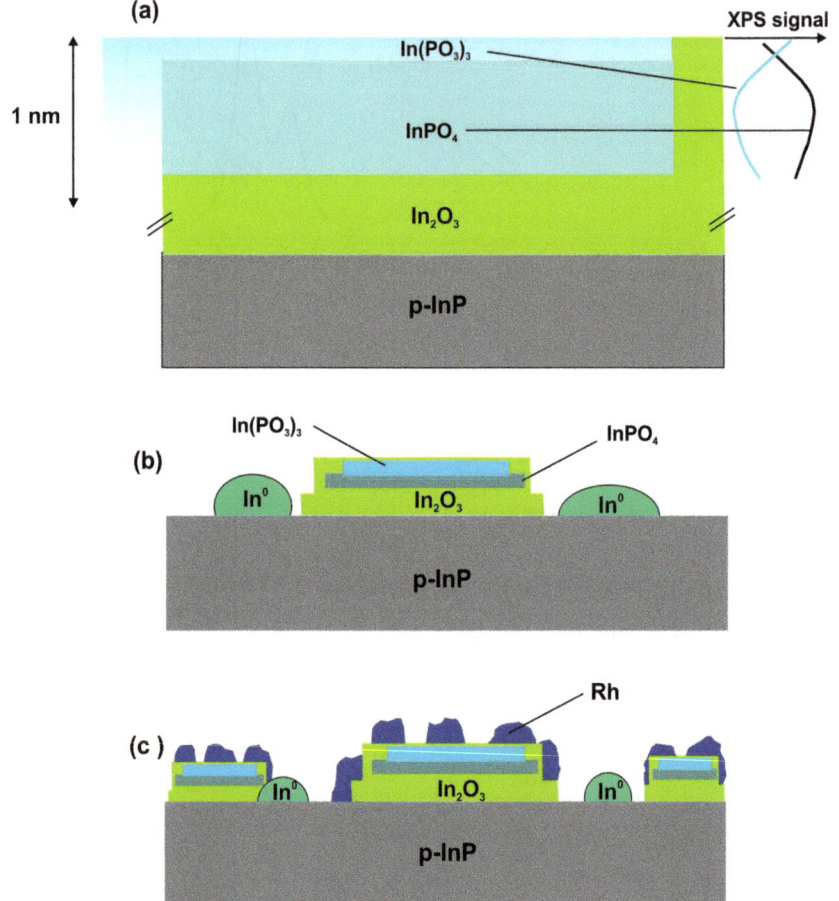

Figure 9.18 Schematic on the surface coverage after photoelectrochemical treatment of p-InP; (a) layer-type arrangement for intelligibility, indicating the relative changes of the P-O related groups with depth after cycling in HCl; (b) more realistic assumption taking into account the findings in Figs. 9.15, 9.16 after HCl treatment; (c) the envisaged situation after Rh photoelectrodeposition (see text).

Therefore, the scheme in Figure 9.17(b) accounts for the basic findings revealed in Figs. 9.15–9.17.

Figure 9.18(a) shows that with photon energies up to 600 eV, the highest depth resolution is about 1 nm. The composition of the oxide film can be extracted from Figs. 9.15–9.17, showing that after HCl cycling, substantial amounts of In metal, InP as well as signals from In_2O_3 and the P-O groups are all found on the surface. This contradicts the planar coverage model shown in Figure 9.18(a), but the figure shows how the phosphates and phosphites are distributed in the topmost part of the film. It cannot be deduced from the data whether the P-O groups are coplanar to indium oxide; here, it was assumed that the film formed after HCl treatment is vertically stacked. Since the corrosion potentials are located between −0.45 and −0.72 V in a rather narrow range compared to the potential cycling range, it is plausible that the oxidation reactions occur simultaneously although one would expect those related to the P-O groups to proceed at higher rate. Figure 9.18(b) shows the assumed arrangement of the components within the film and its basic topography. Since signals from InP, In and the three oxide signals are all present, the film must be patchy as indicated. After Rh deposition, one finds all signals being still present; the relative amounts, however, have changed: the In and InP signals have become smaller with respect to the signals from oxidic products. Therefore, the open surface areas are reduced as well as the top surface In metal islands. The oxide coverage has increased, as indicated in Figure 9.18(c).

9.5.2.3   *Energy Band Relations: Efficiency and Stability Criteria*

The assessment of the energy band relations in the system becomes possible by analysis of UP (ultraviolet photoelectron spectroscopy) spectra and by XPS analysis of the valence band region where the surface sensitivity is reduced, which allows extrapolation of the valence band onsets for weaker signals. Figure 9.19 shows He I UP spectra for HCl treated samples and after Rh deposition. The spectral width S_w, given by the onset of the valence band emission and the secondary electron cut-off, allows to determine the electron affinity and, from the onset of the valence band emission, the band bending on the semiconductor surface. The valence band onset of InP, determined by SRPES at a photon energy of 100 eV ($\lambda_{esc} \sim 0.3$–0.4 nm), is located at about 1.1 eV below

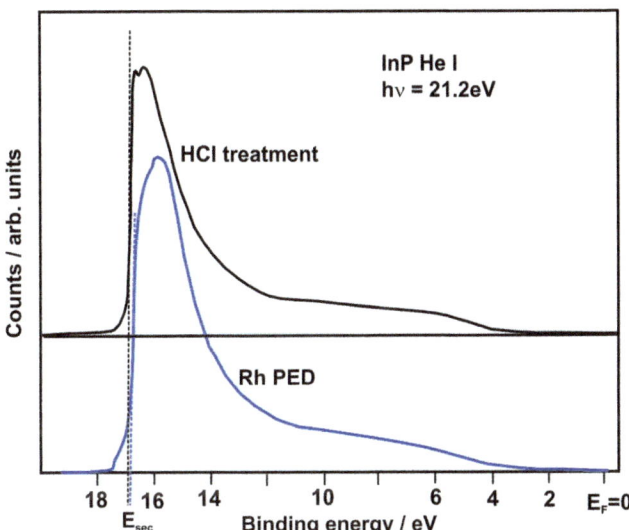

Figure 9.19   He I UP spectra for samples after HCl and Rh treatment as indicated; E_{sec}: secondary electron cut-off (see text).

the Fermi level (binding energy 0). With $E_g = 1.35\,\text{eV}$ for InP and $E_F - E_V \sim 0.2\,\text{eV}$, this indicates a pronounced depletion-type band bending in the absorber surface. The electron affinity is obtained by [13,44]

$$\chi(InP) = h\nu - S_w - E_g \qquad (9.5)$$

For the almost identical cut-offs after HCl treatment and Rh deposition (Figure 9.19), an electron affinity of $\chi = 21.2\,\text{eV} - 15.9\,\text{eV} - 1.35\,\text{eV} \sim 4\,\text{eV}$ is obtained. The pronounced band bending into the limit of strong inversion where the difference $E_C - E_F$ at the surface equals the difference $E_F - E_V$ from the doping in the bulk indicates that a buried p-n junction has been formed. This is discussed further below. The electron affinity of InP, known to be $4.4\,\text{eV}$,[92] has been changed to a smaller value by $0.4\,\text{eV}$, indicating an interfacial dipole that has the positive end pointing outwards of the InP surface. This alignment with the oxide film is advantageous as conduction band electrons, generated as excess carriers in InP upon illumination, have a higher potential energy and can easier cross interfacial barriers.

The InP energy band relations can be extracted more easily from the data on HCl treatment since the density of states near the Fermi level of InP is less than that of Rh with pronounced d-band emission near E_F. One observes that the onset of the emission, due to oxide film coverage, is about 3.5–$4\,\text{eV}$ below E_F. The band gap of In_2O_3 is $2.6\,\text{eV}$[93] but that of $InPO_4$ is considerably larger with $E_g \sim 4.5\,\text{eV}$.[94] Assuming alloying in the film could result in the observed band gap increase. This indicates also that the conduction band edge is likely located close to the oxide film conduction band edge, indicating high n-type doping. We propose that a buried p-n^+ junction has been formed. The energy band diagram in Figure 9.20 shows the situation under short circuit condition. The band edges of InP are shifted downward by $0.4\,\text{V}$, the interfacial dipole at the InP-oxide film interface is indicated by the green arrows. The buried p-n junction between InP and the film is characterized by a reduced contact potential difference (equaling the maximum attainable photovoltage in energy terms, eV_{Ph}^{max}). The resulting cliff in the conduction band promotes minority carrier (electron) transport *via* the film conduction band to Rh and, finally, with a reaction overpotential η to sustain the high photocurrent, to the hydronium ions in solution. Equilibration with the electrolyte resulted also in an upward shift of the Rh Fermi level (i) in order to equilibrate with the redox energy and (ii) to allow passing of the high ($35\,\text{mAcm}^{-2}$) photocurrents.

The device is based on an internal photovoltaic junction formation at the p-InP/n-type film interface. The associated dipole, however, reduces the contact potential difference to about $0.8\,\text{V}$, which is the value actually seen in the experiment. The excited excess electrons move by drift (in the space charge region of InP) through the conduction band of the film and reach unimpeded the Rh catalyst. The rate of thermalization of hot electrons in the Rh film can only be estimated roughly. In Pt, ballistic electrons move in the 5–$10\,\text{nm}$ range at low excess energies.[95,96] The structure therefore allows for excellent stability as the film thickness is not a limitation in the structure. In addition, the concept of band alignment between absorber and passivating film allows the preparation

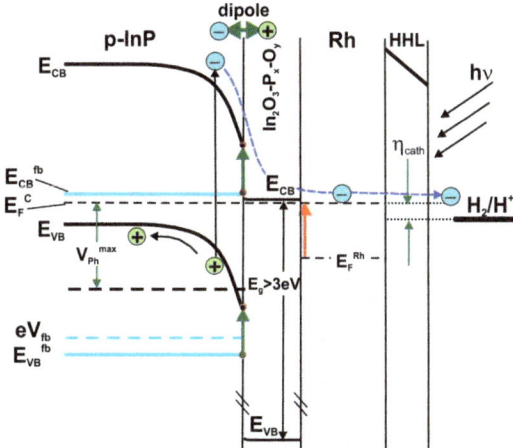

Figure 9.20 Energy schematic of the InP/Film/Rh/electrolyte junction; the sub- or superscripts fb refer to the flatband situation, the superscript C to the Fermi level after contact formation at short circuit condition; green arrows: shift of the InP band edges due to an interfacial dipole at the InP/film contact; red arrow: shift of the Rh Fermi level due to (i) establishing equilibrium with the hydrogen redox couple and (ii) overvoltage η related to high flux of excess minority electrons (see text).

of highly stable devices since the reactive interface has been removed from the absorber surface.

The *in situ* conditioning, a wet process at low temperature which is scalable, has again shown to result in excellent device performance of photoactive structures as already demonstrated for InP, CuInS$_2$, CuInSe$_2$ and Si photo-electrochemical and solid-state photovoltaic systems.[12–14,29,97,98] Issues related to In cost and scarcity could be alleviated by developing specific lift-off techniques for the InP homoepitaxial film.[99]

9.6 GaP(100) and Hetero-Interfaces

GaP represents a model-system for the integration of III-V with Si as substrate. With only a small deviation of the lattice constant (Figure 9.1) and the thermal expansion coefficient, relatively thick layers in the order of 100 nm can be grown pseudomorphically before accumulated strain results in relaxation and hence misfit dislocations. The growth of zinc-blende polar GaP on non-polar, diamond lattice Si(100) substrates is challenging and growth conditions have been studied in detail applying *in situ* RAS control in MOCVD ambients.[100,101] The preparation of well-defined single-domain Si surfaces is a crucial step to avoid the formation of anti-phase disorder, when growing III-V on Si(100).[102] Anti-phase disorder originates from the reduced symmetry of GaP compared to Si. Single- or odd-numbered atomic steps on Si invoke anti-phase domains (APDs) in the GaP film (see Figure 9.21), which in turn create anti-phase boundaries, where homo-atomic bonds have to be formed. Double- or

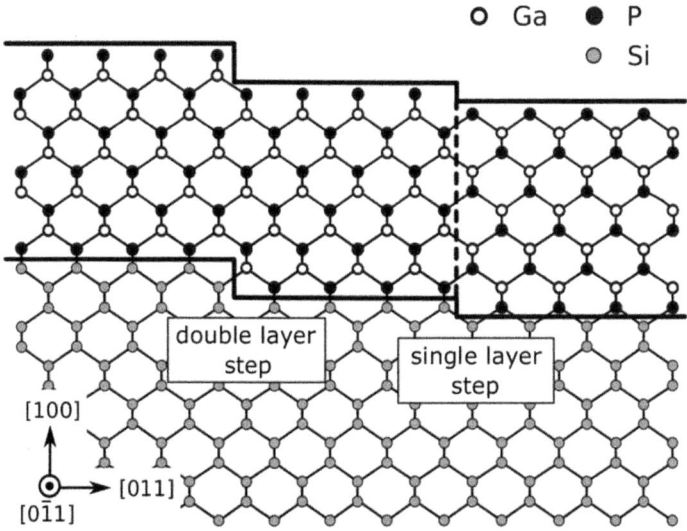

Figure 9.21   Formation of anti-phase disorder due to the reduced symmetry of the III-V semiconductor grown on top. The dashed line indicates an anti-phase boundary at a single layer step.

even-numbered steps on the substrate can avoid the formation of APDs. Therefore, preparation of double-stepped, single-domain Si substrates is employed. Such a surface treatment under RAS in-situ control was developed,[34,70,103] enabling simultaneous quantification as well as prevention of this defect type.

Exact lattice match to Si, however, can be achieved *via* incorporation of nitrogen forming the dilute nitride GaPN.[104] The impact of N on the electronic structure and potential applications will be described in section 9.6.3 below.

For photoelectrochemical applications, the GaP surface has to be functionalized (see section 9.5.2), in order to mitigate limited charge transfer efficiency while simultaneously preventing corrosion. Ideally, the whole process should be monitored *in situ*. RAS could again serve as a powerful tool here, as already demonstrated for metal-electrolyte interfaces.[105] As a first step, we examine the interface between GaP and H_2O in model-experiments outlined in section 9.6.2.

9.6.1   GaP Preparation

The principle of the preparation of GaP on Si(100) is sketched in the following, for details see references 106 and 65. After de-oxidation and homoepitaxial buffer growth with silane, the single-domain, double-stepped, and hydrogen-terminated Si surface is prepared in H_2 atmosphere.[103,107] RAS enables an in-situ quantification of the relative amount of APDs here, as the magnitude of the characteristic RA signal is directly connected to the relative dominance of a single domain. If a high-quality, APD-free Si surface is prepared, the successive GaP nucleation is followed by a heteroepitaxial buffer growth of GaP with the metal-organic precursors TBP and TEGa. After growth, samples are cooled

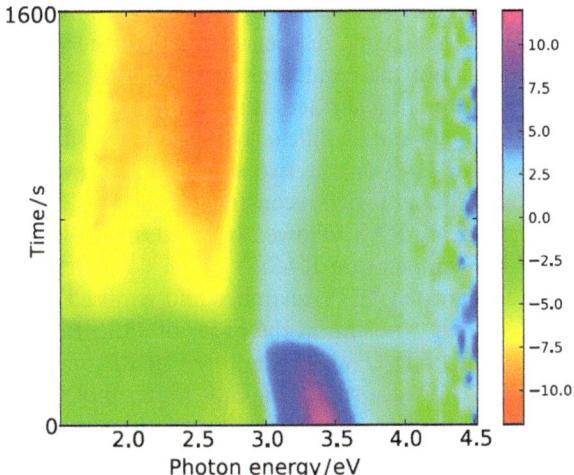

Figure 9.22 Preparation of a III-rich GaP (on Si(100)) surface by annealing. The plot shows color-coded $(\mathrm{Re}(\Delta r/r)/10^{-3})$ subsequent RA spectra over time, details see text.

down to 570 K under TBP stabilization (see 9.4.2 above). The surface then exhibits excess phosphorous which can be desorbed by careful annealing at 690 K without P stabilization.[108]

A Ga-rich surface can be obtained by subsequent annealing of a surface that was initially prepared P-rich as described above. For this purpose, the surface is annealed at 950 K for 5 min without P stabilization. The transition from a P-rich, (2×2) reconstructed surface to a Ga-rich, (2×4) reconstructed surface can be monitored *in situ* with RAS. The typical Ga-rich preparation of an 80 nm heteroepitaxial GaP buffer on a Si(100) substrate (misorientation 2° in the [011] direction) is illustrated in Figure 9.22: The RAS color plot shows successive RA spectra during a temperature ramp of 76 K/min, starting from the P-rich surface at 570 K. At 950 K, the temperature was held constant for 5 min, followed by cooling down again to 570 K. The color indicates the RA intensity $(\mathrm{Re}(\Delta r/r) / 10^{-3})$, time increases on the y-scale. Note that the RA spectra here differ from those of homoepitaxial samples, as the buried interface impacts the RA signal, see ref. 65 for details. For further characterization, samples were directly transferred to UHV from the MOCVD reactor.

9.6.2 Interface Formation with Water

The interface of semiconductors with water is crucial for the understanding of charge-transfer processes, initial oxide formation or corrosion leading to improved (*in situ*) surface conditioning routes. Recently, the interaction of H_2O with GaP has been the subject of theoretical studies, which still need to be put into an experimental context.[109,110] This context could be established by model-experiments using H_2O adsorption in UHV, which have proven to be highly useful to investigate problems of electrochemistry[111] or even questions related to superconductivity.[112] Previously, dissociatively adsorbed water on InP(110) has been investigated,[113] but the results of these experiments are not directly applicable to MOCVD-prepared surfaces since hydrogen is a major ingredient of the MOCVD-process gas and significantly impacts the surfaces.[70,103,114]

To study the impact of H_2O on the two surface reconstructions typical for MOCVD preparation, Ga- and P-rich reconstructed surfaces were transferred to UHV and studied employing RAS followed by studies with PES, FTIR, STM, LEED, LEEM and low-temperature RAS.[64,77,108,115,116] Afterwards, they were also transferred to a dedicated UHV chamber and exposed to H_2O vapor at room temperature, with an exposure in the order of 1 kL. Finally, changes induced by the adsorption of H_2O were monitored in PE and RA spectra. As described in models for H_2O adsorption on GaP, water molecules (or hydroxyl groups) can adsorb and bind to particular sites of the surface reconstruction creating surface motifs such as Ga-[OH]-Ga bridges.[109] Such bond geometries would break the original surface reconstruction, which should be visible to RAS.

For our experiments, we monitored the surfaces *in situ* with RAS during H_2O adsorption. Exposures in the order of several kL were needed to impact the RA signal significantly; see Figure 9.23. In similar experiments reported in the literature, H_2O exposure applied to the closely related InP was significantly lower, *i.e.* in the order of 1 L.[113] This is due to the fact that samples were cleaved in UHV, resulting in a more reactive surface with a higher sticking coefficient and that they were additionally kept at low temperatures.[113] After exposure, samples were transferred in UHV to a PES system.

Figure 9.23 juxtaposes RA spectra of the two surface reconstructions for different H_2O exposures. A general decrease of the signal is observed, indicating a surface reconstruction change. The different contributions of the RA signal indicate a varying influence of H_2O on the different electronic states of the surface reconstruction. The minimum for the P-rich surface around 2.5 eV

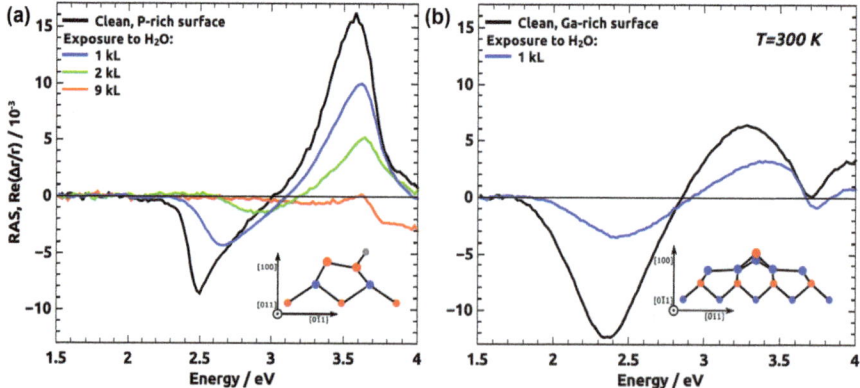

Figure 9.23   RA spectra of GaP(100) before and after exposure to water in UHV at 300 K. a) P-rich and b) Ga-rich surface. Insets show a side-view of the initial P- and Ga-rich surface reconstruction, respectively.

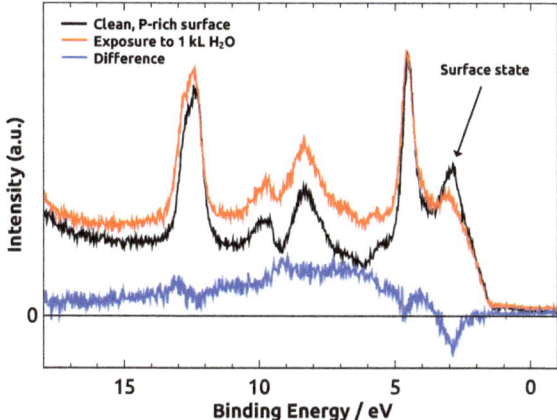

Figure 9.24 He II PE spectrum of P-rich GaP(100) before and after exposure to H_2O. Binding energy relative to Fermi level E_F.

consists of two contributions, where the lower-energetic feature is characteristic for hydrogen termination.[81] This contribution is reduced more quickly upon water adsorption. However, no clear additional RA signal, which would indicate an ordered adsorption on the surface, could yet be observed. This could be due to the finding that the formation energies for different geometries of *e.g.* bonded hydroxyl groups are expected to be very close and a distinction would probably require low temperature experimentation.[109] This is also in line with previous observations of P-dimer flipping at room temperature for P-rich GaP(100).[84]

PES (Figure 9.24) confirms a pronounced reduction of the emission lines around $E_B = 3$ eV, attributed to surface states. At an exposure of 1 kL, the three emission lines typical for molecularly adsorbed H_2O,[117] which should lie in the range of 3 to 13 eV relative to E_F, are absent, indicating dissociative adsorption or very low coverage. This observation suggests a higher stability of the P-rich surface when compared to UHV-cleaved III-V surfaces.[113] More detailed studies and an interpretation in the context of calculated electronic structures are the subject of ongoing investigation.

9.6.3 The GaP(N,As)/Si Tandem System

An epitaxial, lattice-matched tandem structure employing GaP-based III-V compounds on Si constitutes an approach for water splitting, which could enable efficient and bias-free photoelectrolysis. Si substrates, which would serve as bottom cell absorber material, are earth-abundant, available in highest quality and are the semiconductor material that is mainly driving the market for photovoltaics so far. Dilute nitride $GaP_{1-x}N_x$ is supposed to constitute the top cell absorber as it can be grown lattice-matched on Si with a nitrogen content in the range of $x \approx 0.02$, enabling a thin-film application due to its direct-like band gap.[118,119]

Nitrogen atoms are embedded in the crystal's bulk material in non-periodic geometries and therefore, their electronic states are delocalized in k-space. According to the BAC model,[42,120] they interact with the conduction band of GaP creating finally a significantly smaller, "direct-like" band gap of 1.95 eV for $x = 0.02$. A further decrease of the band gap towards the optimum for a tandem (Figure 9.2) can be achieved *via* the incorporation of arsenic and boron, which allow a higher content of N, while maintaining the lattice constant of Si. A photovoltaic GaPNAs tandem, however, which was realized and reported earlier, reached only a V_{OC} of 1.53 V,[62] which would not be sufficient for water splitting. So far, the main deficiency of MOCVD-prepared dilute nitride compounds is a reduced charge carrier lifetime with increasing nitrogen content, effectively decreasing the overall electronic quality of the material.[121] Combining recent advances in the quality of nitrogen precursors along with APD-free epitaxial growth, we aim to significantly improve achievable voltage and overall efficiency.[35] In our first approach, arsenic will not yet be incorporated, enabling higher voltages due to the higher band gap.

Involving nitrogen adds another benefit of improving the band alignment of GaPN, as it increases the electron affinity of GaPN compared to pristine GaP (Figure 9.3). This shifts the conduction band edge closer to the redox potential of hydrogen evolution, possibly improving the electron transfer efficiency to the electrolyte when used as photocathode.[51]

Incorporation of nitrogen already improves the stability in wet-chemical environment,[122] but in addition, one could combine the nano-emitter approach, demonstrated for Si half-cells,[30] with GaPN compounds. In such an approach, the Si surface is electrochemically converted into a nanoporous oxide matrix. The pores of the matrix are then filled with catalysts such as Pt creating nanoemitters. If the distance between the nanoemitters and the carrier diffusion length is properly balanced, the current into the electrolyte flows over the catalytically active nanoemitters, while the semiconductor is protected by the oxide. As the envisaged GaPN compound is lattice-matched to Si, one could grow a 10 nm-thick layer of Si on the dilute nitride and afterwards apply the nanoemitter concept. The quality of the buffer layer would suffer from the growth temperature being too low for optimal Si growth due to the limited temperature stability of the III-V layer. However, possible pinholes due to imperfect growth could be filled and passivated by the catalyst. The average distance between the nanoemitters has to be adapted to the carrier diffusion length in GaPN, which strongly decreases upon the band-crossover.[123] A great benefit of this approach would be a reduction of the required quantity of catalysts to an amount in the order of 100 g per km^2,[30] combined with an effective charge transfer to the electrolyte as well as a passivation of the semiconductor surface.

9.6.4 Micro- and Nanostructured Systems

Absorber layers involving quantum structures hold promise for highly efficient photon energy conversion devices operating beyond the Shockley-Queisser limit of a single band gap material. In size-tunable quantum structures such as quantum wells (see also section 9.3.2), separated electronic states of electrons and holes open opportunities to selectively accumulate charge carriers in well-defined states. In such structures, benefits such as an increased absorption coefficient and slower relaxation, as described in the phonon-bottleneck model,[124] can be considered. The energetics and dynamics of excited electronic states, charge transport and charge separation can be fine-tuned to the specific needs of the envisaged device.

III-V compounds are well suited for designing device structures for exploitation of the photonic excess energy. High-quality III-V p-i-n solar cells involving multiple quantum wells have already demonstrated efficiencies close to 30% under concentrated sunlight light.[125] It has been shown that quantum well structures can be engineered *via* strain-balancing[126] and provide means to engineer the absorption edges of solar cells. In current realizations, it is thought that carriers escape from the strain-balanced quantum wells *via* thermally assisted tunneling. In another approach, illustrated in Figure 9.25, electronic up-conversion of charge carriers is suggested to realize a 2-photon absorption process and to exploit different photon energies more efficiently as possible within the Shockley-Queisser single-band gap limit: the structure schemed in Figure 9.25 allows the absorption of photons with three different energies (typified by red, green and blue arrows) creating a quasi-triple absorber.

In bulk semiconductor material, primary loss mechanisms of electron excess energy occur on time scales of femtoseconds to picoseconds *via* inelastic electron-phonon scattering. This was shown in reference 78, where capture times of optically excited hot electrons were determined in 2-dimensional surface bands of high-purity epitaxial InP bulk material by means of femtosecond-resolved 2-photon photoemission. A 2-photon absorption approach, illustrated in Figure 9.25, implies a strong occupation of the first excited state in the

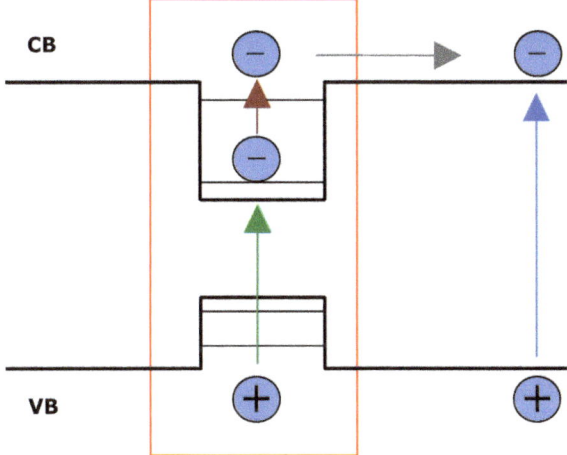

Figure 9.25 Schematic drawing of a quantum well structure and a two-photon (green and red arrow) absorption process.

quantum structure. This enables an efficient absorption of the second photon, *i.e.* the second electronic transition in the quantum structure (upper red arrow in Figure 9.25). Work to explore relaxation and transport mechanisms of hot charge carriers in planar III-V quantum well structures provided a basis for the improvement of meanwhile well-established novel electronic and optoelectronic devices such as high electron mobility transistors,[127] quantum cascade lasers,[128] or quantum well IR photodetectors.[129,130] However, in a planar multiple quantum well device, charge carriers have to pass multiple times through the different quantum wells which increases the probability of capture and to non-radiative recombination. A realization of a corresponding p-i-n core-shell nanowire configuration, sketched in Figure 9.26 would avoid the multiple passages. Nanowire arrays can either be realized by utilizing bottom-up or top-down techniques. In the bottom-up approach, nanowires are grown on a substrate, while the top-down approach creates the nanowires by reducing a given substrate with *e.g.* lithography. Owing to the complexity of the device structure sketched in Figure 9.26 and also for economic reasons (removal of large quantities of the substrate), top-down procedures cannot be considered to be applied. In contrast, it has been shown recently that the catalysis-assisted vapor-liquid-solid growth mode offers the opportunities to prepare corresponding axial and radial III-V nanowire structures in a bottom-up approach with promising photovoltaic performance.[131,132] Surface passivation turned out to be a key issue to reduce the interface recombination and to achieve suitable minority carrier diffusion lengths across the passivated interfaces. In reference 133, it was shown that epitaxial coating of III-V nanowires can basically be prepared almost free of defects and exciton lifetimes, that were quasi equivalent to those of intrinsic bulk material, were measured at $GaAs/Al_xGa_{1-x}As$ nanowire core-shell structures. Currently, intensive studies of preparing these semiconductor structures and measuring charge carrier dynamics on free-standing radial III-V nanowires are the aim of a network research project and ongoing.

Hence, coaxial p-i-n core-shell-shell nanowire arrangements, based on III-V semiconductor compounds and combined with silicon half-cells, are suggested here and illustrated in Figure 9.26 to be considered for highly efficient water splitting devices. The controlled realization of III-V-based radial

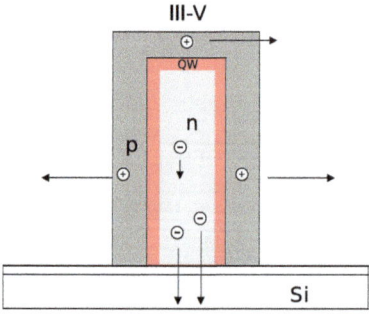

Figure 9.26   Schematic drawing of a p-i-n core-shell-shell III-V device component on a Si substrate for water splitting. QW: quantum well.

core-shell-shell[134] nanowire structures opens new opportunities and advantages in comparison to planar layer structures due to (i) a strongly increased surface area, (ii) a much smaller interface area between nanowires and substrate, and (iii) much shorter charge transfer distances. Based on these merits, a number of additional benefits arise:

1. The so-called critical thickness is usually limiting highly lattice-mismatched, defect-free epitaxial growth.[135] A greatly reduced interface area to the substrate in a nanowire allows for highly mismatched heterostructure growth, when compared to planar structures or micro-wires. Consequently, the freedom in terms of possible III-V material combinations and thus of freely engineered interfaces to stagger multiple junctions, is significantly higher.

2. Strongly increased absorber surface areas give rise to sufficient light absorption, also reinforced by enhanced absorption coefficients in quantum wells. The Drude absorption of metal catalysts, however, can result in substantial losses in the absorbance of the semiconductors. This can be alleviated if electrocatalysts that are semiconducting are used.

3. A highly enlarged surface naturally also increases the reactive interface, where the solar-driven electrochemical reaction for water splitting occurs by orders of magnitude. This broadens the range of utilizable (photo-active) catalysts. It also increases the kinetic parameter exchange current density thus allowing earth abundant materials to be used as heterogeneous catalysts.

4. As charge separation occurs on shorter distances due to the orthogonalization of light absorption and charge carrier separation, the diffusion length is much less critical compared to a planar device structure. Hence, the requirement on material purity is less critical and in consequence allows material selection beyond technologically advanced semiconductors.

5. Such nanowire core-shell structures and their functionality is not only attractive for solar applications (photovoltaics, solar-driven water splitting), but for all electronic and optoelectronic device applications. Nano- and microwire arrays allow integration with membranes in a scalable manner if the array can be prepared on a scalable basis.

Besides the interfacial microscopic understanding of multi-photon absorption and the photoelectrochemical reactions, further aspects such as plasmonic structures for light management and catalytic efficiency, optimized arrangement of nanowires and the interfacial properties between the micro-/nanowires and the substrate as well as between the nanowires' core and shell are of interest to be pursued (see Figure 9.26).

Figure 9.27 shows the electronic structure of an illuminated p-i-n core-shell-shell half-cell. A first step of light absorption is illustrated in the n-doped core material. Here, the increased electron and hole concentration due to light absorption shifts the quasi Fermi levels towards the conduction band minimum

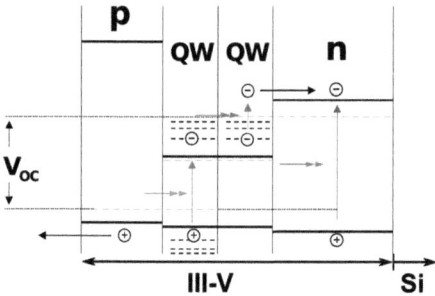

Figure 9.27   Electronic structure of the III-V p-i-n core-shell-shell structure proposed
here for a highly efficient water splitting device.[137,138] Band edges, quasi
Fermi levels, as well as electronic levels and transitions are indicated in
the core material (n-doped), the quantum well (intrinsic) and the large
band gap, charge separating shell (p-doped). QW: quantum well, V_{OC}:
open-circuit voltage. Green, dashed lines: quasi Fermi levels.

(E_{CBM}) and valence band maximum (E_{VBM}), respectively. The two-photon
absorption scenario in the quantum well, suggested here, is displayed in two
separated adjacent diagrams and two electronic transitions (QW left-right). The
first light absorption step occurs in the quantum well as interband transition
(green arrow) from the initial level to an intermediate one. The increased
electron and hole concentration due to light absorption is also illustrated, as the
quasi Fermi levels are shifted towards the band edges of the quantum well. The
photon energy required for that is thought to be smaller than for the band-band
transition of the core material. For the electronic up-conversion of the two-
photon-absorption process and for the second absorption step (red arrow), it is
essential that the intermediate electronic level of the quantum well is strongly
occupied, *i.e.* that the band-band transition of the quantum well is efficiently
pumped. This could be supported by a slowdown of relaxation and recombination rates in the quantum well due to quantum-size and phonon-bottleneck
effect.[124,136] Besides appropriate material parameters and electronic properties in
the quantum well, this will be a critical issue of the quantum well interfaces to the
adjacent material (core and outer shell). Absorption of the second photon of the
two-step absorption in the quantum well and the corresponding electronic
transition is thought to occur with the lowest photon energy also illustrated in the
band diagram of Figure 9.27 (red arrow). The final step of the proposed scenario
is the charge separation in the outer shell. For that, a p-doped large band gap
compound is needed, with a high conductivity for holes and a very low conductivity for electrons. The same concept could be realized by opposite
arrangement of the radial p-i-n structure discussed and illustrated above, *i.e.* a
p-type doped core, intrinsic quantum well, and n- doped outer shell.

9.7   Synopsis

This chapter has shown that the use of III-V absorbers for light-induced water
splitting enables highest solar to-fuel conversion efficiencies. The material class

is, in addition, characterized by an extraordinary flexibility that allows tuning of energy gaps, optical properties (direct *vs.* indirect band gaps) and well-defined homo- and heteroepitaxial growth of thin films. Employing lift-off techniques or nanostructures allows material conscious applications. The preparation of III-V heterostructures with abrupt interfaces by MOCVD applying optical in-situ control has been demonstrated. Based on the high control of growth parameters, it was possible to prepare custom-made thin films for the development of an efficient and stable photocathode starting from an In-rich prepared InP(100) absorber. In-situ photoelectrochemical surface transformation led to the formation of a phosphate containing In_2O_3 interfacial layer that generated a p-n$^+$ buried photovoltaic junction. This enabled both a stabilization of the InP against corrosion and an efficient electron transfer of photo-generated electrons to the Rh catalyst due to appropriate band alignment.

Extending this work on half cells for photoelectrocatalysis to the development of water splitting devices, we will make use of our recent advances in III-V/IV heteroepitaxy. In particular, the dilute nitride absorber GaPN with a direct energy gap around 2 eV will be incorporated into such a structure, where lattice mismatch with the bottom cell is virtually absent. Our approach for surface functionalization regarding stabilization layers and electrocatalyst deposition will employ atomic layer deposition and in-situ photoelectrochemical methods (soft solution processing). Besides the realization of planar tandem structures, the work will also be extended to the development of quantum well-based nanostructures.

Acknowledgements

Many thanks to Peter Würfel and Klaus Schwarzburg for intense and valuable discussions. Part of this work was supported by the BMBF (Project No. 03SF0404A). MM May acknowledges financial support by Studienstiftung des deutschen Volkes. The authors would also like to thank Oliver Supplie and Helena Stange for helpful discussions. This material is based upon work performed by the Joint Center for Artificial Photosynthesis, a DOE Energy Innovation Hub, as follows: the interpretation of the photoelectrochemical and synchrotron radiation photoelectron spectroscopy experiments as well as the manuscript wording and composition were supported though the Office of Science of the US Department of Energy under Award No. DE-SC0004993.

References

1. A. J. Bard and L. R. Faulkner, *Electrochemical methods: fundamentals and applications*, Wiley, New York, 2. edn., 2001.
2. M. G. Walter, E. L. Warren, J. R. McKone, S. W. Boettcher, Q. Mi, E. A. Santori and N. S. Lewis, *Chem. Rev.*, 2010, **110**, 6446–6473.
3. H. Gerischer, *J. Electroanal. Chem. Interfacial Electrochem.*, 1977, **82**, 133–143.

4. H. Tributsch, *Sol. Energy Mater. Sol. Cells*, 1994, **31**, 548–558.
5. H. Wang, T. Deutsch, A. Welch and J. A. Turner, *Int. J. Hydrogen Energy*, 2012, **37**, 14009–14014.
6. A. J. Bard and M. A. Fox, *Acc. Chem. Res.*, 1995, **28**, 141–145.
7. S. Haussener, C. Xiang, J. M. Spurgeon, S. Ardo, N. S. Lewis and A. Z. Weber, *Energy Environ. Sci.*, 2012, **5**, 9922–9935.
8. A. Magnuson and S. Styring, *Aust. J. Chem.*, 2012, **65**, 564–572.
9. A. Fujishima and K. Honda, *Nature*, 1972, **238**, 37–38.
10. R. van de Krol, Y. Liang and J. Schoonman, *J. Mater. Chem.*, 2008, **18**, 2311–2320.
11. S. Fiechter, P. Bogdanoff, T. Bak and J. Nowotny, *Adv. Appl. Ceram.*, 2012, **111**, 39–43.
12. S. Menezes, H.-J. Lewerenz and K. J. Bachmann, *Nature*, 1983, **305**, 615–616.
13. H. J. Lewerenz and K. H. Schulte, *Electrochim. Acta*, 2002, **47**, 2639–2651.
14. T. Stempel, M. Aggour, K. Skorupska, A. Muñoz and H.-J. Lewerenz, *Electrochem. Commun.*, 2008, **10**, 1184–1186.
15. B. Kalanyan and G. Parsons, *ECS Trans.*, 2011, **41**, 285–292.
16. M. Tallarida and D. Schmeisser, *Semicond. Sci. Technol.*, 2012, **27**, 074010.
17. A. Fujishima, K. Kohayakawa and K. Honda, *J. Electrochem. Soc.*, 1975, **122**, 1487–1489.
18. A. J. Bard, *J. Photochem.*, 1979, **10**, 59–75.
19. R. E. Rocheleau, E. L. Miller and A. Misra, *Energy Fuels*, 1998, **12**, 3–10.
20. O. Khaselev and J. A. Turner, *Science*, 1998, **280**, 425–427.
21. H. Gerischer, *Pure Appl. Chem.*, 1980, **52**, 2649–2667.
22. J. R. Bolton, S. J. Strickler and J. S. Connolly, *Nature*, 1985, **316**, 495–500.
23. Y. Sakai, S. Sugahara, M. Matsumura, Y. Nakato and H. Tsubomura, *Can. J. Chem.*, 1988, **66**, 1853–1856.
24. R. Abe, *J. Photochem. Photobiol., C*, 2010, **11**, 179–209.
25. R. R. King, D. C. Law, K. M. Edmondson, C. M. Fetzer, G. S. Kinsey, H. Yoon, R. A. Sherif and N. H. Karam, *Appl. Phys. Lett.*, 2007, **90**, 183516–183516–3.
26. W. Guter, J. Schöne, S. P. Philipps, M. Steiner, G. Siefer, A. Wekkeli, E. Welser, E. Oliva, A. W. Bett and F. Dimroth, *Appl. Phys. Lett.*, 2009, **94**, 223504–223504–3.
27. M. A. Green, K. Emery, Y. Hishikawa, W. Warta and E. D. Dunlop, *Prog. Photovoltaics: Research and Applications*, 2012, **20**, 606–614.
28. A. Heller, H. J. Lewerenz and B. Miller, *Berichte der Bunsengesellschaft für physikalische Chemie*, 1980, **84**, 592–595.
29. K. H. Schulte and H. J. Lewerenz, *Electrochim. Acta*, 2002, **47**, 2633–2638.
30. H. J. Lewerenz, C. Heine, K. Skorupska, N. Szabo, T. Hannappel, T. Vo-Dinh, S. A. Campbell, H. W. Klemm and A. G. Muñoz, *Energy Environ. Sci.*, 2010, **3**, 748.

31. H. Yoneyama, H. Sakamoto and H. Tamura, *Electrochim. Acta*, 1975, **20**, 341–345.
32. A. Heller, B. Miller, H. J. Lewerenz and K. J. Bachmann, *J. Am. Chem. Soc.*, 1980, **102**, 6555–6556.
33. S. Menezes, H. J. Lewerenz, F. A. Thiel and K. J. Bachmann, *Appl. Phys. Lett.*, 1981, **38**, 710–712.
34. H. Döscher, O. Supplie, S. Brückner, T. Hannappel, A. Beyer, J. Ohlmann and K. Volz, *J. Cryst. Growth*, 2011, **315**, 16–21.
35. H. Döscher, O. Supplie, M. M. May, P. Sippel, C. Heine, A. G. Muñoz, R. Eichberger, H.-J. Lewerenz and T. Hannappel, *ChemPhysChem*, 2012, **13**, 2899–2909.
36. A. de Vos and D. Vyncke, *Proc. 5th E.C. Photovoltaic Solar Energy Conf. (Athens)*, 1983, 186.
37. J. M. Olson, S. R. Kurtz, A. E. Kibbler and P. Faine, *Appl. Phys. Lett.*, 1990, **56**, 623–625.
38. R. R. King, D. Bhusari, D. Larrabee, X.-Q. Liu, E. Rehder, K. Edmondson, H. Cotal, R. K. Jones, J. H. Ermer, C. M. Fetzer, D. C. Law and N. H. Karam, *Prog. Photovoltaics: Research and Applications*, 2012, **20**, 801–815.
39. B. E. Sağol, U. Seidel, N. Szabó, K. Schwarzburg and T. Hannappel, *Chimia*, 2007, **61**, 775–779.
40. G. Létay and A. W. Bett, *Proceedings of the 17th European Photovoltaic Solar Energy Conference*, 2001, 178–181.
41. I. Vurgaftman, J. R. Meyer and L. R. Ram-Mohan, *J. Appl. Phys.*, 2001, **89**, 5815–5875.
42. J. Wu, W. Walukiewicz, K. M. Yu, J. W. Ager, E. E. Haller, Y. G. Hong, H. P. Xin and C. W. Tu, *Phys. Rev. B*, 2002, **65**, 241303.
43. W. N. Hansen and D. M. Kolb, *J. Electroanal. Chem. Interfacial Electrochem.*, 1979, **100**, 493–500.
44. H. J. Lewerenz, *Chem. Soc. Rev.*, 1997, **26**, 239.
45. J. Turner, G. Sverdrup, M. K. Mann, P.-C. Maness, B. Kroposki, M. Ghirardi, R. J. Evans and D. Blake, *Int. J. Energy Res.*, 2008, **32**, 379–407.
46. J. Sun, C. Liu and P. Yang, *J. Am. Chem. Soc.*, 2011, **133**, 19306–19309.
47. A. Heller and R. G. Vadimsky, *Phys. Rev. Lett.*, 1981, **46**, 1153–1156.
48. R. L. Anderson, *Solid-State Electron.*, 1962, **5**, 341–351.
49. H. Kroemer, *Proc. IEEE*, 1963, **51**, 1782–1783.
50. Z. Kollonitsch, H.-J. Schimper, U. Seidel, K. Möller, S. Neumann, F.-J. Tegude, F. Willig and T. Hannappel, *J. Cryst. Growth*, 2006, **287**, 536–540.
51. B. Kaiser, D. Fertig, J. Ziegler, J. Klett, S. Hoch and W. Jaegermann, *ChemPhysChem*, 2012, **13**, 3053–3060.
52. J. Mukherjee, B. Erickson and S. Maldonado, *J. Electrochem. Soc.*, 2010, **157**, H487–H495.
53. A. Etcheberry, J. Lou Sculfort and A. Marbeuf, *Sol. Energy Mater.*, 1980, **3**, 347–355.

54. A. G. Munoz, C. Heine, H. W. Klemm, T. Hannappel, N. Szabo and H.-J. Lewerenz, *ECS Trans.*, 2011, **35**, 141–150.

55. M. J. Ludowise, R. A. LaRue, P. G. Borden, P. E. Gregory and W. T. Dietze, *Appl. Phys. Lett.*, 1982, **41**, 550–552.

56. B.-C. Chung, G. F. Virshup and J. G. Werthen, *Appl. Phys. Lett.*, 1988, **52**, 1889–1891.

57. T. Takamoto, E. Ikeda, H. Kurita, M. Ohmori, M. Yamaguchi and M.-J. Yang, *Jpn. J. Appl. Phys.*, 1997, **36**, 6215–6220.

58. W. T. Dietze, M. J. Ludowise and P. E. Gregory, *Appl. Phys. Lett.*, 1982, **41**, 984–986.

59. F. Dimroth, R. Beckert, M. Meusel, U. Schubert and A. W. Bett, *Prog. Photovoltaics: Research and Applications*, 2001, **9**, 165–178.

60. J. Geisz, J. Olson, W. McMahon, T. Hannappel, K. Jones, H. Moutinho, and M. Al-Jassim, in *Progress in Compound Semiconductor Materials Iii - Electronic and Optoelectronic Applications*, eds. D. Friedman, O. Manasreh, I. Buyanova, A. Munkholm,and F. Auret, Materials Research Society, Warrendale, 2004, vol. 799, pp. 27–31.

61. S. F. Fang, K. Adomi, S. Iyer, H. Morkoç, H. Zabel, C. Choi and N. Otsuka, *J. Appl. Phys.*, 1990, **68**, R31–R58.

62. J. Geisz, J. Olson, D. Friedman, K. Jones, R. Reedy and M. Romero, in *Conference Record of the Thirty-First IEEE Photovoltaic Specialists Conference – 2005*, IEEE, New York, 2005, pp. 695–698.

63. W. G. Schmidt, N. Esser, A. M. Frisch, P. Vogt, J. Bernholc, F. Bechstedt, M. Zorn, T. Hannappel, S. Visbeck, F. Willig and W. Richter, *Phys. Rev. B*, 2000, **61**, R16335–R16338.

64. L. Töben, T. Hannappel, K. Möller, H.-J. Crawack, C. Pettenkofer and F. Willig, *Surf. Sci.*, 2001, **494**, L755–L760.

65. O. Supplie, T. Hannappel, M. Pristovsek and H. Döscher, *Phys. Rev. B*, 2012, **86**, 035308.

66. D. E. Aspnes and A. A. Studna, *Phys. Rev. Lett.*, 1985, **54**, 1956–1959.

67. W. Richter and J.-T. Zettler, *Appl. Surf. Sci.*, 1996, **100–101**, 465–477.

68. P. Weightman, D. S. Martin, R. J. Cole and T. Farrell, *Rep. Prog. Phys.*, 2005, **68**, 1251–1341.

69. N. Witkowski, R. Coustel, O. Pluchery and Y. Borensztein, *Surf. Sci.*, 2006, **600**, 5142–5149.

70. S. Brückner, H. Döscher, P. Kleinschmidt and T. Hannappel, *Appl. Phys. Lett.*, 2011, **98**, 211909–211909–3.

71. W. G. Schmidt, F. Bechstedt and J. Bernholc, *Phys. Rev. B*, 2001, **63**, 045322.

72. D. E. Aspnes, *J. Vac. Sci. Technol., B: Microelectron. Nanometer Struct.-Process., Meas., Phenom.*, 1985, **3**, 1498–1506.

73. D. E. Aspnes and N. Dietz, *Appl. Surf. Sci.*, 1998, **130–132**, 367–376.

74. Z. Kollonitsch, K. Möller, H.-J. Schimper, C. Giesen, M. Heuken, F. Willig and T. Hannappel, *J. Cryst. Growth*, 2004, **261**, 289 293.

75. S. M. Scholz, A. B. Müller, W. Richter, D. R. T. Zahn, D. I. Westwood, D. A. Woolf and R. H. Williams, *J. Vac. Sci. Technol., B: Microelectron. Nanometer Struct.--Process., Meas., Phenom.*, 1992, **10**, 1710–1715.
76. K. Möller, L. Töben, Z. Kollonitsch, C. Giesen, M. Heuken, F. Willig and T. Hannappel, *Appl. Surf. Sci.*, 2005, **242**, 392–398.
77. T. Hannappel, S. Visbeck, L. Töben and F. Willig, *Rev. Sci. Instrum.*, 2004, **75**, 1297.
78. L. Töben, L. Gundlach, R. Ernstorfer, R. Eichberger, T. Hannappel, F. Willig, A. Zeiser, J. Förstner, A. Knorr, P. H. Hahn and W. G. Schmidt, *Phys. Rev. Lett.*, 2005, **94**, 067601.
79. R. F. C. Farrow, *J. Phys. D: Appl. Phys.*, 1974, **7**, 2436.
80. K. Haberland, P. Kurpas, M. Pristovsek, J.-T. Zettler, M. Weyers and W. Richter, *Appl. Phys. A: Mater. Sci. Process.*, 1999, **68**, 309–313.
81. W. G. Schmidt, P. H. Hahn, F. Bechstedt, N. Esser, P. Vogt, A. Wange and W. Richter, *Phys. Rev. Lett.*, 2003, **90**, 126101.
82. P. Vogt, A. M. Frisch, T. Hannappel, S. Visbeck, F. Willig, C. Jung, R. Follath, W. Braun, W. Richter and N. Esser, *Appl. Surf. Sci.*, 2000, **166**, 190–195.
83. A. M. Frisch, P. Vogt, S. Visbeck, T. Hannappel, F. Willig, W. Braun, W. Richter, J. Bernholc, W. G. Schmidt and N. Esser, *Appl. Surf. Sci.*, 2000, **166**, 224–230.
84. P. Kleinschmidt, H. Döscher, P. Vogt and T. Hannappel, *Phys. Rev. B*, 2011, **83**, 155316.
85. N. Szabó, B. E. Sağol, U. Seidel, K. Schwarzburg and T. Hannappel, *Phys. Status Solidi RRL*, 2008, **2**, 254–256.
86. U. Seidel, H. Döscher, C. Lehmann, C. Pettenkofer and T. Hannappel, *Surf. Sci.*, 2010, **604**, 2012–2015.
87. T. Bitzer and H. J. Lewerenz, *Surf. Sci.*, 1992, **269–270**, 886–892.
88. S. Rauscher, T. Dittrich, M. Aggour, J. Rappich, H. Flietner and H. J. Lewerenz, *Appl. Phys. Lett.*, 1995, **66**, 3018–3020.
89. R. T. Sanderson, *J. Am. Chem. Soc.*, 1983, **105**, 2259–2261.
90. R. T. Sanderson, *Inorg. Chem.*, 1986, **25**, 1856–1858.
91. P. J. Linstrom and W. G. Mallard, Eds., *NIST Chemistry WebBook, NIST Standard Reference Database Number 69*, National Institute of Standards and Technology, Gaithersburg MD, 2013.
92. T. E. Fischer, *Phys. Rev.*, 1966, **142**, 519–523.
93. C. Janowitz, V. Scherer, M. Mohamed, A. Krapf, H. Dwelk, R. Manzke, Z. Galazka, R. Uecker, K. Irmscher, R. Fornari, M. Michling, D. Schmeißer, J. R. Weber, J. B. Varley and C. G. V. de Walle, *New J. Phys.*, 2011, **13**, 085014.
94. J. F. Wager, C. W. Wilmsen and L. L. Kazmerski, *Appl. Phys. Lett.*, 1983, **42**, 589–591.
95. V. P. LaBella, L. J. Schowalter and J. C. A. Ventrice, in *The 24th conference on the physics and chemistry of semiconductor interfaces*, AVS, 1997, vol. 15, 1191–1195.

96. H. Nienhaus, S. . Weyers, B. Gergen and E. McFarland, *Sens. Actuators, B*, 2002, **87**, 421–424.
97. H. J. Lewerenz, H. Goslowsky, K.-D. Husemann and S. Fiechter, *Nature*, 1986, **321**, 687–688.
98. B. Berenguier and H. J. Lewerenz, *Electrochem. Commun.*, 2006, **8**, 165–169.
99. P. Demeester, I. Pollentier, P. D. Dobbelaere, C. Brys and P. V. Daele, *Semicond. Sci. Technol.*, 1993, **8**, 1124–1135.
100. H. Döscher, T. Hannappel, B. Kunert, A. Beyer, K. Volz and W. Stolz, *Appl. Phys. Lett.*, 2008, **93**, 172110.
101. H. Döscher and T. Hannappel, *J. Appl. Phys.*, 2010, **107**, 123523.
102. H. Kroemer, *J. Cryst. Growth*, 1987, **81**, 193–204.
103. S. Brückner, H. Döscher, P. Kleinschmidt, O. Supplie, A. Dobrich and T. Hannappel, *Phys. Rev. B*, 2012, **86**, 195310.
104. Y. Furukawa, H. Yonezu, K. Ojima, K. Samonji, Y. Fujimoto, K. Momose and K. Aiki, *Jpn. J. Appl. Phys.*, 2002, **41**, 528–532.
105. C. I. Smith, A. Bowfield, G. J. Dolan, M. C. Cuquerella, C. P. Mansley, D. G. Fernig, C. Edwards and P. Weightman, *J. Chem. Phys.*, 2009, **130**, 044702–044702-9.
106. K. Volz, A. Beyer, W. Witte, J. Ohlmann, I. Németh, B. Kunert and W. Stolz, *J. Cryst. Growth*, 2011, **315**, 37–47.
107. A. Dobrich, P. Kleinschmidt, H. Doescher and T. Hannappel, *J. Vac. Sci. Technol., B: Microelectron. Nanometer Struct.--Process., Meas., Phenom.*, 2011, **29**.
108. H. Döscher, K. Möller and T. Hannappel, *J. Cryst. Growth*, 2011, **318**, 372–378.
109. B. C. Wood, T. Ogitsu and E. Schwegler, *J. Chem. Phys.*, 2012, **136**, 064705–064705-11.
110. S. Jeon, H. Kim, W. A. Goddard and H. A. Atwater, *J. Phys. Chem. C*, 2012, **116**, 17604–17612.
111. W. Jaegermann, *Modern Aspects of Electrochemistry 30*, Plenum Press, New York, 1996.
112. M. M. May, C. Brabetz, C. Janowitz and R. Manzke, *Phys. Rev. Lett.*, 2011, **107**, 176405.
113. O. Henrion, A. Klein and W. Jaegermann, *Surf. Sci.*, 2000, **457**, L337–L341.
114. T. Letzig, H.-J. Schimper, T. Hannappel and F. Willig, *Phys. Rev. B*, 2005, **71**, 033308.
115. T. Hannappel, L. Töben, K. Möller and F. Willig, *J. Electron. Mater.*, 2001, **30**, 1425–1428.
116. H. Döscher, B. Borkenhagen, G. Lilienkamp, W. Daum and T. Hannappel, *Surf. Sci.*, 2011, **605**, L38–L41.
117. M. A. Henderson, *Surf. Sci. Rep.*, 2002, **46**, 1–308.
118. H. P. Xin, C. W. Tu, Y. Zhang and A. Mascarenhas, *Appl. Phys. Lett.*, 2000, **76**, 1267.

119. S. V. Dudiy, A. Zunger, M. Felici, A. Polimeni, M. Capizzi, H. P. Xin and C. W. Tu, *Phys. Rev. B*, 2006, **74**, 155303.
120. W. Shan, W. Walukiewicz, K. M. Yu, J. Wu, J. W. Ager, E. E. Haller, H. P. Xin and C. W. Tu, *Appl. Phys. Lett.*, 2000, **76**, 3251.
121. J. F. Geisz, R. C. Reedy, B. M. Keyes and W. K. Metzger, *J. Cryst. Growth*, 2003, **259**, 223–231.
122. T. G. Deutsch, C. A. Koval and J. A. Turner, *J. Phys. Chem. B*, 2006, **110**, 25297–25307.
123. I. A. Buyanova, G. Pozina, J. P. Bergman, W. M. Chen, H. P. Xin and C. W. Tu, *Appl. Phys. Lett.*, 2002, **81**, 52–54.
124. J. H. Van Vleck, *Phys. Rev.*, 1941, **59**, 724–729.
125. J. G. J. Adams, W. Elder, G. Hill, J. S. Roberts, K. W. J. Barnham, and N. J. Ekins-Daukes, *Proc. SPIE*, 2010, **7597**, 759705–759705.
126. N. J. Ekins-Daukes, K. W. J. Barnham, J. P. Connolly, J. S. Roberts, J. C. Clark, G. Hill and M. Mazzer, *Appl. Phys. Lett.*, 1999, **75**, 4195–4197.
127. P. Yuh and K. L. Wang, *Appl. Phys. Lett.*, 1986, **49**, 1738–1740.
128. J. Faist, F. Capasso, D. L. Sivco, C. Sirtori, A. L. Hutchinson and A. Y. Cho, *Science*, 1994, **264**, 553–556.
129. J. Shah, *Solid-State Electron.*, 1989, **32**, 1051–1056.
130. J. Shah, *Ultrafast spectroscopy of semiconductors and semiconductor nanostructures*, Springer, Berlin, 1996.
131. C. Gutsche, A. Lysov, D. Braam, I. Regolin, G. Keller, Z.-A. Li, M. Geller, M. Spasova, W. Prost and F.-J. Tegude, *Adv. Funct. Mater.*, 2012, **22**, 929–936.
132. C. Gutsche, R. Niepelt, M. Gnauck, A. Lysov, W. Prost, C. Ronning and F.-J. Tegude, *Nano Lett.*, 2012, **12**, 1453–1458.
133. S. Perera, M. A. Fickenscher, H. E. Jackson, L. M. Smith, J. M. Yarrison-Rice, H. J. Joyce, Q. Gao, H. H. Tan, C. Jagadish, X. Zhang and J. Zou, *Appl. Phys. Lett.*, 2008, **93**, 053110–053110–3.
134. F. Qian, Y. Li, S. Gradečak,, D. Wang, C. J. Barrelet and C. M. Lieber, *Nano Lett.*, 2004, **4**, 1975–1979.
135. S. C. Jain, M. Willander and H. Maes, *Semicond. Sci. Technol.*, 1996, **11**, 641–671.
136. I.-K. Park, M.-K. Kwon, C.-Y. Cho, J.-Y. Kim, C.-H. Cho and S.-J. Park, *Appl. Phys. Lett.*, 2008, **92**, 253105–253105–3.
137. Personal communication with P. Würfel.
138. Personal communication with K. Schwarzburg.

CHAPTER 10

Photoelectrochemical Water Splitting: A First Principles Approach

ANDERS HELLMAN

Applied Physics and Competence Centre for Catalysis, Chalmers University of Technology, SE-41296, Sweden
Email: ahell@chalmers.se

10.1 Introduction

In the novel, *L'île mystérieuse*, Jules Verne writes, "water will be the coal of the future". This is the passing conclusion of the main characters after a lengthy discussion on whether the known coal reserves of that time would be depleted within the near future. The discussion in the novel is still very relevant today. Crude oil is the backbone of the energy system in modern society. However, owing to increased global energy consumption, expectation of peak oil production and in conjunction with increasing concerns regarding environmental impact, there is a growing awareness that we need to find a renewable source of energy, and thereby decrease our dependence on fossil fuels. Although Verne will never be right about water being used as fuel (due to thermodynamics considerations), water still is a key element in the envisioned future hydrogen-based energy infrastructure. With enough supply of energy, water can be separated into its elementary components, hydrogen and oxygen, and the hydrogen can be used, for example, as fuel in a fuel cell. As a matter of fact, the use of hydrogen and oxygen is also what was envisioned in the story of Jules Verne.

RSC Energy and Environment Series No. 9
Photoelectrochemical Water Splitting: Materials, Processes and Architectures
Edited by Hans-Joachim Lewerenz and Laurence Peter
Published by the Royal Society of Chemistry, www.rsc.org

Continuing this line of thought, the challenge remains to find a source of energy that is abundant, free and sustainable. The most obvious idea is to harvest the energy of the sun, which is constantly bombarding the Earth with enough energy to sustain society over any foreseeable future.[1–4] Energy from the sun can be harvested in many different ways. Sunlight can be transformed into chemical energy by means of photosynthesis, which is crucial for life on earth. For fuel production, plants with low water and fertilizer requirements can be used to produce large quantities of biomass that can be used at a later stage to produce power or fuel. Furthermore, sunlight can be used to drive photovoltaic cells, which produce electricity that can be converted into fuels by electrocatalytic processes. The electricity can also be produced from wind or wave power originating from the energy of the sun. Finally, the absorption of sunlight can be coupled directly with electrochemical processes in photoelectrocatalytic devices that produce fuel directly from electrical currents obtained from absorbed sunlight.

Although Bequerel, with his measurement of photovoltaic effect from an illuminated silver chloride electrode (*ca.*1839) can be viewed as the founder of photoelectrochemistry, it is the measurements by Honda and Fujishima[5] published in 1972 that really showed the potential of photoelectrochemical systems to harvest and store solar energy as chemical energy. In the Honda-Fujishima experiment, titanium dioxide (TiO_2) was used as the photoanode material to produce hydrogen from water using the energy of light.

In order to make use of water oxidation from a sustainable perspective, stable and inexpensive photoanode materials are required.[6,7] To date, titanium dioxide is the benchmarking material.[8,9] It is stable over a range of pHs and potentials, and under favourable circumstances requires no additional bias to run the water splitting reaction. However, the downside of TiO_2 is its large band gap, which limits the photon absorption to only a fraction of the solar spectrum in the ultraviolet. This corresponds to only a few percent of photons that actually reach the surface of the Earth, which limits the long-term potential of TiO_2. Other metal oxides, such as tungsten oxide and hematite have a more favorable band gap, and both are considered as promising candidates for photoanode material. Several other semiconductor photoanodes, such as Si, have even more favorable band gaps, but are not stable under the aqueous conditions required for water oxidation.

The ideal photoanode material[6,7] should meet the following criteria; (i) a band gap ranging between 1.8 and 2.4 eV, (ii) band edge positions that brackets the water redox potentials, (iii) electron-hole mobility and lifetimes that allow the electron-hole pair to reach the active site, (iv) the rate for water-splitting should be faster than any competing recombination reaction. Finally, the material needs to be stable in an aqueous environment under illumination. So far no material has met all these criteria.

The actual processes involved in photoelectrochemistry are many,[10] see Figure 10.1. The first is the capture of photons; the second is the creation of electron-hole pairs that need to be separated. The separated electrons and/or holes then need to be transported to sites, preferable catalytically active, at

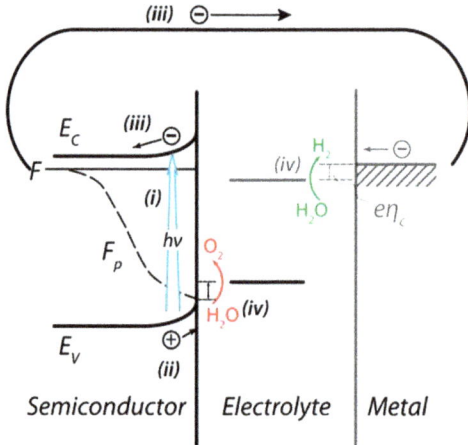

Figure 10.1 A simplified energy diagram for a photoanode (n-type semiconductor). Several important steps are illustrated, namely: (i) light absorption; (ii) charge transfer; (iii) charge transport; and (iv) surface chemical reactions. Reprinted with permission from ref. 10. Copyright 2011 Elsevier.

which the transformation to chemical energy can occur. The multitude of processes has made research in photoelectrochemistry truly cross-disciplinary.[11] For instance, photon absorption and the transport of charge carriers are topics of interest to semiconductor physics, but the chemical transformations occur at surface sites, so surface science and heterogeneous catalysis are also relevant areas. The close connection with electrochemistry is perhaps the most obvious and important area.

This chapter describes some of the processes involved in photoelectrochemistry and demonstrates how first-principles methods can be utilized to provide further understanding of these processes (first-principles are used in conjunction with density functional theory throughout the chapter). The chapter is not intended to be a complete overview, and some aspects do not get the attention they deserve, for which the author can only apologize. Furthermore, the chapter may deviate from the normal description of the processes involved, given that the author's background is not photoelectrochemistry, but rather theoretical surface science[12] and computational aspects of femtosecond spectroscopy,[13] non-adiabatic surface processes[14] and heterogeneous catalysis.[15] It is hoped that this different perspective will provide new angles to approach what is already a very interesting area of research.

10.2 Capture of the Photon

As the photon penetrates a material, the electronic structure of the substance interacts with the propagating electromagnetic wave.[13,16,17] The interaction might result in an electron accruing the photon energy, *i.e.* there is an electronic excitation in which an electron in an occupied band is transferred to an unoccupied band. If the perturbation is small, the process can be described by Fermi's golden rule, where the probability for a transition from state i to j is expressed as follows,

$$T_{i \to j} = \frac{2\pi}{\hbar} \left| \langle i | H.' | j \rangle \right| \rho_i \rho_j \tag{1}$$

Here, ρ_i and ρ_j are the corresponding densities of states of the initial and final states. From equation (1), it is clear that a good light-absorbing material needs to have not only a large optical cross-section, but also a favorable density of state distribution. Here metals, of course, stand out. However, as the goal is to be able to utilize the photon energy to generate something other than heat, the material needs to have a band gap.

In semiconductor and insulating materials, the valence and conduction bands are separated by a band gap. The band gap efficiently hinders the dissipation of energy, as the lack of accessible electronic states quenches many of the common dissipation channels (carrier-carrier interaction and the phonon coupling). Generally one can classify materials with band gaps into two classes; direct and indirect band gap materials. The difference lies in the prerequisites for photon adsorption. In a material with a direct band gap, the momentum of electrons and holes is conserved during the transition. In a material with an indirect band gap, the conservation of momentum requires the creation/annihilation of one (or many) lattice phonon(s). Focusing on the process of capturing the photon, two main (first-principles) research directions can be recognized. The first is to focus on the band gap, *i.e.* how to optimize the positions and densities of states, whereas the second focuses on the photon absorption process.

10.2.1 Band Gap Design

In its simplest form, the power conversion efficiency in a semiconductor depends only on the band gap and the incident light spectrum,[7] As the solar spectrum is fixed, only the band gap can be varied to optimize the conversion efficiency. Density-functional theory (DFT) is supposedly a predicting theory, and there exist several successful examples in the literature. However, in the case of band gap design, there is a well-known problem. The most frequently used approximations for the exchange-correlation functional fail at calculating the band gaps of even the simplest materials. For instance, Si is calculated to have a band-gap of 0.52 eV, whereas the experimental value is 1.17 eV.[18] There are many extensions to DFT designed to circumvent this problem, such as, DFT + U, hybrid functionals and random-phase approximation.[19] However, work has also been done on the simple semi-local functionals[20,21] that describe PBEsol (PBE = Perdew-Burke-Ernzerhof) and GLLB-SC (GLLB = Gritsenko, van Leeuwen, van Lenthe, Baerends). All of these extensions are complicated and lie beyond the scope of this chapter. In many cases the use of first-principles for trend studies are is still valid without the use of these extensions.

Doping has a strong influence on the fundamental and optical band gaps of semiconductors, and there are numerous examples in the literature where first-principles have been used to guide how to include dopant atoms in order to improve the band gap of a given material. Asahi *et al.*[22] showed that nitrogen doped into substitutional sites of TiO_2 resulted in a band-gap narrowing, since the *p* states of N contribute to the narrowing by mixing with O $2p$ states. Teeffelen *et al.*[23] investigated the shifts of the fundamental and optical band gap energies as functions of dopant concentration in heavily n-type and p-type doped $Si_{1-x}Ge_x$ in order to improve solar cell efficiency. The calculated band gap narrowing of Si and of $Si_{0.82}Ge_{0.18}$ was found to be in good agreement with values derived from photoluminescence measurements. The above examples show that first-principles methods can be used to provide explanations of what is happening at the atomic level. However, first-principles methods should also be able to provide guidelines for the search for new materials with improved band gaps.

The next example indicates what we can expect in a near future. Recently, Castelli *et al.*[24] demonstrated the power of computational screening for the discovery of new light harvesting materials for water splitting. More than 2700 oxides with the cubic perovskite structure where investigated with respect to stability and band-gap. In the end, 15 potential candidates were identified (see Figure 10.2). Unfortunately, these candidates are already known to be suitable for water splitting, but the successful outcome of the theoretical work appears to promise that identification of new material formulations will be reliable if more complex structures and compounds are included in the screening.

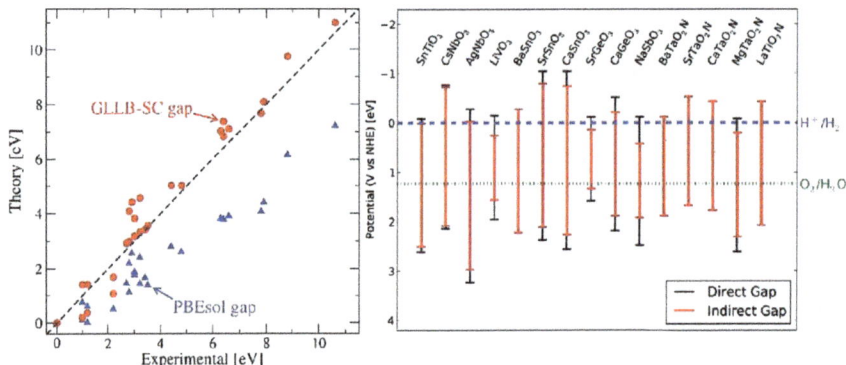

Figure 10.2   (Right) Calculated and measured band-gap of a number of oxides. The first-principles are based on two different state-of-the-art functionals, namely, PBESol[20] and GLLB-SC.[21] (Left) The identified oxides and oxynitrides in the cubic perovskite structure with potential for splitting water in visible light. The figure shows the calculated band edges for both the direct (red) and indirect (black) gaps. The levels for hydrogen and oxygen evolution are also indicated.

Reprinted with permission from ref. 24. Copyright 2011 Royal Society of Chemistry.

10.2.2 Plasmon-Assisted Photon Absorption

When light interacts with a (metallic) surface, resonant collective oscillations in the electronic system can be excited. If these oscillations are localized in a nano-sized entity such as a noble metal nanoparticle, they are called localized surface plasmon resonances (LSPR).[25] The frequency of the LSPR can be tuned across the electromagnetic spectrum through the choice of metal size, shape and dielectric environment. LSPR results in a strong coupling between the photon and the nanoparticle, which gives rise to an enhanced photon absorption and light scattering and strongly enhances electromagnetic fields around the metal nanoparticle.

All of these effects are interesting from a solar harvesting perspective, also for photoelectrochemical devices.[26] Besides the obvious benefit of enhanced photon absorption, light scattering can be utilized to engineer light management in order to decrease the amount of material required.[27–29] The near-field can stimulate light absorption in the proximity of the metal nanoparticle. This is particularly interesting if the minority carriers have a short lifetime, since the inclusion of LSPR particles at the interface should lead to carrier generation closer to the active site for water splitting. If the LSPR particles were placed at the bottom of the photoanode instead, this would help in the generation of the majority carriers. Since they are situated at the photoanode/electrolyte interface, special consideration must be given to the influence of the nanoparticles on (i) stability, (ii) Fermi level and (iii) band-bending. Furthermore, catalytic effects and the formation of trap states may also be critical.

There are several recent studies reported in the literature that provide evidence for plasmon-assisted enhancement of light-driven reactions. Linic *et al.*[30,31] measured an increase in reaction rate in the case of Ag nanoparticles, where the enhancement was attributed to energy transfer from the metal to the semiconductor arising from overlap with the LSPR frequency of the Ag (see Figure 10.3). Overall, the work of Ingram and Linic provides a strong indication that there needs to be an absorption overlap between the semiconductor and LSPR for energy transfer to result in preferential excitation of the semiconductor in the vicinity of the metal nanoparticle. This is in agreement with theoretical studies of similar systems.[32]

Plasmon-enhanced photon absorption is particularly interesting if the photoanode material has a weak optical absorption or if the intrinsic properties of the material make some of the necessary processes to slow. For instance, hematite suffers from a short hole diffusion length.[33–35] Here, the idea is that the plasmon particle will enhance photon absorption, creating the electron-hole pair closer to the anode-electrolyte interface. Several successful measurements demonstrating this effect have appeared in the recent literature.[36–38]

In principle, the collective motion of electrons that constitutes the LSPR phenomenon can be described theoretically by time-dependent density functional theory.[39] However, owing to the high computational cost of such an approach, a linear response formalism is much more favorable. Therefore, the idea here is to introduce a small perturbation that can be evaluated as a Dyson

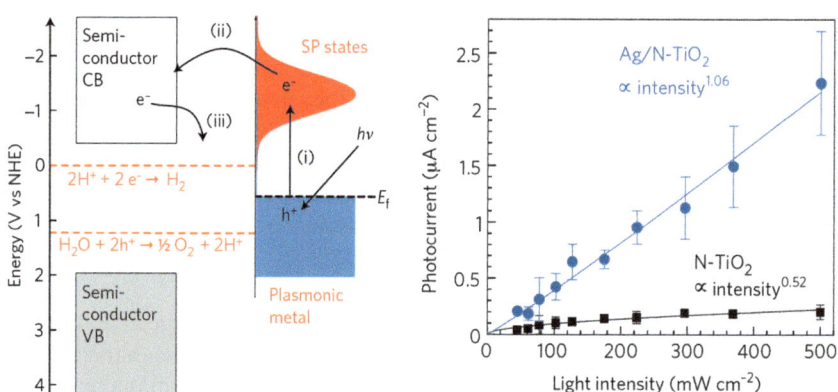

Figure 10.3 (Left) Mechanism of plasmon induced charge transfer with approximate energy levels on the NHE scale. Dashed red lines refer to the water-splitting redox potentials. As the plasmon decays the energy is transferred to an electron–hole pair where the electron can transfer to a nearby semiconductor particle. Depending on the position of the plasmon and the valance and conduction band this process can drive the water oxidation reaction. (Right) Photocurrent as a function of broadband visible-light intensity for samples of TiO_2, with and without the plasmon active particles.
Reprinted with permission from ref. 31. Copyright 2011 American Chemical Society and from ref. 30. Copyright 2011 Nature Materials.

equation. The poles of the solution then basically yield the plasmonic response of the considered system. Thus, there exists a theoretical framework with roots in DFT that can be developed and used to evaluate the plasmonic response from various materials at the atomic level.[40–42]

From a photoelectrochemical point of view, the decay of plasmons is interesting because the decay of plasmons at finite momentum transfer is dominated by Landau damping, *i.e.* the decay into electron–hole (e–h) pairs. The damping of surface plasmons is purely a quantum mechanical process and is governed by the coupling between the surface plasmons and e–h pairs. Unfortunately, this process is not well understood, even for the simplest crystalline surfaces. In a recent study, Gao and coworkers[43] presented a semiclassical model of plasmon–electron coupling and Landau damping for metal thin films and surfaces, based on the quantization of the plasmon hybridization (PH) model of Nordlander and coworkers,[44,45] which describes this process qualitatively. Desirable developments for the near future are (i) to connect first-principles results with the nanostructure of the photoactive material, and (ii) to investigate how the plasmon decay can be incorporated into first-principles studies.

10.3 Electron–Hole Separation

Once the electron–hole pair is created, it needs to be separated, but this is not a straightforward process. Figure 10.4 shows some of the different fundamental

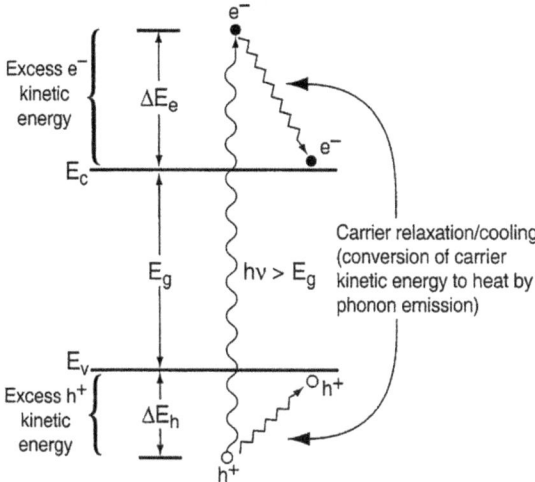

Figure 10.4   The relaxation process of a hot electron–hole pair. There exist several pathways for dissipation of the energy from the electron–hole pair. More details of this are given in the text.
Reprinted by permission from *Annual Review: Annual Review of Physical Chemistry*, copyright 2001.[46]

processes that can happen after the photon is absorbed. Depending on the material at hand, some of the pathways will be more useful than others.

If the photon energy is larger than the minimum band gap, states deeper down in the valence band and/or higher in the conduction band are available to create the electron–hole pair. As a consequence, the newly created electron–hole pair is effectively in a non-equilibrium state, but the additional energy is dissipated rapidly owing to carrier-carrier interaction and phonon coupling. The cooling process normally goes through the following steps.[46–48] First electrons and the holes reach equilibrium *via* their respective carrier-carrier collisions, resulting in two different temperatures defined by the distribution of energy of the respective carrier distributions. This temperature is always higher than the phonon temperature, giving rise to the terms "hot electrons" and "hot holes". This initial relaxation process is very rapid (1–10 fs). Next, the hot carriers equilibrate with phonons *via* carrier-phonon interactions, transferring excess energy to the heating of the photoanode material. This second relaxation process occurs on the time scale of 1–100 ps. The last step involves electron-hole recombination, which can be either radiative (luminescence) or non-radiative (heat). Clearly, the last step is undesirable in photoelectrochemistry.

The electron–hole pair feels a coulomb attraction, and if it is strong enough, the electron–hole pair can be referred as an exciton. The attraction will affect the spatial distribution and transport properties of both the electron and the hole. However, as the wavefunction of the electron–hole pair becomes un-correlated, electrons and holes can be viewed as free carriers. The charge carrier with the lowest effective mass (as determined by the second derivative of the dispersion relation) will exhibit a larger root-mean-square motion as compared to the heavier charge carrier. This implies that the lighter carrier will effectively create a charge separation, known as the photo-Dember effect.[10] This is the main effect that determines carrier transport under zero-field conditions.

In semiconductor physics, an interface is often used to separate electron–hole pairs. At such an interface, a space charge layer is formed owing to the difference in the electrochemical potentials of electrons in the two phases. As the system equilibrates, electron flows from higher free energy to lower free energy until the compensating field is sufficient to stop the flow. This is the physics behind p-n junctions. In a photoelectrochemical system, the interface is created between the semiconductor and the liquid phase (see Figure 10.5). Owing to the

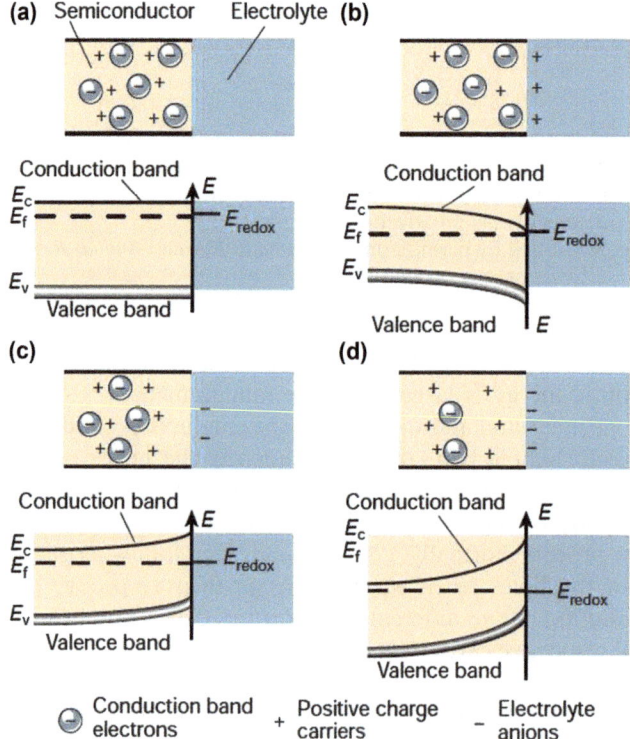

Figure 10.5 Schematic showing the electronic energy levels at the interface between an n-type semiconductor and an electrolyte containing a redox couple. The four cases indicated are: (a) flat band potential, where no space-charge layer exists in the semiconductor; (b) accumulation layer, where excess electrons have been injected into the solid producing a downward bending of the conduction and valence band towards the interface; (c) depletion layer, where electrons have moved from the semiconductor to the electrolyte, producing an upward bending of the bands; and (d) inversion layer where the electrons have been depleted below their intrinsic level, enhancing the upward band bending and rendering the semiconductor p-type at the surface.[49]
Reprinted by permission from Macmillan Publishers Ltd: *Nature* 414 338, copyright 2001.

mobile charge carriers in the liquid, a Schottky junction is formed that involves a redistribution of the charge on the electrolyte side that corresponds to formation of the Helmholtz layer. The field in the semiconductor can extend over several hundred nm, depending on the doping level of the semiconductor.

The field in the space charge region will assist in the separation of the photogenerated electron–hole pairs and drive the minority carrier to the surface. The minority carriers will pick up excess energy from the field. Furthermore, carrier-carrier interaction is much less in the space-charge layer, so that only the phonon channel is open for dissipation of energy. This implies that the carriers can arrive at the surface with an excess energy corresponding to the degree of bend banding, *i.e.* hot carriers. In contrast, the majority carrier will experience both the carrier-carrier and phonon-scattering processes during its propagation through the bulk semiconductor material.

The inclusion of plasmon-active metal nanoparticles can have pronounced effects on bend bending because the work functions of metals generally differ from the electrochemical potentials of electrolyte solutions. This implies that the band bending is different in different regions of the semiconductor. The most common plasmon-active nanoparticles used are gold and silver, where the metal Fermi level is positioned at a higher energy than the redox Fermi level corresponding to water oxidation. This results in less band bending at the metal-semiconductor contact, thereby hindering charge separation. Furthermore, the metal-semiconductor interface can generate surface states and cause Fermi level pinning, which will have a deleterious effect on photoelectrochemical performance. However, if surface states are removed or if the work function of the metal is positioned more suitably, improvements in charge separation can be observed.[26,49]

As the electron or hole is transported through the lattice, the lattice responds to this charge by delocalizing it over many atoms (a large polaron) or by localizing it over just a few atoms (a small polaron). As the extra charge carrier will fill or empty states with the bonding or antibonding characteristics of the lattice atoms, the charge influences bond lengths and angles in the material. If the charge is delocalized, the influence is small because it is spread over many atoms, while a highly localized charge distorts the lattice significantly. One way to view the process is to consider a carrier dragging a cloud of phonons along as it propagates through the lattice. This gives rise to an understanding of the phenomenon of self-trapping. A small polaron will shift the surrounding lattice to its new position within a few lattice vibrations. This increases the stability of the state, creating a deeper potential well that must be overcome to transfer the polaron to a neighboring lattice site, *i.e.* there is a large reorganization energy associated with small polarons.

Deskins *et al.*[50] calculated the electronic structure of one excess electron in bare and singly hydroxylated rutile (110) surfaces. According to their calculations, the excess electron behaves as a small polaron with its spin density and associated lattice distortion localized around a single site. The study also showed that the surface hydroxyl group only perturbs the electronic potential slightly and that both clean and hydroxylated surfaces exhibit similar polaron stability. This is an important conclusion because hydroxyls and photoinduced e-polarons give rise to excess charges, which can affect the reactivity of surface absorbates and other photochemical processes. As for the formation and migration of hole polarons, Deskins et al.[51] used DFT in combination with the Marcus/Holstein theory of electron/polaron transfer[52–54] to show that holes were formed by removal of an $O(2p)$ valence electron, and that hole hopping (where reorganization energy and electronic coupling was taken into account) in most directions in rutile and along one direction in anatase was adiabatic in character, *i.e.* thermal processes coupled to phonons. Lattice distortions around hole polarons were found to be larger than around electron polarons. The study also showed that holes are thermodynamically more stable in the rutile phase, while electrons are more stable in the anatase phase.[51] Furthermore, Pacchioni et al.[55,56] showed that in order to get the correct description of the localized defect states on reduced and hydroxylated $TiO_2(110)$, it is necessary to go beyond semi-local description of the exchange-correlation functional and use hybrid exchange functionals instead. Even though the electron trapping nature of Ti(OH) groups was verified, no evidence that these defects also act as hole traps was found. The results show that the Ti(OH) defect can be a good electron trap, and that the lattice distortion is essential in the electron trapping process.

A similar approach was used by Sicolo et al.[57] to describe the electronic structure and spectral properties of self-trapped holes in SiO_2, where two classical variants of self-trapped holes were studied, namely, (i) a hole trapped at the $2p$ nonbonding orbital of an O atom bridging two Si atoms, and (ii) a metastable defect where the hole is delocalized over the $2p$ orbitals of two bridging O atoms. The first-principles results showed that the ground state of the first type of self-trapped hole allows an unpaired electron to occupy a nonbonding $2p$ level of a bridging oxygen, where the $2p$ level is normal to the Si-O-Si plane. This results in a strong elongation of one Si-O bond, thereby classifying this center as a small polaron. The description of the second self-trapped hole required modification of the normal lattice structures to make the structure flexible enough to allow the O-Si-O angle to shrink from $110°$ to $80°–90°$. This latter condition actually reflects the fact that the electronic structure of the second-type of self-trapped hole uses a bonding combination between the $2p$ levels on two O atoms, resulting in a net bonding interaction that closes the O-Si-O angle and decreases the O-O distance. Other angles around the defect also assume values that deviate substantially from those of quartz, while the Si-O distances are only moderately elongated. In this respect, the center has the typical characteristics of a molecular polaron.[57]

Kleiman-Shwarsctein et al.[58] introduced strain in a Fe_2O_3 film by substitutional doping of Al, which resulted in a 2- to 3-fold increase of the incident photon conversion efficiency. By means of first-principles, it was shown that there was no substantial change to the electronic structure. Instead, the doping benefits small polaron migration, resulting in an improvement in conductivity compared to the undoped sample. In a similar study (although based on unrestricted Hartree-Fock calculations), Liao and Carter[59] investigated how the activation energies for hole diffusion is affected by different dopants. The study suggests that hole hopping occurs *via* oxygen anions for hematite, and hole carriers are predicted to be attracted to O anions near the dopants.

The examples above indicate that DFT is able to characterize electron/hole polarons, and that the first-principles results can also be used as input into modeling of polaron transport. However, further developments will be necessary before DFT can be used to calculate all the details involved in the electron-hole transport process. The difficulty in obtaining an adequate description of the energy difference between a localized and a delocalized state is particularly crucial, and the predictive power of DFT is hampered by the need to compare with experiment. Furthermore, in the case of issues such as non-equilibrium systems and nonadiabatic transitions, the correct handling of different length and time scales are all important.[60] In a recent review, Shluger *et al.*[61] give more examples of modeling electron and hole trapping in metal oxides and also discussion concerning the different challenges involved.

10.4 Charge Transfer

Assuming a four-electron transfer water oxidation mechanism, the photo-generated holes that have reached the photoanode surface need to transfer electrons from the water molecule (or the intermediate products). Transfer of an electron in one direction is equivalent to transfer of a hole in the other, which is convenient because it allows ET and HT (hole transfer) to be treated within the same theoretical framework. Now electron transfers (ET) are very common and important for chemical reactions, but are of course crucial for all electrochemical reactions. It is very common to classify an ET into either an adiabatic or a non-adiabatic process (sometimes equivalently termed *diabatic*).

Both adiabaticity and non-adiabticity are sometimes described in quantum-mechanics where the of atomic and electronic motions are separated in the adiabatic approximation, also called the Born–Oppenheimer approximation (BOA).[62] An early description of the terms adiabatic and non-adiabatic was given by O'Malley:[63] the adiabatic states are simply the eigenstates of the electronic Hamiltonian, and the adiabatic PESs are the corresponding eigenvalues of the same Hamiltonian defined for each nuclear configuration, $\mathbf{R}$. The diabatic representation provides an alternative description that includes many of the so-called non-adiabatic transitions in a natural and straightforward way, particularly for processes in which fast electronic transitions either occur within a spatially localized region of configuration space or they do not occur at all. In many areas of physics, there are real highlights in the breakdown of the BOA. Recent experiments[64] and calculations[65] have shown the importance of electron–hole pair excitation in gas–surface dynamics by analyzing the *chemi-current* – a current due to direct transformation of chemical into electrical energy – of electrons and holes in a Schottky diode induced by the adsorption of molecules. For more details, see reference 14.

Electron transfer plays a prominent role in key processes in all areas of physics, chemistry, and biology. For instance, bond making and bond breaking

involve electron transfer from one electronic state to another. In adiabatic reaction rates described by transition-state theory, the potential energy surface
for nuclear motion is well separated from higher potential energy surfaces so
that the transitions to the latter surfaces are negligible. However, sometimes
this approximation comes into question.[66] Even though electron transfer is a
common phenomenon, its modeling is often primitive and intended more to
make use of simple analytical models than to represent the system accurately.
However, such models do have considerable interpretive value. The Landau–
Zener model is widely used to describe non-adiabatic processes occurring in the
gas and liquid phases. It was originally constructed to calculate the probability
of non-adiabatic transitions in a two-level system, and as such it is a good
starting point to study non-adiabticity in electron transfer from one electronic
state to another. However, the continuum of states that are present in the
valence and conduction bands of semiconductor materials may cause the two-
state approximation to fail, but nevertheless – with proper modification – the
model can still be very useful for photoelectrochemical processes.[67] For further
details, see Figure 10.6.

In the chemistry and electrochemistry community, ET processes are usually
described in the context of the Marcus theory.[54] In short, the basic idea of
the classical electron transfer theory of Marcus can be summarized as (i) the
complete system starts off from the equilibrium state of the donor state, (ii) the
donor and acceptor states are described by two separate potential energy surfaces, (iii) electron transfer occurs at the intersection of both potential energy
curves. Thermal fluctuations are needed for the system to reach the intersection.
Fluctuations in the vibrational coordinates and the orientational coordinates
are important ingredients. The rate constant for electron transfer then depends
on the probability of reaching the intersection, a frequency factor, and the
probability of crossing the surface. The analogy with the Landau–Zener model

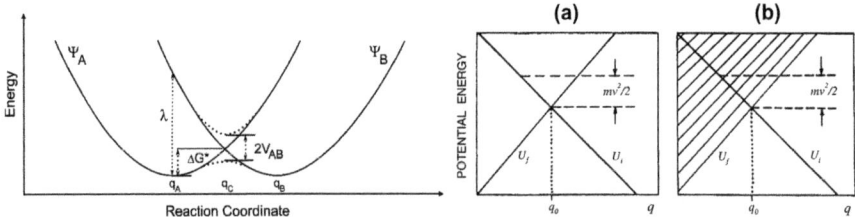

Figure 10.6   (Left) A simple picture of Marcus theory for symmetric polaron transfer.
The potential energy surfaces of the initial state and final state are shown.
The adiabatic energy curves are shown as dashed lines, with the electronic coupling matrix element, given as half the energy difference
between the two adiabatic states. Reprinted with permission from reference 51.[51] Copyright 2008 American Chemical Society. (Right) The
ordinary and modified Landau-Zener model for single and multiple
crossing of potential energy surface.
Reprinted with permission from ref. 67. Copyright 1997 American
Physical Society.

is clear. The direct application of the simple models to photoelectrochemistry is not without problems. For instance, the processes of transferring charge (electrons, protons, *etc.*) across the interface between the semiconductor and the electrolyte are difficult to model correctly. There are often significant dipole moments, which affect the description of the donor and acceptor states. Moreover, the dynamics of screening in the semiconductor and electrolyte complicate the picture further.[60]

10.5 Surface Reaction (Electrochemical Conversion)

Although the holes that drive the water oxidation reaction are generated by photons and the driving force comes from the valence band position and not from an external potential, the surface reaction is the same as in an electrochemical cell. This is a real advantage, since the recent developments in computational electrochemistry have been quite remarkable. The frameworks formulated by, *e.g.* the Anderson group,[68–72] the Neurock group,[73–77] and the Rossmeisl/Norskov group[78–84] have provided a molecular-level insight into the atomic-scale processes that occur at the vicinity of the anode/cathode surface.

Recently, Valdes *et al.*[85–87] suggested a novel, theoretical framework in which the photo-oxidation of water can be described by first-principles methods. It is an extension of the electrochemical framework suggested by Rossmeisl *et al.*[78–84] Although the framework only treats the thermodynamics of the reaction mechanism, it provides a methodology for a detailed atomistic understanding of photoelectrochemical water splitting. The framework assumes that the driving force for the reaction at the anode originates from the photogenerated hole at the edge of the valence band. Hence, in the scale obtained by aligning the energy levels of oxide semiconductors with the redox level of the standard hydrogen electrode, a deeper valence band edge energy level will result in a larger thermodynamic driving force.

Within the computational electrochemical framework, the potential of the standard hydrogen electrode (SHE) is used as reference point. The SHE is zero by definition, where the chemical potential of the $H^+(aq) + e^-$ pair is equal to that of $1/2\ H_2$ in the gas-phase. This solves the problem of calculating the energy of solvated protons and electrons, and instead the gas-phase value of the energy of H_2, which is easily described by first-principles, can be used. Furthermore, the effect of the electrode potential on the adsorption energies is simple to include by addition of a stabilization energy of $+eU$ when appropriate. In principle, the adsorption energy of reaction intermediates can depend on the electrode potential. However, first-principles studies indicate that this effect is small, *e.g.* the adsorption energies of *O, *H, and *OH are only changed slightly when an electric field in the range of $-0.3\ \text{V/Å}$ to $0.3\ \text{V/Å}$ is used. Assuming a double layer thickness of 3 Å, the range corresponds to a potential of $-0.9\ \text{V}$ and $0.9\ \text{V}$ with respect to the zero potential.[88] Therefore, the primary effect of the electrode potential is to change the (free) energy of the electrons.

At the photoanode/electrolyte interface, the solvent water molecules may play an important role. Contributions from the liquid phase are normally approximated by including several of water layers in the simulation.[89,90] The water network, with its many hydrogen bonds, will have a large effect on the stability of the reaction intermediates, especially the ones that can contribute with more hydrogen bonds. First-principles results show that OH-containing reaction intermediates, such as, *OH and *OOH, can be stabilized up to 0.6 eV on Pt (111), whereas those of H and O are affected to a lesser extent, *ca.* 0.1 eV.[91] Furthermore, entropy contributions from the liquid phase are approximated by reference to the equilibrium pressure in contact with liquid water at 298.15 K and vapor pressure 0.0317 bar, where the free energy of gas phase water is equal to the free energy of liquid water. This permits the use of gas-phase water to calculate the binding energies and its transformation into liquid-phase water when adding the entropy correction.[82] The free energy change of the reaction step involving the formation of O_2 is set to the experimentally obtained value of 4.92 eV per O_2 molecule. The free energy of the reaction intermediates is calculated *via* DFT by also including the zero-point energy (ZPE) and vibrational contributions. Normally the entropy contribution is low as the temperatures are not so high under photoelectrochemical conditions.

The main difference between the frameworks for electrochemistry and photoelectrochemistry is the origin of the driving potential.[85–87,92] In electrochemistry, the driving potential can be varied externally, whereas the photoelectrochemical driving force is the redox potential originating from the photoinduced hole in the valence band. It should be noted that the energy position of the valence band of oxide semiconductors depends on pH, but the same dependence applies to the free energy of each reaction step for water oxidation. Thus, to a first approximation, the thermodynamics of the reaction is unchanged by changes in the pH.

10.5.1 Pourbaix Surface Diagrams

From the photoelectrochemical framework it is now possible to establish a surface phase diagram, which gives an estimate of which reaction intermediates are adsorbed on the surface for given pH values and under dark or light conditions, see Figure 10.7. In electrochemistry these phase diagrams are called Pourbaix diagrams.[83,93,94] Although Pourbaix diagrams were originally constructed for bulk transitions, the use of first-principles has shown that they can accurately describe which reaction intermediates are present and which are unstable under electrochemical conditions. The phase diagrams are only applied under stable conditions, which implies that a photostationary concentration of holes builds up at interface in order to generate a driving force for the photoanode reaction. Furthermore, it should be noted that the potential experienced by a metal nanoparticle at the surface of a photoelectrode during illumination differs from the externally applied potential because the quasi Fermi level of holes shifts to positive potentials, leading to a shift in the Fermi level of the metal particle.

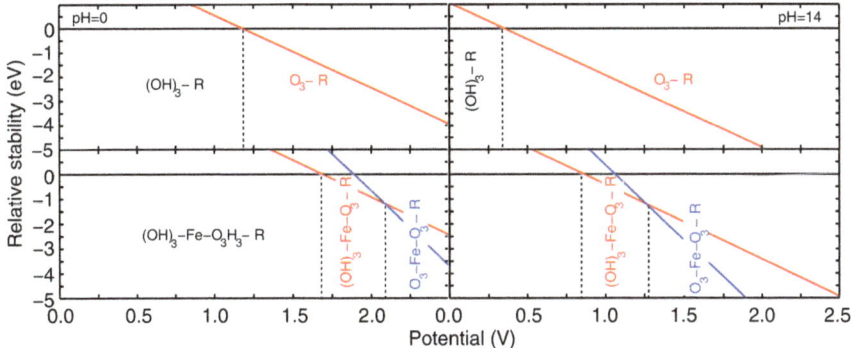

Figure 10.7 The relative stability of all considered surface terminations as a function of applied potential and at two different pH, namely, pH = 0 (left) and pH = 14 (right).
Reprinted with permission from ref. 92. Copyright 2011 American Chemical Society.

10.5.2 Reaction Mechanism

A four proton and electron transfer water oxidation mechanism could consist of the following elementary steps,

$$H_2O + * \rightarrow OH_* + H^+ + e^-$$

$$OH_* \rightarrow O_* + H^+ + e^-$$

$$O_* + H_2O \rightarrow OOH_* + H^+ + e^-$$

$$OOH_* \rightarrow O_2 + * + H^+ + e^-$$

Once the reaction mechanism has been proposed, the photoelectrochemical framework allows the calculation of the reaction landscape of all reaction intermediates from first-principles. Since all steps only involve one proton and electron transfer, the height of the different steps scales linearly with the redox potential. See Figure 10.8 for an example of the thermodynamics of water-oxidation at different potentials.

The described photoelectrochemical framework has been used to study photo induced water-oxidation on rutile $TiO_2(110)$,[86] WO_3 (various facets),[87] and $Fe_2O_3(0001)$.[92] In most cases, the redox potential arising from the valence band edge is sufficiently positive to make the reaction thermodynamically favorable. Only in the case of Fe_2O_3 was water oxidation predicted to be prohibited on some surface terminations. However, these terminations were not the most stable ones, which implies that they do not play a part under normal operation conditions.

The electrochemical framework that has been described so far does not lend itself easily to the calculation of activation barriers; hence, information on the kinetics of any reaction is still missing. However, there exist extensions that deal with this issue. For instance, by varying the number of protons/electrons in the

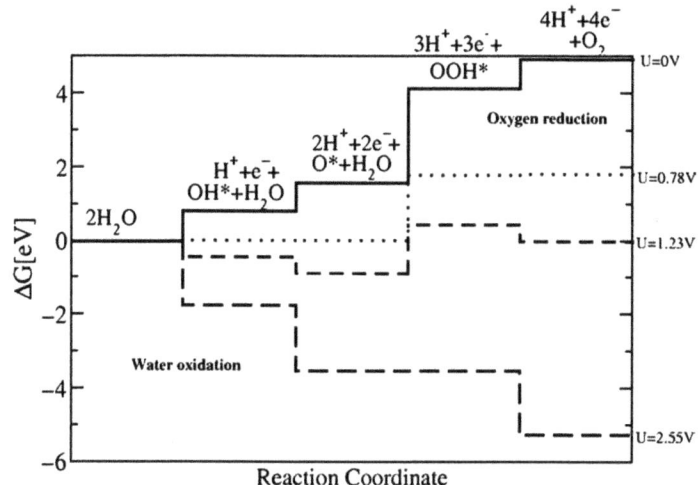

Figure 10.8   The water splitting reaction at different potentials. At potentials between 0 and 0.78 V all steps in the oxygen reduction are exothermic. For potentials beyond 2.55 V all water splitting reaction steps become exothermic. This variation is obtained by varying the term eU in the free energy per electron transferred to the electrode.
Reprinted from *Chemical Physics* 319 (2005) 178, Copyright 2005, with permission from Elsevier.[95]

electrolyte, Skulasson *et al.*[80,90] were able to determine the activation energy for the hydrogen evolution reaction as a function of electrode potential. One important conclusion from the study is the manifestation of a Brønsted–Evans–Polanyi-type relationship between activation energy and reaction energy found throughout surface chemistry.[95–97] This result implies that theoretical electrochemistry can use a lot of the knowledge gained in heterogeneous catalysis.[98,99] Since the driving force in photoelectrochemistry comes – at least as a first approximation – from the hole (electron) at the edge of the valence (conduction) band, it still remains to be confirmed whether the same technique can be used to find new catalyst formulations in photoelectrochemistry. However, with the envisioned development of band gap design it might still be possible.

10.5.3   Overpotential

Recently, Abild-Pedersen *et al.*[100] demonstrated the existence of approximately linear relationships between the adsorption energy of hydrogen and non-hydrogen containing species, *e.g.* OH and O, over many different materials. In combination with Brønsted-Evans-Polanyi relations,[95] this finding has provided a breakthrough in the computational screening of heterogeneous catalysts. In electrochemistry, the same linear relations have been used to show that in the oxygen evolution reaction (OER) and the oxygen reduction reaction (ORR) there exists a fundamental overpotential.[101–103] This was done by calculating the difference in Gibbs free energy for each reaction step, and by use of the linear scaling relations, expressing each reaction intermediate as a function of one of these differences or a linear combination. Especially the combination $\Delta G_{O*} - \Delta G_{OH*}$ has proved universal for the description of OER and ORR activities of a large set of data.[102] The OER volcano plots for different substrates are shown in Figure 10.9. Furthermore, the fundamental overpotential, which originates front the fixed distance between the binding of OH and OOH, is also shown. This observation provides an upper limit on how good an OER electrocatalyst can be expected to be.

The origin of the fundamental overpotential provides directions of how to overcome this limitation. If a catalyst is able to stabilize *OOH with respect to *OH (*i.e.* make the free energy difference between *OOH and *OH to values closer to 2.46 eV), it will circumvent the fundamental overpotential. One possible pathway would be to use 3D structures that are able to differentiate between the intermediates by, *e.g.* confining the OOH group. Another suggestion is the use of the so-called Hangman porphyrins,[103] which have indeed

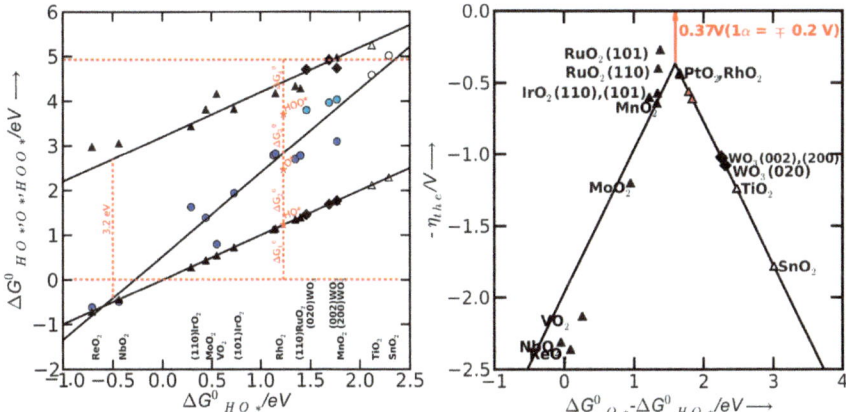

Figure 10.9   The Gibbs free energy of adsorbed HOO*, O*, HO* on rutile surfaces (110) and (101) and for WO$_3$ (200,020,002). Filled symbols represent the adsorption energies on the surfaces with a high coverage of oxygen. The hollow symbols represent adsorption energies on the clean surfaces with no nearest neighbors. Triangles are for HOO* and HO* species, while the circles are for O* species. The difference between the two red dashed horizontal lines is the standard free energy for oxygen molecule to be formed. (Left) The activity trends for oxygen evolution (OER) on the rutile surfaces (black line). The negative value of theoretical overpotential is plotted against the descriptor for OER (the standard free energy of HO* oxidation). Solid black triangles include the effect of the interaction with the oxygen from the neighboring sites, while the red triangles include the effect of the interaction with the HO* species. The diamond symbols represent the overpotentials for WO$_3$. The minimum possible overpotential for any oxide is shown by the red arrow (the difference between the peak of the volcano and the zero line).
Reprinted from ref. 11, Copyright 2012, with permission from Royal Society of Chemistry.

been shown to be good catalysts for OER. However, there is no support of this claim from first-principles as of yet.[104]

10.6 Conclusion and Outlook

This chapter has discussed a selection of the many processes involved in photoelectrochemistry and shown how first-principles methods can help in interpretation and future improvements. The many success stories reveal a promising future, *i.e.* soon first-principles calculations will not only be used for reproducing the experimentally known facts about a given photoelectrochemical reaction on a given photoanode (cathode) material, but could also become the standard starting point when a new photoelectrocatalyst for a known reaction is desired, or even when an unknown reaction to obtain a given product is needed. However, some issues remain that should be addressed and solved before this promise can become reality. The use of more accurate exchange-correlation functionals seems to be able to push first-principles from being explanatory to predicative. This development will certainly continue, and hopefully band gap design will soon be reality. As for the plasmon-assisted photon capture process, the clear link between the microscopic description provided by first-principles and the nanostructured plasmon active particle remains to be settled. The problem associated with charge transfer and the connection between DFT calculations of photoelectrocatalysis and dynamical simulations of electron transfer has not yet fully matured. The use of the standard hydrogen electrode as the reference point has opened up the possibility of using first-principles methods for some straightforward applications to electrochemical and photoelectrochemical systems. However, the approach can only be used for electrochemical steps, *i.e.* elementary reactions in which a proton and an electron are simultaneously transferred. In the case of reactions where only a proton or only an electron is transferred, the standard hydrogen electrode approximation cannot be applied. In spite of these remaining unresolved issues, the future looks bright for first-principles and its impact on photoelectrochemistry.

Acknowledgements

The author wants to acknowledge the financial support of the Swedish Research Council. Furthermore, Chris Cornwall is acknowledged for proofreading the chapter.

References

1. N. S. Lewis, *Science*, 2007, **315**(5813), 798–801.
2. T. R. Cook, D. K. Dogutan, S. Y. Reece, Y. Surendranath, T. S. Teets and D. G. Nocera, *Chem. Rev.*, 2010, **110**(11), 6474–6502.
3. D. G. Nocera, *Energy Environ. Sci.*, 2010, **3**(8), 993–995.

4. E. L. Miller, *Solar Hydrogen Production by Photoelectrochemical Water Splitting: The Promise and Challenge*, in *On Solar Hydrogen & Nanotechnology*, 2010, John Wiley & Sons, Ltd, pp. 1–35.

5. A. Fujishima and K. Honda, *Nature*, 1972, **238**(5358), 37–38.

6. M. G. Walter, E. L. Warren, J. R. McKone, S. W. Boettcher, W. Shannon, M. Qixi and E. A. Santori, *Chem. Rev.*, 2010, **110**(11), 6446–6473.

7. T. Bak, *et al.*, *Int. J. Hydrogen Energy*, 2002, **27**, 991.

8. M. A. Henderson, *Surf. Sci. Rep.*, 2011, **66**(6–7), 185–297.

9. A. L. Linsebigler, G. Lu and J. T. Yates, *Chem. Rev.*, 1995, **95**(3), 735–758.

10. R. Memming, and R. J. Dwayne Miller, in *Nanostructured And Photoelectrochemical Systems For Solar Photon Conversion*, eds. M. D. Archer, A. J. Nozik. World Scientific Press, 2008, pp. 39–145.

11. A. Valdes, J. Brillet, M. Grätzel, H. Gudmundsdottir, A. Heine, H. Jonsson, P. Klupfel, G.-J. Kroes, F. Le Formal, I. C. Man, R. S. Martins, J. K. Norskov, J. Rossmeisl, K. Sivula, A. Vojvodic and M. Zach, *Phys. Chem. Chem. Phys.*, 2012, **14**(1), 49–70.

12. A. Groß, *Theoretical Surface Science: A Microscopic Perspective*, 2003, Springer.

13. H. L. Dai, and W. Ho, *Laser Spectroscopy and Photochemistry on Metal Surfaces*, 1995, World Scientific.

14. B. I. Lundqvist, A. Hellman, I. Zoric, in *Handbook of surface science Dynamics*, E. Hasselbrink and B. I. Lundqvist, 2008, Elsevier Science.

15. G. A. Somorjai, and Y. Li, *Introduction to Surface Chemistry And Catalysis*, 2010, John Wiley & Sons.

16. P. A. M. Dirac, Proceedings of the Royal Society of London. Series A, 1927, **114**(767), pp. 243–265.

17. A. Liebsch, *Electronic Excitations at Metal Surfaces,* 1997, Plenum Press.

18. R. W. Godby, M. Schlüter and L. J. Sham, *Phys. Rev. B*, 1987, **36**(12), 6497–6500.

19. J. Hafner, *J. Comput. Chem.*, 2008, **29**(13), 2044–2078.

20. J. P. Perdew, A. Ruzsinszky, G. I. Csonka, O. A. Vydrov, G. E. Scuseria, L. A. Constantin, X. Zhou and K. Burke, *Phys. Rev. Lett.*, 2008, **100**(13), 136406.

21. M. Kuisma, J. Ojanen, J. Enkovaara and T. T. Rantala, *Phys. Rev. B*, 2010, **82**(11), 115106.

22. R. Asahi, T. Morikawa, T. Ohwaki, K. Aoki and Y. Taga, *Science*, 2001, **293**(5528), 269–271.

23. S. v. Teeffelen, C. Persson, O. Eriksson and B. Johansson, *J. Phys.: Condens. Matter*, 2003, **15**(3), 489.

24. I. E. Castelli, T. Olsen, S. Datta, D. D. Landis, S. Dahl, K. S. Thygesen and K. Jacobsen, *Energy Environ. Sci.*, 2012, **5**(2), 5814–5819.

25. J. M. Pitarke, V. M. Silkin, E. V. Chulkov and P. M. Echenique, *Rep. Prog. Phys.*, 2007, **70**(1), 1.

26. S. C. Warren and E. Thimsen, *Energy Environ. Sci.*, 2012, **5**(1), 5133–5146.

27. C. Hägglund and S. P. Apell, *J. Phys. Chem. Lett.*, 2012, **3**(10), 1275–1285.
28. H. A. Atwater and A. Polman, *Nat. Mater.*, 2010, **9**(3), 205–213.
29. C. Hägglund, S. P. Apell and K. Basemo, *Nano Letters*, 2010, **10**(8), 3135–3141.
30. S. Linic, P. Christopher and D. B. Ingram, *Nat. Mater.*, 2011, **10**(12), 911–921.
31. D. B. Ingram and S. Linic, *J. Am. Chem. Soc.*, 2011, **133**(14), 5202–5205.
32. W. Ni, T. Ambjörnsson, S. P. Apell, H. Chen and J. Wang, *Nano Letters*, 2009, **10**(1), 77–84.
33. B. Klahr, S. Gimenez, F. Fabregat-Santiago, J. Bisquert and T. W. Hamann, *Energy Environ. Sci.*, 2012, **5**(6), 7626–7636.
34. C. Gleitzer, J. Nowotny and M. Rekas, *Appl. Phys. A: Mater. Sci. Process.*, 1991, **53**(4), 310–316.
35. K. Sivula, F. Le Formal and M. Grätzel, *ChemSusChem*, 2011, **4**(4), 432–449.
36. I. Thomann, B. A. Pinaud, Z. Chen, B. M. Clemens, T. F. Jaramillo and M. L. Brongersma, *Nano Letters*, 2011, **11**(8), 3440–3446.
37. B. Iandolo and M. Zäch, *Aust. J. Chem.*, 2012, **65**(6), 633–637.
38. H. Gao, C. J. Liu, H. E. Hoon and P. Yang, *ACS Nano*, 2011, **6**(1), 234–240.
39. C. Alberto, A. L. M. Miguel, A. A. Julio and R. Angel, *J. Comput. Theor. Nanosci.*, 2004, **1**(3), 231–255.
40. A. G. Marinopoulos, L. Reining, A. Rubio and V. Olevano, *Phys. Rev. B*, 2004, **69**(24), 245419.
41. J. Yan, K. W. Jacobsen and K. S. Thygesen, *Phys. Rev. B*, 2011, **84**(23), 235430.
42. J. Yan, K. W. Jacobsen and K. S. Thygesen, *Phys. Rev. B*, 2011, **83**(24), 245122.
43. Y. Gao, Z. Yuan and S. Gao, *J. Chem. Phys.*, 2011, **134**(13), 134702.
44. E. Prodan, C. Radloff, N. J. Halas and P. Nordlander, *Science*, 2003, **302**(5644), 419–422.
45. E. Prodan and P. Nordlander, *J. Chem. Phys.*, 2004, **120**(11), 5444–5454.
46. A. J. Nozik, *Annu. Rev. Phys. Chem.*, 2001, **52**(1), 193–231.
47. A. C. Luntz, in *Chemical Bonding at Surfaces and Interfaces*, ed. A. Nilsson, L. G. M. Pettersson, J. K. Nørskov, Amsterdam: Elsevier.
48. R. Franchy, *Rep. Prog. Phys.*, 1998, **61**(6), 691.
49. P. V. Kamat, *J. Phys. Chem. Lett.*, 2012, **3**(5), 663–672.
50. N. A. Deskins, R. Rousseau and M. Dupuis, *J. Phys. Chem. C*, 2009, **113**(33), 14583–14586.
51. N. A. Deskins and M. Dupuis, *J. Phys. Chem. C*, 2008, **113**(1), 346–358.
52. T. Holstein, *Ann. Phys.*, 1959, **8**(3), 325–342.
53. T. Holstein, *Ann. Phys.*, 1959, **8**(3), 343–389.
54. R. A. Marcus, *Rev. Mod. Phys.*, 1993, **65**(3), 599–610.
55. C. Di Valentin, G. Pacchioni and A. Selloni, *Phys. Rev. Lett.*, 2006, **97**(16), 166803.
56. G. Pacchioni, *J. Chem. Phys.*, 2008, **128**(18), 182505.
57. S. Sicolo, G. Palma, C. Di Valentin and G. Pacchioni, *Phys. Rev. B*, 2007, **76**(7), 075121.

58. A. Kleiman-Shwarsctein, M. N. Huda, A. Walsh, Y. Yan, G. D. Stucky, Y.-S. Hu, M. W. Al-Jassim and E. W. McFarland, *Chemistry of Materials*, 2009, **22**(2), 510–517.
59. P. Liao and E. A. Carter, *Journal of Applied Physics*, 2012, **112**(1), 013701.
60. M. Stoneham, *Modell. Simul. Mater. Sci. Eng.*, 2009, **17**(8), 084009.
61. A. L. Shluger, K. P. McKenna, P. V. Sushko, D. Muñoz Ramo and A. V. Kimmel, *Modell. Simul. Mater. Sci. Eng.*, 2009, **17**(8), 084004.
62. M. Born and R. Oppenheimer, *Ann. Phys.*, 1927, **84**, 457.
63. T. F. O'Malley, *Adv. Atom. Mol. Phys.*, 1971, **7**, 223.
64. B. Gergen, H. Nienhaus, W. H. Weinberg and E. W. McFarland, *Science*, 2001, **294**(5551), 2521–2523.
65. S. N. Maximoff and M. P. Head-Gordon, *Proceedings of the National Academy of Sciences*, 2009, **106**(28), 11460–11465.
66. V. P. Zhdanov, *Elementary Physicochemical Processes on Solid Surfaces,* 1991, New York, Plenum.
67. V. P. Zhdanov, *Phys. Rev. B*, 1997, **55**(11), 6770–6772.
68. A. B. Anderson and T. V. Albu, *J. Am. Chem. Soc.*, 1999, **121**(50), 11855–11863.
69. A. B. Anderson, *Phys. Chem. Chem. Phys.*, 2012, **14**(4), 1330–1338.
70. A. B. Anderson, *Electrochimica Acta*, 2003, **48**(25–26), 3743–3749.
71. A. B. Anderson, N. M. Neshev, R. A. Sidik and P. Shiller, *Electrochimica Acta*, 2002, **47**(18), 2999–3008.
72. R. Jinnouchi and A. B. Anderson, *J. Phys. Chem. C*, 2008, **112**(24), 8747–8750.
73. C. D. Taylor, R. G. Kelly and M. Neurock, *J. Electroanal. Chem.*, 2007, **607**(1–2), 167–174.
74. C. Taylor, R. G. Kelly and M. Neurock, *J. Electrochem. Soc.*, 2006, **153**(12), E207–E214.
75. C. Taylor, R. G. Kelly and M. Neurock, *J. Electrochem. Soc.*, 2007, **154**(3), F55–F64.
76. C. D. Taylor, R. G. Kelly and M. Neurock, *J. Electrochem. Soc.*, 2007, **154**(12), F217–F221.
77. M. J. Janik, C. D. Taylor and M. Neurock, *J. Electrochem. Soc.*, 2009, **156**(1), B126–B135.
78. J. Rossmeisl, K. Dimitrievski, P. Siegbahn and J. K. Nørskov, *The Journal of Physical Chemistry C*, 2007, **111**(51), 18821–18823.
79. E. Skúlason, V. Tripkovic, M. E. Björketun, S. Gudmundsdóttir, G. Karlberg, J. Rossmeisl, T. Bligaard, H. Jónsson and J. K. Nørskov, *J. Phys. Chem. C*, 2010, **114**(50), 22374–22374.
80. E. Skúlason, V. Tripkovic, M. E. Björketun, S. Gudmundsdóttir, G. Karlberg and J. Rossmeisl, *J. Phys. Chem. C*, 2010, **114**(42), 18182–18197.
81. A. S. Bondarenko, I. E. L. Stephens, H. A. Hansen, F. J. Pérez-Alonso, V. Tripkovic, T. P. Johansson, J. Rossmeisl, J. K. Nørskov and I. B. Chorkendorff, *Langmuir*, 2011, **27**(5), 2058–2066.
82. J. K. Nørskov, J. Rossmeisl, A. Logadottir, L. Lindqvist, J. R. Kitchin, T. Bligaard and H. Jónsson, *J. Phys. Chem. B*, 2004, **108**(46), 17886–17892.

83. J. Rossmeisl, J. K. Nørskov, C. D. Taylor, M. J. Janik and M. Neurock, *J. Phys. Chem. B*, 2006, **110**(43), 21833–21839.
84. J. Rossmeisl, Z. W. Qu, H. Zhu, G. J. Kroes and J. K. Nørskov, *J. Electroanal. Chem.*, 2007, **607**(1–2), 83–89.
85. A. Valdés and G. J. Kroes, *J. Phys. Chem. C*, 2010, **114**(3), 1701–1708.
86. A. Valdés, Z. W. Qu, G. J. Kroes, J. Rossmeisl and J. K. Nørskov, *J. Phys. Chem. C*, 2008, **112**(26), 9872–9879.
87. A. Valdes and G.-J. Kroes, *J. Chem. Phys.*, 2009, **130**(11), 114701.
88. G. S. Karlberg, J. Rossmeisl and J. K. Nørskov, *Phys. Chem. Chem. Phys.*, 2007, **9**(37), 5158–5161.
89. I. Hamada and Y. Morikawa, *J. Phys. Chem. C*, 2008, **112**(29), 10889–10898.
90. E. Skulason, G. S. Karlberg, J. Rossmeisl, T. Bligaard, J. Greeley, H. Jonsson and J. K. Norskov, *Phys. Chem. Chem. Phys.*, 2007, **9**(25), 3241–3250.
91. M. T. M. Koper, *Electrochimica Acta*, 2011, **56**(28), 10645–10651.
92. A. Hellman and R. G. S. Pala, *J. Phys. Chem. C*, 2011, **115**(26), 12901–12907.
93. H. A. Hansen, J. Rossmeisl and J. K. Nørskov, *Phys. Chem. Chem. Phys.*, 2008, **10**(25), 3722–3730.
94. K. A. Persson, B. Waldwick, P. Lazic and G. Ceder, *Phys. Rev. B*, 2012, **85**(23), 235438.
95. B. I. Lundqvist, T. Bligaard and J. K. Nørskov, in *Understanding Heterogeneous Catalysis from the Fundamentals*, in *Dynamics*, ed. E. Hasselbrink, BI. Lundqvist, 2008, p. 269.
96. T. Bligaard, J. K. Nørskov, S. Dahl, J. Matthiesen, C. H. Christensen and J. Sehested, *J. Catal.*, 2004, **224**(1), 206–217.
97. A. Vojvodic, F. Calle-Vallejo, W. Guo, S. Wang, A. Toftelund and F. Studt, *J. Chem. Phys.*, 2011, **134**(24), 244509.
98. J. K. Nørskov, T. Bligaard, J. Rossmeisl and C. H. Christensen, *Nature Chemistry*, 2009, **1**, 37.
99. J. K. Nørskov, M. Scheffler and H. Toulhoat, *MRS Bull.*, 2006, **31**, 669.
100. F. Abild-Pedersen, J. Greeley, F. Studt, J. Rossmeisl, T. R. Munter, P. G. Moses, E. Skullason, T. Bligaard and J. K. Nørskov, *Phys. Rev. Lett.*, 2007, **99**, 016105.
101. V. Viswanathan, H. A. Hansen, J. Rossmeisl and J. K. Norskov, *ACS Catalysis*, 2012.
102. I. C. Man, H.-Y. Su, F. Calle-Vallejo, H. A. Hansen, J. I. Martínez and N. G. Inoglu, *ChemCatChem*, 2011, **3**(7), 1159–1165.
103. F. Calle-Vallejo, J. I. Martinez and J. Rossmeisl, *Phys. Chem. Chem. Phys.*, 2011, **13**(34), 15639–15643.
104. J. D. Baran, H. Grönbeck, A. Hellman, To be published in 2013.

CHAPTER 11

Electro- and Photocatalytic Reduction of CO_2: The Homogeneous and Heterogeneous Worlds Collide?

DAVID BOSTON, KAI-LING HUANG,
NORMA DE TACCONI, NOSEUNG MYUNG,
FREDERICK MACDONELL AND
KRISHNAN RAJESHWAR*

Department of Chemistry and Biochemistry, University of Texas at
Arlington, Arlington Texas 76019-0065, USA
*Email: rajeshwar@uta.edu

11.1 Introduction and Scope

The catalytic reduction of carbon dioxide (CO_2) to fuels and organic compounds using light, electricity, or a combination of both, is not a new topic. References to this topic date back to the 1800s,[1–3] although rapid progress was made only since the 1970s. As elaborated below, a major challenge relates to the fact that the CO_2 molecule is extremely stable and is kinetically inert. A number of review articles and book chapters already summarize what has been accomplished on this challenging R&D topic.[4–14] This chapter contains an overview of recent developments in molecular catalysts for CO_2 reduction, summarized in Table 11.1 and Table 11.2, as well as a review of the progress made in our own laboratories against the backdrop of the rather vast body of

RSC Energy and Environment Series No. 9
Photoelectrochemical Water Splitting: Materials, Processes and Architectures
Edited by Hans-Joachim Lewerenz and Laurence Peter

Table 11.1 Electrochemical systems for CO_2 reduction with reduction potentials, electrolyte, electrodes, electrolysis potentials with all potentials are reported in NHE except where noted. (2-m-8-Hq = 2-methyl-8-hydroxyquinoline, 2-Qui. = 2-quinoxalinol, Hiq = Hydroxyisoquinoline, 4-m-1,10-Phen = 4-methyl-1,10-phenanthroline, dmbpy = 4,4'-dimethyl-2,2'-bipyridyl, phen = 1,10-phenanthroline, bpy = 2,2'-bipyridine, salophen = (4-acetamidophenyl) 2-hydroxybenzoate, dophen = 2,9-bis(2-hydroxyphenyl)-1,10-phenanthroline, tpy = 2,2';6',2''-terpyridine, TPP = 5,10,15,20-Tetraphenylporphin, N-MeIm = 1-methyl-imidazole, tBu-bpy = 4,4'-tertbutyl-2,2'-bipyridine, dppm = 1,1-Bis(diphenylphosphino)methane, dppe = 1,1-Bis(diphenylphosphino)ethane, dmg = dimethylglyoxime, cyclam = 1,4,8,11-tetraazacyclotetradecane, COD = 1,5-Cyclooctadiene, tmdnTAA = 5,7,12,14-tetramethyldinaphtho[b,i][1,4,8,11]tetraaza[14]annulene, HACD = 1,3,6,9,11,14-hexaazacyclohexadecane, decyclam = 1,8-diethyl-1,3,6,8,10,13-hexaazacyclotetradecane, TBA = tetra-N-butylammonium, TEtA = tetra-N-ethylammonium, TMA = tetra-N-methylammonium).

	Catalyst	WE	Electrolyte	Solvent	Product	Efficiency	Redox Couple	CO_2 reduction	pH	Temp	Notes	Refs.
1	$[Co(salophen)]^{2+}$	Hg	Li(ClO$_4$)	MeCN	CO, CO$_3^{2-}$		-1.02 V	-1.29 V			TON >20	23, 115, 116
2	$[Fe^{3+}(cophen)Cl]_2$	GC	TBAPF$_6$	DMSO	CO, HCOO$^-$, C$_2$O$_4^{2-}$	18.5%/ 67.2%/ 9.8%	-1.75 V	-1.69 V			improved by Li$^+$ and CF$_3$CH$_2$OH	117
3	$[Fe^{3+}(dophen)(N\text{-}MeIm)_2]_2$	GC	TBAPF$_6$	DMSO	CO, HCOO$^-$, C$_2$O$_4^{2-}$	13.3%/ 73.6%/ 7.3%	-1.72 V	-1.69 V			improved by Li$^+$ and CF$_3$CH$_2$OH	117
4	$[Fe^{3+}(dophen)Cl]_2$	GC	TBAPF$_6$	DMF	CO, HCOO$^-$, C$_2$O$_4^{2-}$	22.5%/ 57.2%/ 13.4%	-1.71 V	-1.69 V			improved by Li$^+$ and CF$_3$CH$_2$OH	117
5	$[Fe^{3+}(dophen)(N\text{-}MeIm)_2]_2$	GC	TBAPF$_6$	DMF	CO, HCOO$^-$, C$_2$O$_4^{2-}$	23.9%/ 58.9%/ 11.1%	-1.72 V	-1.69 V			improved by Li$^+$ and CF$_3$CH$_2$OH	117
6	$[Ni(cyclam)]^{2+}$	Hg	KNO$_3$	H$_2$O	CO	99%	-1.33 V	-1.0 V	4.10		4 h, 18 TOF/77.5 TON	118–121
7	$[Ni(tmdnTAA)]^{2+}$	GC	TEtA(ClO$_4$)	DMF:H$_2$O 1:1	CO		-0.84 V	-1.60 V				122
8	$[Ni(HACD)]^{2+}$	Hg (HMD)	Li(ClO$_4$)	H$_2$O	CO		-1.12 V	-1.36 V				123
9	$[Ni(decyclam)]^{2+}$	Hg (HMD)	Li(ClO$_4$)	H$_2$O	CO, HCOO$^-$, H$_2$		-1.23 V	-1.36 V	5.00			124

#	Complex	Electrode	Electrolyte	Solvent	Products	FE (%)	E	E	Conditions	Notes	Ref
10	[CHx(Ni(cyclam))2]	HMD	TBAPF$_6$	MeCN/H$_2$O	CO, H$_2$		−1.21 V	−1.46 V			125
11	[Co(dmg)$_2$(H$_2$O)Py]	GC	TMACl	EtOH	CO	10%	−0.53 V	−0.65 V			126
12	[Co(TPP)]	GC/Pt	TBAF	DMF	HCOO$^-$			−1.26 V	20–22 C		84, 127–129
13	[Fe(TPP)]	Hg	TEtA(ClO$_4$)	DMF	CO	94%	−1.41 V	−1.46 V		Mg^{2+}, or CF$_3$CH$_2$OH detected by chromotropic assay	130–134
14	[Co(tpy)$_2$]$^{2+}$	GC	TBA(ClO$_4$)	DMF	HCOOH		−1.46 V	−1.46 V			135
15	[Ni(tpy)$_2$]$^{2+}$	GC	TBA(ClO$_4$)	DMF	CO, CO$_3^{2-}$		−0.96 V	−0.96 V			135, 136
16	[Ni(bpy)$_3$]$^{2+}$	GC	TBA(ClO$_4$)	MeCN			−0.9 V	−0.90 V			137
17	[Ru(bpy)$_2$(CO)$_2$]$^{2+}$	Hg (HMD)	TBA(ClO$_4$)	H$_2$O:DMF 9:1	HCOO$^-$, CO	34%/	−0.79 V	−1.26 V	9.5/6	16 TON/12 TON	37, 138
18	[Ru(bpy)$_2$(CO)$_2$]$^{2+}$	Hg	TBA(ClO$_4$)	MeOH	HCOO$^-$, CO, H$_2$	52.5%/ 32.0%	−0.79 V	−1.26 V	30 C		35, 38, 138
19	[Ru(bpy)$_2$(CO)$_2$]$^{2+}$	Hg	TBA(ClO$_4$)	MeCN	HCOO$^-$, CO, H$_2$	84.2%/ 2.4%/ 6.8%	−0.79 V	−1.06 V		Me$_2$NH·HCl, Effiecency of HCOO$^-$ increases with increasing pk$_a$	35, 38, 43, 138
20	[Ru(dmbpy)(bpy)(CO)$_2$]$^{2+}$	Hg	TBA(ClO$_4$)	MeCN:H$_2$O 4:1	CO	71.80%	−0.89 V	−1.06 V			35
21	[Ru(dmbpy)(bpy)(CO)$_2$]$^{2+}$	Hg	TBA(ClO$_4$)	MeOH	CO, HCOO$^-$	34.2%/ 39.8%	−0.89 V	−1.06 V			35
22	[Ru(dmbpy)$_2$(CO)$_2$]$^{2+}$	Hg	TBA(ClO$_4$)	MeCN:H$_2$O 4:1	CO	65.30%	−0.89 V	−1.06 V			35
23	[Ru(dmbpy)$_2$(CO)$_2$]$^{2+}$	Hg	TBA(ClO$_4$)	MeOH	CO, HCOO$^-$	44.7%/ 32.5%	−0.89 V	−1.06 V			35
24	[Ru(phen)$_2$(CO)$_2$]$^{2+}$	Hg	TBA(ClO$_4$)	MeCN:H$_2$O 4:1	CO	61.5	−0.82 V	−1.06 V			35
25	[Ru(phen)$_2$(CO)$_2$]$^{2+}$	Hg	TBA(ClO$_4$)	MeOH	CO, HCOO$^-$	34.7%/ 24.5%	−0.82 V	−1.06 V			35
26	[Ru(bpy)(Cl)$_2$(CO)$_2$]$^{2+}$	Hg	TBA(ClO$_4$)	MeCN:H$_2$O 4:1	CO	87.80%		−1.06 V			35
27	[Ru(bpy)(Cl)$_2$(CO)$_2$]$^{2+}$	Hg	TBA(ClO$_4$)	MeOH	CO, HCOO$^-$	27.3%/ 37.7%	−1.06 V	−1.06 V			35

Table 11.1 (Continued)

	Catalyst	WE	Electrolyte	Solvent	Product	Efficiency	Redox Couple	CO$_2$ reduction	pH	Temp	Notes	Refs.
28	[Ru(dmbpy)(Cl)$_2$(CO)$_2$]$^{2+}$	Hg	TBA(ClO$_4$)	MeCN:H$_2$O 4:1	CO	66.00%		−1.06 V				35
29	[Ru(dmbpy)(Cl)$_2$(CO)$_2$]$^{2+}$	Hg	TBA(ClO$_4$)	MeOH	CO, HCOO$^-$	39.2%/ 26.8%		−1.06 V				35
30	cis-[Os(bpy)$_2$H(CO)]$^+$	Pt$_{mesh}$	TBAPF$_6$	MeCN	CO	90%	−1.10, −1.36 V	−1.16 to −1.36 V				44
31	cis-[Os(bpy)$_2$H(CO)]$^+$	Pt	TBAPF$_6$	MeCN 0.3M H$_2$O	CO, HCOO$^-$	x/25%						44
32	[Re(CO)$_3$(Cl)(bpy)]	GC/Pt	TEtACl	DMF 10% H$_2$O	CO	98%	−1.47 V	−1.25 V				27, 29, 139
33	[Re(CO)$_3$(ClO$_4$)(bpy)]	GC	TBAPF$_6$	DMF 10% H$_2$O	CO	99%	−1.12 V	−1.25 V				26, 27, 29, 64
34	[Re(CO)$_3$Cl(dmbpy)]	GC	TEtA(BF$_4$)	MeCN	CO		−1.30 V	−1.52 V				28, 30
35	[Re(CO)$_3$Cl(pbmbpy)]	Pt$_{mod}$	TBA(ClO$_4$)	MeCN	CO, CO$_3$$^{2-}$	81%	−1.72 V vs. Ag/ 10 mM Ag$^+$	−1.85 V vs. Ag/ 10 mM Ag$^+$			14% oxalate	140
36	Re(tBu-bpy)(CO)$_3$Cl	GC	TBAPF$_6$	MeCN	CO	99%	−1.59 V	−1.76 V				28
37	[Rh(COD)(bpy)]$^+$	Pt		MeCN	CO, HCOO$^-$							141
38	[(η^6-C$_6$H$_6$)Ru(bpy)Cl]$^+$	Pt		MeCN	CO, HCOO$^-$							141
39	cis-[Rh(bpy)$_2$(CF$_3$SO$_3$)$_2$]$^+$	Pt	TBAPF$_6$	MeCN	HCOO$^-$		−0.98, −1.27 V				40 to 100 minute run, 12.3 TON	141, 142
40	cis-[Ir(bpy)$_2$(CF$_3$SO$_3$)$_2$]$^+$	Pt	TBAPF$_6$	MeCN	HCOO$^-$			−0.96 to −1.36 V				141, 142
41	[Ni(MeCN)$_4$(PPh$_3$)]$^{2+}$	GC	TBA(ClO$_4$)	MeCN	CO, CO$_3$$^{2-}$							137
42	[Ni$_3$(μ-CNMe)(μ$_3$-I)(dppm)$_3$]$^+$	Hg	NaPF$_6$	THF	CO, CO$_3$$^{2-}$		−0.89 V	−0.89 V				143
43	[Ru(terpy)(dppe)Cl]$^+$	Pt		MeCN	CO, HCOO$^-$							141
44	[RhCl(dppe)]	Hg	TEtA(ClO$_4$)	MeCN	HCOO$^-$	42%	−1.52 V	−1.21 V	−1.52		MeCN proton source	144

No.	Complex	Electrode	Electrolyte	Solvent	Products	Yield	E_1	E_2	Temp	Ref
45	[Ir(CO)Cl(PPh$_3$)$_2$]	Hg	TBABF$_4$	DMF 10% H$_2$O	CO		-1.70 V	-1.06 V	20 C	145
46	[Pd(PPh$_3$)(PPh$_3$)]	GC	TEtA(BF$_4$)	MeCN +H$^+$	CO, H$_2$		-0.33 V			146
47	[Pd(PPh$_3$)(PEt$_3$)]	GC	TEtA(BF$_4$)	MeCN +H$^+$	CO, H$_2$		-0.69 V			146
48	[Pd(PPh$_3$)(P(OMe)$_3$)]	GC	TEtA(BF$_4$)	MeCN +H$^+$	CO, H$_2$		0.37 V			146
49	[Pd(PPh$_3$)(P(CH$_2$OH)$_3$)]	GC	TEtA(BF$_4$)	MeCN +H$^+$	CO, H$_2$		-0.51 V			146
50	[Pd(PPh$_3$)(MeCN)]	GC	TEtA(BF$_4$)	MeCN +H$^+$	CO, H$_2$		-0.48 V			146
51	[Pd(PPh$_3$)$_2$(2-m-8-Hq)]Cl	Pt	TBAPF$_6$	MeCN	CO	60.2%		-0.94 V		147
52	[Pd(PPh$_3$)$_2$(2-m-8-Hq)]Cl	Pt	TBAPF$_6$	MeCN:H$_2$O 25:2	CO, HCOO$^-$	25.2%/44.8%		-0.94 V		147
53	[Pd(PPh$_3$)$_2$(2-Qui)]Cl	Pt	TBAPF$_6$	MeCN	CO	56.7%		-0.94 V		147
54	[Pd(PPh$_3$)$_2$(2-Qui)]Cl	Pt	TBAPF$_6$	MeCN:H$_2$O 25:2	CO, HCOO$^-$	24.0%/37.7%		-0.94 V		147
55	[Pd(PPh$_3$)$_2$(3-Hiq)]Cl	Pt	TBAPF$_6$	MeCN	CO	73%		-0.94 V		147
56	[Pd(PPh$_3$)$_2$(3-Hiq)]Cl	Pt	TBAPF$_6$	MeCN:H$_2$O 25:2	CO, HCOO$^-$	31.7%/25.2%		-0.94 V		147
57	[Pd(PPh$_3$)$_2$(1-Hiq)]Cl	Pt	TBAPF$_6$	MeCN	CO	74.5%		-0.94 V		147
58	[Pd(PPh$_3$)$_2$(1-Hiq)]Cl	Pt	TBAPF$_6$	MeCN:H$_2$O 25:2	CO, HCOO$^-$	31.1%/25.8%		-0.94 V		147
59	[Pd(PPh$_3$)$_2$(2-m-1,10-phen)](ClO$_4$)$_2$	Pt	TBAPF$_6$	MeCN	CO	60.9%		-0.94 V		147
60	[Pd(PPh$_3$)$_2$(2-m-1,10-phen)](ClO$_4$)$_2$	Pt	TBAPF$_6$	MeCN:H$_2$O 25:2	CO, HCOO$^-$	31.5%/39.5%		-0.94 V		147
61	[Pd(PPh$_3$)$_2$(dmbpy)](ClO$_4$)$_2$	Pt	TBAPF$_6$	MeCN	CO	81.0%		-0.94 V		147
62	[Pd(PPh$_3$)$_2$(dmbpy)](ClO$_4$)$_2$	Pt	TBAPF$_6$	MeCN:H$_2$O 25:2	CO, HCOO$^-$	44.2%/30.0%		-0.94 V		147
63	[Co(PPh$_3$)$_2$(2-m-1,10-phen)](ClO$_4$)$_2$	Pt	TBAPF$_6$	MeCN	CO	62.6%		-0.94 V		147
64	[Co(PPh$_3$)$_2$(2-m-1,10-phen)](ClO$_4$)$_2$	Pt	TBAPF$_6$	MeCN:H$_2$O 25:2	CO, HCOO$^-$	32.6%/41.0%		-0.94 V		147
65	[Co(PPh$_3$)$_2$(dmbpy)](ClO$_4$)$_2$	Pt	TBAPF$_6$	MeCN	CO	83.4%		-0.94 V		147
66	[Co(PPh$_3$)$_2$(dmbpy)](ClO$_4$)$_2$	Pt	TBAPF$_6$	MeCN				-0.94 V		147

Table 11.1 (Continued)

Catalyst	WE	Electrolyte	Solvent	Product	Efficiency	Redox Couple	CO_2 reduction	pH	Temp	Notes	Refs.
67 $[Co(PPh_3)_2(2\text{-}m\text{-}8\text{-}Hq)]Br$	Pt		MeCN:H_2O 25:2	CO, $HCOO^-$	44.8%/ 29.1%		−0.94 V				147
68 $[Co(PPh_3)_2(2\text{-}m\text{-}8\text{-}Hq)]Br$	Pt	TBAPF$_6$	MeCN	CO	61.4%		−0.94 V				147
		TBAPF$_6$	MeCN:H_2O 25:2	CO, $HCOO^-$	25.8%/ 43.9%						
69 $[Rh_2(PhCHOHCOO)_2\text{-}(phen)_2(H_2O)_2]^{2+}$	Pt	TBA(BF$_4$)	DMF:H_2O 10:1	CO, $HCOO^-$	85-90%	−0.55 V	−0.74 V				148
70 $[Fe_4S_4(SCH_2Ph)_4]^{2-}$	Hg	TBA(BF$_4$)	DMF	$HCOO^-$	40% /		−1.76 V				149
71 $[Fe_4S_4(SXN^-)_4]^{2-}$	Hg	TBA(BF$_4$)	DMF	$HCOO^-$	23%		−1.80 V			X = − COCMe^{2-}, COC$_6$H$_4$CH$_2$	150
72 Pyridine	Pt/Pd/p-GaP	Na(ClO$_4$)	H_2O	HCOOH, MeOH	10.8%/ 22%		−0.34 V	5.00		30–50 µA	25, 110, 114

Table 11.2 Photochemical systems for CO$_2$ reductions with reaction conditions, catalysts, chromophore, and products(tb-cabinol = tris(4'-methyl-2,2'-bipyridyl-4-methyl)carbinol, dmb = 4,4'-dimethyl-2,2'-bipyridyl, TEA = triethylamine, TEOA = triethanolamine, BNAH = 1-Benzyl-1,4-dihydronicotinamide, H$_2$A = Ascorbic Acid, bpz = 2,2'-bipyrazine, HMD = 5,7,7,12,14,14-hexamethyl-1,4,8,11-tetraazacyclotetradeca-4,11-diene, cyclam = 1,4,8,11-tetraazacyclotetradecane, pr-cyclam = 6-((p-methoxybenzyl)pyridin-4-yl)methyl-1,4,8,11 -tetraazacyclotetradecan, MV = methyl viologen, phen = 1,10-phenanthroline, bpy = 2,2'-bipyridine, EDTA = ethylenediaminetetraacetate, TPA = tripropylamine, TMA = trimethylamine, TPP = 5,10,15,20-Tetraphenylporphin, TBtA = tributylamine, TPtA = tripentylamine, TiBA = triisobutylamine, TMEDA = N,N,N,N-tetramethylethylenediamine, {[Zn(TPP)]/[Re(CO)$_3$(pic)(bpy)]} = 5-[4-[(2-methoxy-4-([rhenium (I) tricarbonyl (3-picoline)]4-methyl-2,2'-bipyridine-4'-carboxyamidyl) carboxyamidyl) phenyl] phenyl]-10,15,20-triphenyl porphyrinatozinc(II)).

	Chromophore	Cat/Relay	Donor	Solvent	Product	Φ(mol/einsteins)	TON/TOF	pH	irradiation time	λ/nm	Refs.
1	Ru(bpy)$_3^{2+}$		TEOA	15%H$_2$O in DMF	HCOO$^-$	0.049	19/9.5				151
2	Ru(bpy)$_3^{2+}$		TEOA	15%H$_2$O in DMF	HCOO$^-$	0.096	43/21.5				151
3	Ru(bpy)$_3^{2+}$	MV^{2+}	TEOA, EDTA	H$_2$O	HCOO$^-$	0.01	75/18.8		4 hr		152
4	Ru(bpy)$_3^{2+}$	Co^{2+}/bpy	TEA, TPA, TEOA, TMA	MeCN/donor/H$_2$O, 3:1:1 (vol/vol)	CO, H$_2$		9/0.4		22 hr		153
5	Ru(bpy)$_3^{2+}$	Co^{2+}/2,9-Me$_2$phen	TEA, TPA, TMA, TBA, TPtA, TiBA, TMEDA	MeCN or DMF/donor/ H$_2$O, 3:1:1 (vol/vol), DMF/ H$_2$O 3:2	CO, H$_2$	0.012 (CO), 0.065 (H$_2$)		8.6	26 hr		154
6	Ru(bpy)$_3^{2+}$	Ru(bpy)$_2$(CO)$_2^{2+}$	TEOA	H$_2$O /DMF 1:9 and DMF	HCOO$^-$	2% , 1%	50, 125	6/9.5	10 hr		41, 42, 138
7	Ru(bpy)$_3^{2+}$	Ru(bpy)$_2$(CO)$_2^{2+}$	BNAH	H$_2$O /DMF 1:9	HCOO$^-$, CO	0.03 (HCOO$^-$), 0.15 (CO)		6/9.5	10 hr		41, 42, 138
8	Ru(bpy)$_3^{2+}$	Ru(bpy)$_2$(CO)H$^+$	TEOA		HCOO$^-$	0.15	161/80.5 (X=Cl)				151
9	Ru(bpy)$_3^{2+}$	Ru(bpy)$_2$(CO)X^{n+}	TEOA		HCOO$^-$		163/81.5, 54/27 (X=CO)				151
10	Ru(bpy)$_3^{2+}$	Co(HMD)$^{2+}$	H$_2$A		CO, H$_2$						77

Table 11.2 (*Continued*)

	Chromophore	Cat/Relay	Donor	Solvent	Product	ϕ (mol/einsteins)	TON/TOF	pH	irradiation time	λ/nm	Refs.
11	$Ru(bpy)_3^{2+}$	$Ni(cyclam)^{2+}$	H_2A	H_2O	CO, H_2	0.001 (CO)		4	22 hr		76, 78
12	$Ru(bpy)_3^{2+}$	$Ni(Pr\text{-}cyclam)^{2+}$	H_2A	H_2O	CO, H_2		ca. 0.005 (CO)	5.1	4 hr		79
13	$Ru(bpy)_3^{2+}$	$Bipyridinium^+$, Ru/OS Colloid	TEOA	H_2O	CH_4, H_2	10^{-4} (CH_4), $10^{-3}(H_2)$		7.8	2 hr		87
14	$Ru(phen)_3^{2+}$	$Ni(cyclam)^{2+}$	H_2A	H_2O	CO, H_2		<0.1	4	22 hr		78
15	$Ru(phen)_3^{2+}$	Pyridine	H_2A	H_2O	CH_3OH	7.22×10^{-7}	0.9	5	6 hr	470	86
16	$Ru(bpz)_3^{2+}$	Ru colloid	TEOA	H_2O/EtOH 2:1	CH_4	0.04%	15	9.5	2 hr		87, 88
17	$Ru(dmb)_3^{2+}$	$ReCl(dmb)(CO)_3$	BNAH	DMF:TEOA 5:1	CO	0.062	101/6.3		16 hr	≥500	69
18	$[Ru(phen)_2\text{-}(phenC_1cyclam)Ni]^{2+}$		H2A	H_2O	CO, H_2		<0.1	5.1	4 hr		79
19	$[(dmb)_2Ru(MebpyC_3OHMebpy)\text{-}Re(CO)_3Cl]^{2+}$		BNAH	DMF:TEOA 5:1	CO	0.12	170/10.7		16 hr	≥500	69
20	$[(dmb)_2Ru(MebpyC_nH_{2n}Mebpy)\text{-}Re(CO)_3Cl]^{2+}$		BNAH	DMF:TEOA 5:1	CO	0.13 (n = 2), 0.11 (n = 4,6)	180/15 (n − 2), 120/10 (n = 4,6)		16 hr	<500	73
21	$[(dmb)_2Ru(MebpyC_3OHMebpy)\text{-}Re(CO)_3\{P(POEt)_3\}]^{3+}$		BNAH		CO	0.21	232/19.3				70
22	$[Ru\{(Mebpy C_3OHMebpy)\text{-}Re(CO)_3Cl\}_3]^{2+}$		BNAH	DMF:TEOA 5:1	CO	0.093	240/15		16 hr	≥500	69
23	$[(dmb)_2Ru(MebpyC_2Mebpy)\text{-}Re(CO)_2\{P(p\text{-}FPh)\}_2]^{2+}$		BNAH	DMF:TEOA 5:1	CO	0.15	207/281		20 hr	>500	74
24	$[(dmb)_2Ru(tb\text{-}carbinol)\{Re(CO)_3Cl\}_2]^{2+}$		BNAH	DMF:TEOA 5:1	CO		190/11.8		16 hr	≥500	71, 72
25	$[[(dmb)_2Ru]_2(tb\text{-}carbinol)Re(CO)_3Cl]^{2+}$		BNAH	DMF:TEOA 5:1	CO		110/6.9		16 hr	≥500	52,53
26	p-terphenyl	$Co(cyclam)^{3+}$	TEOA	MeOH/MeCN 1:4	$CO, H_2, HCOO^-$	0.25 ($CO + HCOO^-$)			1 hr	290	81
27	p-terphenyl	$Co(HMD)^{2+}$	TEOA	MeOH/MeCN	$CO, H_2, HCOO^-$				1 hr	313	81, 83
28	Phenazine	$Co(cyclam)^{3+}$	TEA	MeOH/MeCN/TEA 10:1:0.5	$CO, H_2, HCOO^-$	0.07 ($HCOO^-$)			3 hr	313	82

No.	Catalyst	Reductant	Solvent	Product	Φ	TN	Time	λ (nm)	Ref.
29	$Fe^{III}(TPP)$	TEA	DMF	CO		70	180 hr	UV	85
30	$Co^{III}(TPP)$	TEA	MeCN	$HCOO^-$, CO		>300 (total)	200 hr	<320	84
31	{[Zn(TPP)]/[Re(CO)$_3$(pic)(bpy)]}	TEOA	DMF:TEOA 5:1	CO		30		>520	155
32	[Re(4,4'-(MeO)$_2$-bpy)(CO)$_3$-(P(OEt)$_3$)]$^+$	TEOA	DMF:TEOA 5:1	CO	0.59		25 hr	<330	65
33	ReCl(bpy)(CO)$_3$	TEOA	DMF:TEOA 5:1	CO		27	4 hr	>400	64
34	ReCl(bpy)(CO)$_3$	TEA	DMF:TEA 0.8 M TEA	CO		8.2(Cl) 42(COO)	25 hr		62
35	ReBr(bpy)(CO)$_3$	TEOA	TEOA:DMF 1:2	CO	0.15	20/5	11.7 min	436	57, 59, 64
36	ReOCHO(bpy)(CO)$_3$	TEOA	TEOA:DMF 1:5	CO	0.05	12	20 min	>330	64, 68
37	[Re(bpy)(CO)$_3$(PR$_3$)]$^+$	TEOA	DMF:TEOA 5:1	CO	0.38 (R=Oet), 0.013 (R=nBu), 0.024(R=Et), 0.2 (OiPr), 0.17 (R-Ome)	7.5/0.5 (R=Oet), <1/<0.1 (R=nBu), 6.2/0.5 (OiPr), 5.5/0.4 (R-Ome)	13 hr	365	66, 156
38	[Re(bpy)(CO)$_3$(P(Ohex)$_3$)]$^+$	TEA	CO$_2$(liquid)	CO		2.2/1.1	2 hr	365	157
39	[Re(bpy)(CO)$_3$(P(OiPr)$_3$)]$^+$	TEOA	DMF:TEOA	CO		15.6/0.7	24 hr	365	158
40	[Re(bpy)(CO)$_3$(4-X-py)]$^+$	TEOA	TEOA:DMF 1:5	CO	0.03 (x=tBu, Me,H), 0.04 (x=C(O)Me), 0.13 (X=CN)	1/0.1 (x=tBu, Me,H, C(O)Me, 3.5/0.4 (X=CN)	8.5 hr	365	159
41	[Re(4,4'-(CF$_3$)$_2$-bpy)(CO)$_3$(P(OEt)$_3$)]$^+$	TEOA	DMF:TEOA 5:1	CO	0.005	<1/<0.1	17 hr	365	156
42	[Re(dmb)(CO)$_3$(P(OEt)$_3$)]$^+$	TEOA	DMF:TEOA 5:1	CO	0.18	4.1/0.2	17 hr	365	156

literature. Similar compilations for semiconductor-based studies appear in reference 5 and other sources cited above. The focus here is on the use of inorganic molecules and/or semiconductor electrode materials to sustain the reduction of CO_2. Biochemical or bioelectrochemical approaches (for example, see reference 10), which are clearly of interest and importance from a comparative perspective of the artificial photosynthesis approach under discussion, are not addressed specifically.

At the outset it seems prudent to review critically the various terminologies used in the literature, in order to place the present discussion in the proper context. Thus a *homogeneous* CO_2 reduction system consists of an assembly of dissolved (molecular) catalyst that may be present in addition to a light absorber, sacrificial electron donor, and/or electron relay all in the same solution. In some cases, the light-absorbing function may be built into the same catalyst molecule, but the key is that all participating components are present in the same phase. A *heterogeneous* system, on the other hand, has the catalyst present in a different (*i.e.* solid) phase. The catalyst may be a metal that is anchored to a support, which in turn may also be a metal or an inorganic semiconductor. We believe that this distinction based on phase is less important (and may even be misleading) when applied to a situation such as CO_2 reduction; thus this terminology is avoided in the discussion that follows. The terminology problem is illustrated by approaches based on the tethering (or strong adsorption) of (catalyst) *molecules* on metal or semiconductor electrode surfaces. Does the CO_2 reduction occur in such cases in the solution phase or at the solid/liquid phase boundary? Clearly the distinction between "homogeneous" and "heterogeneous" becomes much fuzzier here.

The term "photoelectrochemical" has been largely applied in the literature to situations involving a semiconductor electrode, whereas we apply this terminology in the present context to denote situations involving either the traditional semiconductor/liquid junctions or catalyst molecules that serve the dual functions of both light absorption and electron transfer mediation. Alternate descriptions based on "electrocatalytic" and "photocatalytic" systems are synonymous and denote approaches wherein the CO_2 reduction is driven electrochemically and with the assistance of light, respectively. On the other hand, the concept of *photochemical* systems is best reserved for approaches based on colloidal suspensions of metal or inorganic semiconductor nanoparticles or purely homogeneous systems with molecular catalysts in solution. While approaches using colloids or nanoparticles have been reviewed[6] (and indeed one example of it is discussed below), we believe that they are problematic in terms of process scale-up and product separation. Systems based on colloidal suspensions also are prone to low conversion efficiencies stemming largely from the prevalence of back reactions. These issues are largely circumvented in *photoelectrocatalytic* systems, in whcih the colloidal particles are anchored onto a solid electronically conducting support (*e.g.* conducting transparent glass) so that a negative electrode potential can be applied to bias forward electron transfer and thus inhibit the back recombination pathway.

11.2   Thermodynamics of CO$_2$ Reduction

Equations (11.1) to (11.6), below show the various products resulting from the reduction of CO$_2$, ranging from a one-electron reduction to the radical anion all the way to an 8-electron deep reduction to methane. Multiple proton-coupled electron transfer (PCET) steps occur in Equations (11.2) to (11.6), and herein lies the rich electrochemistry inherent with this system. Given that these electrochemical processes are pH-dependent, the potentials below are given at pH 7 in aqueous solution versus the normal hydrogen electrode (NHE), 25 °C, 1 atm gas pressure, and 1M for the solutes.[5,15]

$$CO_2 + e^- \rightarrow CO_2{}^{\bullet-} \qquad\qquad E° = -1.90V \qquad (11.1)$$

$$CO_2 + 2e^- + 2H^+ \rightarrow CO + H_2O \qquad\qquad E° = -0.53V \qquad (11.2)$$

$$CO_2 + 2e^- + 2H^+ \rightarrow HCOOH \qquad\qquad E° = -0.61V \qquad (11.3)$$

$$CO_2 + 4e^- + 4H^+ \rightarrow H_2CO + H_2O \qquad\qquad E° = -0.48V \qquad (11.4)$$

$$CO_2 + 6e^- + 6H^+ \rightarrow H_3COH + H_2O \qquad\qquad E° = -0.38V \qquad (11.5)$$

$$CO_2 + 8e^- + 8H^+ \rightarrow CH_4 \qquad\qquad E° = -0.24V \qquad (11.6)$$

While progress on the concerted $2e^- - 2H^+$ reduction to CO or formate has been impressive (see below), the formation of more useful fuel products such as methanol and methane necessitates multiple electron and proton transfers. The kinetic barriers associated with these are formidable, as briefly discussed next.

11.3   Energetics of CO$_2$ Reduction

11.3.1   General Remarks

The terms "electrocatalytic" and "photocatalytic" are used herein in a generic sense with the implicit and important recognition that the reactions above are endergonic with ΔG values ranging from 1.90 eV to 8.31 eV respectively. Putting an electron into the linear and inert CO$_2$ molecule (Reaction (11.1)) entails a steep energy cost because of the resultant structural distortion.[8] This is reflected in the very negative reduction potential for Reaction (11.1) above. Thus this radical formation step is very energy-inefficient and the steep activation barrier associated with it, must be avoided *via* the use of a catalyst.[16,17] From an electrochemical perspective, this translates to sizeable "overpotentials" (spanning several hundred mV) for driving this reduction process.[16,17] Thus a catalyst molecule, by interacting strongly with the radical anion, can reduce this energy barrier. This is the essence of many of the catalysis-based approaches to be discussed below.

11.3.2   Band Energy Positions of Selected Oxide Semiconductors and CO_2 Redox Potentials

In a discussion of the energetics of CO_2 photoreduction, it is convenient to display the relevant solution redox potentials in an energy diagram on the same scale as the semiconductor band-edges. The latter are experimentally derived from flat-band potential measurements.[18] Figure 11.1 shows an example of this diagram for several inorganic oxide semiconductors. As with solution redox couples involving proton transfer, the oxide band-edges are pH-dependent, shifting at a Nernstian rate of -59 mV/pH unit at 25 °C. Therefore, the particular situation illustrated in Figure 11.1 pertains to a solution pH of 7. A similar diagram appears in reference [5] for TiO_2, Cu_2O, and eight other non-oxide semiconductors (see Figure 11.4 therein).

Such a representation is useful for assessing whether the photogenerated carriers in the semiconductor are thermodynamically capable of reducing (or oxidizing) a given species in the solution phase. We assume at the outset that these carriers are thermalized such that they possess average energies close to the semiconductor band edges. A further assumption is that Fermi level pinning does not occur, so that the band-edges are fixed relative to the energies of solution redox couples.[18] Therefore, any reduction reaction whose potential

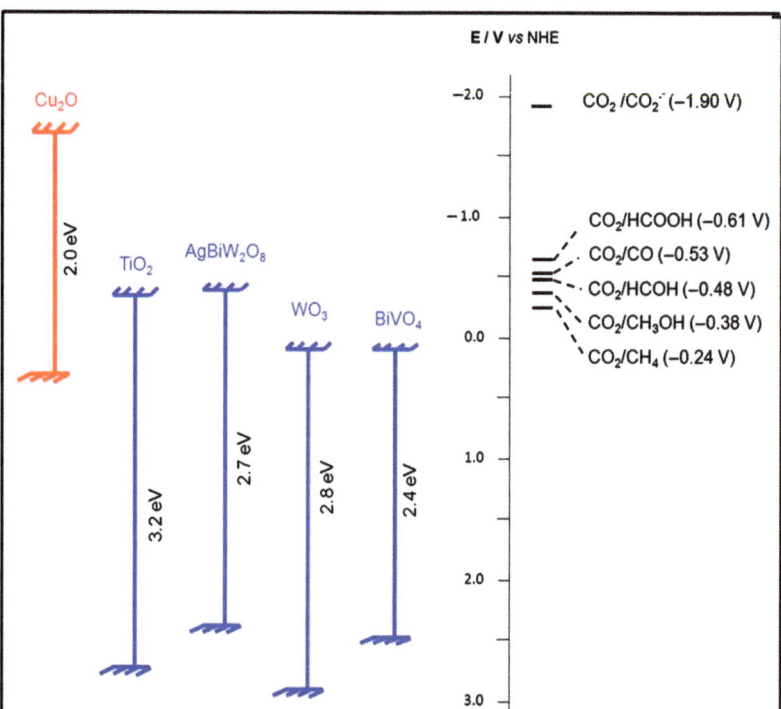

Figure 11.1   Comparison between the band-edges of selected semiconductors and relevant thermodynamic potentials for CO_2 reduction. All data are for pH 7 and versus a normal hydrogen electrode (NHE).

falls above the conduction band-edge (for a given semiconductor) would be thermodynamically prohibited because the photogenerated electrons would simply not have enough energy to sustain the reduction.

Given that the redox potentials for most of the CO$_2$-derived reduction processes lie negative of the hydrogen evolution reaction (HER) in water (the two exceptions being the deep reduction of CO$_2$ to methanol and methane), the more negative the conduction band-edge location is, the better is the corresponding semiconductor candidate. The thermodynamic driving force for the reduction is roughly given by the difference between the semiconductor conduction band-edge and the corresponding redox potential. Thus the more negative the semiconductor conduction band edge is, the greater the driving force (all other factors being equal). However, it is worth emphasizing that such thermodynamic arguments are only *starting points*. Whether the reduction occurs at a fast rate depends on the inherent kinetics at that semiconductor/ solution interface and the associated overpotentials. It is clear from Figure 11.1 that the use of new-generation semiconductors such as AgBiW$_2$O$_8$ and even materials such as Cu$_2$O rather than the well-studied TiO$_2$ prototype may be advantageous because of the relatively negative location of their conduction band edges. Results on both these oxide semiconductors are presented below.

11.3.3 Molecular Orbital Energy Diagram for Ru(phen)$_3$$^{2+}$ Compared with CO$_2$ Redox Potentials

For photochemical reduction in homogeneous system, the chromophores [Ru(phen)$_3$]$^{2+}$ and [Ru(bpy)$_3$]$^{2+}$ still represent one of the most widely used systems for driving highly endogonic redox reactions, due to their excited state energetics and good chemical stability.[19] We make a particular point of introducing this chromophore as it has been the primary one used in our own studies on homogeneous photochemical reduction of CO$_2$. As shown schematically in Figure 11.2, the reduction potential for both the photoexcited state [Ru(phen)$_3$]$^{2+}$* or the reductively quenched chromophore [Ru(phen)$_3$]$^+$ are negative of the key CO$_2$ reduction couples, meaning that these species are thermodynamically capable of reducing CO$_2$. The difficulty in using them is that they themselves lack the chemical functionality to lower the activation barriers involved and are only capable of delivering a single electron each towards these multi-electron reactions. It is also worth noting that the initial conversion of CO$_2$ to CO is the energy "hog" in the overall process and consumes a minimum of 1.33 eV.[20] Much of the progress associated with the conversion of CO$_2$ to CO and formate has revolved around electro- and photocatalytic strategies for minimizing the additional overpotential over and above this minimum threshold.

11.4 Electrocatalytic CO$_2$ Reduction with Molecular Catalysts

Many homogeneous catalysts have been developed for both electrochemical and photochemical systems; however, few are capable of deeper reduction than the

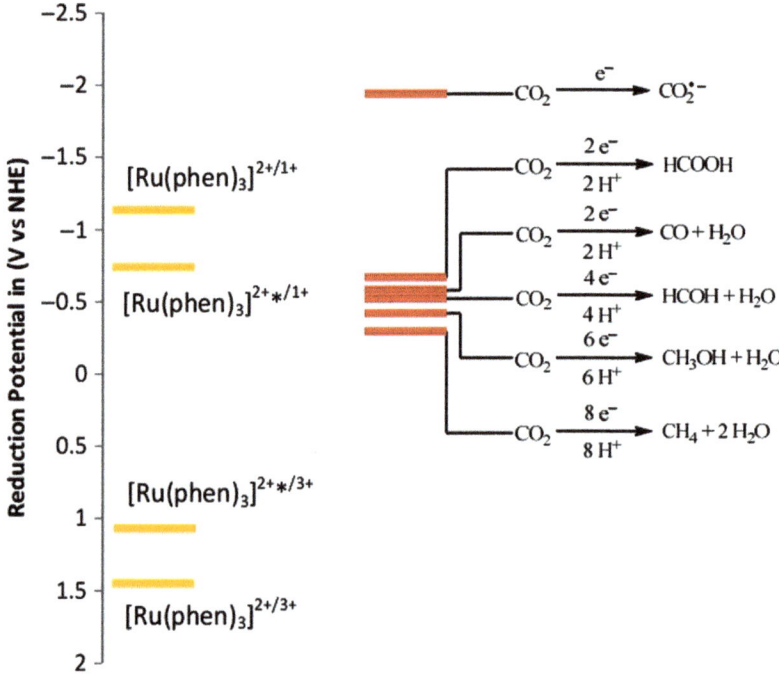

Figure 11.2   Reduction potentials of carbon dioxide reduction as compared with the HOMO and LUMO levels of $[Ru(phen)_3]^{2+}$.

two-electron reduced products of CO_2, such as CO and formic acid. Table 11.1 contains a collection of all (at least to the best of our knowledge) reported molecular electrocatalytic systems for CO_2 reduction in which the actual products of reduction were identified. The table includes information on the catalyst, electrochemical conditions, and products. Of the 72 entries in Table 11.1, 71 are metal complexes, and the final product represents a net two-electron reduction of CO_2. The sole organic entry (#72) is pyridine, and this simple catalyst is able to catalyze even deeper reduction to products including methanol as described in the next section. Metal phthalocyanines have also been reported to catalyze the even deeper reduction of CO_2 to CH_4, but these are known to form electrochemically active films and thus are more of a heterogeneous catalyst.[21]

11.4.1   Pyridine for Electrocatalytic Reduction of CO_2

Bocarsly *et al.* have shown it is possible to reduce CO_2 to methanol by using a very simple electrocatalyst, pyridinium, which upon reduction can bind CO_2 to form carbamate-type adducts and, *via* redox cycling, shuttles six electrons to ultimately form methanol, as shown in Figure 11.3.[22]

Through simulation of experimental results and kinetic studies, they were able to deduce the possible mechanism of the reduction of CO_2 to methanol.[23,24] The electron transfer in this process proceeds through an inner sphere electron

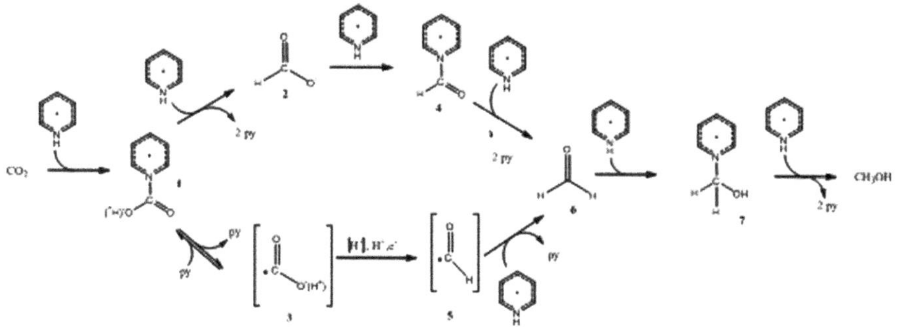

Figure 11.3   Pyridinium catalyzed reduction of CO$_2$.[3,25,114]

transfer as was shown by $^{13}C^{15}N$ coupling in NMR and by gas-phase photo-electron spectroscopy.[25] Based on the calculated bond distance and bond angles, the nature of the N-CO$_2$ bond was found to be primarily of π-character as opposed to σ-character. Reductions beyond the first electron transfer were found to depend on the electrodes being used.[25] For electrodes with low hydrogen over-potential, such as Pt or Pd, it was found that dissociation of the pyridine-formate radical adduct occurs, allowing the next reduction to formate or formic acid to take place on the electrode surface.[25] For electrodes with high hydrogen over-potential, the reaction is catalyzed entirely by the pyridinium with no dissociation of the formate radical, but a second pyridinium radical passes an electron to the pyridine-formate radical adduct instead.[24,25] With low hydrogen overpotential electrodes, formic acid adsorbs onto the surface to produce the hydroxyformyl radical that reacts with a surface hydrogen atom to make the formyl radical, which is reduced finally to the pyridinium radical to formaldehyde, as shown at the bottom of Figure 11.3.[25] For high hydrogen overpotential electrodes, formic acid reacts with the pyridinium radical to make the pyridinium-formyl adduct, which is reduced further by a second pyridinium radical to form free for-maldehyde and two equivalents of pyridine, as shown at the top of Figure 11.3.[25] For both electrode types, the reduction of formaldehyde results from the reaction with a pyridinium radical to form a pyridinium-formyl radical adduct, and this species reacts with a second pyridinium to produce methanol and two equiva-lence of pyridine.[25] There is some debate about the mechanism proposed by Bocarsly *et al.*: some groups claim a non-innocent role of the surface in the process and thus a process which is necessarily heterogeneous.[23,25]

11.4.2   Rhenium Polypyridyl Complexes for Electrocatalytic Reduction of CO$_2$

The highly selective and efficient nature of the *fac*-Re(bpy)(CO)$_3$X (X = Cl, Br) has driven the large amount of research activity of this complex towards electrochemical CO$_2$ reduction.[13,15,26–32] The catalytic reduction can proceed through two different pathways: a one-electron or a two-electron pathway (Figure 11.4), both of which yield CO.[27] In the one-electron pathway

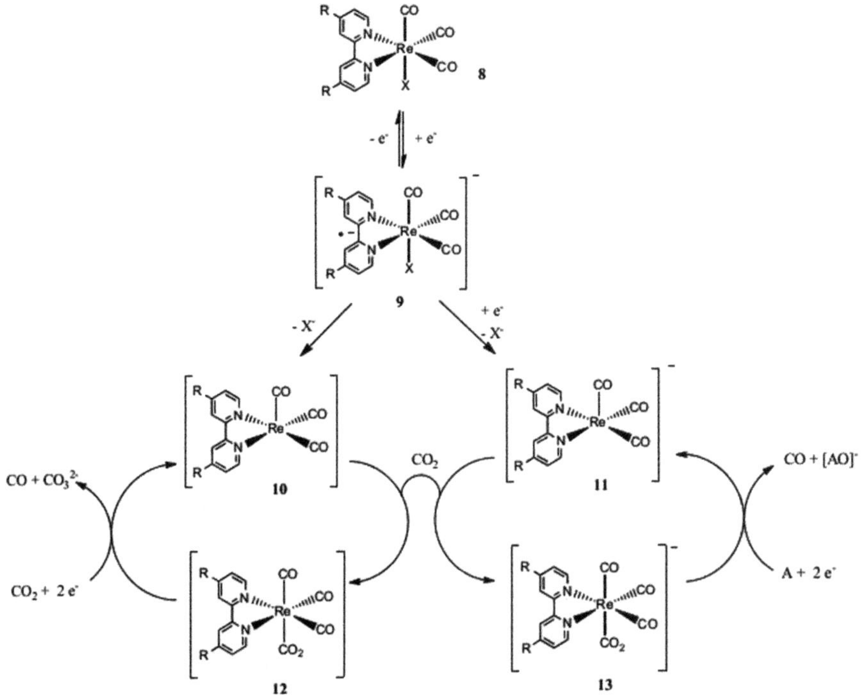

Figure 11.4 The proposed one-electron and two-electron pathways.

(Figure 11.4, left), the coordinatively unsaturated Re(0) species binds CO_2 to give the equivalent of a bound $CO_2^{\bullet-}$ radical anion, which upon reaction with another $CO_2^{\bullet-}$ radical anion (CO_2 and a second electron) disproportionates to yield CO and CO_3^{2-}.[33,34] In the two electron pathway (Figure 11.4, right), the starting complex is reduced twice, and following loss of the halide yields a coordinatively unsaturated Re(-1) complex. Binding of CO_2 and in the presence of some oxide acceptors, such as H^+, yields CO and H_2O.[13,27]

Both of these pathways, the one-electron and the two-electron, are accessible with the same complex but at different potentials. Based on the work published by Meyer *et al.*,[27] species **9**, the one-electron reduced species, is formed at -1.11 V versus NHE. Under Ar, this couple is also associated with the formation of a Re-Re dimer, [*fac*-Re(bpy)(CO)$_3$]$_2$, which has been implicated as the reactive species in some studies.[27,30] The second reduction of *fac*-[Re(bpy)-(CO)$_3$X]$^-$ is observed at -1.26 V *vs.* NHE and can also result in halide loss and generation of the active catalyst.[13] At a more negative potential of -1.56 V *vs.* NHE, the reaction proceeds through the two-electron reduction pathway as shown on the right side of Figure 11.4 to give CO without the formation of carbonate.[27] Studies by Kubiak *et al.*[28] have shown that by changing the 4,4$'$ substituents on the bpy-ligand in *fac*-Re(bpy)(CO)$_3$X it was possible to enhance the electrocatalysis reaction rate from 50 $M^{-1}s^{-1}$ for H to 1000 $M^{-1}s^{-1}$ for the *t*-Bu derivative as well as to increase the Faradaic efficiency ($\sim$99%).

11.4.3 Ruthenium Polypyridyl Complexes for Electrocatalytic Reduction of CO₂

Other than rhenium-based complexes, ruthenium polypyridyl complexes are the next most well explored. $Ru(phen)_2(CO)_2{}^{2+}$ and $Ru(bpy)_2(CO)_2{}^{2+}$ are reported to reduce CO_2 electrocatalytically. These complexes typically make CO, H_2, and formate as products of reduction,[35–42] with the ratio of these products being pH dependent.[39,40] At pH 6, the products are CO and H_2; however at pH 9.5, formate is produced in addition to CO and H_2.

Two different catalytic pathways have been proposed: the one proposed by Tanaka *et al.* is shown in Figure 11.5,[39,40,43] and the other proposed by Meyer *et al.*[9] is shown in Figure 11.6. Both schemes involve the reduction and loss of CO to form a neutral coordinatively unsaturated $Ru(L\text{-}L)_2(CO)$ (**16**) species. Tanaka *et al.* can start with the dicarbonyl, species **14**, or the monocarbonyl monochloride, species **15**,[39,40,43] where the electrons are thought to sit on the bpy ligands.[9] Tanaka *et al.* propose that **16** can react with either CO_2 or H^+ to form one of two intermediates, the formato species **17** or the hydride species **19**.[39,40,43] The species **17** reacts with a proton to form the metallo-carboxylic acid (species **18**), which at pH 6 and reforms the dicarbonyl complex **14** but at pH 9.5, adds two electrons to produce formate and **16**.[39,40,43] The generation of hydrogen is explained *via* the formation of the hydride in a competing side reaction.

As shown in Figure 11.6, Meyer *et al.* propose that the hydride species **19** is involved directly in the CO_2 fixing cycle and that – after reduction to **21** – insertion of CO_2 into the metal hydride bond forms the formato-complex (species **22**).[9,44–46] A further reduction step releases formate and generates a solvate complex, which reacts with water to reform the hydride **19**.[46] Although there is

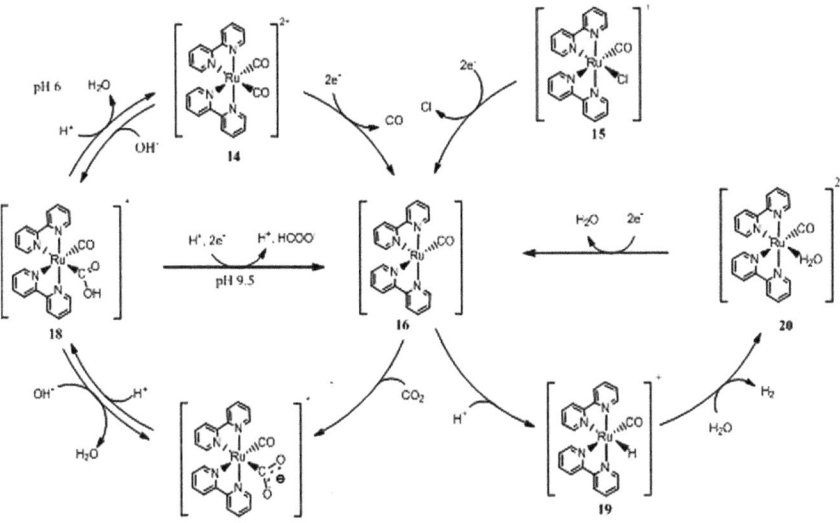

Figure 11.5 Electrocatalytic reduction of CO_2 by $[Ru(bpy)_2(CO)_2]^{2+}$ with possible pathways for CO, formate and H_2 formation.

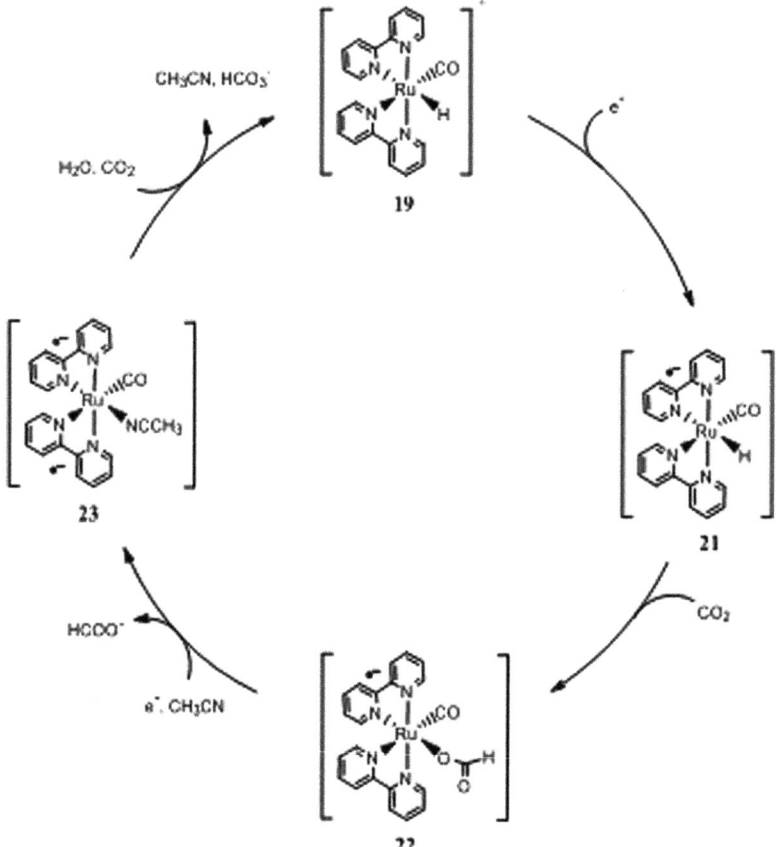

Figure 11.6   Electrochemical reduction of CO$_2$ *via* hydride bond insertion to form formate.

no direct evidence for the formation of species **23**, the catalysis could proceed through a process that goes directly from species **22** to species **19** with loss of formate and reduction of water in one step.[46]

11.5   Electrocatalytic Materials Inspired by Fuel Cell Electrode Nanocomposites

11.5.1   Pt-carbon Black-TiO$_2$ Nanocomposite Films Containing Highly Dispersed Pt Nanoparticles as Applied to CO$_2$ Electroreduction

The proton electrolyte membrane (PEM) fuel cell and water electrolysis technologies have obvious synergies, especially as they pertain to the proton exchange membrane (*e.g.* Nafion) separating the two compartments. Therefore, it is not surprising that the electrochemical engineering underpinning both these technologies shares many common aspects.[19] It occurred to us that many of the design aspects related to fuel cell electrode materials are equally relevant to the CO$_2$ electroreduction system. This led us to evaluate the applicability of oxygen reduction reaction (ORR) cathode materials for the CO$_2$ electroreduction application.[47] For this purpose, we utilized a Pt-carbon black-TiO$_2$ nano-composite film electrode in conjunction with a pyridinium ion co-catalyst.[47] The latter is further elaborated in the next section of this chapter.

The photocatalytic route for modifying a C-TiO$_2$ support with Pt (or another target metal or bimetallic Pt-Pd electrocatalyst) hinges on the fact that TiO$_2$ is a semiconductor and thus is capable of absorbing light of wavelengths equal to or greater than those corresponding to its band gap energy, E$_g$.[48,49] As elaborated elsewhere, this in situ catalyst photodeposition strategy first consisted of ultrasonically dispersing: (i) a suitable carbon black; (ii) the metal oxide (in desired proportions); (iii) the catalyst precursor salt (*e.g.* H$_2$PtCl$_6$, PdCl$_2$); (iv) an electron donor or hole scavenger (*e.g.* formate); and (v) a surfactant (as needed). This dispersion was introduced into a custom-built photochemical reactor[50] equipped with a medium pressure Hg arc lamp as a UV irradiation source. A crucial aspect of this synthetic approach is that photoinduced electron transfer from the TiO$_2$ phase to the carbon support results in the uniform deposition of the metal catalyst throughout the composite as a whole rather than localized deposition on the titania surface.[48,49]

For preparation of Pt-Pd/C-TiO$_2$, an essentially similar procedure was used, except for addition of the Pd precursor salt to the photocatalytic deposition medium. The two metals (Pt and Pd) were photodeposited sequentially, with the total loading of the noble metal in the nanocomposite being maintained the same in both cases. Other details may be found in refernces 48 and 49.[49,48]

Figure 11.7 contains representative high-resolution transmission electron micrographs of Pt/C-TiO$_2$ (Figure 11.7a) and Pt-Pd/C-TiO$_2$ (Figure 11.7b) nanocomposite powders. In both images, the dark spots are the metallic nanoclusters that are seen highly dispersed on the carbon-oxide support (showing corrugated appearance). Metal cluster sizes are seen to be in the range 3–5 nm, with slighter larger sizes and elongated shapes in the case of the Pd-Pt nanocomposite (Figure 11.7b). It is worth noting that photocatalytic deposition at a neat TiO$_2$ surface (*i.e.* without carbon black in electronic contact) affords Pt nanoparticles that are significantly larger than those obtained with the C-TiO$_2$ support.[49] Consistent with the essentially similar nanocomposite morphology in both cases, cathodes derived from either Pt/C-TiO$_2$ or Pt-Pd/C-TiO$_2$ exhibited comparable electrocatalytic activity for CO$_2$ reduction, and consequently a further distinction is not made between the two materials in the data trends presented next.

Cyclic voltammetry (data not shown) of 0.2 M NaF solutions (pH = 5.3) containing 50 mM pyridinium cation (PyH$^+$) revealed the onset of the 1e$^-$ reduction of the cation to the neutral pyridinium radical at *ca.* −0.5 V on both Pt and Pd surfaces. Remarkably, an unmodified glassy carbon surface is completely inactive toward electroreduction of the pyridinium cation. Figure 11.8 compares hydrodynamic voltammograms (at 1200 rpm rotation speed) for a Pt/C-TiO$_2$

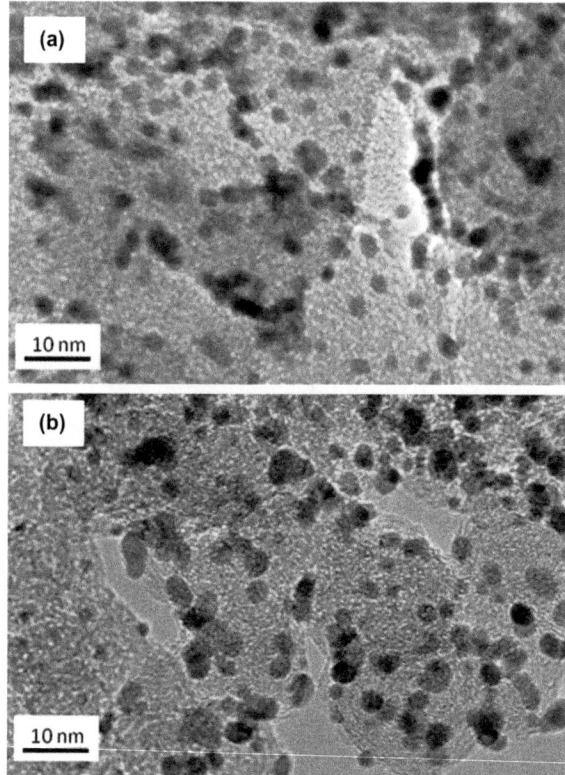

Figure 11.7 HR-TEM images of (a) Pt/C-TiO$_2$ and (b) Pd-Pt/C-TiO$_2$ nanocomposite surfaces.
Reproduced from *Electrochem and Solid-State Lett.*, 2012, **15**, B5–B8 with copyright permission of ECS, The Electrochemical Society.

RDE and a comparably-sized Pt RDE in 10 mM pyridine-loaded NaF supporting electrolyte saturated with either N$_2$ or CO$_2$. The Pt/C-TiO$_2$ nanocomposite contained 10 μg Pt dispersed on 0.196 cm^2 of the glassy carbon rotating disk electrode (RDE) surface, whereas the comparably-sized Pt RDE translated to a Pt content of *ca.* 3–4 mg, depending on the metal thickness.

As seen in Figure 11.8, addition of CO$_2$ led to marked enhancement of the cathodic current for both cathode materials, attesting to the catalytic role played by both Pt and the pyridine co-catalyst toward CO$_2$ reduction. In particular, note the *ca.* 30% enhancement in cathodic current flow for the Pt/C-TiO$_2$ nanocomposite relative to the massive Pt electrode case. This enhancement occurred in spite of the fact that the loading of Pt in the Pt/C-TiO$_2$ electrode is approximately a *million-fold* less than that corresponding to a Pt RDE. Thus the mass activity of Pt/C-TiO$_2$ as expressed by the current density normalized by the mass of platinum at −0.55 V were 46.4 and 76.6 mA.mg^{-1} in N$_2$ and CO$_2$ respectively, whereas for the Pt disk (assuming ∼1 g of metal content) the corresponding values were 1.44×10^{-4} and 2.51×10^{-4} mA.mg^{-1}, respectively.

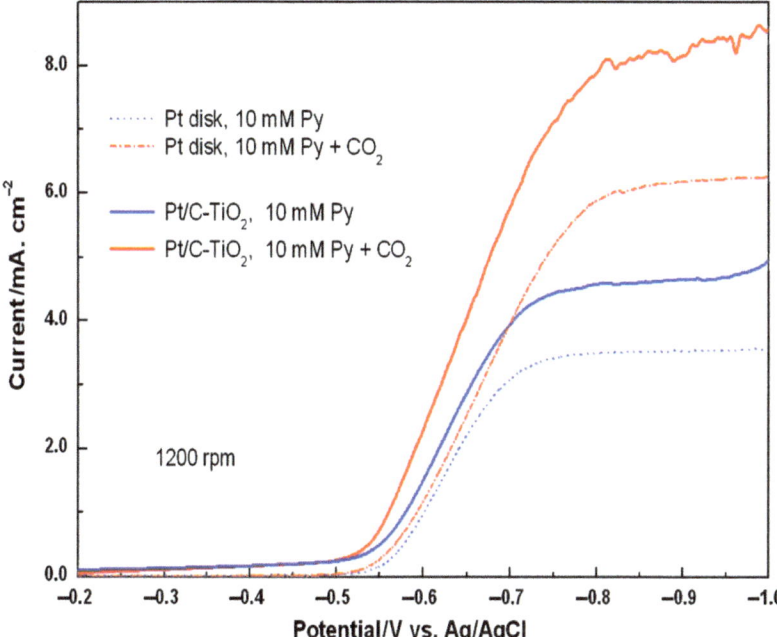

Figure 11.8 Comparison of hydrodynamic voltammograms for a Pt/C-TiO$_2$ nano-composite RDE with a Pt RDE in PyH$^+$-loaded NaF solutions saturated with N$_2$ and CO$_2$ respectively.
Reproduced from *Electrochem and Solid-State Lett.*, 2012, **15**, B5–B8 with copyright permission of ECS, The Electrochemical Society.

Significantly, close examination of the hydrodynamic voltammogram in the plateau region in Figure 11.8 for the Pt/C-TiO$_2$ case in the presence of CO$_2$, showed noise that is consistent with the visual observation of gas bubbles (presumably CO and/or H$_2$) on the electrode surface. Comparable data trends were seen in the Pt-Pd/C-TiO$_2$ *bimetallic* case, although Pt outperformed Pd alone in this regard (data not shown). The hydrodynamic voltammetry trends shown in Figure 11.8 were elaborated further for Pt/C-TiO$_2$ using variable rotation rates, and the corresponding data are presented in Figure 11.9. Once again, the cathodic current enhancement was seen at all rotation speeds, and noise in the plateau region is unmistakable when CO$_2$ is present in the electrolyte (compare Figure 11.9 a and b). The increase in current flow observed at potentials more negative than −0.9 V in Figure 11.8 and is associated with proton reduction (and hydrogen evolution).

Figure 11.10 shows Levich plots[51] (constructed from the hydrodynamic voltammetry data in Figure 11.9) of the limiting current density, j_L *vs.* the square root of the electrode rotation speed for the Pt/C-TiO$_2$ nanocomposite cathode in N$_2$ and CO$_2$ saturated solutions.

$$j_L = 0.62 \, n \, FD^{2/3} \nu^{-1/6} C \, \omega^{1/2} \qquad (11.7)$$

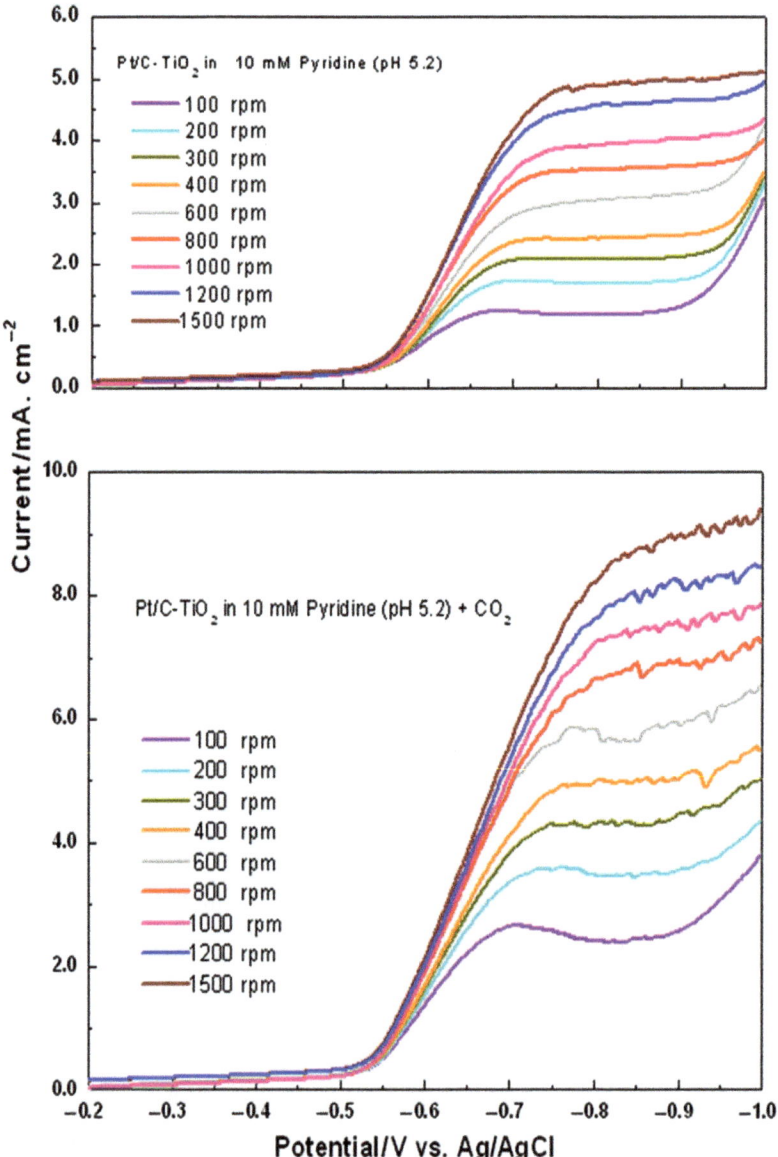

Figure 11.9   Hydrodynamic voltammetry data for a Pt/C-TiO$_2$ RDE in 10 mM PyH$^+$ - loaded aqueous solutions (NaF, pH 5.3) saturated with N$_2$ (top) and CO$_2$ (bottom), respectively.
Reproduced from *Electrochem and Solid -State Lett*, 2012, **15**, B5–B8 with copyright permission of ECS, The Electrochemical Society.

In Equation (11.7), n is the number of electrons transferred, F is the Faraday constant, D is the diffusion coefficient of the electroactive species (PyH$^+$), ν is the kinematic viscosity of the electrolyte ($\sim$0.01 cm^2 s^{-1}),[51] C is the PyH$^+$

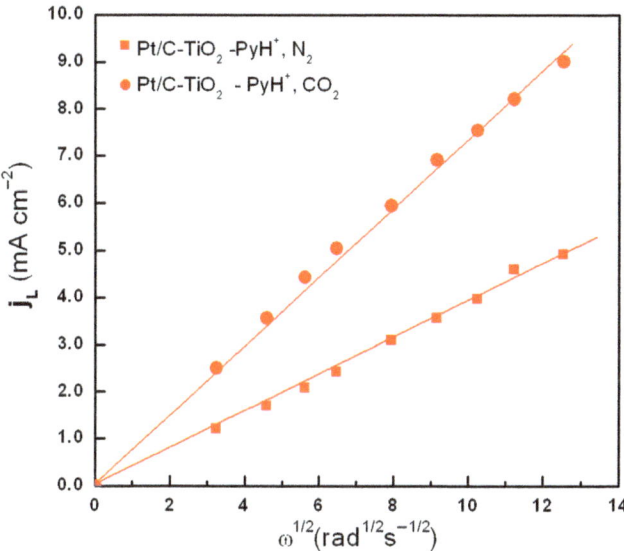

Figure 11.10   Comparison of Levich plots for the electroreduction of PyH⁺ in N₂ and
CO₂ saturated solutions using a Pt/C-TiO₂ RDE. Slopes and correlation
coefficients from least-squares fits are 0.401 mAcm⁻²s¹/² and 0.9986 and
0.718 mAcm⁻²s¹/² and 0.9987 for the N₂- and CO₂ saturated solutions
respectively. Data for the Levich plots are taken from Figure 11.9.
Reproduced from *Electrochem and Solid-State Lett.*, 2012, **15**, B5–B8.
with copyright permission of ECS, The Electrochemical Society.

concentration (mol cm⁻³) at the respective solution pH and ω is the electrode
rotation rate in radians s⁻¹. The values used for D and C were 7.6×10^{-6} cm² s⁻¹
and 0.8×10^{-5} mol cm⁻³, respectively.[52]

   As seen in Equation (11.7), a plot of j_L *vs.* $\omega^{\frac{1}{2}}$ (Levich plot) should be linear
for diffusion-controlled electrochemical processes. Both sets of data in
Figure 11.10 show good linearity; the slopes of the two Levich plots afford
values for the electron stoichiometry (n) which are very close to 1.0 and 1.8 in
N₂ and CO₂ saturated solutions respectively. Clearly, the pyridinium-catalyzed
CO₂ reduction is sustained by the initial diffusion-controlled one-electron reduction of PyH⁺ to PyH• radical on Pt/C-TiO₂. Importantly, the nanocomposite electrocatalyst layer was thin enough (0.05–0.08 μm range) to not
contribute to film diffusion-limited behavior.[49,53] Further implications of the
data in Figure 11.10 require follow-up work.

   The good stability of the nanocomposite electrodes were confirmed by re-
cording the electrode potential during a 50-h galvanostatic electrolysis run.
Only a minor and gradual potential drift (from −0.50 V to −0.58 V) was observed, attesting to the good stability of the nanocomposite material. It is worth
underlining that as these galvanostatic electrolyses were performed at low
current density, the electrode potential remained near the bottom of the
hydrodynamic current–potential regime (see Figure 11.8 and Figure 11.9), thus
involving mainly the reduction of pyridine without much interference from H₂

formation at the nanocomposite surface. As the solution was intentionally not buffered to avoid interference in alcohol detection, this small potential drift probably reflects a progressive increase of pH with time due to the consumption of protons associated with product generation.

Gas chromatography (GC) analyses of the electrolysis solution revealed the formation of both methanol and isopropanol as solution products at the Pt/C-TiO_2 nanocomposite cathode Figure 11.11(a). As shown in Figure 11.11 a, at

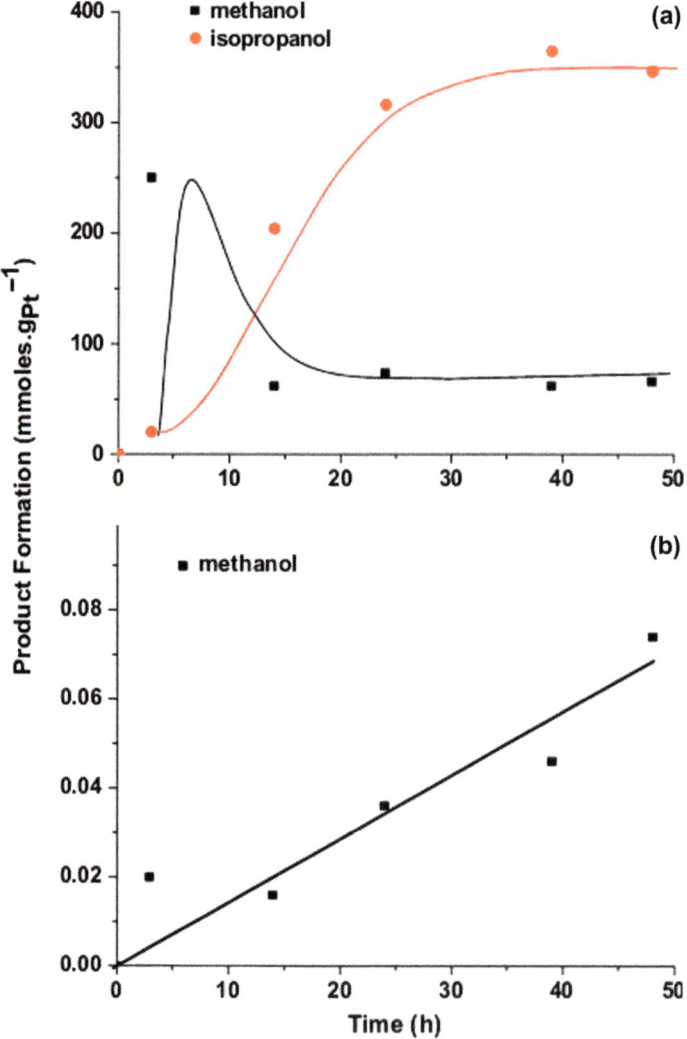

Figure 11.11   GC analyses as a function of time for constant-current electrolysis of CO_2-saturated solutions using: (a) Pt/C-TiO_2 and (b) Pt foil cathodes. Electrolysis conditions specified in the text.
Reproduced from *Electrochem and Solid-State Lett.*, 2012, **15**, B5–B8 with copyright permission of ECS, The Electrochemical Society.

short reaction times (t < 1 h), methanol is the major product with lesser amounts of isopropanol. Subsequently, the amount of methanol drops precipitously, whereas the isopropanol content increases regularly. This drop in methanol content could arise from volatilization and entrainment of methanol in the CO_2 stream (due to constant CO_2 bubbling). This would be less of an issue with isopropanol due to its higher vapor pressure, and as seen in Figure 11.11a, the concentration of isopropanol increases regularly up to 40 h, after which production is flat throughout the electrolysis.

Only methanol was seen as the CO_2 electroreduction product in the Pt foil case (Figure 11.11(b)). Importantly, no CO was seen in the final reduction products in either case in Figure 11.11a and b. This is especially significant in that metals such as Pt or Pd are known to produce CO selectively on electroreduction of CO_2.[54] Clearly, the nanocomposite matrix facilitates deeper reduction (to alcohols) beyond the 2e$^-$ CO stage. The mechanistic factors underlying this trend will be probed further in follow-on work that is planned in the Rajeshwar laboratory. Specifically, one intriguing aspect of these data is that *multi-carbon* products such as isopropanol are generated in the Pt/C-TiO$_2$ nanocomposite case, whereas Pt generates only methanol. Clearly the intermediate products from the initial reduction remain in close proximity to the interface in the former case so that subsequent product coupling can occur.

Finally, we note that the production of alcohol per unit mass of Pt for the nanocomposite matrix case (Figure 11.11(a)) is *three orders of magnitude* larger than the Pt foil case (Figure 11.11(b)). This represents substantial reduction in Pt usage and is an important practical advance toward improving the economics of electrochemical reduction of CO_2 to liquid fuels. Thus, the data discussed above show for the first time that nanocomposite cathode matrices derived from either Pt/C-TiO$_2$ or Pt-Pd/C-TiO$_2$ are effective electrocatalysts for aqueous CO_2 reduction in the presence of a solution co-catalyst such as the pyridinium cation. It is worth noting in this regard that matrices such as carbon black have good adsorption affinity for CO_2[55] pointing to synergistic factors such as the site-proximity effect[56] being operative for nanocomposite matrices such as the ones developed in this particular proof-of-concept study.

11.6 Transition Metal Complexes for Photocatalytic CO$_2$ Reduction

11.6.1 Catalysts for Reduction of CO$_2$ to CO or HCOO$^-$

Compared to the number of electrocatalysts for CO_2 reduction, photochemical catalysts are far more limited and are largely limited to Re and Ru complexes. Table 11.2, to the best of our knowledge, contains a collection of all reported molecular photocatalytic systems for CO_2 reduction. Of the 42 systems reported, only 5 systems do not involve Re or Ru. Three utilize the organic chromophores phenazine and para-terphenyls and two utilize Fe- and Co-based porphyrins. In most cases, the chromophore is coupled with a CO_2-reducing co-catalyst which is a known electrocatalyst for CO_2 reduction. CO,

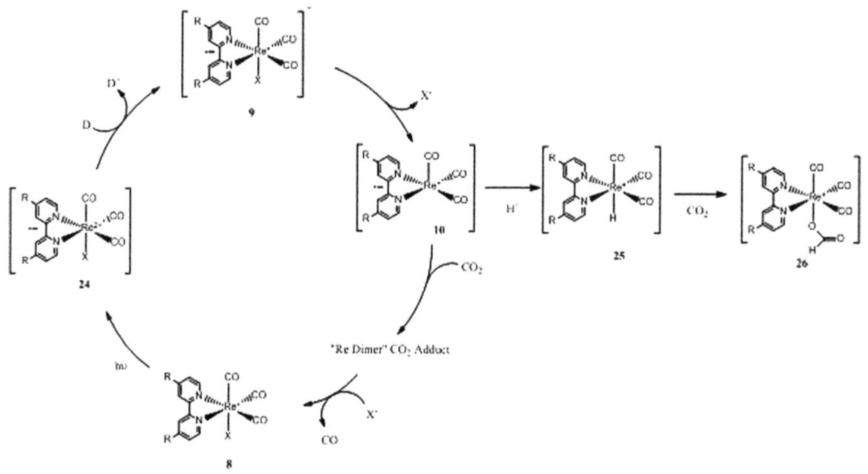

Figure 11.12   Photocatalytic cycle of *fac*-Re(bpy)(CO)$_3$X with the formation of the formate adduct.

formate, and H$_2$ are the only reported products for these systems, with the sole exception of entry #15, indicating that deeper reduction has remained an elusive goal. In most of these systems, CO formation is proposed to occur *via* the disproportionation of the CO$_2$$^{\bullet-}$ radical anion[34] or a two-electron reduction to produce CO.[57–61]

One of the better studied photocatalysts is *fac*-[Re(bpy)(CO)$_3$X]$^+$ which is also one of the few systems in which the chromophore is also the CO$_2$ reducing catalyst. Hori[62] and Lehn[29,63,64] have proposed the mechanism shown in Figure 11.12 for the production of CO. After reductive quenching of the photoexcited state (**24**), dissociation of the halide forms the coordinatively unsaturated species **10** which then can react with CO$_2$. While the exact structure of the CO$_2$ complex is not fully known, one proposed structure is the μ_2-η^2-CO$_2$ bridged binuclear Re adduct.[29,62–64] In any case, the CO$_2$ adduct is unstable and decomposes to yield CO and **8**. Ishitani proposes a similar mechanism except that **10** adds CO$_2$ and instead of dimerization as a method to provide a second reducing equivalent, the CO$_2$ adduct is reduced a second time by an outer-sphere mechanism to yield CO and complex **8**.[65] At present, both mechanisms have reasonable data to support their claims.[27] Among a related rhenium photocatalyst family, *fac*-[Re(bpy)(CO)$_3$P(OEt)$_3$]$^+$ has been demonstrated to be most efficient.[60,66–68] The one electron reduced species is almost quantitatively produced and is unusually stable in solution because of the strong electron-withdrawing property of the P(OEt)$_3$ ligand.[60,66–68]

A competing reaction in this system is the reaction of species **10** with a proton to give the metallo-hydride (species **25**).[29] This species can react with a proton to give hydrogen gas,[29] or it can insert CO$_2$ to give the metallo-formate complex (species **26**), which kills the catalyst as the formate is not released.[29,64] Hori and coworkers noted that it was possible to prevent this deactivation pathways, *i.e.* inhibit hydride formation, by increasing the CO$_2$ partial pressure.[62] Catalyst turnover exhibited a 5-fold improvement at 54 atm over a 1 atm system.

One issue with the Re photocatalysts is the limited range of absorption in the visible region, which is typically limited to wavelengths below 440 nm. Multinuclear metal supramolecular complexes were developed for CO$_2$ reduction photocatalysts for this reason. These complexes were composed of a photosensitizer part, a ruthenium(II) bpy-type complex, and a catalyst part based on the of rhenium(I) tricarbonyl complexes. A number of bi-, tri-, and tetranuclear complexes linked by several types of bridging ligands have been investigated in the literature (Figure 11.13).[60,69–74] The bridging ligands strongly influence photocatalytic ability, including selectivity of CO over H$_2$, high quantum yields and large turn over numbers, of the complexes. In all of these complexes, the ^{3}MLCT excited state of the ruthenium moiety was reductively quenched by a sacrificial reducing agent and the one electron reduced Re complex was formed through intramolecular electron transfer from the reduced Ru chromophore.[60] Ishitani *et al.* have shown that the catalytic activity of these mixed Re/Ru assemblies improves upon increasing the Re/Ru ratio, presumably due to the ready availability of Re sites (which catalyzes the slow step) to pick up electrons from the photoreduced Ru site (the fast step).[69,71,72]

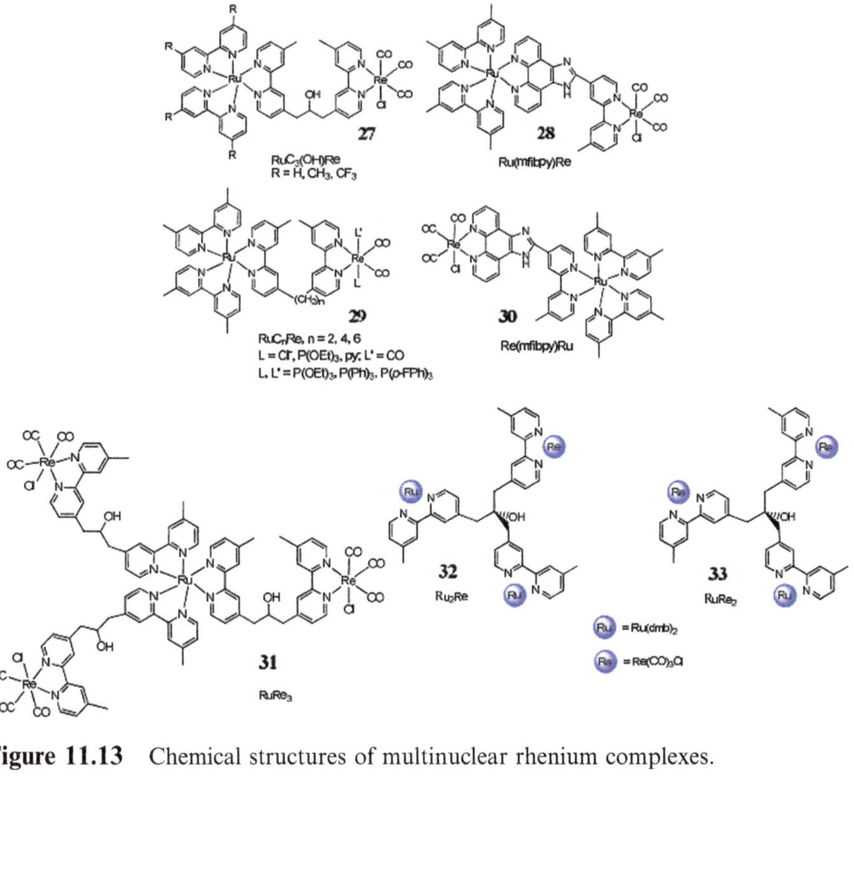

Figure 11.13 Chemical structures of multinuclear rhenium complexes.

Table 11.3 TON$_{CO}$ of different numbers of active sites and their comparison with appropriate monometallic model complexes containing different ratios of [Ru(dmb)$_3$]$^{2+}$ and [(dmb)Re(CO)$_3$Cl]. TON is calculated after 16 h of irradiation based on Ru(II) moieties with the complex concentration of 0.05 M.

Complexes	Ru + Re (1:1)	Ru-Re	Ru + Re (1:2)	Ru-Re$_2$	Ru + Re (1:3)	Ru-Re$_3$	Ru + Re (2:1)	Ru$_2$-Re
TON$_{CO}$	100	160	89	190	60	240	55	110

Moreover, a comparison of each multinuclear system to their corresponding system composed of the individual components shows that the intramolecular system is superior as shown in Table 11.3.[69–71] The electronic structure of the bridging ligands is important and systems in which there is strong electronic communication between metal sites perform less well than those with weak electronic coupling.[69,70] Thus, the more conjugated bridging is not better for supramolecular architecture for photochemical CO$_2$ reduction.[73] In order to direct the electron towards the Re center in the Ru/Re assemblies, it is important to adjust the π* orbital energy on peripheral bpy ligands of the Ru chromophores, such that they do not become electron traps.[69] Ideally, the acceptor orbital of the bpy-like portion of the bridging ligand should be equal or lower in energy than that of the peripheral ligands to help direct the electron to the Re site.[60,69]

Cyclam-based macrocyclic ligands with either cobalt or nickel ions are one of the most commonly used co-catalysts for the photochemical reduction of CO$_2$ in the presence of Ru(bpy)$_3$$^{2+}$.[75–83] As seen in Figure 11.14,[78] the production of CO is shown to proceed through the reduction of the macrocycle by the singly reduced Ru species, Ru(bpy)$_3$$^+$, followed by formation of metal hydride intermediate (species **36**). The next step is insertion of CO$_2$ into the metal hydride bond to form the metallo-formate (**37**). This species then decomposes to form CO and water. This species can rearrange to for the oxygen-bound formate species **38** which can be lost as formate by simple protonation.[75,77,78,82,84] In both cases, the catalyst is regenerated by a reducing agent. A competing pathway in this system is the reaction between the metal hydride and another proton to form H$_2$. The use of a different photosensitizer yields an additional product with these catalysts. When p-terphenyl[81,83] or phenazine[82] is used in place of Ru(bpy)$_3$$^{2+}$ the formation of formate is also observed in addition to CO and H$_2$.

Another class of macrocycle complexes that are used in the reduction of CO$_2$ are porphyrin-based complexes using iron and cobalt. The photochemical reduction of CO$_2$ with iron porphyrins, follows a similar catalytic process as the electrochemical reduction, shown in Figure 11.15, with the key difference being the method in which the active species, [Fe(0)TPP]$^{2-}$, is formed.[85] In the photochemical method, the active species is formed through a three-step process which involves four photons and two porphyrin rings. The first step (Equation (11.8)) is photoreduction of the iron(III) porphyrin to iron(II)

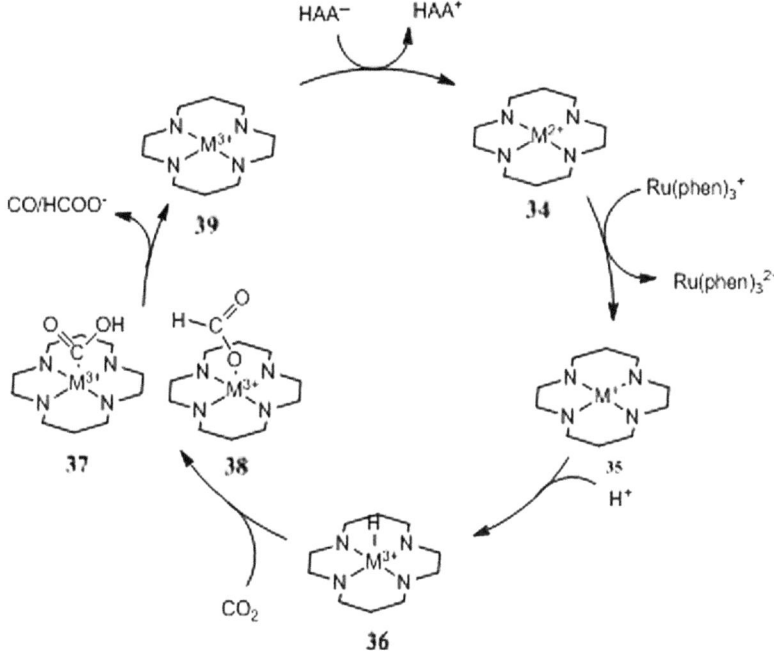

Figure 11.14 Acid assisted CO_2 electroreduction by metal cyclam where M is cobalt or nickel.

porphyrin in the presence of a reducing agent, TEA, which is coordinated to the axial site of the complex.

The second step is further illumination of the iron(II) species to result in iron(I) species with a mechanism similar to first step (Equation (11.9)). This process is far less efficient than the previous one and is affected by the concentration of TEA.

$$\text{TEA-Fe(III)TPP} \xrightarrow{h\nu} \text{Fe(II)TPP} + \text{TEA}^{\bullet+} \tag{11.8}$$

$$\text{TEA-Fe(II)TPP} \xrightarrow{h\nu} \text{Fe(I)TPP}^- + \text{TEA}^{\bullet+} \tag{11.9}$$

The last step to make the active species of this complex is a disproportionation reaction of two iron(I) species in solution to give one iron(0) species and one iron(II) species (Eqn. 11.10). CoTPP, was demonstrated to perform similarly to FeTPP with Co(0)TPP^{2-} as the active catalytic species.[84]

$$2\text{Fe(I)TPP}^- \rightarrow \text{Fe(0)TPP}^{2-} + \text{Fe(TPP)} \tag{11.10}$$

11.6.2 Deeper Reduction Using a Hybrid System

As mentioned previously, molecule-based photochemical systems capable of photoreducing CO_2 beyond formic acid or CO are hard to come by. In recent

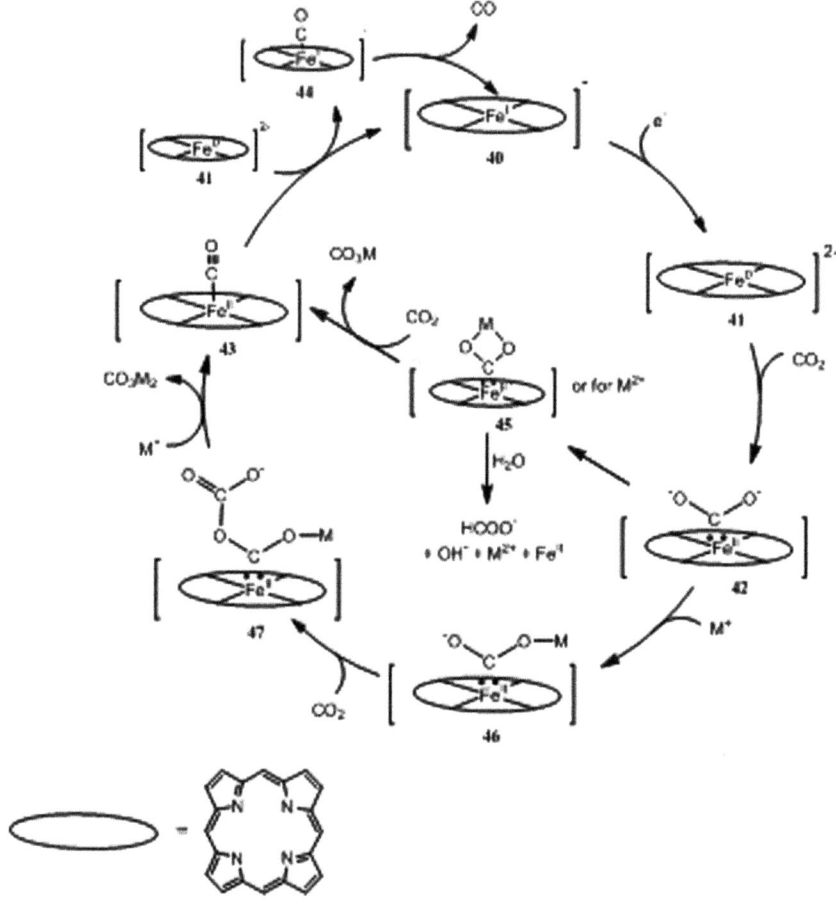

Figure 11.15 Metal ion-assisted CO_2 reduction where the metal ion is Ca^{2+}, Mg^{2+}, Na^+, Li^+, with Mg^{2+} working the best.

work, we examined the use of pyridine and the $[Ru(phen)_3]^{2+}$ chromophore for reduction of CO_2 to methanol, basing the deeper reduction on the prior work of Bocarsly *et al.* (see section 11.4.1).[86] As shown in Figure 11.16, methanol was produced in a system composed of $[Ru(phen)_3]^{2+}$, pyridine, ascorbic acid, and CO_2 in water (pH 5.0) irradiated at 470 nm over a period of 6 h.[86] The proposed mechanism of reduction is indicated in Figure 11.17; the entire system is seen to be simply a combination of the CO_2-reducing ability of pyridine and the reducing power of $[Ru(phen)_3]^+$ formed upon reductive quenching of the ^{3}MLCT state in the complex. Interestingly, this is the first report of pyridine functioning to produce methanol in a purely molecular system. In the electrocatalytic systems, it has been reported that the nature of the working electrode is integral to the observed catalytic activity. For example, methanol production is only

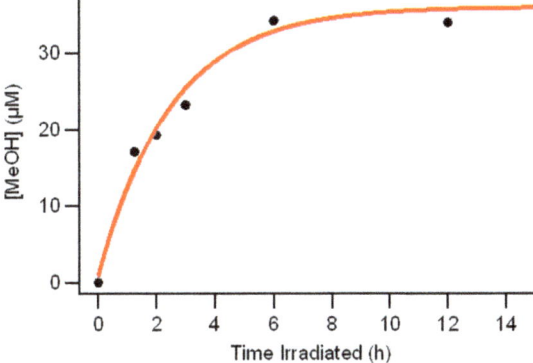

Figure 11.16 Temporal evolution of methanol from reduction of CO_2 with time for irradiation with 470 nm light.

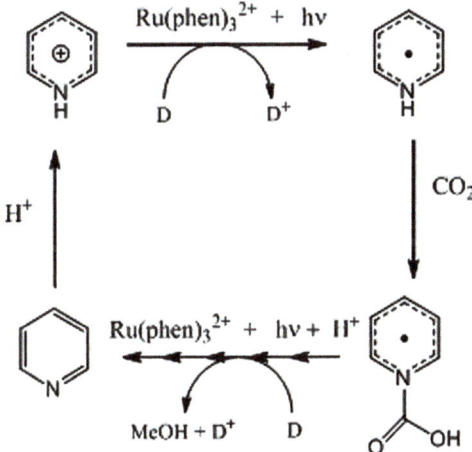

Figure 11.17 Proposed mechanism of the reduction of CO_2 by pyridine in the presence of $[Ru(phen)_3]^{2+}$ in water.

observed with Pt working electrodes, and theoretical studies have suggested that the Pt surface is required for methanol production.[86] At present, the activity of the $[Ru(phen)_3]^{2+}$/pyridine system is modest, with 0.15 methanol molecules produced per Ru chromophore over a 6 h period, which – when adjusted for the number of electrons involved – corresponds to 0.9 TON.[86] The loss of activity after 6 h is attributed to decomposition of the $[Ru(phen)_3]^{2+}$ chromophore, which is known to occur *via* ligand labilization in the excited state, a problem than can exacerbated by the presence of pyridine in the system. Reports of deeper reductions of CO_2 to methane are quite rare in the literature, although the process has been observed for the $[Ru(bpy)_3]^{2+}$ chromophore and colloidal metals.[87,88]

11.7 Photocatalytic CO$_2$ Reduction using Semiconductor Nanoparticles

11.7.1 Syngas as Precursor in Fisher-Tropsch Process for Production of Synthetic Fuels

Syngas (or "synthesis gas") is the name given to a gas mixture that contains varying amounts of CO and H$_2$. Common examples of producing syngas include the steam reforming of methane or liquid hydrocarbons and the gasification of coal or biomass. Thermally mild (*i.e.* low temperature) alternatives for producing this transportable chemical fuel mixture would obviously be significant from an energy perspective. Value is added even further if the production process can be driven *via* a renewable solar energy source from a source greenhouse gas material such as CO$_2$.

11.7.2 Photogeneration of Syngas Using AgBiW$_2$O$_8$ Semiconductor Nanoparticles

Figure 11.1 depicts the band-edges for AgBiW$_2$O$_8$ superimposed on the relevant redox levels for CO$_2$ reduction. Unlike for semiconductors such as WO$_3$ and BiVO$_4$ that are otherwise excellent photocatalysts for oxidative processes (*e.g.* water photooxidation), the conduction band-edge of AgBiW$_2$O$_8$ lies far enough negative to sustain CO$_2$ reduction.

Double tungstates of mono- and trivalent metals with composition: AB(WO$_4$)$_2$ are known to have structural polymorphism.[89] Here, A is a monovalent alkali metal (or Ag, Tl) and B represents a tri-valent element such as Bi, In, Sc, Ga, Al, Fe, or Cr. To date, very little is known on the exact crystal structure of AgBiW$_2$O$_8$. Our density functional theory calculations[90] indicate that the wolframite structural modification should be the most stable, while the scheelite and fergusonite structures are 0.69 eV and 0.28 eV higher in energy, respectively. Further details are given in reference 90. Synthesis, structural, surface, and optical characterizations of this oxide semiconductor are also given elsewhere.[90,91] We focus here on the use of nanosized particles of Pt-modified AgBiW$_2$O$_8$ for mild syngas photogeneration.

Formic acid was used as a precursor for CO$_2$ in these proof-of-concept experiments. The UV-photogenerated holes in AgBiW$_2$O$_8$ were then used to generate CO$_2$ in situ at the oxide semiconductor-solution interface. It is worth noting here that the direct electrochemical reduction of CO$_2$ in aqueous media is hampered both by the low partial pressure of CO$_2$ in the atmosphere (3.9×10^{-4} atm) and by its low solubility in water (1.5 g/L at 298 K).[92] On the other hand, formate species have high solubility in water (945 g/L at 298 K).[93] Further, they have high proclivity for being adsorbed on oxide semiconductor surfaces[94] and are easily oxidized by the photogenerated holes in the oxide. Figure 11.18 presents the data.[90,91] The GC results clearly show the formation of CO and H$_2$ *via* photocatalytic reactions. Control experiments conducted without

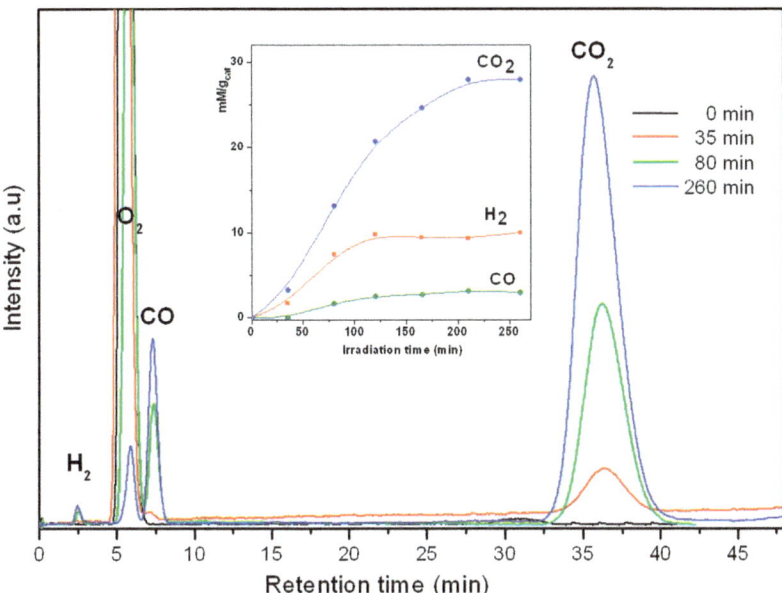

Figure 11.18   GC data for the simultaneous evolution of CO and H_2 from a platinized $AgBiW_2O_8$ suspension.
Reproduced with copyright permission from *J. Nano Research,* 2012, **17**, 185–191, Trans Tech Publications, Switzerland.

the oxide semiconductor did not yield any of the products shown in Figure 11.18, indicating that direct photolysis of the organic acid was not a factor here.

One surprising aspect of these experiments was that product generation could be sustained for periods up to at least several minutes without a noticeable fall-off in the rate. This indicates that CO-poisoning of the Pt surface was not a factor, at least in the initial stages of the photocatalytic process. The site specificity of product evolution, namely HER on the Pt sites and CO_2 photoreduction on the unmodified oxide surface, is an interesting aspect that deserves further study. The results in Figure 11.18 are also non-optimized in terms of product evolution efficiency and quantum yield. Practical application of this syngas photogeneration approach will necessitate modification of the oxide semiconductor so that wavelengths in the visible part of the solar spectrum (rather than the UV portion in Figure 11.18) can be accessed.

11.7.3   Photoreduction of CO_2 using Cu_2O as Semiconductor and Pyridine as Solution Co-Catalyst

Copper (I) oxide, Cu_2O, is an attractive p-type semiconductor with a band gap energy of ∼ 2.0 eV, a conduction band-edge lying far enough negative to sustain CO_2 reduction, and a high absorption coefficient over the visible wavelength range in the solar spectrum. One limitation of using Cu_2O photoelectrodes is their poor stability in aqueous electrolytes because the redox potentials for

Cu_2O/Cu^0 and CuO/Cu_2O lie within the band gap of thee semiconductor. However, the performance of these photoelectrodes can be enhanced by metal and/or semiconductor coatings[95, 96] as well as by solution electron shuttles such as methyl viologen.[96] We reported the stabilization of electrodeposited Cu_2O photoelectrodes by a Ni surface protection layer and also by the use of an electron shuttle in solution.[96] High photoactivity of electrodeposited Cu_2O was obtained under a surface protection consisting of nanolayers of Al-doped ZnO and TiO_2 to avoid photocathodic decomposition.[95] These films were also activated with Pt nanoparticles to enhance solar hydrogen generation.[95] Photocurrents for electrodeposited $Cu_2O(111)$ were optimized separately for solar H_2 generation without using any surface treatment.[97]

Cathodic electrodeposition of copper oxide from a copper sulfate bath containing lactic acid is a versatile and low-cost technique. The bath pH is known to determine the grain orientation and crystallite shape of the resulting film.[97–100] Thus, films electrodeposited from an alkaline bath of pH ~ 9 are highly oriented along the (100) plane, while a preferential (111) orientation occurs in films grown at more alkaline pHs (pH ~ 12).[98] Interestingly, as the bath pH is varied from 7.5 to 12, a third preferred orientation, (110), can be identified in a narrow pH range of 9.4 to 9.9.[101] The preferred grain orientation and crystallite shape were also found to depend on the applied potential (in the potentiostatic growth mode),[102] while layered Cu_2O/Cu nanostructures were formed by galvanostatic electrodeposition.[103]

Although there are quite a few studies on the use of Cu_2O photocathodes for HER,[96,97,104–106] to the best of our knowledge there are no reports of their use for CO_2 photoreduction. An intriguing aspect for study hinges in the capability of preferentially-oriented electrodeposited films for CO_2 photoreduction. Among the (100), (110) and (111) crystal faces of Cu_2O, the (111) orientation is the best one for photoreduction reactions,[107] and therefore the data presented below correspond to electrodeposited $Cu_2O(111)$ films. Besides, the observation of high cathodic photocurrents ($\sim$ 6–8 mA/cm^2) for HER on $Cu_2O(111)$ films electrodeposited from a copper lactate solution (pH 12)[95–97] is very encouraging because it demonstrated that the highest cathodic photocurrent were associated with electrodeposited Cu(111) films.[97]

The use of a p-type photoelectrode for CO_2 reduction was reported in the late eighties for the case of CdTe and GaP electrodes in aqueous solutions in the presence of tetraalkylammonium salts.[108] However, much more recently, pyridine was incorporated into the electrolyte as a co-catalyst for the reduction of CO_2 at illuminated p-type electrodes (GaP and FeS_2).[109,110]

Our data below correspond to electrodeposited $Cu_2O(111)$ films grown on gold-coated quartz crystal working electrodes polarized at –0.4 V in 0.2 M $CuSO_4 + 3$ M lactic acid $+ 0.5$ M K_2HPO_4, pH 12 at 25 C to reach a total charge of 180 mC/cm^2 which corresponds to 0.25 µm film thickness. The associated total mass of the film was 140 µg/cm^2 as measured during its growth by electrochemical quartz crystal microgravimetry (EQCM).

To analyze the stability of electrodeposited $Cu_2O(111)$ electrodes, cathodic photocurrent density and mass change (Δm) were recorded as a function of potential under chopped simulated AM 1.5 solar illumination in a solution just containing the supporting electrolyte (Figure 11.19).[111] Prior to defining the optimal film thickness, careful analyses were performed (data not shown) to verify compliance of the EQCM set-up with the Sauerbrey equation.[111–113]

Figure 11.19 compares the effect of solar light intensity on the photocurrent and Δm values for the photoelectroreduction of water to hydrogen. The data in Figure 11.19 (top) are for illumination from a 1.5 AM solar simulator, and those in Figure 11.19 (bottom) are for the case when a 10% neutral density filter was placed in between the light source and the electrochemical cell. Clearly, although significantly higher photocurrents (up to $2\,\mathrm{mA\,cm^{-2}}$) are reached under full illumination, there is a beneficial effect on the relative photocurrent and mass changes brought about by decreasing the light intensity. The total photocurrent is mainly associated with photoelectron transfer to water, photogenerating H$_2$ as represented in Equation (11.11):

$$H_2O + 2e^- \rightarrow H_2 + 2OH^- \qquad (11.11)$$

Nonetheless, there are also measurable mass changes associated principally with the photocathodic decomposition of the oxide as indicated in Equation (11.12):

$$Cu_2O + 2e^- + H_2O \rightarrow 2Cu + 2OH^- \qquad (11.12)$$

Reaction (11.12) has the net effect of decreasing the photoactivity of the electrode.

The above data show that electrodeposited Cu$_2$O(111) films are more stable and reached higher photon to current conversion ratios than under the full output of simulated AM 1.5 illumination. This effect on performance seems to be rooted in the better capability of the photoelectrode/electrolyte interface to deal with a lower photon flux because: (i) there is less photoconversion of Cu$_2$O to Cu^0 (Equation (11.12)) as indicated by a lower mass loss and (ii) a more efficient photocurrent is associated with the transfer the photogenerated electrons to the electrolyte, *i.e.* conversion of water to H$_2$ (Equation (11.11)) when the light intensity is low. It is useful to recall here that the minority-carriers (photoexcited electrons) are collected only over a distance corresponding to the sum of the space charge layer thickness and the electron diffusion length. For electrodeposited Cu$_2$O(111) electrodes, this distance is always less than 100 nm in the potential range under study,[97] while the photon absorption depth near the band gap is much larger and in the micrometer range (1–2 μm).

For CO$_2$ reduction on illuminated Cu$_2$O(111), chopped (0.2 Hz) and continuous AM 1.5 simulated solar illumination were used (Figure 11.20). The pH of the electrolyte was adjusted to pH 5 to compare CO$_2$ reduction in the absence and in the presence of protonated pyridine (PyH$^+$, pKa = 5.5) as solution co-catalyst. Chopped illumination at a potential of −0.4 V (a rather low potential for Reaction 7 to be dominant) shows that the presence of PyH$^+$ enhances the cathodic photocurrent associated to CO$_2$ reduction only by *ca.* 15 % (compare Figure 11.20 a and b). While the photocurrent enhancement by PyH$^+$ is modest, the PyH$^+$ contribution by draining the photoelectrons to the solution

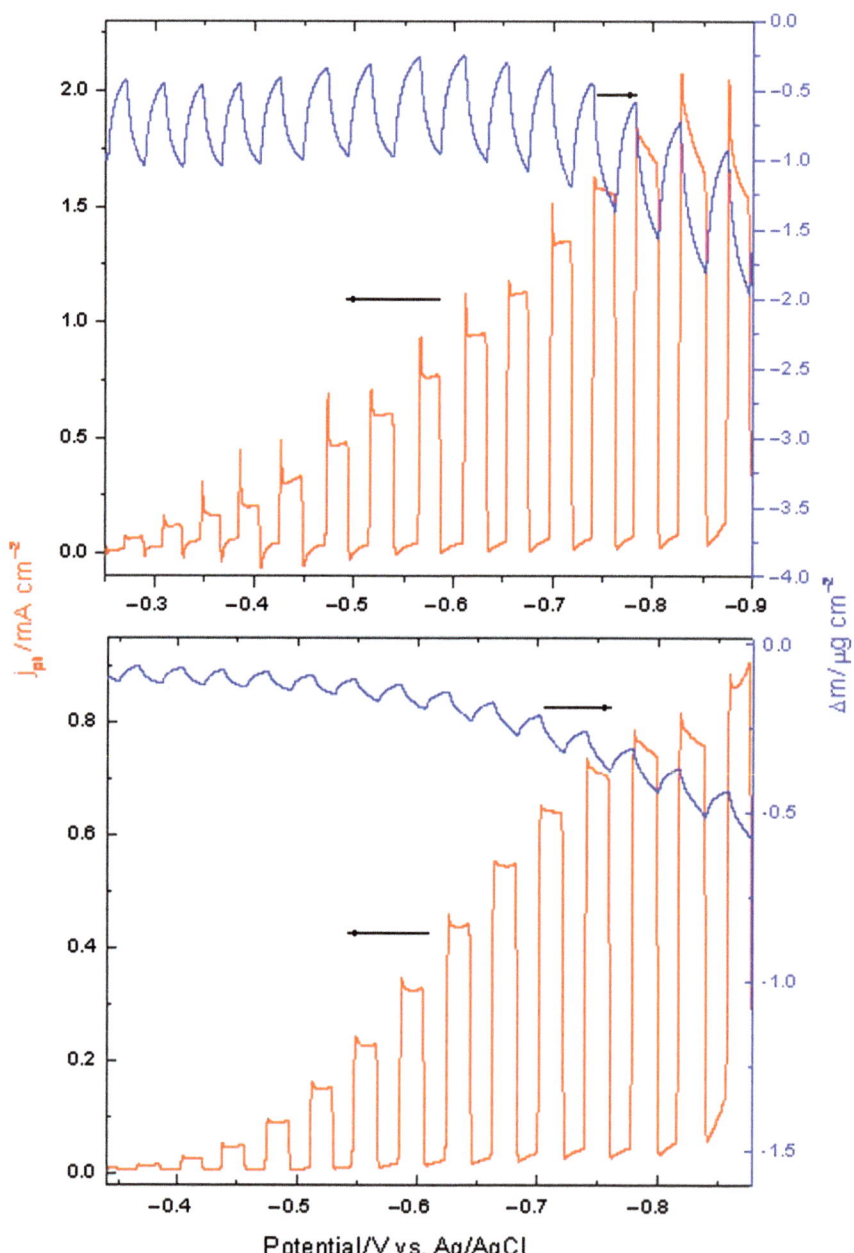

Figure 11.19 Effect of light intensity on the photocurrent/potential and Δm/potential profiles of $Cu_2O(111)$ films in N_2 saturated $0.1\,M\,Na_3PO_4$ under chopped simulated AM 1.5 illumination at full output (top) and at 10% intensity with a neutral filter (bottom).

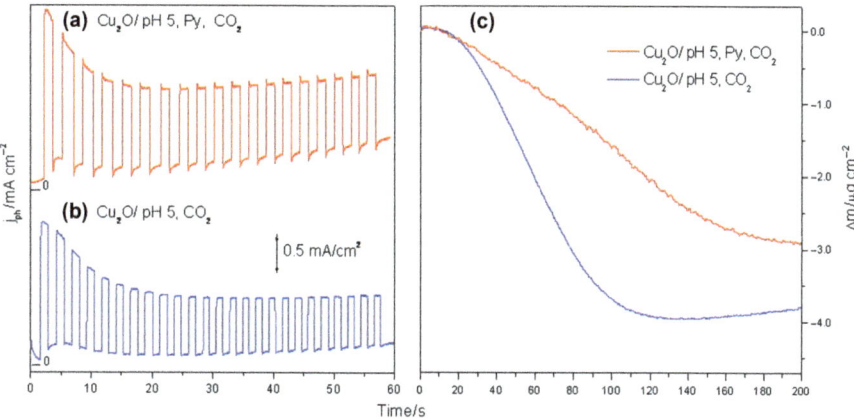

Figure 11.20 Photocurrent flow (under chopped 105 AM 1.5 illumination) at −0.4 V (*vs.* Ag/AgCl) for a Cu$_2$O(111) film in pH 5, CO$_2$ saturated solution without (a, blue line) and with (b, red line) 10 mM pyridine. (Curves were shifted for better comparison and each one includes the zero current). Frame (c) contains a comparison of mass changes (as measured by EQCM) for the respective (a) and (b) scenarios except that continuous instead of chopped irradiation was used.

side, thus avoiding photocorrosion of the electrode, is quite clear as shown by the comparative mass changes under continous irradiation (Figure 11.20c). It can be seen that the mass decrease is faster in the absence of PyH$^+$, reaching a total mass loss of 4.0 µg/cm^2 after 120 s of irradiation, while in the presence of PyH$^+$ the mass loss was reduced to only 2.9 µg/cm^2 after 200 s.

Although these mass changes indicate the instability of the Cu$_2$O photoelectrodes, longer times of irradiation show only relatively insignificant mass changes, which in comparison to the total film mass (140 µg/cm^2) are only in the 2–3% range. Further, the presence of PyH$^+$ enhances the film inertness as indicated by the absence of mass increase associated with adsorption of solution species on the Cu$_2$O(111) surface. Verification of these ideas is being carried out currently by analyzing the surface composition after performing CO$_2$ reduction as well as in the product detection from reactions such as those represented in Equation (11.12). Other Cu$_2$O film morphologies with different preferential orientations are also under scrutiny in our laboratory.

11.8 Concluding Remarks and Future Directions

Progress in the realm of molecular catalysts in the last 40 years toward the reduction of CO$_2$ has been substantial, but primarily limited to reduction to CO or formate. CO does have value in that it can be used in a Fischer–Tropsch reaction to produce higher carbon fuel products, but there is a general recognition that deeper reduction to more value-added products such as methanol is needed. Semiconductor-based photocatayst systems have also shown promise for both the above reduction pathways.

Although there are promising examples in the literatures, some obvious problems remain. The current technology does not meet the grand goal for industrial large-scale operation. The challenge to overcome is the overpotential for the electrochemical systems and short-lived one-electron reduced species for the photochemical systems. It is also critical to replace sacrificial reducing agents with more practical donors such as water, so as to close the loop in a practical fuel cycle. The use of sunlight to drive photoreduction is a sustainable method for the use of CO_2 as a C1 feedstock. Examples of incorporation of a chromophore with the real catalyst in either intermolecular or intramolecular photochemical systems have demonstrated the feasibility of CO_2 reduction. However, photoinduced electron transfer from the chromophore to the catalyst or from the semiconductor to the solution still account for much of the inefficiency in these systems. To address this issue will require sustained efforts by scientists in the rational design of molecule- or semiconductor-based assemblies to reduce these inefficiencies. It is the authors' hope that this book chapter will contribute to further progress and stimulate future generations of scientists to dvelop new electro-/photocatalyst design paradigms.

Acknowledgements

The authors would like to thank the National Science Foundation (CHE 0911720 to FMM, NT) and the Robert A. Welch Foundation (Y-1301 to FMM) for financial support.

References

1. M. E. Royer, *Comptes rendus hebdomadaires des séances de l'Académie des sciences*, 1870, **70**, 731.
2. A. d. s. (France), *Comptes rendus hebdomadaires des séances de l'Académie des sciences*, 1893, **116**, 1145.
3. H. H. Fenton, *J. Chem. Soc., Trans*, 1907, **91**, 687–693.
4. N. S. Spinner, J. A. Vega and W. E. Mustain, *Catal. Sci. Technol.*, 2012, **2**.
5. B. Kumar, M. Llorente, J. Froelich, T. Dang, A. Sathrum and C. P. Kubiak, *Annu. Rev. Phys. Chem.*, 2012, **63**, 541.
6. M. R. Hoffmann, J. A. Moss and M. M. Baum, *Dalton Trans.*, 2011, **40**, 5151.
7. E. B. Cole and A. B. Bocarsly, *Carbon Dioxide as Chemical Feedstock*, Wiley-VCH, Weinheim, 2011.
8. A. J. Morris, G. J. Meyer and E. Fujita, *Acc. Chem. Res.*, 2009, **42**, 1983.
9. E. E. Benson, C. P. Kubiak, A. J. Sathrum and J. M. Smieja, *Chem. Soc. Rev.*, 2009, **38**, 89–99.
10. T. Reda, C. M. Plugge, N. J. Abram and J. Hirst, *Proceedings of the National Academy of Sciences*, 2008, **105**, 10654.
11. Y. Hori, *Modern Aspects of Electrochemistry*, Springer, New York, 2008.
12. D. L. DuBois, in *Encyclopedia of Electrochemistry*, A. J. Bard and M. Stratmann, Wiley-VCH, Weinheim, 2007, vol. 7a , p. 202.

13. I. Taniguchi, *Mod. Aspects Electrochem.*, 1989, **20**, 327–400.
14. J. P. Collin and J. P. Sauvage, *Coord. Chem. Rev.*, 1989, **93**, 245–268.
15. M. D. Doherty, D. C. Grills, J. T. Muckerman, D. E. Polyansky and E. Fujita, *Coord. Chem. Rev.*, 2010, **254**, 2472–2482.
16. J. O. M. Bockris and J. C. Wass, *J. Electrochem. Soc.*, 1989, **136**, 2521.
17. K. Chandrasekharan and J. O. M. Bockris, *Surface Science*, 1987, **185**, 495.
18. K. Rajeshwar, in *Encyclopedia of Electrochemistry*, S. Licht, Wiley-VCH, Weinheim, 2001, pp. 3–53.
19. K. Rajeshwar, R. McConnell and S. Licht, *Solar Hydrogen Generation: Toward a renewable Energy Future*, Kluwer Academic, New York, 2008.
20. R. F. Service, *Science*, 2011, **334**, 925–927.
21. N. Furuya and K. Matsui, *J. Electroanal. Chem.*, 1989, **271**, 181–191.
22. G. Seshadri, C. Lin and A. B. Bocarsly, *J. Electroanal. Chem.*, 1994, **372**, 145–150.
23. A. Gennaro, A. A. Isse and E. Vianello, in *NATO ASI Series*, A. J. L. Pombeiro and J. A. McCleverty, Kluwer, The Netherlands, 1992, vol. 385 , pp. 311–316.
24. A. J. Morris, R. T. McGibbon and A. B. Bocarsly, *ChemSusChem*, 2011, **4**, 191–196.
25. E. Barton-Cole, P. S. Lakkaraju, D. M. Rampulla, A. J. Morris, E. Abelev and A. B. Bocarsly, *J. Am. Chem. Soc.*, 2010, **132**, 11539–11551.
26. J. Hawecker, J.-M. Lehn and R. Ziessel, *J. Chem. Soc., Chem. Commun.*, 1984, 328–330.
27. B. P. Sullivan, C. M. Bolinger, D. Conrad, W. J. Vining and T. J. Meyer, *J. Chem. Soc., Chem. Commun.*, 1985, 1414–1416.
28. J. M. Smieja and C. P. Kubiak, *Inorganic Chemistry*, 2010, **49**, 9283–9289.
29. J. Hawecker, J. M. Lehn and R. Ziessel, *Helv. Chim. Acta*, 1986, **69**, 1990–2012.
30. P. Christensen, A. Hamnett, A. V. G. Muir and J. A. Timney, *J. Chem. Soc., Dalton Trans.*, 1992, 1455–1463.
31. R. Ziessel, *Nato ASI Series, Series C*, 1987, **206**, 113–138.
32. M. A. Scibioh and B. Viswanathan, *Photo/electrochemistry & photobiology in environment, energy and fuel*, 2002, 1–46.
33. C. Amatore, L. Nadjo and J. M. Saveant, *Nouveau Journal de Chimie*, 1984, **8**, 565–566.
34. P. A. Christensen, A. Hamnett and A. V. G. Muir, *J. Electroanal. Chem. Interfacial Electrochem.*, 1988, **241**, 361–371.
35. H. Ishida, K. Fujiki, T. Ohba, K. Ohkubo, K. Tanaka, T. Terada and T. Tanaka, *J. Chem. Soc., Dalton Trans.*, 1990, 2155–2160.
36. H. Ishida, K. Tanaka, M. Morimoto and T. Tanaka, *Organometallics*, 1986, **5**, 724–730.
37. H. Ishida, K. Tanaka and T. Tanaka, *Chemistry Letters*, 1985, 405–406.
38. H. Ishida, H. Tanaka, K. Tanaka and T. Tanaka, *J. Chem. Soc., Chem. Commun.*, 1987, 131–132.
39. K. Tanaka, *Kagaku (Kyoto)*, 1989, **44**, 570–575.

40. K. Tanaka, *Denki Kagaku Oyobi Kogyo Butsuri Kagaku*, 1990, **58**, 989–996.
41. H. Ishida, T. Terada, K. Tanaka and T. Tanaka, *Inorganic Chemistry*, 1990, **29**, 905–911.
42. H. Ishida, K. Tanaka and T. Tanaka, *Chemistry Letters*, 1988, 339–342.
43. H. Tanaka, H. Nagao, S. M. Peng and K. Tanaka, *Organometallics*, 1992, **11**, 1450–1451.
44. M. R. M. Bruce, E. Megehee, B. P. Sullivan, H. Thorp, T. R. O'Toole, A. Downard and T. J. Meyer, *Organometallics*, 1988, **7**, 238–240.
45. M. R. M. Bruce, E. Megehee, B. P. Sullivan, H. H. Thorp, T. R. O'Toole, A. Downard, J. R. Pugh and T. J. Meyer, *Inorganic Chemistry*, 1992, **31**, 4864–4873.
46. J. R. Pugh, M. R. M. Bruce, B. P. Sullivan and T. J. Meyer, *Inorganic Chemistry*, 1991, **30**, 86–91.
47. N. R. de Tacconi, W. Chanmanee, B. H. Dennis, F. M. MacDonnell, D. J. Boston and K. Rajeshwar, *Electrochem. Solid-State Lett.*, 2012, **15**, B5–B8.
48. N. R. de Tacconi, K. Rajeshwar, W. Chanmanee, V. Valluri, W. A. Wampler, W.-Y. Lin and L. Nikiel, *J. Electrochem. Soc.*, 2010, **157**, B147.
49. N. R. de Tacconi, C. R. Chenthamarakshan, K. Rajeshwar, W.-Y. Lin, T. F. Carlson, L. Nikiel, W. A. Wamper, S. Sambandam and V. Ramani, *J. Electrochem. Soc.*, 2008, **155**, B1102.
50. W. Y. Lin, C. Wei and K. Rajeshwar, *J. Electrochem. Soc.*, 1993, **140**, 2477.
51. A. J. Bard and L. R. Faulkner, *Electrochemical Methods Fundeamentals and Applications*, Second edn., John Wiley & Sons, Inc, New York, New York, 2001.
52. *GSI Chemical Data Base*, http://www.gsi-net.com/en/publications/gsi-chemical-database/single/474.html, Accessed 2012.
53. T. Ikeda, P. Denisevich, K. Willman and R. W. Murray, *J. Am. Chem. Soc.*, 1982, **104**, 2683.
54. Y. Hori, H. Wakebe, T. Tsukamoto and O. Koga, *Electrochimica Acta*, 1994, **39**, 1833.
55. V. Y. Gavrilov and R. S. Zakharov, *Kinetics and Catalysis*, 2010, **51**, 633.
56. N. R. de Tacconi, H. Wenren, D. McChesney and K. Rajeshwar, *Langmuir*, 1998, **14**, 2933.
57. C. Kutal, A. J. Corbin and G. Ferraudi, *Organometallics*, 1987, **6**, 553–557.
58. Y. Hayashi, S. Kita, B. S. Brunschwig and E. Fujita, *J. Am. Chem. Soc.*, 2003, **125**, 11976–11987.
59. C. Kutal, M. A. Weber, G. Ferraudi and D. Geiger, *Organometallics*, 1985, **4**, 2161–2166.
60. H. Takeda and O. Ishitani, *Coord. Chem. Rev.*, 2010, **254**, 346–354.
61. A. Inagaki and M. Akita, *Coord. Chem. Rev.*, 2010, **254**, 1220–1239.

62. H. Hori, Y. Takano, K. Koike and Y. Sasaki, *Inorg. Chem. Commun.*, 2003, **6**, 300–303.
63. J. Hawecker, J. M. Lehn and R. Ziessel, *J. Chem. Soc., Chem. Commun.*, 1985, 56–58.
64. J. Hawecker, J. M. Lehn and R. Ziessel, *J. Chem. Soc., Chem. Commun.*, 1983, 536–538.
65. H. Takeda, K. Koike, H. Inoue and O. Ishitani, *J. Am. Chem. Soc.*, 2008, **130**, 2023–2031.
66. H. Hori, F. P. A. Johnson, K. Koike, O. Ishitani and T. Ibusuki, *J. Photochem. Photobiol., A*, 1996, **96**(171–174).
67. H. Takeda, K. Koike, T. Morimoto, H. Inumaru and O. Ishitani, in *Advances in Inorganic Chemistry*, E. Rudi vanand S. Grażyna, Academic Press, 2011, vol. 63 , pp. 137–186.
68. H. Hori, F. P. A. Johnson, K. Koike, K. Takeuchi, T. Ibusuki and O. Ishitani, *J. Chem. Soc., Dalton Trans.*, 1997, 1019–1023.
69. B. Gholamkhass, H. Mametsuka, K. Koike, T. Tanabe, M. Furue and O. Ishitani, *Inorganic Chemistry*, 2005, **44**, 2326–2336.
70. S. Sato, K. Koike, H. Inoue and O. Ishitani, *Photochem. Photobiol. Sci.*, 2007, **6**, 454–461.
71. Z.-Y. Bian, K. Sumi, M. Furue, S. Sato, K. Koike and O. Ishitani, *Inorganic Chemistry*, 2008, **47**, 10801–10803.
72. Z.-Y. Bian, K. Sumi, M. Furue, S. Sato, K. Koike and O. Ishitani, *Dalton Trans.*, 2009, 983–993.
73. K. Koike, S. Naito, S. Sato, Y. Tamaki and O. Ishitani, *J. Photochem. Photobiol., A*, 2009, **207**, 109–114.
74. Y. Tamaki, K. Watanabe, K. Koike, H. Inoue, T. Morimoto and O. Ishitani, *Faraday Discussions*, 2012, **155**, 115–127.
75. E. Fujita, B. S. Brunschwig, D. Cabelli, M. W. Renner, L. R. Furenlid, T. Ogata, Y. Wada and S. Yanagida, *Stud. Surf. Sci. Catal.*, 1998, **114**, 97–106.
76. C. A. Craig, L. O. Spreer, J. W. Otvos and M. Calvin, *J. Phys. Chem.*, 1990, **94**, 7957–7960.
77. A. H. A. Tinnemans, T. P. M. Koster, D. H. M. W. Thewissen and A. Mackor, Recl, *J. R. Neth. Chem. Soc.*, 1984, **103**(288–295).
78. J. L. Grant, K. Goswami, L. O. Spreer, J. W. Otvos and M. Calvin, *J. Chem. Soc., Dalton Trans.*, 1987, 2105–2109.
79. E. Kimura, S. Wada, M. Shionoya and Y. Okazaki, *Inorganic Chemistry*, 1994, **33**, 770–778.
80. E. Kimura, X. Bu, M. Shionoya, S. Wada and S. Maruyama, *Inorganic Chemistry*, 1992, **31**, 4542–4546.
81. S. Matsuoka, K. Yamamoto, T. Ogata, M. Kusaba, N. Nakashima, E. Fujita and S. Yanagida, *J. Am. Chem. Soc.*, 1993, **115**, 601–609.
82. T. Ogata, Y. Yamamoto, Y. Wada, K. Murakoshi, M. Kusaba, N. Nakashima, A. Ishida, S. Takamuku and S. Yanagida, *J. Phys. Chem*, 1995, **99**, 11916–11922.

83. T. Ogata, S. Yanagida, B. S. Brunschwig and E. Fujita, *J. Am. Chem. Soc.*, 1995, **117**, 6708–6716.

84. D. Behar, T. Dhanasekaran, P. Neta, C. M. Hosten, D. Ejeh, P. Hambright and E. Fujita, *J. Phys. Chem. A*, 1998, **102**, 2870–2877.

85. J. Grodkowski, D. Behar, P. Neta and P. Hambright, *J. Phys. Chem. A*, 1997, **101**, 248–254.

86. D. J. Boston, C. Xu, D. W. Armstrong and F. M. MacDonnell, Manuscript in Preparation.

87. I. Willner, R. Maidan, D. Mandler, H. Duerr, G. Doerr and K. Zengerle, *J. Am. Chem. Soc.*, 1987, **109**, 6080–6086.

88. R. Maidan and I. Willner, *J. Am. Chem. Soc.*, 1986, **108**, 8100–8101.

89. P. V. Klevtsov and R. F. Klevtsova, *J. Struct. Chem.*, 1977, **18**, 339.

90. N. R. de Tacconi, H. K. Timmaji, W. Chanmanee, M. N. Huda, P. Sarker, C. Janaky and K. Rajeshwar, *ChemPhysChem*, 2012, **13**, 2945.

91. K. Rajeshwar, N. R. de Tacconi and H. K. Timmaji, *J. Nano Res.*, 2012, **17**, 185.

92. D. A. Palmer and R. Vaneldik, *Chemical Reviews*, 1983, **83**, 651.

93. CRC *Handbook of Chemistry and Physics*, 82 edn., CRC Press, Boca Raton, FL, 2001–2001.

94. C. R. Chenthamarakshan and K. Rajeshwar, *Electrochem. Commun.*, 2000, **2**, 527.

95. A. Paracchino, V. Laporte, K. Sivula, M. Gratzel and E. Thimsen, *Nature Materials*, 2011, **10**, 456.

96. S. Somasundaram, C. R. Chenthamaraksha, N. R. de Tacconi and K. Rajeshwar, *Int. J. Hydrogen Energy*, 2007, **32**, 4661.

97. A. Paracchino, J. C. Brauer, J.-E. Moser, E. Thimsen and M. Gratzel, *J. Phys. Chem. C*, 2012, **116**, 7341.

98. Y. Zhou and J. A. Switzer, *Mater. Res. Innovations*, 1998, **2**, 1731.

99. A. E. Rakhshani and J. Varghese, *J. Mater. Sci.*, 1988, **23**, 2847.

100. A. E. Rakhshani, A. A. Al-Jassar and J. Varghese, *Thin Solid Films*, 1987, **148**, 191.

101. L. C. Wang, N. R. de Tacconi, C. R. Chenthamarakshan, K. Rajeshwar and M. Tao, *Thin Solid Films*, 2007, **515**, 3090.

102. Y. Zhou and J. A. Switzer, *Scripta Materialia*, 1998, **38**, 1731.

103. E. W. Bohannan, L.-Y. Huang, F. S. Miller, M. G. Shumsky and J. A. Switzer, *Langmuir*, 1999, **15**, 813.

104. C. C. Hu, J. N. Nian and H. Teng, *Sol. Energy Mater. Sol. Cells*, 2008, **92**, 1071.

105. W. Siripala, A. Ivanovskaya, T. F. Jaramillo, S. H. Baeck and E. W. McFarland, *Sol. Energy Mater. Sol. Cells*, 2003, **77**, 229.

106. P. E. de Jongh, D. Vanmaekelbergh and J. J. Kelly, *J. Electrochem. Soc.*, 2000, **147**, 486.

107. Z. Zheng, B. Huang, Z. Wang, M. Guo, X. Qin, X. Zhang, P. Wang and Y. Dai, *J. Phys. Chem. C*, 2009, **113**, 14448.

108. H. Yoneyama, K. Sugimura and S. Kuwabata, *J. Electroanal. Chem.*, 1988, **249**, 143.

109. A. B. Bocarsly, Q. D. Gibson, A. J. Morris, R. P. L'Esperance, Z. M. Detweiler, P. S. Lakkaraju, E. L. Zeitler and T. W. Shaw, *ACS Catalysis*, 2012, **2**, 1684.

110. E. E. Barton, D. M. Rampulla and A. B. Bocarsly, *J. Am. Chem. Soc.*, 2008, **130**, 6342–6344.

111. N. Myung, N. R. de Tacconi and K. Rajeshwar, To be submitted.

112. K. K. Kanazawa and J. G. Gordon, *Analytical Chemistry*, 1985, **57**, 1770.

113. G. Sauerbrey, *Zeitschrift für Physik*, 1959, **155**, 206.

114. G. Seshadri, C. Lin and A. B. Bocarsly, *J. Electroanal. Chem.*, 1994, **372**, 145–150.

115. A. A. Isse, A. Gennaro, E. Vianello and C. Floriani, *J. Mol. Catal.*, 1991, **70**, 197–208.

116. D. J. Pearce and D. Pletcher, *J. Electroanal. Chem. and Interfacial Electrochemistry*, 1986, **197**, 317–330.

117. S.-N. Pun, W.-H. Chung, K.-M. Lam, P. Guo, P.-H. Chan, K.-Y. Wong, C.-M. Che, T.-Y. Chen and S.-M. Peng, *J. Chem. Soc., Dalton Trans.*, 2002, 575–583.

118. J. Costamagna, J. Canales, J. Vargas and G. Ferraudi, *Pure and Applied Chemistry*, 1995, **67**, 1045–1052.

119. S. Sakaki, *J. Am. Chem. Soc.*, 1992, **114**, 2055–2062.

120. M. Beley, J. P. Collin, R. Ruppert and J. P. Sauvage, *J. Am. Chem. Soc.*, 1986, **108**, 7461–7467.

121. M. Beley, J.-P. Collin, R. Ruppert and J.-P. Sauvage, *J. Chem. Soc., Chem. Commun.*, 1984, 1315–1316.

122. A. Rios-Escudero, M. Isaacs, M. Villagran, J. Zagal and J. Costamagna, *J. Argent. Chem. Soc.*, 2004, **92**, 63–71.

123. M. A. Scibioh and V. R. Virayaraghavan, *Bull. Electrochem.*, 1997, **13**, 275–279.

124. M. A. Scihioh and V. R. Virayaraghavan, *Bull. Electrochem.*, 2000, **16**, 376–381.

125. A. C. De, J. A. Crayston, T. Cromie, T. Eisenblatter, R. W. Hay, Y. D. Lampeka and L. V. Tsymbal, *Electrochim. Acta*, 2000, **45**, 2061–2074.

126. I. S. Adaev, T. V. Korostoshevskaya, V. T. Novikov and T. V. Lysyak, *Russ. J. Electrochem.*, 2005, **41**, 1125–1129.

127. T. Atoguchi, A. Aramata, A. Kazusaka and M. Enyo, *J. Chem. Soc., Chem. Commun.*, 1991, 156–157.

128. I. M. B. Nielsen and K. Leung, *J. Phys. Chem. A*, 2010, **114**, 10166–10173.

129. M. Tezuka and M. Iwasaki, *Chemistry Letters*, 1993, 427–430.

130. I. Bhugun, D. Lexa and J.-M. Saveant, *J. Am. Chem. Soc.*, 1996, **118**, 1769–1776.

131. I. Bhugun, D. Lexa and J.-M. Savéant, *J. Am. Chem. Soc.*, 1994, **116**, 5015–5016.

132. I. Bhugun, D. Lexa and J.-M. Savéant, *J. Phys. Chem.*, 1996, **100**, 19981–19985.

133. M. Hammouche, D. Lexa, M. Momenteau and J. M. Saveant, *J. Am. Chem. Soc.*, 1991, **113**, 8455–8466.

134. M. Hammouche, D. Lexa, J. M. Savéant and M. Momenteau, *J. Electroanal. Chem. Interfacial Electrochem.*, 1988, **249**, 347–351.

135. C. Arana, M. Keshavarz, K. T. Potts and H. D. Abruña, *Inorganica Chimica Acta*, 1994, **225**, 285–295.

136. J. A. Ramos Sende, C. R. Arana, L. Hernandez, K. T. Potts, M. Keshevarz-K and H. D. Abruna, *Inorganic Chemistry*, 1995, **34**, 3339–3348.

137. S. Daniele, P. Ugo, G. Bontempelli and M. Fiorani, *J. Electroanal. Chem. Interfacial Electrochem.*, 1987, **219**, 259–271.

138. H. Ishida, K. Tanaka and T. Tanaka, *Organometallics*, 1987, **6**, 181–186.

139. A. I. Breikss and H. D. Abruna, *J. Electroanal. Chem.*, 1986, **201**, 347–358.

140. S. Cosnier, A. Deronzier and J.-C. Moutet, *J. Electroanal. Chem. Interfacial Electrochem.*, 1986, **207**, 315–321.

141. C. M. Bolinger, B. P. Sullivan, D. Conrad, J. A. Gilbert, N. Story and T. J. Meyer, *J. Chem. Soc., Chem. Commun.*, 1985, 796–797.

142. C. M. Bolinger, N. Story, B. P. Sullivan and T. J. Meyer, *Inorganic Chemistry*, 1988, **27**, 4582–4587.

143. K. S. Ratliff, R. E. Lentz and C. P. Kubiak, *Organometallics*, 1992, **11**, 1986–1988.

144. S. Slater and J. H. Wagenknecht, *J. Am. Chem. Soc.*, 1984, **106**, 5367–5368.

145. A. Szymaszek and F. P. Pruchnik, *J. Organomet. Chem.*, 1989, **376**, 133–140.

146. D. L. DuBois and A. Miedaner, *J. Am. Chem. Soc.*, 1987, **109**, 113–117.

147. H. A. G. M. Mostafa, T. Nagaoka and K. Ogura, *Electrochim. Acta*, 1997, **42**, 2577–2585.

148. A. Szymaszek and F. Pruchnik, *Rhodium Express*, 1994, **5**, 18–22.

149. M. Nakazawa, Y. Mizobe, Y. Matsumoto, Y. Uchida, M. Tezuka and M. Hidai, *Bull. Chem. Soc. Jpn.*, 1986, **59**, 809–814.

150. T. Tomohiro, K. Uoto and H. Okuno, *J. Chem. Soc., Chem. Commun.*, 1990, 194–195.

151. J.-M. Lehn and R. Ziessel, *J. Organomet. Chem.*, 1990, **382**, 157–173.

152. N. Kitamura and S. Tazuke, *Chemistry Letters*, 1983, 1109–1112.

153. J.-M. Lehn and R. Ziessel, *Proceedings of the National Academy of Sciences*, 1982, **79**, 701–704.

154. R. Ziessel, J. Hawecker and J.-M. Lehn, *Helv. Chim. Acta*, 1986, **69**, 1065–1084.

155. R. Perutz, C. D. Windle, M. V. Campian, E. Gibson, A. Duhme-Klair and J. Schneider, *Chem. Commun.*, 2012.

156. K. Koike, H. Hori, M. Ishizuka, J. R. Westwell, K. Takeuchi, T. Ibusuki, K. Enjouji, H. Konno, K. Sakamoto and O. Ishitani, *Organometallics*, 1997, **16**, 5724–5729.

157. H. Hori, K. Koike, K. Takeuchi and Y. Sasaki, *Chemistry Letters*, 2000, 522–523.

158. H. Hori, K. Koike, Y. Suzuki, M. Ishizuka, J. Tanaka, K. Takeuchi and Y. Sasaki, *J. Mol. Catal. A: Chem.*, 2002, **179**, 1–9.

159. H. Hori, J. Ishihara, K. Koike, K. Takeuchi, T. Ibusuki and O. Ishitani, *J. Photochem. Photobiol., A*, 1999, **120**, 119–124.

CHAPTER 12

Key Intermediates in the Hydrogenation and Electrochemical Reduction of CO₂

KLAAS JAN SCHOUTEN AND MARC KOPER*

Leiden Institute of Chemistry, Leiden University, Einsteinweg 55,
PO Box 9502, 2300 RA, Leiden, The Netherlands
*Email: m.koper@chem.leidenuniv.nl

12.1 Introduction

Carbon dioxide is the main product of the oxidation of hydrocarbons. Since hydrocarbon-based fuels are the world's most important energy source, the use of fossil fuels has led to significant increases of atmospheric CO_2 levels that are not expected to level out in the coming decades unless drastic measures are taken.[1] The increasing presence of CO_2 in the atmosphere is causing widespread concern about its possible consequences. On the other hand, from a more positive perspective, CO_2 is a vast and sustainable carbon feedstock that could partly replace the widespread use of petroleum-based hydrocarbons as chemical building blocks. Therefore, converting carbon dioxide into hydrocarbons would not only limit the emission of carbon dioxide but also supply us with a sustainable carbon feedstock, provided the conversion is performed using sustainable energy and without much additional CO_2 production. In this way,

RSC Energy and Environment Series No. 9
Photoelectrochemical Water Splitting: Materials, Processes and Architectures
Edited by Hans-Joachim Lewerenz and Laurence Peter
© The Royal Society of Chemistry 2013
Published by the Royal Society of Chemistry, www.rsc.org

the re-usage of the carbon dioxide caused by human emissions would enable a sustainable carbon cycle. If the hydrocarbons produced can be used as fuels, a carbon energy cycle is created. Such a carbon-based energy cycle has two main advantages compared to other proposed energy cycles that are, for example, based on storing energy in hydrogen or batteries. Firstly, hydrocarbons have a higher energy density, and secondly, storage is easier and there will be no need to change the existing fuel infrastructure, provided the generated fuel is a liquid.

Photosynthesis is one of the most important processes through which CO_2 is recycled in the earth's natural carbon cycle. CO_2 is inserted in carbon chains using the energy from sunlight to create carbohydrates, which are used in nature as chemical building blocks and energy carriers. Although these fuels used by nature are oxygen-rich, in contrast to fossil fuels that are oxygen poor and therefore more energy rich, mimicking photosynthesis would still be an attractive way to close our carbon-based energy loop and create a sustainable carbon energy cycle.

One of the promising ways to convert carbon dioxide into hydrocarbons is to do this electrochemically and ultimately to integrate such a process in a photo-electrochemical device. An auspicious discovery in this area was made by Hori in 1985, who showed that CO_2 can be converted directly to hydrocarbons on copper electrodes.[2] Only copper electrodes catalyze this reaction to a significant extent, and the main carbon products are methane and ethylene.[3] Ample research has been performed to understand the electrochemical reduction on the molecular level, but in spite of the extensive literature, the molecular mechanism is still a matter of debate.[4,5] With the renewed interest in solar fuels and CO_2 reduction and recycling, the mechanistic details of the electrochemical CO_2 reduction have become a topical subject of interest again in recent years.[6,7] Understanding the mechanism of this reaction is important, as it would open up routes to the production of high-energy fuels by the (photo-)electrochemical reduction of CO_2.

To further our understanding of the mechanistic aspects of the electrochemical conversion of CO_2, it is worthwhile to compare this mechanism with other hydrogenation reactions of CO_2. Both homogeneous and heterogeneous catalysis have been used to convert CO_2 to various hydrocarbons, *i.e.* carbon monoxide, methanol, methane, and formic acid.[8] This chapter is focussed on the reaction mechanisms of CO_2 reduction on a molecular level using metal catalysts. In order to obtain more insights into the key intermediates that determine the selectivity of CO_2 reduction to the various products, we compare the electrochemical reduction of CO_2, using copper and other metal electrodes in solution, with the metal-catalyzed hydrogenation and reduction of CO_2, both homogeneously in solution and heterogeneously in the gas phase. This chapter does not discuss the technical and economical feasibility studies of the various ways in which CO_2 can be converted: for such discussions we refer to other recent literature.[9–12]

We begin with a brief mechanistic overview of CO_2 fixation as it takes place in nature. Next, we give an overview of the reaction mechanisms for the various processes that are based on the hydrogenation of CO_2 including (i) the synthesis of carbon monoxide *via* the reverse water-gas shift (RWGS) reaction, (ii) the methanation of CO_2, (iii) methanol synthesis, (iv) hydrocarbon synthesis, and (v) the hydrogenation of CO_2 to formic acid using homogeneous catalysis. We then discuss the latest insights into the mechanisms of the electrochemical CO_2 reduction using metal electrodes and metal complexes, and finally compare the various mechanisms in the concluding section.

12.2 CO$_2$ Fixation in the Calvin Cycle

The conversion of CO_2 to carbohydrates in plants is called the Calvin cycle, named after Melvin Calvin who discovered the cycle in the 1950s and was awarded the Nobel Prize for Chemistry in 1961. In this cycle, the energy of sunlight is used to fix CO_2 and convert it into triose phosphates.[13] The carbon fixation in the Calvin cycle can be broken down in four steps, as shown in Figure 12.1. First, a proton is removed from ribulose-1,5-biphosphate, resulting in the formation of an enediolate intermediate. CO_2 binds to this enediolate to form a C_6 intermediate through a carboxylation reaction. This intermediate is hydrated in the next step, after which it breaks into two 3-phosphoglycerates.

After the carboxylation, the two 3-phosphoglycerates are converted into glyceraldehyde 3-phosphate, a three-carbon sugar phosphate. Five of these glyceraldehyde 3-phosphates are regenerated by converting them into three ribulose-1,5-biphosphates. So overall, three turns of the Calvin cycle yield one C_3 product, the glyceraldehyde 3-phosphate. These triose phosphates are used to synthesize hexose phosphates, which can be converted to (i) sucrose for transport, (ii) starch for energy storage, (iii) cellulose for cell wall synthesis, and (iv) pentose phosphates for metabolic intermediates.

The enzyme that catalyzes the carboxylation reaction is ribulose-1,5-biphosphate carboxylase oxygenase (rubisco). The active site of this enzyme contains an Mg^{2+} ion, that brings together and orients the reactants, as shown in Figure 12.2[14] Deprotonation of ribulose-1,5-biphosphate results in the formation of the enediolate, shown in Figure 12.1a. CO_2, polarized by the Mg^{2+} ion, is then added to the double C–C bond of the enediolate, resulting in the formation of a carboxylate, as shown in Figure 12.2b.[13]

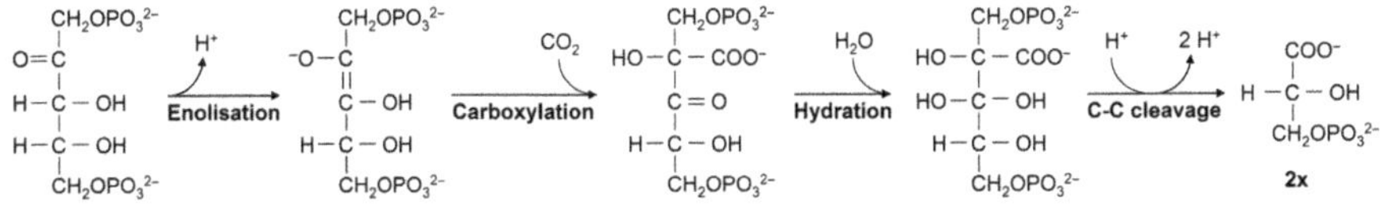

Figure 12.1 Carbon dioxide fixation in the Calvin cycle.

(a)   **(b)**

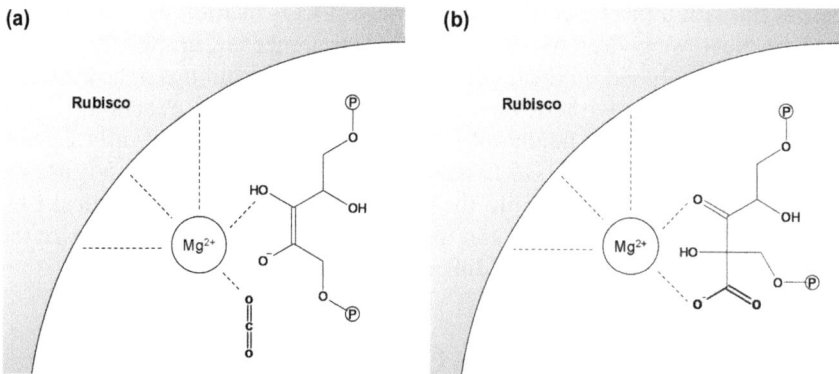

Figure 12.2   Carboxylation in the active site of Rubisco.

12.3   The Mechanisms of CO_2 Reduction Using Heterogeneous Catalysis

12.3.1   Carbon Monoxide Synthesis *via* the Reverse Water-Gas Shift Reaction

The formation of carbon monoxide from CO_2 *via* the reverse water-gas shift (RWGS) reaction,

$$CO_2 + H_2 \rightarrow CO + H_2O \tag{1}$$

is one of the most promising ways to convert CO_2, for several reasons.[15] First, since the RWGS is an endothermic reaction, the reaction product CO is a way to store energy, *i.e.* the conversion of CO_2 to CO can be used to store hydrogen energy. Next, this reaction can be used to change the ratio of H_2/CO in syngas, allowing for selective hydrocarbon formation. Furthermore, the RWGS occurs as a side reaction in many processes where CO_2 and H_2 are present, for example, in methanol synthesis. Finally, CO can be converted to various useful chemicals such as formic acid, methanol, formaldehyde, and long hydrocarbon chains.

Since the RWGS is a reversible reaction, catalysts active in the water-gas shift (WGS) reaction are often also active in the reverse reaction.[11]

$$CO + H_2O \rightleftharpoons CO_2 + H_2 \tag{2}$$

Since Cu-based catalysts are the most studied for the WGS reaction, they are also applied for the RWGS reaction.[8] Examples of catalysts used for the RWGS are Cu-Ni/γ-Al$_2$O$_3$ and Cu-ZnO/Al$_2$O$_3$, the latter is used for methanol synthesis as well.[8] Cerium-based catalysts are also active in the (R)WGS reaction.[8,11]

The mechanism of the RWGS reaction is still controversial. According to two recent review articles, two main reaction mechanisms have been proposed: the redox-mechanism and the associative formate-mechanism.[8,10] In the redox-mechanism, the CO_2 dissociates directly to CO and O, followed by the reduction of the oxide by hydrogen, resulting in the formation of water. On a Cu-based catalyst, this reaction can by modeled by:

$$CO_2 + 2Cu^0 \rightarrow Cu_2O + CO \tag{3}$$

$$H_2 + Cu_2O \rightarrow 2Cu^0 + H_2O \tag{4}$$

Both reaction (3) and (4) have been suggested as the rate determining step (RDS) for the (R)WGS.[16–19] Since it is a continuous process, the reduction of the oxidized Cu has to be faster than the oxidation process, and the RDS is probably the dissociation of CO_2.[8,18,20]

In the formate-mechanism, formate is formed by the association of hydrogen with CO_2. CO is formed subsequently by the decomposition of formate into CO and OH.

$$CO_2 \rightarrow CO_{2,ads} \tag{5}$$

$$H_2 \rightarrow 2H_{ads} \tag{6}$$

$$CO_{2ads} + H_{ads} \rightarrow HCO_{2,ads} \tag{7}$$

$$HCO_{2,ads} \rightarrow CO + OH_{ads} \tag{8}$$

$$OH_{ads} + H_{ads} \rightarrow H_2O \tag{9}$$

The dissociation of formate is assumed to be the RDS.[20] However, considering the mechanistic studies of the WGS reaction,[21,22] and taking into account the microscopic reversibility of the models used, it cannot be excluded that also in the RWGS hydroxycarboxyl, COOH, is the intermediate to CO instead of formate. One indication why formate is unlikely as an intermediate to CO formation, is that formate binds bidentate with two O atoms to the surface, whereas CO binds with its C atom, which makes the reaction step from HCOO to CO + OH – reaction (8) – very difficult.[21] Although adsorbed formate has been observed during the (R)WGS with several techniques, it could be only a spectator species.[21] Next to formate and hydroxycarboxyl, carbonate has also been proposed as an intermediate for the RWGS reaction.[23]

12.3.2   Methanation of Carbon Dioxide

The hydrogenation of CO_2 to methane is an important process. This reaction, called the Sabatier reaction, can be used for the production of syngas (*via* steam reforming), and is a way to store hydrogen energy and use this in the existing natural gas network.

$$CO_2 + H_2 \rightarrow CH_4 + 2H_2O \tag{10}$$

Supported Ni is the most studied catalyst material; other catalytic systems that are used for the Sabatier reaction are mainly based on Ru.[8,24]

Although differences in the rate and the selectivity for the methanation of CO_2 and CO are observed,[25] the general proposed mechanism of CO_2 hydrogenation is that CO_2 reacts to CO first, and subsequently follows the reaction mechanism of CO methanation.[8,24,26] For the first step, the formation of CO, it is proposed that, similar to the RWGS reaction and methanol synthesis, CO is formed *via* the decomposition of formate[8,24,27] or hydroxycarboxyl.[26] Interestingly, it was proposed recently that the hydrogenation of CO_2 to formate (HCOO) is a dead-end in the reaction, but that instead the hydrocarboxyl (COOH) leads to the formation of CO.[28]

The mechanism of CO methanation was proposed in the seventies to occur *via* an CH_xO intermediate,[29,30] but the generally accepted mechanism nowadays assumes the formation of surface carbon by CO dissociation *via* the Boudouard reaction, with subsequent hydrogenation.[8,24,26]

12.3.3 Methanol Synthesis

The synthesis of methanol is an important industrial process, since methanol is an alternative fuel and an important building block for synthesis in the chemical industry; over 40 million tons of methanol are produced per year. Most methanol plants are fed by natural gas which is converted to syngas by steam reforming, and the syngas subsequently is converted to methanol using copper-based catalysts. It is a promising process since CO_2 and H_2 can be used as a starting material using the same catalysts.

$$CO_2 + 3H_2 \rightarrow CH_3OH + H_2O \qquad (11)$$

The mechanistic details of the reaction of CO_2 to methanol are still a matter of debate. Two possible reaction pathways have been proposed for methanol synthesis from CO_2 and H_2 over Cu-based catalysts.[8,17,32] One pathway is the formate-pathway, in which the reaction to methanol proceeds through the formation of formate (HCOO), dioxomethylene (H_2COO), formaldehyde (CH_2O), and methoxy (CH_3O). On Cu, this is usually considered as the predominant pathway.[17] An example of this pathway, calculated by Nakatsuji and Hu on Cu(100),[31] is shown in Figure 12.3. There is no agreement in the literature on the RDS for this particular pathway. Usually the hydrogenation of adsorbed formate, as shown in Figure 12.3, is considered to be the RDS,[31,33–36] but the hydrogenation of dioxomethylene[17,39] or methoxy[38] have also been proposed as the RDS.

The other pathway involves the RWGS reaction (1), where CO_2 is first converted to CO, which is then hydrogenated to form methanol.

$$CO + 2H_2 \rightarrow CH_3OH \qquad (12)$$

This RWGS-pathway can explain the formation of CO as the major by-product during methanol synthesis from CO_2.[8,17,39] In this pathway, the CO will be hydrogenated to formyl (HCO) and formaldehyde (H_2CO) after which it will

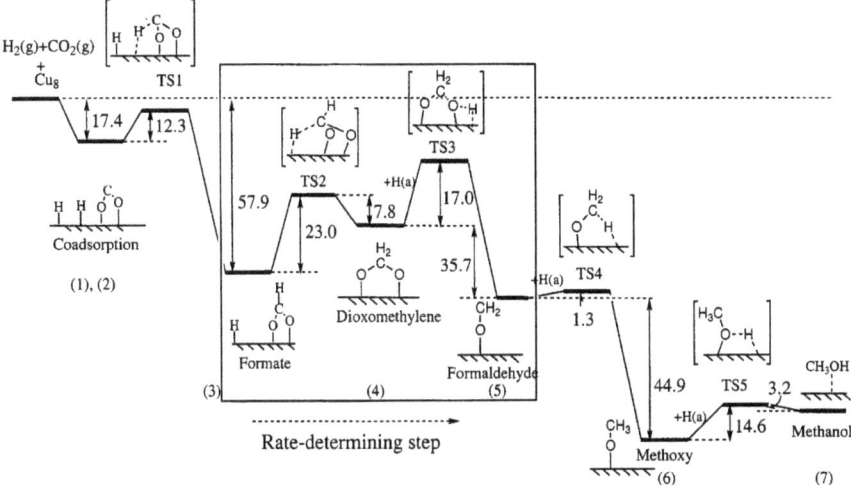

Figure 12.3   Hydrogenation of CO_2 to methanol *via* the formate-pathway on Cu(100). Reprinted with permission from reference 31. Copyright 2000 John Wiley & Sons, Inc.

follow the same path as shown in Figure 12.3. The overall RDS for the RWGS-pathway is the recombination of water from H and OH on the surface.[17] Concerning the reduction of CO to methanol only, the hydrogenation of methoxy has the highest activation energy.[17,32,38]

To avoid undesired by-products, a highly selective catalyst is needed. Although many kinds of metal-based catalysts have been examined for the synthesis of methanol, modified Cu, in particular Cu/ZnO, remains the main active catalyst. Although this Cu/ZnO system has been studied extensively, there is no consensus in the literature on the promotional role of ZnO and on the active site of the catalyst. One proposition is that the active site is metallic Cu,[34,37] and that the ZnO improves the dispersion and stabilization of metallic copper.[8,17] On the other hand, it has been proposed that the active site in methanol synthesis is a Cu^+ species, stabilized by the ZnO phase.[40–42] Recent work by Behrens *et al.* suggests that the active sites are steps at the Cu surface, alloyed with Zn.[43] The $Zn^{\delta+}$ at the steps increases the binding strength of oxygenated intermediates, which decreases the energy barriers in the reaction.

Other catalysts that have been used for methanol synthesis are Pd/β-Ga_2O_3, Cu/ZrO_2, and molybdenum sulfide.[8]

12.3.4   Synthesis of Hydrocarbons

Hydrogenation of CO_2 to hydrocarbon chains would be the ultimate goal of CO_2 utilization. Investigations into this subject can be divided into two categories: methanol-mediated and non-methanol mediated.[8,44]

The hydrogenation of CO_2 *via* methanol is performed on a composite catalyst, a combination of the standard Cu-Zn catalyst used for methanol synthesis with a zeolite used for the methanol-to-gasoline (MTG) process.[45,46] However, this method usually gives light alkanes as major hydrocarbon products because the methanol synthesis catalyst further hydrogenates the intermediate alkenes formed in the zeolites.[45,47]

The non-methanol mediated hydrogenation of CO_2 is usually performed using Fischer-Tropsch (FT) catalysts.[44] The most common metals in the FT process are cobalt and iron. Cobalt is the catalyst of choice in FT when long carbon chains are needed. However, the product distribution changes significantly when switching the feed gas from syngas to a gas mixture containing CO_2 and H_2. In the presence of CO_2, cobalt acts as a methanation catalyst. Iron has a higher catalytic activity for the WGS reaction than cobalt, which means that under FT conditions CO_2 is formed from syngas, making it a less attractive FT catalyst. However, for the hydrogenation of CO_2 this is an advantage, since iron also catalyzes the RWGS reaction. CO_2 hydrogenation on Fe has been shown to occur in two steps: first CO_2 is converted to CO *via* the RWGS reaction, followed by chain propagation *via* the FT mechanism resulting in alkenes in the range C_2–C_5.[44,47]

It would go beyond the scope of this chapter to discuss the controversial FT mechanism in detail. Several different mechanisms have been proposed since the discovery in the 1920s.[48] Currently, the proposed mechanisms can be divided into two classes: according to the first class, CO or CH_xOH species are inserted into the growing hydrocarbon chain, after which the C–O or C–OH bond is broken. In the other class, the C–O or C–OH bond is broken first, leading to the formation of CH_x species that are incorporated into the growing chain. The currently generally accepted mechanism is the latter, where CO is dissociated first resulting in the formation of C_1 (CH_2) species, followed by FT chain growth.[48,49] The breaking of the C–O bond can be activated by the assistance of hydrogen through the formation of adsorbed CHO or COH. Calculations for single-crystal surfaces have shown that this so-called "hydrogen assisted CO activation", is the optimum pathway on flat surfaces whereas at defects on the surface the direct dissociation of CO is favored.[49]

12.4   The Mechanisms of CO_2 Reduction Using Homogeneous Catalysis

12.4.1   Synthesis of Formic Acid

The main product of CO_2 hydrogenation using homogeneous catalysis is formic acid. Formic acid has a wide range of applications, for example, in the leather industry and for food preservation, and is a starting material for the production of various chemicals. Formic acid has also been proposed as a way to store hydrogen,[8,50] and as a fuel for formic acid fuel cells.[51] Complexes of several transition-metals are used to catalyze this reaction, *i.e.* rhodium,

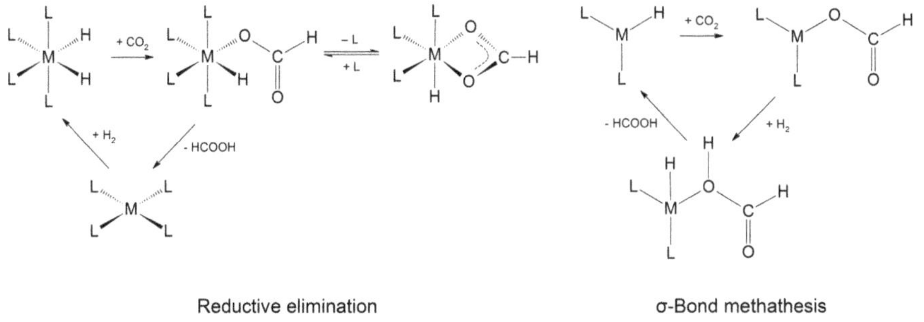

Reductive elimination   σ-Bond methathesis

Figure 12.4   Reaction mechanisms of the hydrogenation of CO_2 to formic acid. "M" represents a metal atom and "L" a ligand group, which could also be the solvent.

ruthenium and iridium.[8,52,53] Ru complexes often show the best activity and selectivity.[8]

$$CO_2 + H_2 \rightarrow HCOOH \tag{13}$$

The key step in the reduction of CO_2 with H_2 to formic acid is the formation of the C–H bond. For the formation of this bond, formate has always been detected as an intermediate.[52] Formate is formed by the insertion of CO_2 into the metal-hydride bond. The binding of formate to the active site can be bidentate, ionic, or monodentate.[54,55] Various reaction mechanisms for the different metal complexes have been proposed,[8,53,54] but two fundamental different reaction mechanisms for the formation of formic acid are distinguished, as shown in Figure 12.4.[56] In the mechanism shown in the left-hand panel of Figure 12.4, the formate is formed upon oxidative addition of the CO_2, followed by reductive elimination of the formate by a hydride in the complex. On Rh, a different pathway is observed, with a smaller energy barrier compared to the reductive elimination mechanism, where the formic acid is formed from the formate directly from a dihydrogen complex by σ-bond metathesis.[54,56,57]

The reaction is often performed in organic solvents, but the addition of small amounts of water or alcohols has been shown to improve the catalytic hydrogenation of CO_2 to formic acid.[8,53,58] A proposed explanation is that hydrogen-bonding to the oxygen atom of CO_2 enhances the electrophilicity of carbon, thereby facilitating its insertion into the metal-hydride bond of the metal complex.[8,58]

12.4.2 Synthesis of Other Products

While active homogeneous catalysts for the production of formic acid have been discovered, there has only been preliminary development of catalysts for the production of other products like methanol and CO.[53] The formation of CO is thermodynamically more favorable at elevated temperatures.[53] Using Ru complexes, the formation of CO, methanol and methane has been observed.[53,59] In this case, CO is observed as initial product, followed by the formation of methanol and methane, suggesting that methanol is formed from CO. For the formation of ethanol, a bimetallic catalyst of Ru and Co is used, where the Ru-complex is believed to be primarily responsible for the reduction of CO_2 to CO and methanol, while the Co-complex is responsible for the formation of ethanol from methanol and CO.[53,59] Interestingly, the formation of ethanol was only observed in the presence of iodide.

12.5 Mechanisms of the Electrochemical Reduction of CO_2

The electrochemical reduction of carbon dioxide has attracted sustained attention in the past decades, for the synthesis of organic molecules as well as a possible means of energy storage, *e.g.* of high-energy electrons generated by photo-excitation. The most common reaction products are those that require the transfer of 2 electrons,[60] *i.e.* formic acid, carbon monoxide, and oxalic acid, but examples of 6 and 8 electron conversions into *e.g.* methanol, ethylene and methane have also been reported:

$$CO_2 + 2H^+ + 2e^- \rightarrow CO + H_2O \tag{14}$$

$$CO_2 + 2H^+ + 2e^- \rightarrow HCO_2H \tag{15}$$

$$2CO_2 + 2H^+ + 2e^- \rightarrow H_2C_2O_4 \tag{16}$$

$$CO_2 + 4H^+ + 4e^- \rightarrow H_2CO + H_2O \tag{17}$$

$$CO_2 + 6H^+ + 6e^- \rightarrow CH_3OH + H_2O \tag{18}$$

$$CO_2 + 8H^+ + 8e^- \rightarrow CH_4 + 2H_2O \tag{19}$$

$$2CO_2 + 12H^+ + 12e^- \rightarrow C_2H_4 + 4H_2O \tag{20}$$

Figure 12.5 shows a Pourbaix diagram of the equilibrium potentials for the reduction of CO_2 to various products in water as a function of pH.[5,61,62] The formation of these products usually proceeds through proton-coupled multi-electron steps, which are generally more favorable than single electron reductions since thermodynamically more stable molecules are formed.[61] The standard potential of the outer-sphere single electron reduction of CO_2 to $CO_2^{-\bullet}$ is -1.90 V *vs.* SHE in water, due to the large reorganizational energy needed for the formation of the bent radical anion.[61]

$$CO_2 + e^- \rightarrow CO_2^{-\bullet} \tag{21}$$

Catalysts (partially) overcome the high overpotential for outer-sphere CO_2 reduction by binding and protonating the CO_2 species such that its stability is improved. Nevertheless, high overpotentials may lead to a broad product distribution, especially on heterogeneous catalysts. Therefore, considerable

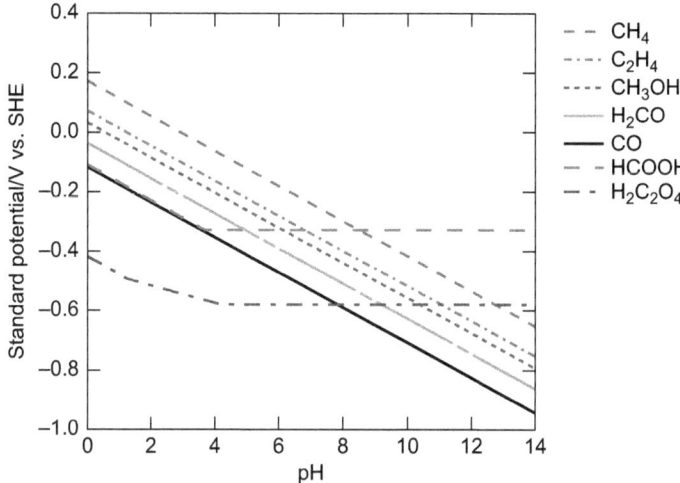

Figure 12.5   Standard potentials of various products of CO$_2$ reduction *vs.* pH.

efforts have been made to find catalysts that not only lower the overpotential but also steer the selectivity of the reaction. Both homogeneous molecular catalysis and heterogeneous catalysis, mostly metals, have been investigated for these purposes.[63]

12.5.1   Homogeneous Electrocatalysis

The field of CO$_2$ (electro)reduction using transition metal complexes in non aqueous media started in the 1970s. In the homogeneously catalyzed reduction of CO$_2$, the reduced form of a reversible couple with an equilibrium potential negative to the reduction potential of CO$_2$ reacts with the CO$_2$ and generates the oxidized form. The reduction of the metal complex at the electrode starts a new catalytic cycle. With respect to homogeneous electrocatalysis, Savéant differentiates between redox catalysis and chemical catalysis.[62] In redox catalysis, the reduced form of the catalyst is only an outer-sphere electron donor, whereas in chemical catalysis the interactions between the catalyst and substrate are stronger and involve the formation of an addition product between the catalyst and (a group of atoms initially belonging to) the substrate.[62]

Two families of transition metal complexes have been reported, namely Ag and Pd porphyrins, as well as some Ni macrocycles, that only form oxalate at potentials close to the CO$_2$/CO$_2^{-\bullet}$ couple, and therefore likely involve a redox or quasi-redox catalysis, in which electron transfer step acts as a pre-equilibrium to the rate determining dimerization of CO$_2^{-\bullet}$. A quasi redox catalysis mechanism has also been reported for the reduction of CO$_2$ to oxalate by anion radicals of aromatic nitriles and esters.[62,64]

Electrocatalytic CO$_2$ reduction with metal complexes in solution typically proceeds through chemical catalysis.[62] The reported catalysts can be divided into different categories, as shown in Figure 12.6: metal catalysts with

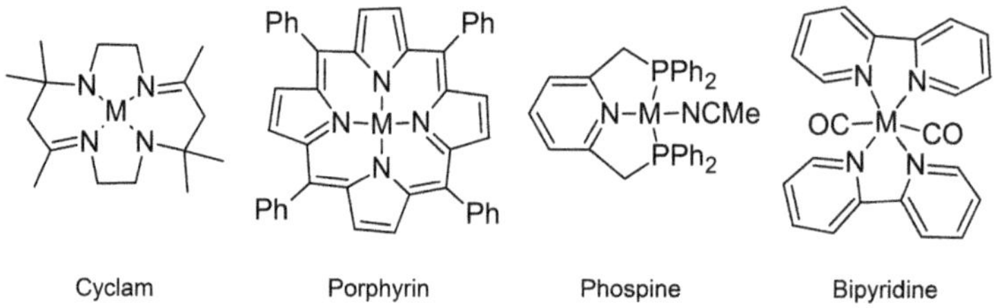

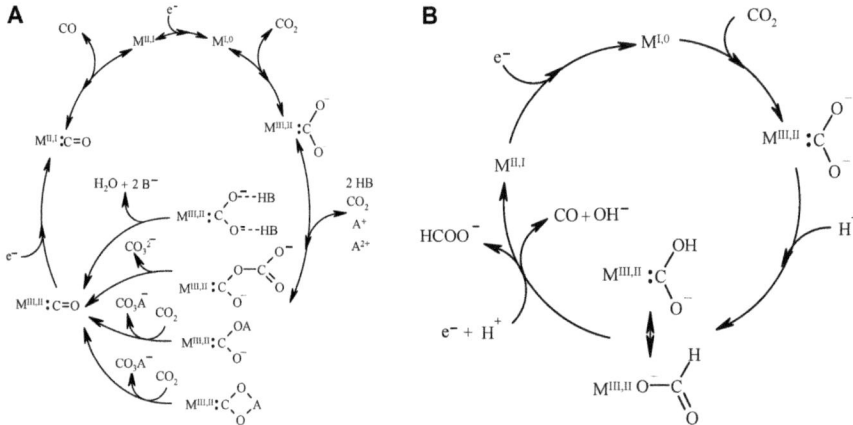

| Cyclam | Porphyrin | Phospine | Bipyridine |

Figure 12.6   Examples of the types of metal complexes that are used to reduce CO_2.

Figure 12.7   (A) Various mechanisms for the reduction of CO_2 to CO in the presence of Lewis acids "A" and weak Brønsted acids "B". (B) Proposed mechanism for the reduction of CO_2 to CO and formate.
Reprinted with permission from reference 62. Copyright 2008 American Chemical Society.

macrocyclic ligands, which can be divided into cyclam-like and porphyrin-like complexes, metal catalysts with phosphine ligands, and metal catalysts with polypyridyl ligands.[61,62,65] The best efficiencies and/or selectivities are obtained with Ni (cyclams), Fe (porhyrins), Re and Ru (polypyridyls), and Pd (phosphines).

The most common reduction product using these catalysts is CO, which is formed at potentials much less negative than the $CO_2/CO_2^{-\bullet}$ couple, which clearly excludes a redox mechanism. Therefore, the first step in CO_2 reduction is most likely the coordination of CO_2 to the previously reduced metal complex.[62] On Ni cyclams this complexation has been shown to be stabilized by a strong back donation from Ni to CO_2. This causes an increase in negative charge at the O atoms of the CO_2, in a similar configuration to the $CO_2^{-\bullet}$ radical.[5,66] The presence of weak Brønsted acids and Lewis acids has been shown to stabilize the coordination of CO_2 and facilitate the breaking of the C–O bond to form CO (see reference 62 and references therein). Also water and CO_2 itself may have the same role as these acids. This has led to the proposed mechanisms for the formation of CO shown in Figure 12.7a.[62,65,67,68]

The other common product obtained is formate. The mechanism suggested for the formation of formate, shown in Figure 12.7b, is a proton-coupled electron transfer to the coordinated CO_2 that, depending on the catalyst, will lead to the formation of CO or formate.[62] Another mechanism, comparable to the reductive elimination reaction shown in Figure 12.4, is *via* an internal hydride transfer to the coordinated CO_2.[65,69]

In some cases, higher reduction products have been obtained with Ru complexes, *i.e.* formaldehyde, methanol, and even some C_2 species such as $CHOCO_2^-$ and $CH_2OHCO_2^-$.[70] These C_2 products, which are the result of 4 and 6 electron transfer reactions, are only possible if CO is a stable intermediate ligand that can be further reduced, and coupled with another CO_2. The nature of the catalyst is very important in this stabilization, and lower temperatures are used to stabilize the intermediates.[62,70] Interestingly, it has been shown that pyridinium is able to reduce CO_2 to methanol through six sequential electron transfers.[71] This is probably the first case in which sequential one-electron transfers provide the low energy pathway for catalysis, in contrast to multi-electron transfer pathways. The mechanism of the reaction was studied on Pt electrodes with pyridinium in solution. Formic acid and formaldehyde were observed as intermediate products in the formation of methanol, and the reaction has been shown be first order in CO_2 and pyridinium.[71,72] The pyridinium radical is proposed as the actual catalyst[71,72] (although recent work by Keith and Carter suggests differently).[73] This radical can bind CO_2 and reduced intermediates through a coordinative interaction that stabilizes the intermediate species. The first step is the coordination of CO_2 to the pyridinium radical resulting in the formation of a carbamate species, suggesting a covalent N–C bond.[72] Subsequent electron transfer results in the formation of formic acid. Formic acid again is coordinated to a pyridinium radical and is, *via* the formyl-radical, reduced to formaldehyde. Coordination of formaldehyde results in hydroxymethyl, which is reduced to methanol. This mechanism suggests an inner-sphere-type electron transfer from the pyridinium radical to the intermediates for the various mechanistic steps, where the pyridinium radical is able to covalently bind the (radical) intermediate species and transfer the electron.

Electrochemical Carboxylation

The application of CO_2 as a C_1 source in organic synthesis might be a way to CO_2 fixation and could yield various useful carboxylic acids, including pharmaceuticals. In organic chemistry, low-valent Ni and Pd species are generated *in situ* from Ni^{II} or Pd^{II} precursors, and facilitate C–C coupling reactions.[74] Electrochemistry can provide an easy way to generate a desired oxidation state of a metal complex that becomes the active catalytic species for an organic reaction at potentials that avoid the direct reduction of the organic compound. An example is the carboxylation of aryl halides.[74–76] The reaction was shown to proceed through a sequence involving Ni^0, Ni^I, Ni^{II} and Ni^{III} intermediates, as shown in Figure 12.8. Another route to electrochemical carboxylation is the direct reduction of the organic compound, followed by the

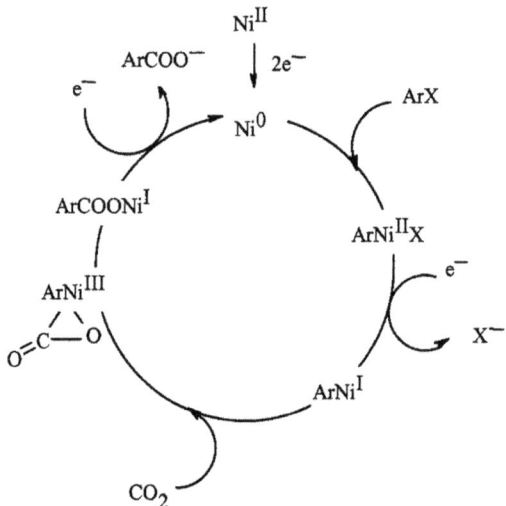

Figure 12.8 Mechanism of the carboxylation of aryl halides (ArX; X = Br, Cl, I). Reprinted with permission from reference 74. Copyright 2002 WILEY-VCH.

carboxylation.[76,77] An example is the carboxylation of α-chloroethylbenzene.[78] The α-chloroethylbenzene is reduced at the Pt cathode, which effectively removes the chlorine. The reduced intermediate reacts with CO_2 to form the carboxylate. An Mg anode enhances the reaction: the Mg^+ cations created by dissolution of the anode stabilize the carboxylate anions by producing an insoluble Mg salt that prevents oxidation or protonation.[78] Hindering the competing protonation reaction is also the reason why this direct carboxylation is usually performed in organic electrolytes or ionic liquids.

12.5.2 Heterogeneous Electrocatalysis

The conversion of CO and CO_2 to hydrocarbons using heterogeneous catalysis is usually performed at elevated temperatures and pressures. The direct electrochemical conversion of CO_2 would allow a process that avoids high temperatures, and where the production rate can be controlled directly, depending on the availability of surplus electricity. Only a few metals are active in this process, and the product distribution is broad, depending heavily on the electrode material and electrolyte used.

Aqueous Media

In aqueous media, metals can be divided roughly into 4 groups, based on the products formed during the electrochemical CO_2 reduction.[3,5] The first group includes the metals that evolve hydrogen at low potentials and with a high CO adsorption strength, such as Ni, Fe, Pt, and Ti. On these metals, CO_2 is reduced to strongly bound CO that blocks further reduction. Therefore, the main product on these metals is hydrogen.

The second group of metals are those with a high hydrogen overvoltage and a very low CO adsorption strength, such as Sn, In, Tl, Pb, Hg, Bi, and Cd. Since the reduced intermediates are not (or only very weakly) adsorbed on the surface, these metals are not able to catalyze the breaking of the C–O bond in CO$_2$. These metals facilitate the conversion of CO$_2$ to formic acid with high current efficiencies: formic acid is formed with 100% current efficiency on Hg. The first step in the formation of formic acid is the formation of CO$_2$$^{-\bullet}$. The formation of this radical is observed in aqueous and non-aqueous solutions, and during its formation and subsequent reduction only a very small fraction of the electrode is covered by adsorbates (see reference 5 and references therein). The subsequent reduction and protonation of CO$_2$$^{-\bullet}$ to form formate does not depend on pH, which shows that the proton donor is not H$^+$ but H$_2$O. This indicates a mechanism in which CO$_2$ is reduced to CO$_2$$^{-\bullet}$ in solution, followed by the protonation by water to form HCO$_2^\bullet$ and its subsequent reduction to HCOO$^-$.

$$CO_2 + e^- \rightarrow CO_2^{-\bullet} \tag{22}$$

$$CO_2^{-\bullet} + H_2O \rightarrow HCO_2^\bullet + OH^- \tag{23}$$

$$HCO_2^\bullet + e^- \rightarrow HCO_2^- \tag{24}$$

The third group of metals produces mainly CO. These include Au, Ag, Zn, and Ga; metals with a weak CO adsorption and a medium hydrogen overvoltage. The potentials at which CO is formed on these metals is less negative compared to the potentials where formic acid is formed, especially on Au.

Hori has shown that the heat of fusion of the various metal electrodes correlates well with the potentials of CO$_2$ reduction.[3] The heat of fusion is related to the *d*-electron contribution to the metallic bond, and may be taken as a measure of *d*-electron availability, which affects the strength of CO$_2$ adsorption. Therefore, metals with a higher heat of fusion adsorb CO$_2$ more strongly, and reduce it at lower potentials. In this context, the CO and HCOOH forming metals are well separated; CO is formed at less negative potentials on metals with a higher heat of fusion. This suggests a different mechanism for the formation of CO, where the intermediate(s) are much more stabilized, *i.e.* adsorbed at the electrode.[5]

A similar relation between adsorption strength and overpotential has been suggested by Peterson and Nørskov, who investigated the correlation between the binding energies of the intermediates of CO$_2$ reduction *vs.* CO (for intermediates binding to the surface through carbon) and *vs.* OH (for intermediates binding to the surface through oxygen) for various transition metals.[79] The limiting potential at which each elementary step of a reaction becomes exergonic can be derived from these binding energies. The protonation of adsorbed CO is singled out as the most important step dictating the overpotential, with Cu having the lowest overpotential compared to other transition metals.

Hori suggested a mechanism for the formation of CO where the CO_2 is bound to the surface, coordinated in a similar way to the CO_2 shown in Figure 12.7b,[5] as has has been calculated for the Ni cyclams discussed in section 12.5.1.[66] The negative charge on the O atoms then facilitates the protonation, and formation of adsorbed COOH. The next step is the breaking of the C–O bond, and the formation of CO and OH^-

$$CO_2 + e^- \rightarrow CO_{2,ads}^- \tag{25}$$

$$CO_{2,ads}^- + H_2O \rightarrow COOH_{ads} + OH^- \tag{26}$$

$$COOH_{ads} + e^- \rightarrow CO + OH^- \tag{27}$$

The fundamental possibility of the reversible conversion of CO_2 to CO and formate has been illustrated by Armstrong and Hirst.[80] They have discussed the immobilization of enzymes, *i.e.* carbon monoxide dehydrogenase (CODH) and formate dehydrogenase (FDH), on electrodes and shown that the electrocatalytic conversion of CO_2 on these electrodes is reversible, *i.e.* CO_2 is converted to CO or formate and vice versa around the equilibrium potential of the corresponding redox couple. During the interconversion of CO_2 and CO by CODH, CO_2 binds at reducing potentials as a bridging ligand between the Ni and the dangling Fe atom in the [Ni4Fe-4S] active site; then, CO migrates to Ni, and OH forms on Fe.[81] In contrast to metal surfaces, these highly efficient active sites can bind and stabilize the intermediate due to the precisely positioned functional groups and thereby lower the overpotential, or kinetically couple the formation and onward reaction of the intermediates. Moreover, enzymes seem to avoid the formation of poisons, such as CO, which frustrate the development of metallic formic acid oxidation catalysts.

Only a few metals can catalyze the conversion of CO_2 to hydrocarbons. The most interesting metal is Cu, with a moderate CO adsorption, where methane and even C_2 species such as ethylene are formed in significant amounts.[82] Some other metals such as Mo and Ru are able to convert CO_2 to methanol and methane, but with low efficiencies.[83,84]

Copper Electrodes

In 1985, Hori discovered that on copper electrodes CO_2 can be reduced to hydrocarbons, mainly methane and ethylene.[2,82] In addition to these hydrocarbons, CO and formic acid are also formed during CO_2 reduction. Also oxygenates such as ethanol and propanol are observed, although usually only in trace amounts.[4,5] Recently, up to 16 different reduction products have been observed.[85]

In spite of the extensive literature on carbon dioxide reduction on copper electrodes, the detailed mechanism of this reaction is still unclear.[4] It is known that ethylene and methane are formed through a different reaction mechanism and that carbon monoxide is a key intermediate in the formation of both

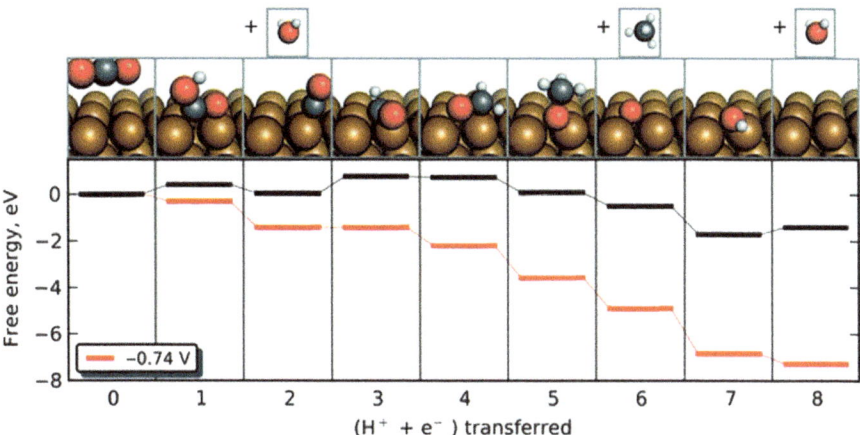

Figure 12.9   Free energy diagram for the lowest energy pathway to CH_4. The black pathway represents the free energy at 0 V *vs.* RHE and the red pathway the free energy at the indicated potential.
Reprinted with permission from reference 6. Copyright 2010 Royal Society of Chemistry.

ethylene and methane.[3,86,87] The exact reaction mechanism of carbon monoxide reduction to either ethylene or methane is still debated. Interestingly, methanol is not or produced only in very small amounts on a metallic Cu electrode, suggesting that the C–O bond is broken early in the reaction.[5]

The formation of methane from CO depends on pH, in such a way that the rate determining step must involve the transfer of a proton and an electron.[86] Recently, DFT calculations by Peterson *et al.* (see Figure 12.9), as well as our own experiments, suggested that the key intermediate to form methane is CHO_{ads}.[6,7] A similar path, but *via* adsorbed COH has been suggested by Hori.[86]

The formation of ethylene from CO, on the other hand, does not depend on pH.[86] Therefore, a dimer of carbon monoxide, whose formation does not involve the transfer of an hydrogen atom but does depend on potential (*i.e.* involves electron transfer), has been suggested as the key intermediate in the C–C coupling.[4,7] A Fischer-Tropsch-like mechanism where CO is coupled to CH_x species cannot explain the observed selectivity to ethylene. Schouten *et al.* showed that the only C_2 species that can be reduced to ethylene is ethylene oxide, suggesting a shared oxametallacycle intermediate for the reduction reactions of CO_2 and ethylene oxide, as shown in Figure 12.10.[7] Enol-type species have also been proposed as key intermediates in the formation of C_2 species by Kuhl *et al.*[85]

Hori *et al.* showed that the extent of methane and ethylene formation sensitively depends on the surface orientation of the copper electrode.[88] Formation of methane is favored on the (111) facet of the copper fcc crystal, whereas the formation of ethylene is dominant on the (100) facet. Recent DFT calculations have predicted that the limiting potential for the formation of the intermediates

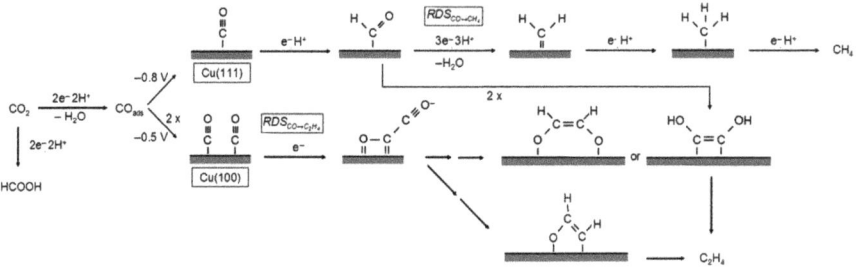

Figure 12.10 Proposed mechanisms for the formation of methane and ethylene from CO$_2$ on copper electrodes.[7]

of the CO$_2$ reduction to CH$_4$ is lower on the (211) surface compared to the Cu(111) and Cu(100) surface.[89] Using Online Electrochemical Mass Spectrometry (OLEMS), it has been shown that CO can be selectively reduced to ethylene on Cu(100) at low overpotentials, whereas at higher potentials ethylene and methane are formed simultaneously both on Cu(100) and Cu(111), suggesting two different pathways for the formation of ethylene from CO, shown in Figure 12.10.[90]

Interestingly, methanol has been observed as a product of CO$_2$ reduction on intentionally oxidized Cu electrodes.[83,91] The electronic properties of Cu(I) in *p*-Cu$_2$O are thought to play an important role in the adsorption of CO$_2$, causing a stronger binding of CO$_2$ and CO on *p*-Cu$_2$O and other Cu(I) centers.[83] It has been shown that *p*-Cu$_2$O-covered Cu can absorb atomic O into the bulk Cu.[83] This might facilitate the dissociation of CO$_2$, but the exact mechanism leading to methanol is still unclear. Recently, Li and Kanan have shown that thick Cu$_2$O films catalyze the reduction of CO$_2$ to CO and HCOOH with high faradaic efficiencies at low overpotentials.[92]

Non-aqueous Media

The electrochemical reduction of CO$_2$ in non-aqueous solutions, *e.g.* methanol, propylene carbonate or dimethyl sulfoxide, has several advantages compared to the reduction in water. The solubility of CO$_2$ is higher, and hydrogen evolution is heavily suppressed. The main products are CO, HCOOH and (COOH)$_2$.[5] CO is the main product in non-aqueous media on Cu, Ag, Au, Zn, In, Sn, Ni, and Pt. (COOH)$_2$ is formed on Cu, Sn, Ag, Zn, In and Au. Some metals like Fe, Cr, Mo, Pd, and Cd form both CO and (COOH)$_2$. HCOOH is formed on Pt, Pb, Hg, Ag and Au. On Cu, hydrocarbons such as CH$_4$ and C$_2$H$_4$ have also been obtained.

Product selectivity is mainly determined by whether or not the reduced CO$_2$ is stabilized at the electrode surface. Metals like Hg and Pb reduce CO$_2$ at potentials close to the potential of the CO$_2$/CO$_2^{-\bullet}$ couple, reaction (21).[5] The main product is oxalate or formate, depending on the concentration of water.[5,93] Therefore, Savéant *et al.* proposed that oxalate is formed by the dimerization of $CO_2^{-\bullet}$, similar to the oxalate formation in homogeneous catalysis.[93]

$$2CO_2^{-\bullet} \rightarrow C_2O_4^{2-} \qquad (28)$$

An alternative route to oxalate is the coupling of CO_2 to $CO_2^{-\bullet}$, resulting in the formation of $(CO_2)_2^{-\bullet}$, which is further reduced to oxalate.[5,94]

Formate is formed in a similar way as suggested for aqueous media and for homogeneous catalysis, where the small amount of water present reacts as a Lewis acid and protonate the $CO_2^{-\bullet}$, reaction (23).[5,62,93] Increasing water concentrations in non-aqueous media increases formate and decreases oxalate formation.

On the metals like Au, Zn and Ag, which stabilize the reduced intermediates to a much greater extent, the main product is CO. In the absence of water, CO_2 reacts as a Lewis acid with adsorbed CO_2^- to form $OCOCO_2^-$ in a way comparable with that which occurs in homogeneous catalysis, depicted in Figure 12.7A.[5,62] The breaking of the C–O bond then results in the formation of CO and CO_3^{2-}.

$$CO_2 + e^- \rightarrow CO_{2,ads}^- \qquad (29)$$

$$CO_{2,ads}^- + CO_2^- \rightarrow OCOCO_2^- \qquad (30)$$

$$OCOCO_2^- + e^- \rightarrow CO + CO_3^{2-} \qquad (31)$$

12.5.3 Photoelectrochemical CO$_2$ Reduction

The ultimate goal of CO_2^- reduction would be to couple the CO_2^- electro-catalyst to light harvesting by photo-excitation. This would yield a kind of artificial photosynthesis device to store the energy of sunlight in hydrocarbons. The basic steps always involve (i) the absorption of light to generate an excited state, (ii) charge separation of the created electron-hole pair, (iii) the energy of this charge separated state is used to reduce CO_2, (iv) catalyst regeneration.

The photoreduction of CO_2 can be divided into to general categories: the first category are the homogeneous systems, which are entirely molecule based.[95] The molecular light absorber and the catalyst could be the same molecule, or a molecular light absorber and a transition metal catalyst could work in concert. In the second category, a semiconductor is used for the light absorption and charge separation steps, after which this energy is transfered to a homogeneous or heterogeneous catalyst.

The most frequently used photosensitizers in homogeneous systems are Ru(II) polypyridyl complexes, such as [Ru(bpy)$_3$]$^{2+}$. This photosensitizer (P) is irradiated by light and forms an excited state, P*. This excited state is quenched by eletron transfer from a donor (D), typically triethylamine, to form P$^-$. The reduced photosensitizer in turn reduces the catalyst (cat), which then reduces the CO_2.[96]

$$P + h\nu \rightarrow P^* \qquad (32)$$

$$P^* + D \rightarrow P^- + D^+ \tag{33}$$

$$P^- + \text{cat} \rightarrow P + \text{cat}^- \tag{34}$$

$$\text{cat}^- + CO_2 \rightarrow \text{cat} + \text{products} \tag{35}$$

Cobalt and nickel macrocylic compounds are often used as catalysts. Sometimes the catalyst is linked to the photosensitizer in a supramolecular complex to increase the efficiency of the electron transfer to the catalyst.[96] Metalloporphyrins with Fe and Co can react both as a light absorber and a catalyst.[96] The products obtained are mainly CO and formate, formed in a similar way as described in section 12.5.1.[96,97]

Semiconductors have been shown to be efficient in the conversion of incident photon energy into electrical energy.[95] The semiconductor can be an electrode or a colloid. If the semiconductor surface is not electrocatalytically active, the separated charge has to be transfered to a catalytic species, which could be adsorbed on the surface or a species in solution. On these electrodes, not only the usual two-electron reduction products, formate and CO, are observed, but also formaldehyde and methanol.[97] A special case is p-GaAs: the (photo)-electrochemical reduction of CO_2 to methanol on the (111) faces of this electrode has been reported with a current efficiency close to 100%.[83,97] The mechanism for methanol formation is unclear. Arsenic-rich surfaces of GaAs spontaneously produce CH_3OH, even at open cicuit in the dark. This is attributed to dissolution of the semiconductor in carbonic acid, resulting in the formation of Ga and As hydroxides and methanol.[98]

12.6   Discussion and Conclusions

In this chapter, we have compared various mechanisms for the catalytic reduction of carbon dioxide, with particular emphasis on the low-temperature electrocatalytic reduction. The reduction of carbon dioxide may yield a variety of products, and most of these reactions tend to suffer from slow kinetics and/or poor selectivity. From the electrochemical point of view, only the conversion of carbon dioxide to carbon monoxide or to formic acid have been shown to be potentially reversible.[80] This is indeed expected for two-electron transfer reactions. In electrocatalysis, this feat has only been accomplished thus far by using enzymes, whereas in heterogeneous catalysis the (reverse) water gas shift reaction is also known as a reversible catalytic reaction, although at higher temperature. No synthetic room-temperature electrocatalyst has yet been developed which can do the same. Such a catalyst would be extremely interesting for formic acid fuel cells, and its reversible counterpart, *i.e.* carbon dioxide and hydrogen storage in formic acid. Higher-energy fuels from carbon dioxide require the transfer of more than 2 electrons, and this invariably leads to overpotential losses.[60,79]

We believe that there are four main pathways for CO_2 reduction to high-energy fuels. The first pathway is methanation, which is the thermodynamically most favorable process. A key intermediate, not only in this pathway but also in the FT process and the electrochemical reduction of CO_2 on copper electrodes, is carbon monoxide. In all processes described in this chapter, the intermediate leading to CO is hydroxycarboxyl, COOH, with the only exception being the RWGS reaction for which a pathway *via* formate, HCOO, has been proposed. However, as argued in section 12.3.1, there is evidence that hydroxycarboxyl rather than formate is also the intermediate to CO in the RWGS. A next important step in the methanation is the breaking of the C–O bond. This dissociation can be either directly, forming surface carbon, or hydrogen assisted *via* the formation of CH_xO species. The latter has also been suggested for the electrochemical reduction of $CO_{(2)}$, where CHO is considered to be the key intermediate in the formation of methane. It is not just the nature of the metal catalyst that determines whether or not CO can be dissociated. Cu and Ru are methanation catalysts, both under electrochemical and heterogeneous gas phase conditions. Under electrochemical conditions, the methane formation rate is much higher on Cu, whereas under gas phase conditions Ru is far more active.[83] Frese has attributed the higher activity of Ru in the gas phase to a higher hydrogen coverage compared to Cu (chemisorption of H_2 is more favorable on Ru) and to Ru being able to dissociate CO on defect sites, whereas Cu binds CO normally in a non-dissociated form. The higher methanation rate on Cu under electrochemical conditions has been explained by the ability of Cu to allow large overpotentials without hydrogen evolution overwhelming the CO_2 reduction reaction. The high potential, in combination with the HCO formation, is here the driving force for the dissociation of CO.

The second pathway is the formation of methanol. It seems that methanol is always formed *via* the formate-formaldehyde route, as has been demonstrated for methanol synthesis and for the (photo)-electrochemical reduction using pyridinium. Stabilization of the various intermediates, by adding ZnO to the Cu catalyst or as observed on the pyridinium by covalent bonding to the radical, seems to play an important role. Oxidized copper also seems to offer this (structural) stabilization, since the formation of CO and HCOOH is enhanced on oxidized copper electrodes,[92] and even the formation of methanol has been observed.[91] The formate-formaldehyde route also explains why methanol is not observed when CO_2 is reduced on metallic copper electrodes, as formate cannot be further reduced on copper.[4,5] On the other hand, C_2 oxygenates are observed (in low quantities) on copper electrodes. This is consistent with a reaction mechanism involving early breaking of the C–O bond, followed by C–C bond formation between $CO_{(2)}$ and the reduced intermediate. Interestingly, heterogeneous gas phase catalysts that are effective for methanol synthesis are in general ineffective for C_2 oxygenate formation, and metals that are active in C_2 oxygenate formation are known to favor CO dissociation.[99]

The third pathway is the pathway that leads to ethylene, observed during the electrochemical reduction on copper electrodes. This reaction has been discovered and described in great detail by Hori. Ethylene is normally formed simultaneously with methane, although at lower overpotentials the formation of ethylene is favored, and formed *via* a carbon monoxide intermediate. Its formation has been suggested to involve reductive CO coupling,[7] and ene(di)ol(ate) intermediates.[7,85] The dehydroxylation of enol-like surface species also explains the formation of C_2 and C_3 oxygenated species, such as ethanol. An enol is also the intermediate to the C–C bond formation in the Calvin cycle.

The fourth pathway is closest to the way CO_2 is fixed in nature through the Calvin cycle, namely through CO_2 insertion into an existing carbon chain, and subsequent carboxylate reduction. In electro-organic synthesis, this strategy is known as electrocarboxylation, but it has not gained much popularity yet as a sustainable solution for fuel production. It is interesting that the same Mg^{2+} ion that plays an important role in the active site of the carboxylation enzyme in the Calvin cycle, strongly enhances the electrocarboxylation reaction.[78]

In conclusion, the (electrochemical) reduction of CO_2 to interesting products, such as potential fuels, can take place through a variety of different pathways, and the pathway selected is highly sensitive to the catalyst material, electrode (over)potential, pH, electrolyte composition, solvent, *etc.* Although some common patterns can be observed, as discussed in some detail in this chapter, it appears that many pathways are similar in energy, involving intermediates with similar stability. This makes the search for or development of active and selective catalysts highly interesting but also highly challenging. Catalyst stability is an additional issue, that we have not touched upon in any detail, but that will clearly be crucial for any future deployment of this technology.

References

1. P. Friedlingstein, R. A. Houghton, G. Marland, J. Hackler, T. A. Boden, T. J. Conway, J. G. Canadell, M. R. Raupach, P. Ciais and C. Le Quere, *Nature Geosci.*, 2010, **3**, 811–812.
2. Y. Hori, K. Kikuchi and S. Suzuki, *Chem. Lett.*, 1985, 1695–1698.
3. Y. Hori, H. Wakebe, T. Tsukamoto and O. Koga, *Electrochim. Acta*, 1994, **39**, 1833–1839.
4. M. Gattrell, N. Gupta and A. Co, *J. Electroanal. Chem.*, 2006, **594**, 1–19.
5. Y. Hori, in *Modern Aspects of Electrochemistry*, ed. C. G. Vayenas, R. E. White and M. E. Gamboa-Aldeco, Springer, New York, 2008, vol. 42, pp. 89–189.
6. A. A. Peterson, F. Abild-Pedersen, F. Studt, J. Rossmeisl and J. K. Nørskov, *Energy Environ. Sci.*, 2010, **3**, 1311–1315.
7. K. J. P. Schouten, Y. Kwon, C. J. M. van der Ham, Z. Qin and M. T. M. Koper, *Chem. Sci.*, 2011, **2**, 1902–1909.
8. W. Wang, S. Wang, X. Ma and J. Gong, *Chem. Soc. Rev.*, 2011, **40**, 3703–3727.
9. C. Graves, S. D. Ebbesen, M. Mogensen and K. S. Lackner, *Renew. Sust. Energ. Rev.*, 2011, **15**, 1–23.
10. D. T. Whipple and P. J. A. Kenis, *J. Phys. Chem. Lett.*, 2010, **1**, 3451–3458.
11. G. Centi and S. Perathoner, *Catal. Today*, 2009, **148**, 191–205.
12. X. Xiaoding and J. A. Moulijn, *Energy Fuels*, 1996, **10**, 305–325.

13. G. Schneider, Y. Lindqvist and C. I. Branden, *Annu. Rev. Biophys. Biomol. Struct.*, 1992, **21**, 119–143.
14. D. L. Nelson and M. M. Cox, *Lehninger principles of biochemistry*, W. H. Freeman, 5th edn., 2008, pp. 773–801.
15. Y. Liu and D. Liu, *Int. J. Hydrogen Energy*, 1999, **24**, 351–354.
16. M. Ginés, A. Marchi and C. Apestegua, *Appl. Catal., A*, 1997, **154**, 155–171.
17. Y. Yang, J. Evans, J. A. Rodriguez, M. G. White and P. Liu, *Phys. Chem. Chem. Phys.*, 2010, **12**, 9909–9917.
18. G.-C. Wang and J. Nakamura, *J. Phys. Chem. Lett.*, 2010, **1**, 3053–3057.
19. S.-I. Fujita, M. Usui and N. Takezawa, *J. Catal.*, 1992, **134**, 220–225.
20. C. S. Chen, J. H. Wu and T. W. Lai, *J. Phys. Chem. C*, 2010, **114**, 15021–15028.
21. A. A. Gokhale, J. A. Dumesic and M. Mavrikakis, *J. Am. Chem. Soc.*, 2008, **130**, 1402–1414.
22. P. Liu and J. A. Rodriguez, *J. Chem. Phys.*, 2007, **126**, 164705.
23. A. Goguet, F. C. Meunier, D. Tibiletti, J. P. Breen and R. Burch, *J. Phys. Chem. B*, 2004, **108**, 20240–20246.
24. W. Wang and J. Gong, *Front. Chem. Sci. Eng.*, 2011, **5**, 2–10.
25. S.-I. Fujita, M. Nakamura, T. Doi and N. Takezawa, *Appl. Catal., A*, 1993, **104**, 87–100.
26. A. Lapidus, N. Gaidai, N. Nekrasov, L. Tishkova, Y. Agafonov and T. Myshenkova, *Petrol. Chem.*, 2007, **47**, 75–82.
27. M. Marwood, R. Doepper and A. Renken, *Appl. Catal., A*, 1997, **151**, 223–246.
28. E. Vesselli, M. Rizzi, L. De Rogatis, X. Ding, A. Baraldi, G. Comelli, L. Savio, L. Vattuone, M. Rocca, P. Fornasiero, A. Baldereschi and M. Peressi, *J. Phys. Chem. Lett.*, 2010, **1**, 402–406.
29. C. Huang and J. Richardson, *J. Catal.*, 1978, **51**, 1–8.
30. J. Sehested, S. Dahl, J. Jacobsen and J. R. Rostrup-Nielsen, *J. Phys. Chem. B*, 2005, **109**, 2432–2438.
31. H. Nakatsuji and Z.-M. Hu, *Int. J. Quantum Chem.*, 2000, **77**, 341–349.
32. Q.-L. Tang, Q.-J. Hong and Z.-P. Liu, *J. Catal.*, 2009, **263**, 114–122.
33. M. Bowker, R. A. Hadden, H. Houghton, J. N. K. Hyland and K. C. Waugh, *J. Catal.*, 1988, **109**, 263–273.
34. J. Yoshihara, S. C. Parker, A. Schafer and C. T. Campbell, *Catal. Lett.*, 1995, **31**, 313–324.
35. H.-W. Lim, M.-J. Park, S.-H. Kang, H.-J. Chae, J. W. Bae and K.-W. Jun, *Ind. Eng. Chem. Res.*, 2009, **48**, 10448–10455.
36. Q. Sun, C.-W. Liu, W. Pan, Q.-M. Zhu and J.-F. Deng, *Appl. Catal., A*, 1998, **171**, 301–308.
37. P. Rasmussen, M. Kazuta and I. Chorkendorff, *Surf. Sci.*, 1994, **318**, 267–280.
38. L. C. Grabow and M. Mavrikakis, *ACS Catal.*, 2011, **1**, 365–384.
39. Y. Nitta, O. Suwata, Y. Ikeda, Y. Okamoto and T. Imanaka, *Catal. Lett.*, 1994, **26**, 345–354.

40. J. Szanyi and D. W. Goodman, *Catal. Lett.*, 1991, **10**, 383–390.
41. J. Yoshihara and C. T. Campbell, *J. Catal.*, 1996, **161**, 776–782.
42. J. Nakamura, Y. Choi and T. Fujitani, *Top. Catal.*, 2003, **22**, 277–285.
43. M. Behrens, F. Studt, I. Kasatkin, S. Kühl, M. Hävecker, F. Abild-Pedersen, S. Zander, F. Girgsdies, P. Kurr, B.-L. Kniep, M. Tovar, R. W. Fischer, J. K. Nørskov and R. Schlögl, *Science*, 2012, **336**, 893–897.
44. R. W. Dorner, D. R. Hardy, F. W. Williams and H. D. Willauer, *Energy Environ. Sci.*, 2010, **3**, 884–890.
45. M. Fujiwara, R. Kieffer, H. Ando and Y. Souma, *Appl. Catal., A*, 1995, **121**, 113–124.
46. K. Fujimoto and T. Shikada, *Appl. Catal.*, 1987, **31**, 13–23.
47. P. Sai Prasad, J. Bae, K.-W. Jun and K.-W. Lee, *Catal. Surv. Asia*, 2008, **12**, 170–183.
48. H. Schulz, *Appl. Catal., A*, 1999, **186**, 3–12.
49. R. A. van Santen, I. M. Ciobîcă, E. van Steen and M. M. Ghouri, in *Advances in Catalysis*, B. C. Gates and H. Knözinger, Elsevier, Amsterdam, 2011, vol. 54, pp. 127–187.
50. S. Enthaler, J. von Langermann and T. Schmidt, *Energy Environ. Sci.*, 2010, **3**, 1207–1217.
51. X. Yu and P. G. Pickup, *J. Power Sources*, 2008, **182**, 124–132.
52. W. Leitner, *Angew. Chem., Int. Ed.*, 1995, **34**, 2207–2221.
53. P. G. Jessop, F. Jo and C.-C. Tai, *Coord. Chem. Rev.*, 2004, **248**, 2425–2442.
54. Y. Gao, J. K. Kuncheria, H. A. Jenkins, R. J. Puddephatt and G. P. A. Yap, *J. Chem. Soc., Dalton Trans.*, 2000, 3212–3217.
55. J. C. Tsai and K. M. Nicholas, *J. Am. Chem. Soc.*, 1992, **114**, 5117–5124.
56. F. Hutschka, A. Dedieu, M. Eichberger, R. Fornika and W. Leitner, *J. Am. Chem. Soc.*, 1997, **119**, 4432–4443.
57. Y. Musashi and S. Sakaki, *J. Am. Chem. Soc.*, 2002, **124**, 7588–7603.
58. C. Yin, Z. Xu, S.-Y. Yang, S. M. Ng, K. Y. Wong, Z. Lin and C. P. Lau, *Organometallics*, 2001, **20**, 1216–1222.
59. K.-I. Tominaga, Y. Sasaki, M. Saito, K. Hagihara and T. Watanabe, *J. Mol. Catal.*, 1994, **89**, 51–55.
60. M. T. Koper, *J. Electroanal. Chem.*, 2011, **660**, 254–260.
61. E. E. Benson, C. P. Kubiak, A. J. Sathrum and J. M. Smieja, *Chem. Soc. Rev.*, 2009, **38**, 89–99.
62. J.-M. Savéant, *Chem. Rev.*, 2008, **108**, 2348–2378.
63. B. P. Sullivan, K. Krist and H. E. Guard, *Electrochemical and electrocatalytic reactions of carbon dioxide*, Elsevier, Amsterdam, 1993.
64. A. Gennaro, A. A. Isse, J.-M. Savant, M.-G. Severin and E. Vianello, *J. Am. Chem. Soc.*, 1996, **118**, 7190–7196.
65. D. L. Dubois, *Electrochemical Reactions of Carbon Dioxide*, Wiley-VCH Verlag GmbH & Co. KGaA, 2007, pp. 202–225.
66. S. Sakaki, *J. Am. Chem. Soc.*, 1992, **114**, 2055–2062.
67. M. Beley, J. P. Collin, R. Ruppert and J. P. Sauvage, *J. Am. Chem. Soc.*, 1986, **108**, 7461–7467.

68. M. Hammouche, D. Lexa, M. Momenteau and J.-M. Saveant, *J. Am. Chem. Soc.*, 1991, **113**, 8455–8466.
69. J. R. Pugh, M. R. M. Bruce, B. P. Sullivan and T. J. Meyer, *Inorg. Chem.*, 1991, **30**, 86–91.
70. H. Nagao, T. Mizukawa and K. Tanaka, *Inorg. Chem.*, 1994, **33**, 3415–3420.
71. B. E. Cole, P. S. Lakkaraju, D. M. Rampulla, A. J. Morris, E. Abelev and A. B. Bocarsly, *J. Am. Chem. Soc.*, 2010, **132**, 11539–11551.
72. A. J. Morris, R. T. McGibbon and A. B. Bocarsly, *ChemSusChem.*, 2011, **4**, 191–196.
73. J. A. Keith and E. A. Carter, *J. Am. Chem. Soc.*, 2012, **134**, 7580–7583.
74. E. Dunach, D. Franco and S. Olivero, *Eur. J. Org. Chem.*, 2003, 1605–1622.
75. C. Amatore and A. Jutand, *J. Am. Chem. Soc.*, 1991, **113**, 2819–2825.
76. J. Grimshaw, *Electrochemical Reactions and Mechanisms in Organic Chemistry*, Elsevier Science B.V., Amsterdam, 2000, pp. 147–150.
77. M. M. Baizer, *Tetrahedron*, 1984, **40**, 935–969.
78. Y. Hiejima, M. Hayashi, A. Uda, S. Oya, H. Kondo, H. Senboku and K. Takahashi, *Phys. Chem. Chem. Phys.*, 2010, **12**, 1953–1957.
79. A. A. Peterson and J. K. Nørskov, *J. Phys. Chem. Lett.*, 2012, **3**, 251–258.
80. F. A. Armstrong and J. Hirst, *Proc. Natl. Acad. Sci. USA*, 2011, **108**, 14049–14054.
81. J.-H. Jeoung and H. Dobbek, *Science*, 2007, **318**, 1461–1464.
82. Y. Hori, M. Murata and R. Takahashi, *J. Chem. Soc. Faraday Trans. 1*, 1989, **85**, 2309–2326.
83. K. W. Frese, in *Electrochemical and electrocatalytic reactions of carbon dioxide*, ed. B. P. Sullivan, K. Krist and H. E. Guard, Elsevier, Amsterdam, 1993, pp. 145–216.
84. D. P. Summers, S. Leach and K. W. F. Jr., *J. Electroanal. Chem.*, 1986, **205**, 219–232.
85. K. P. Kuhl, E. R. Cave, D. N. Abram and T. F. Jaramillo, *Energy Environ. Sci.*, 2012, **5**, 7050–7059.
86. Y. Hori, R. Takahashi, Y. Yoshinami and A. Murata, *J. Phys. Chem. B*, 1997, **101**, 7075–7081.
87. Y. Hori, A. Murata, R. Takahashi and S. Suzuki, *J. Am. Chem. Soc.*, 1987, **109**, 5022–5023.
88. Y. Hori, I. Takahashi, O. Koga and N. Hoshi, *J. Mol. Catal. A: Chem.*, 2003, **199**, 39–47.
89. W. J. Durand, A. A. Peterson, F. Studt, F. Abild-Pedersen and J. K. Nørskov, *Surf. Sci.*, 2011, **605**, 1354–1359.
90. K. J. P. Schouten, Z. Qin, E. P. Gallent and M. T. M. Koper, *J. Am. Chem. Soc.*, 2012, **134**, 9864–9867.
91. J. Karl and W. Frese, *J. Electrochem. Soc.*, 1991, **138**, 3338–3344.
92. C. W. Li and M. W. Kanan, *J. Am. Chem. Soc.*, 2012, **134**, 7231–7234.
93. C. Amatore and J. M. Savéant, *J. Am. Chem. Soc.*, 1981, **103**, 5021–5023.
94. V. S. Bagotzky and N. V. Osetrova, *Russ. J. Electrochem.*, 1995, **31**, 409–425.

95. N. S. Lewis and G. A. Shreve, in *Electrochemical and electrocatalytic reactions of carbon dioxide*, ed. B. P. Sullivan, K. Krist and H. E. Guard, Elsevier, Amsterdam, 1993, pp. 263–289.
96. A. J. Morris, G. J. Meyer and E. Fujita, *Acc. Chem. Res.*, 2009, **42**, 1983–1994.
97. B. Kumar, M. Llorente, J. Froehlich, T. Dang, A. Sathrum and C. P. Kubiak, *Annu. Rev. Phys. Chem.*, 2012, **63**, 541–569.
98. W. M. Sears and S. R. Morrison, *J. Phys. Chem.*, 1985, **89**, 3295–3298.
99. A. F. Sammells and R. L. Cook, in *Electochemical and electrocatalytic reactions of carbon dioxide*, ed. B. P. Sullivan, K. Krist and H. E. Guard, Elsevier, Amsterdam, 1993, pp. 217–262.

CHAPTER 13

Novel Approaches to Water Splitting by Solar Photons

ARTHUR J. NOZIK

National Renewable Energy Laboratory, Golden, CO 80401 and Department of Chemistry and Biochemistry, University of Colorado, Boulder, Boulder, CO 80309, USA
Email: Arthur.Nozik@nrel.gov

13.1 Introduction and Previous History of Photoelectrochemical Water Splitting/ Photoelectrolysis

The efficient and cost-effective splitting of H_2O into H_2 and O_2 to produce solar H_2 as a renewable energy carrier has been a major objective under continuous investigation since the mid-1970 s. The research was spurred by the oil crisis of that period, which occurred 2 years after the famous paper by Fujishima and Honda[1] that showed for the first time that H_2O could be split in H_2 and O_2 by near UV light using a photoelectrochemical cell. In this cell, a single crystal of rutile TiO_2 (a wide band gap (3 eV) semiconductor), was used as a photoanode in contact with aqueous electrolyte to oxidize H_2O to O_2, and the cathode counter-electrode was platinum metal that reduced H_2O to H_2. This process, which has been referred to as photoelectrolysis or photoelectrochemical energy conversion, is based on the science of semiconductor-liquid junctions or semiconductor-molecule interfaces. The latter field of science (photoelectrochemistry) had been studied and developed earlier through pioneering work[2–20] by Gerischer, Memming, Brattain and Garrett, Dewald, Williams,

RSC Energy and Environment Series No. 9
Photoelectrochemical Water Splitting: Materials, Processes and Architectures
Edited by Hans-Joachim Lewerenz and Laurence Peter

Morrison, and Gomes; comprehensive reviews and books with references to the early work and to the later energy applications of photoelectrochemistry are available.[21-34] Figure 13.1 shows the basic features of a semiconductor-electrolyte junction in which an n-type semiconductor is in equilibrium in the dark with the H^+/H_2 redox couple in acid solution.[21]

A feature of the initial Fujishima–Honda cell, not generally recognized initially, was that a potential bias was present between the two electrodes connected by a salt bridge in the cell since the electrolyte in the photoanode region was a strong base (NaOH) and the Pt cathode was in strong acid (H_2SO_4). The pH difference between the photoanode and the cathode could generate an electrochemical bias between the two electrodes of 0.82 V (14×0.059 V), and hence the H_2O splitting process in this case should more properly be called photo-assisted electrolysis, especially when an external electrical potential is applied to the cell. The need for an applied voltage of at least 0.3 V when using rutile as the photoanode together with a Pt counter electrode was reported by Nozik in a follow-up publication in *Nature* in 1975.[35] Notwithstanding the fact that an applied voltage for H_2O splitting was required in these early cells and thus reduced the practical prospects for solar H_2O splitting, the papers stimulated a large, new global research effort in photoelectrochemical energy

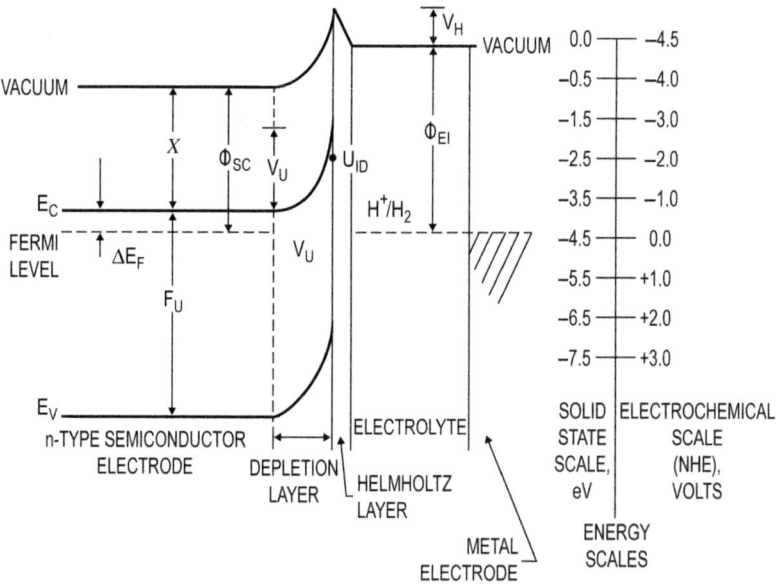

Figure 13.1 Energy level diagram for semiconductor-electrolyte junction showing the relationships between the electrolyte redox couple (H^+/H_2), the Helmholtz layer potential drop (V_h), and the semiconductor band gap (Eg), electron affinity (χ), work function (ϕ_{sc}), band bending (V_B), and flatband potential (U_{fb}). The electrochemical and solid state energy scales are shown for comparison is ϕ_{El} is the electrolyte work function (from reference 21).

conversion to produce low-cost solar hydrogen as well as low-cost and efficient electrochemical photovoltaic cells.

True photoelectrolysis, that is to say, splitting H_2O without any external bias, was first reported by Nozik in 1976, who showed that the need for an external bias could be eliminated by illuminating both electrodes; the two electrodes were an n-type TiO_2 photoanode and a p-type GaP photocathode.[36] In such a cell with two-photoelectrodes, the photopotentials generated in each photoelectrode are added together to provide a high enough voltage (1.23 V + overvoltage) to drive the H_2O splitting reaction spontaneously. Subsequently, it was shown[37] that the two photoelectrodes could be simply formed into a monolithic bilayer structure which requires no external bias wherein the two n-and p-type materials are sandwiched through a common ohmic contact layer. This monolithic structure, shown in Figure 13.2, was termed a "photochemical diode".[37] Its operation is analogous to the Z-scheme of biological photosynthesis; the n-type region of the photochemical diode is analogous to the oxygen-evolving Photosystem II (PSII), the p-type region is analogous to the CO_2-reducing Photosystem I (PSI), and the ohmic contact between the two n-and p-type regions is analogous to the cytochrome electron transport chain in photosynthesis that combines electrons from PSII with the positive holes in PSI (see Figure 13.3).

The Fermi level of a semiconductor electrode is equivalent to its electro-chemical potential, and in electrochemistry it is referenced to a standard oxidation-reduction (redox) potential in an electrolyte, such as the H^+/H_2 redox potential at pH 0 and 1 atm H_2, the NHE (normal hydrogen electrode) with a value defined to be equal to zero, or the standard calomel electrode (SCE), which has a value 0.24 V positive with respect to NHE. When a single n-type photoelectrode is used for H_2O splitting, an external bias is necessary if the Fermi level (electrochemical potential) of the photoanode is below the redox Fermi level of the H^+/H_2 redox couple. This means that when the anode and cathode are short circuited under these conditions so that their Fermi levels equilibrate and become equal, photogenerated free electrons in the conduction band of the n-type photoanode are unable to drive the electrochemical

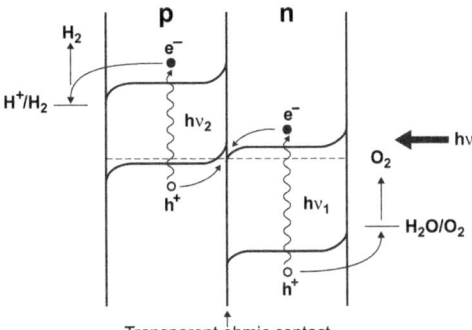

Figure 13.2 Energy level diagram for a photochemical diode (from references 25 and 37).

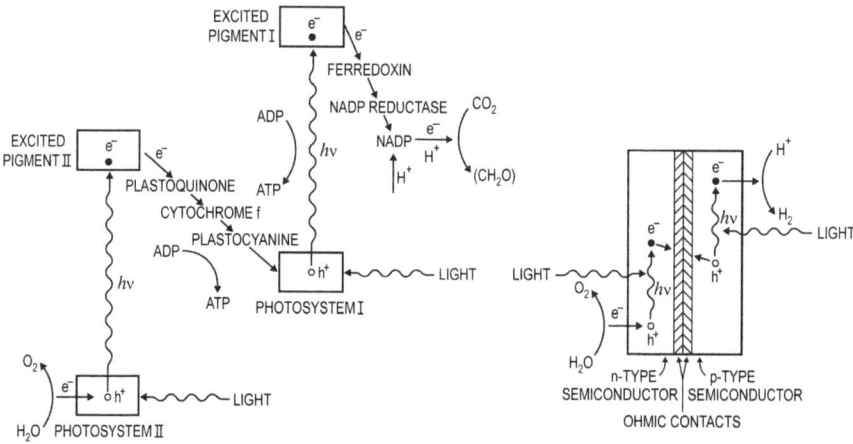

Figure 13.3 Analogy between the Z scheme of biological photosynthesis for reacting CO_2 and H_2O to produce glucose and O_2 and the operation of a photochemical diode to split H_2O into H_2 and O_2.

reduction reaction to generate H_2 when they are transferred through external contacts to the cathode because their free energy is insufficient. Thus, an external bias must be applied to the cell to raise the Fermi level of electrons in in the cathode above the H^+/H_2 redox Fermi level, *i.e.* to a more negative potential value on the standard electrochemical scale (see Figure 13.4). The higher the Fermi level of the cathode is relative to the H^+/H_2 redox Fermi level, the faster the rate of the reduction reaction; the corresponding potential difference is termed the overvoltage of the electrode reduction reaction. At the photoanode, the photogenerated holes drive the oxidation of H_2O to O_2 to complete the cell reaction of splitting H_2O into H_2 and O_2. In this case, the overvoltage can be considered to be the difference between the H_2O/O_2 redox couple and the potential of the valence band edge. Figure 13.4 illustrates the sequence of energetic conditions from the initial state to the final state for a photoelectrolysis cell configured with a single n-type photoanode and Pt cathode.

In a single electrode cell, it is possible to have the illuminated semiconductor electrode as the cathode (photocathode). In this case, photogenerated electrons are injected from the p-type semiconductor photocathode to the H^+/H_2 redox couple to generate H_2, and photogenerated holes move through the external circuit to a metal anode to bring about the oxidation of H_2O to O_2. The energy level diagrams for this case are analogous to those in Figure 13.4, except that the signs and directions of interfacial electric fields and charge flow in the semiconductor electrodes are reversed.

The position of the Fermi level of a photoanode or photocathode can be determined by measuring the flatband potential (U_{fb}) of the electrode from a Mott-Schottky plot ($1/C_{sc}^2$ *vs.* electrode potential), where C_{sc} is the space charge capacitance of the electrode. U_{fb} is the potential of the semiconductor electrode at which the conduction and valence bands are flat when in contact

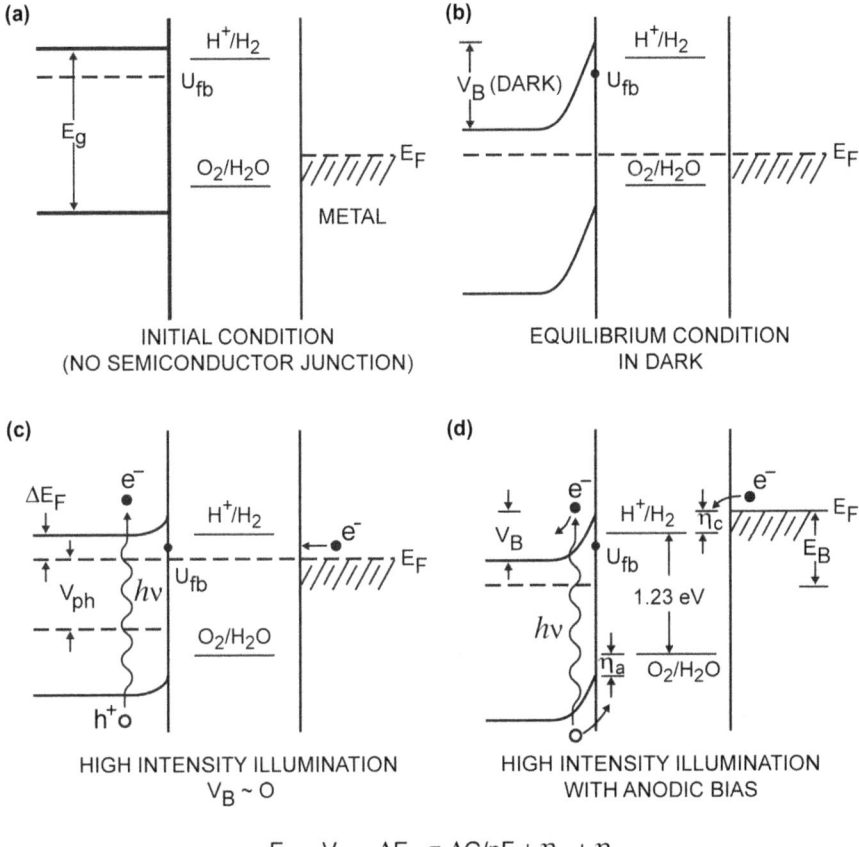

$$Eg - V_B - \Delta E_F = \Delta G/nF + \eta_a + \eta_c$$

Figure 13.4 Sequence of energy level diagrams for photoelectrolysis cell with an n-type semiconductor anode and metal cathode from the initial condition in the dark (a) to the final condition of photoelectrolysis with light and bias (d) (from reference 35).

with an electrolyte. When a semiconductor forms an ideal junction with an electrolyte (*i.e.* there is no surface state charging and hence no band edge movement when the electrode potential changes because of equilibration of the Fermi level with the electrolyte or *via* application of an external voltage), the U_{fb} and the semiconductor band edges remain constant and fixed. What changes with applied potential is the value (and sign) of the electric field at the semiconductor-liquid interface; this electric field is distributed over a finite distance into the semiconductor (called a space charge layer or depletion layer because the majority carrier population in this region decreases) where it causes both semiconductor band edges to either bend up or down depending on the sign of the change of electrode potential (Fermi level) after equilibration is achieved. This situation occurs if the potential drop across the double layer of charge at the semiconductor-electrolyte interface – the Helmholtz double layer – remains constant with changes in the Fermi level of the semiconductor electrode. If surface states become charged during illumination, either through trapping of photogenerated electrons or holes or through charge transfer from donor or acceptor species in the electrolyte, then the potential drop in the Helmholtz layer can change and cause the position of the U_{fb} and the band edges to move. Furthermore, if the Fermi level is initially equilibrated with a high density of surface states, and not with the dominant redox couple in the electrolyte, then the band edges can also move with changes in electrode potential; this situation is referred to as Fermi level pinning. Finally, the pH of the electrolyte affects the potential drop in the Helmholtz layer when ionic surface charges are created by the equilibrium between H_2O, H^+, and OH^- species on the surface. This equilibrium is affected by pH and hence will control the U_{fb}. For photoelectrolysis cells, where the surface charge is determined by the H_2O dissociation equilibrium, the U_{fb} and redox potentials for water oxidation and reduction will change with pH by the same amount, and hence the relative band edge positions with respect to the $H + /H_2$ and H_2O/O_2 redox potentials are independent of pH. All these issues are important for water splitting because it is the positions of the conduction and valence band edges of the semiconductor electrodes with respect to the redox potentials of the desired water oxidation and reduction reactions that determine whether the desired two redox reactions, resulting in H_2 and O_2 evolution, can occur.

Another very important criterion for a viable H_2O splitting system is that the photomaterials and redox catalysts must be stable against photocorrosion and photodegradation for a very long period (10–25 years), depending on the initial capital cost of the H_2O splitting system. For semiconductor electrodes in direct contact with an aqueous electrolyte, this means that photooxidation and photoreduction of the semiconductor electrode materials should be thermodynamically forbidden (*i.e.* the decomposition potentials must lie outside the bracket of the oxidation and reduction potentials defined by the band positions of the semiconductor electrode). If this is not the case, then the anodic and cathodic decomposition rates must be many orders of magnitude slower than the rates of the desired oxidation and reduction of H_2O, respectively in order to ensure adequate long-term stability.

Thus, three factors must be satisfied simultaneously for conventional PEC cells based on semiconductor-electrolyte junctions for H_2O splitting: (1) optimum band gap to maximize photocurrent and generate net photovoltages above 1.8 V; (2) flatband potentials that allow the conduction and valence band edges of the semiconductor electrode to straddle the redox potentials of the H^+/H_2 and H_2O/O_2 redox couples; and (3) photoelectrode and catalytic materials that are (photo)stable for 10–25 years, depending upon the initial capital cost of the photoelectrolysis system. All of these basic scientific issues and factors are presented and discussed in detail in several earlier published reviews, book chapters, and books cited above in Section 13.1.[21–34] Much experimental and theoretical research has been conducted over the years to discover and develop PEC photoelectrodes based on semiconductor-aqueous electrolyte junctions that satisfy the three requirements presented above; these include the other chapters of this book, as well other reviews.[38-40] This specific topic will not be discussed here. However, as discussed in section 13.2.3, two of the three requirements for PEC H_2O splitting cells can be relaxed for certain architectures for H_2O splitting cells, and this is one of the focal areas of this chapter. However, we begin with a discussion of the thermodynamic limits on the maximum possible power conversion efficiency (PCE) of H_2O splitting cells.

13.2 Detailed Balance Thermodynamic Calculations of Solar Water-Splitting PCEs

13.2.1 Conventional Solar Cells at One Sun Intensity

The detailed balance model of Shockley and Queisser[41] is used to calculate the PCE of solar photoconversion devices for H_2O splitting.[42] This model is used first to calculate the PCE of single band gap and multiple band gap tandem PEC devices, applying the usual assumption made for all present day solar cells that just one electron–hole pair is generated per absorbed photon. In later sections, the calculations are expanded to include the effects of allowing more than one electron–hole pair to be generated per absorbed photon when the photon energy is at least twice the band gap (or HOMO-LUMO energy of the photoconverter) in order to satisfy energy conservation. Finally, the effects of combining exciton multiplication process (termed multiple exciton generation, MEG in semiconductor nanocrystals and singlet fission, SF, in molecules) with solar concentration are analyzed. As discussed later, MEG occurs efficiently in semiconductor nanocrystals, and SF occurs efficiently in unique molecular chromophores. Both systems can be incorporated into solar cells for efficient H_2O splitting as well as for more efficient photovoltaic (PV) cells.

In general, the current versus voltage dependence for a single threshold (band gap) photoconversion device is written as[42]

$$j(V, E_g) = j_G(E_g) - j_R(V, E_g) \tag{13.1}$$

where j_G is the photogenerated current, j_R is the recombination current associated with radiative emission, E_g is the absorption threshold or band gap (or HOMO-LUMO gap) of the absorber and V is the photovoltage generated by the cell. Expressions for the photogenerated current, j_G, and recombination current, j_R for a single band gap cell under one sun intensity are written as:

$$j_G(E_g) = q \int_{E_g}^{E_{max}} QY(E)\Gamma(E)dE \tag{13.2}$$

$$j_R(V, E_g) = qg \int_{E_g}^{\infty} \frac{QY(E)E^2}{\exp\left(\frac{E-qQY(E)V}{k_BT}\right) - 1} dE \tag{13.3}$$

where E is the photon energy, q is the electronic charge, $\Gamma(E)$ is the photon flux, k_B is Boltzmann's constant, T is the absolute temperature of the device ($T = 300$ K in this work), $g = 2\pi/c^2h^3$, c is the speed of light in vacuum and h is Planck's constant. The quantum yield, $QY(E)$, generally allows for the generation and recombination of multiple charge pairs per absorbed photon (*i.e.* MEG or SF) over the appropriate energy range. The ASTM G-173-3 Reference AM1.5G solar spectrum[43] is used as the illumination source, $\Gamma(E)$ is the photon flux associated with the AM1.5G spectrum and E_{max} is the maximum photon energy in the solar spectrum, (for AM1.5G, $E_{max} = 4.428$ eV). For practical purposes, $E_{max} \sim 4$ eV, because the integrated solar current above 4 eV in the standard AM1.5G spectrum is only $\sim 5\,\mu A/cm^2$. In equation (13.2), carrier generation from ambient blackbody radiation becomes important for E_g less than ~ 0.2 eV. The assumptions implicit in equations (13.1) to (13.3) are those of the detailed balance model: all photons with energy greater than the absorption threshold are absorbed, the quasi-Fermi level separation (in semiconductors) is constant and equal to V across the device (equivalent to infinite carrier mobility), and radiative recombination is the only active recombination mechanism. The chemical potential of the emitted photons is $qVQY(E)$, as required by thermodynamics.[44]

The PCE for the production of stored chemical energy in the form of H_2 from water splitting is written as

$$\eta_{H_2}(V) = j(V)E_{H_2} / P_{IN} \qquad (13.4)$$

where $E_{H_2} = 1.23$ V is the minimum thermodynamic potential (*i.e.* no overvoltage) required for water splitting at 300 K. In actual water splitting devices, the operating or bias point of the cell, V, will be larger than E_{H_2} by the sum of the anode and cathode overpotentials and the resistive potential drop of the electrolyte. V_o is used to denote the sum of these overpotentials (energy losses). Then, the operating voltage is

$$V = V_O + E_{H_2} \qquad (13.5)$$

The maximum efficiency for a single band gap device with a given band gap and QY can be found from the above equations by maximizing the efficiency of equation (13.4) with respect to the operating voltage V.[42] The results are shown in Figure 13.5, where the maximum possible efficiency for H_2O splitting is plotted *vs.* the single band gap energy for various overvoltages ranging from $V_0 = 0$ to $V_0 = 0.8$ V. The figure shows that the maximum efficiency for H_2O splitting drops rapidly as the overvoltage increases; thus the efficiency drops from $\sim 30\%$ at $V_0 = 0.0$ V, to 17% at $V_0 = 0.4$ V, to 8% at $V_o = 0.8$ V. This is because the band gap required for H_2O splitting increases with increasing V_o and therefore reduces the amount of absorbed solar photons.

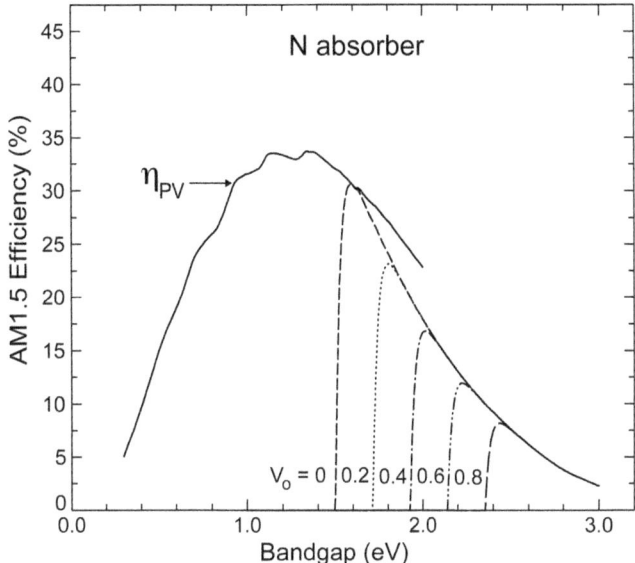

Figure 13.5   Power Conversion Efficiency (PCE) *vs.* bandgap for photoelectrolysis using a single conventional (M1) photoelectrode for different values of the total cell overvoltage (V_o) (sum of anodic and cathodic overvoltages). Also shown for comparison is the PCE for a conventional single band gap PV cell (from reference 42).

13.2.2   Tandem Semiconductor Structures for Water Splitting

As discussed in the introduction, the use of two band gaps arranged in tandem in H_2O-splitting cells can increase the PCE.[42] In the equations. below, the subscript 1 is used to denote parameters for the top cell and subscript 2 for the bottom cell. The current voltage equations analogous to equation (13.1) for the top and bottom cells in a tandem device are

$$j_1(V_1, E_1) = j_{G1}(E_1) - j_{R1}(V_1, E_1) \qquad (13.6)$$

and

$$j_2(V_2, E_2) = j_{G2}(E_2) - j_{R2}(V_2, E_2) \qquad (13.7)$$

The generation and recombination currents, j_{G1}, j_{G2}, j_{R1} and j_{R2} have the same form as equations (13.2) and (13.3) with E_1 and E_2 substituted for E_g and V_1 and V_2 substituted for V. For the bottom cell, E_{max} in the upper integral limit of equation (13.2) is replaced with the band gap of the top absorber, E_1, because the top cell is assumed to absorb all photons with energy greater than E_1. For a series connected tandem device, the total current must be the same in each cell, while the photovoltage across the device is the sum of the voltages developed across each cell.

$$j(V) = j_1(V_1, E_1) = j_2(V_2, E_2) \tag{13.8}$$

$$V = V_1 + V_2 \tag{13.9}$$

The $j(V)$ curve for the tandem cell is found by solving equations (13.8) and (13.9) simultaneously for V_1 and V_2. The current is then calculated using equations (13.6) and (13.7). The maximum PCE of the tandem water splitting device is found using equations (13.4) and (13.5) with j and V given by equations (13.8) and (13.9).

For water splitting tandem devices, a restriction on the possible combinations of E_1 and E_2 arises from the requirement that a portion of the solar spectrum must be absorbed in the bottom cell for a tandem device to function. For all absorber combinations of top and bottom cell, the band gap for the top cell must be larger than the band gap of the bottom cell. Additionally, a water-splitting tandem device must have a combination of band gap energies (E_1, E_2) that generates an open circuit voltage greater than $V_o + E_{H_2}$ for the device to be able to split water.

The efficiency of a tandem H_2O-splitting cell is plotted versus the bottom cell band gap, E_2, in Figure 13.6a.[42] The top cell band gap, E_1 was chosen to maximize the efficiency. The top cell band gap giving the maximum efficiency, E_{1max}, is plotted in Figure 6b. The maximum efficiency under the ideal condition where $V_o = 0\,V$ is 40.0% and occurs with top and bottom cell gaps of 1.40 eV and 0.52 eV, respectively. As the overpotential increases to 0.4 and 0.8 V, the maximum efficiency decreases to 33.2% and 27.1%, respectively, while the optimum top and bottom cell gaps move to higher energies.

We have seen above that the PCE of H_2O splitting can be greatly increased if two photosystems, optically in tandem, are used instead of one. The use of two photosystems in the tandem cell is analogous to the use of two photosystems in biological photosynthesis to boost the photopotential to the values needed for the photosynthetic reaction: $CO_2 + H_2O \rightarrow (CH_2O) + O_2$, except that in the case of the Z-scheme, the two photosystems have about the same band gap (HOMO-LUMO).[45] Figure 13.7 is an isoefficiency contour plot for H_2O splitting at a fixed overvoltage of $V_0 = 0.4\,V$; for this overvoltage, the values of constant PCE are indicated for the various possible combinations of the two band gaps in the tandem cell. A summary of these calculations of maximum PCE *vs.* V_o is shown in Figure 13.8.

As can be readily seen in Figure 13.8, tandem cells have the potential to greatly increase the efficiency of solar driven water splitting at all overvoltage values. Thus, the limiting maximum efficiency at zero overvoltage is $\sim 41\%$ for a tandem cell and $\sim 31\%$ for a single junction cell. At an overvoltage of 0.5 V, the maximum values are $\sim 30\%$ and 15%, respectively, and at 1.0 V overvoltage the maximum values are 20% and 5%, respectively. Thus, depending upon overvoltage, the 2-junction tandem cell increases the relative efficiency by 33% at zero overvoltage to 400% at 1 V overvoltage. Calculations show that there is no benefit to be gained regarding higher PCEs if three tandem junctions are used.[42]

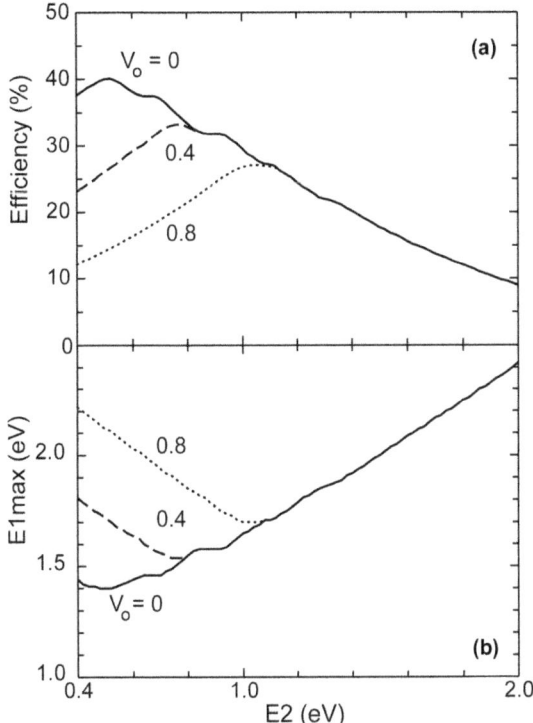

Figure 13.6 Maximum water splitting PCE *vs.* band gap of the bottom cell *E2* for a two band gap series connected tandem device with M1 top and bottom absorbers (a), and the corresponding value of the top cell band gap E_1max for maximum PCE (b). Efficiency and E_1max vs. E_2 curves are shown for three values of cell overpotential ($Vo = 0$, 0.4, and 0.8 V) (from reference 42).

The strong dependence of PCE on cell overvoltage shown in Figures 13.6 and 13.8 demonstrates the great importance of catalysis to reduce overvoltage and maximize efficiency. The maximum PCE results calculated here thus far are for the usual absorbers which have a maximum QY = 1 (*i.e.* the absorber produces 1 electron-hole pair per absorbed photon, and is labeled M1). Higher thermodynamic conversion efficiencies are possible when 2 or more electron-hole pairs are created per absorbed photon *via* MEG (a process that occurs efficiently in quantum dots) and *via* SF in molecular chromophores.

13.2.3 Power Conversion Efficiencies Based on Utilization of Hot Electron-Hole Pairs and Singlet Fission

13.2.3.1 PCE Using Hot Carriers in Bulk Semiconductor Electrodes

One ubiquitous feature of the majority of current solar cells is that they are based on macroscale semiconductor materials in which absorbed photons with energies greater than the band gap create free electrons and holes with excess kinetic energy ("hot carriers") that rapidly (ps to sub-ps time scales) lose their excess energy by electron-phonon scattering, converting the excess kinetic energy into heat. Subsequently, the "cold" free carriers occupy the lowest available energy levels (the bottom and top of the conduction and valence bands, respectively), where they can be removed to do electrical work in the external circuit or where they are lost through radiative or non-radiative recombination. In 1961, Shockley and Queisser[41] (S-Q) calculated the maximum possible thermodynamic efficiency of converting solar irradiance into electrical free energy in a PV cell assuming: (1) complete carrier cooling, (2) equilibrium between photogenerated electron and holes and the phonons they interact with, (3) the only other free energy loss mechanism is radiative recombination, and (4) sub-band gap photons are not absorbed. This detailed balance calculation (called the radiative limit) yields a maximum thermodynamic efficiency of 31–33% (depending upon the selected solar spectrum), corresponding to optimum band gaps between about 1.1 and 1.4 eV (the band gaps of Si and GaAs are close to these optimum values).

One way used currently to reduce energy loss due to carrier cooling is to stack a series of semiconductors with different band gaps in tandem with the largest band gap irradiated first followed by decreasing band gaps. In the limit

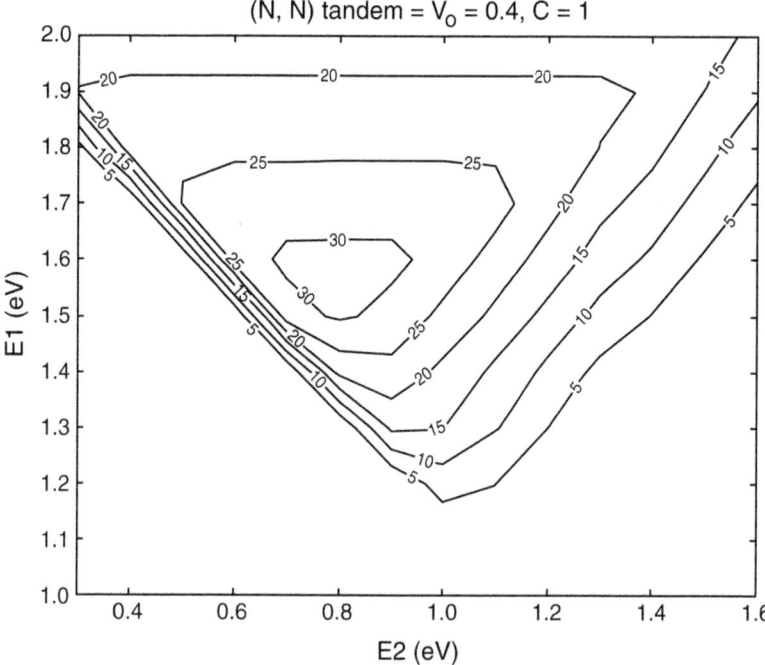

Figure 13.7   Contour plot of the possible PCEs as a function of the two bandgaps (E_1 and E_2) in a tandem cell with two conventional M1 photoelectrodes for an overvoltage $V_o = 0.4$ eV.

of a large number of different multiple band gaps matched to the solar spectrum, the conversion efficiency can reach 67% at one-sun intensity. However, only 2 to 3 band gaps are used in practice because most of the gain in efficiency for these multijunction PV cells is obtained with 3 band gaps; after that the returns diminish. Detailed balance calculations show that with 2 band gaps, the maximum efficiency is 43%, with 3 it is 48%, with 4 it is 52%, and with 5 it is 55%. The present record laboratory efficiency of a multijunction solar cell based on 3 junctions ($GaInP_2$/GaAs/Ge (or GaInAs)) is 43.5% under a solar concentrations of 418 suns.[46] The present highest one-sun efficiency of single junction PV cells in the laboratory is 28.3%.[47] The efficiency of commercial PV modules is about 75–80% of the maximum values measured in the laboratory.

In 1982, thermodynamic calculations[48] first showed that the same high conversion efficiency as a tandem stack of different band gaps can be obtained by utilizing the excess energy of hot photogenerated carriers before they cool to the lattice temperature through electron-phonon scattering; in the limit of a carrier temperature of 3000 K, the conversion efficiency reaches 67%. One way to achieve this is to transport the hot carriers to carrier-collecting contacts with appropriate work functions (either into an electrolyte redox system in a photoelectrochemical fuel-producing cell or to a solid state ohmic contact in a PV cell),[48–52] before the carriers cool. These cells are called hot carrier solar cells.[48–58]

A second approach is to use the excess kinetic energy of the hot carriers to produce additional electron–hole pairs (EHPs). In bulk semiconductors, this process is called impact ionization,[59–63] and it is an inverse Auger type of process. However, impact ionization (II) cannot contribute to improved quantum yields in current solar cells based on bulk Si, CdTe, $CuIn_xGa_{1-x}Se_2$, or III-V semiconductors because the QY for II does produce extra carriers (QYs > 100%) until photon energies reach the ultraviolet region of the spectrum, where almost no solar photons are present. For smaller band gap bulk semiconductors, where II could occur in the visible region, the photovoltages would be too small to make the PCE with II efficient for solar photon conversion. In bulk semiconductors, the threshold photon energy for II exceeds that required for energy conservation alone because crystal momentum (k) must also be conserved.[59] Additionally, the rate of II must compete with the rate of energy relaxation by phonon emission through electron-phonon scattering. It has been shown that the rate of II becomes competitive with phonon scattering rates only when the kinetic energy of the electron is many multiples of the band gap energy.[60–62] In bulk semiconductors, the observed transition between inefficient and efficient II occurs slowly; for example, the II efficiency in Si was found to be only 5% (*i.e.* total quantum yield = 105%) at $h\nu \approx 4\,eV$ ($3.6E_g$), and 25% at $h\nu \approx 4.8\,eV$ ($4.4E_g$).[63]

13.2.3.2 PCE Enhancement by Using Hot Excitons in Nanocrystals

Nanostructures of semiconductor materials exhibit quantization effects when the electronic carriers in these materials are confined by potential barriers to very small regions of space. The confinement can be in 1 dimension (quantum films, also termed 1-D quantum wells), 2 dimensions (quantum rods or wires), or in 3 dimensions (quantum dots: QDs). Nanostructures of crystalline materials are also referred to as nanocrystals; this term encompasses a variety of nanoscale shapes with the three types of spatial confinement, including spheres, cubes, rods, wires, tubes, tetrapods, ribbons, cups, discs, and platelets.[64]

As a consequence of quantum confinement of electrons and holes in nanostructures, the following new physics obtains:[64-66] (1) e^--h^+ pairs are correlated and thus exist as excitons rather than free carriers; (2) the rates of hot electron and hole (*i.e.* exciton) cooling can be slowed because of the formation of discrete electronic states; (3) momentum is not a good quantum number and thus the need to conserve crystal momentum is relaxed so that the minimum threshold energy of a photon to produce MEG (hv_{th}) is $2E_g$ which satisfies energy conservation only; and (4) Auger processes are greatly enhanced because of increased e^--h^+ Coulomb interaction. On the basis of these factors, it had been predicted[64-66] and confirmed experimentally [67-74] that the production of multiple e^--h^+ pairs (excitons) will be enhanced in QDs compared to the II process in bulk semiconductors. The threshold energy for electron–hole pair multiplication (EHPM) is greatly reduced, and the EHPM efficiency, η_{EHPM} (defined as the number of excitons produced per additional band gap of photon energy above the EHPM threshold energy) is greatly enhanced; this has been demonstrated in reference 74. The formation of multiple EHPs (excitons) in QDs is termed Multiple Exciton Generation (MEG),[67] and η_{EHPM} is termed η_{MEG}. Free carriers are formed upon the dissociation of the excitons, for example in various types of solar cells.

The possibility of enhanced MEG in QDs was first proposed by Nozik in 2001–2002.[65,66] The first experimental verification of exciton multiplication was reported by Schaller and Klimov[70] in 2004 for PbSe QDSs; they found an excitation energy threshold for the efficient formation of two excitons per photon at $3E_g$. Soon after, it was shown that efficient MEG also occurs in PbS, PbTe PbSe, CdSe, PbTe InAs, Si, InP, CdTe and CdSe/CdTe core-shell QDs (these results are reviewed in reference 64); the time scale for MEG was initially reported to be $< 100\,fs$.[67] Several reviews of recent work, including both experimental and theoretical MEG and carrier multiplication results have been published recently.[64,70-73] The reported ultrafast MEG rates are much faster than the hot exciton cooling rate produced by electron-phonon interactions, and MEG can therefore beat exciton cooling and become efficient.

Multi-excitons can be detected using several spectroscopic measurements. The first method used was to monitor the signature of multi-exciton generation using transient (pump-probe) absorption (TA) spectroscopy. In one type of TA experiment, the probe pulse monitors the interband bleach dynamics with excitation across the QD band gap; whereas in a second type of experiment, the probe pulse is in the mid-IR and monitors the intraband transitions (*e.g.* $1\,S_e$–$1\,P_e$) of the newly created excitons. Additional spectroscopic techniques use time-resolved photoluminescence and THz spectroscopy. The most direct measure of MEG is photocurrent QY in a QD-based solar cell device; in this approach, the electrons are counted rather than deriving their population from an analysis of spectroscopic data. A recent publication[75] has shown the photo-current measurement of MEG confirms earlier spectroscopic measurements when the latter experiment is done properly to eliminate the possible con-founding effects of surface photocharging and variations in surface chemistry evident in several earlier research publications.[71-76]

Regarding the application of MEG and exciton multiplication to solar water splitting, detailed balance S-Q type calculations have been made to determine the PCE with photoelectrodes that exhibit MEG.[42] The ideal MEG process produces N excitons (where N is a integer) when the photon energy is $N \times E_g$, this yields a threshold photon energy $h\nu_{th}$ for MEG of $2E_g$. This ideal case produces a staircase characteristic for a plot of quantum yield (QY) *vs.* photon energy divided by band gap $h\nu/E_g$, such that 2 EHPs are produced at $2E_g$, 3 EHPs at $3E_g$, and so forth (see Figure 13.8). However, most MEG results to date do not show a staircase characteristic but rather a linear dependence of QY on after $h\nu_{th}$ is passed, and for these linear characteristics a recent analysis[74] shows that $h\nu_{th}$ and η_{MEG} are related by the simple expression:

$$\frac{h\nu_{th}}{E_g} = 1 + \frac{1}{\eta_{MEG}} \tag{13.10}$$

For example, a threshold of $2E_g$ yields a MEG efficiency of 1, $3E_g$ an efficiency of 0.5, *etc.* For the linear characteristic, the slope of the QY *vs.* $h\nu/E_g$ plot is η_{MEG} and QY $= (h\nu/E_g - 1)\eta_{MEG}$. In the following presentation, we considered various MEG linear characteristics that are labeled $L(n)$ where L indicates the

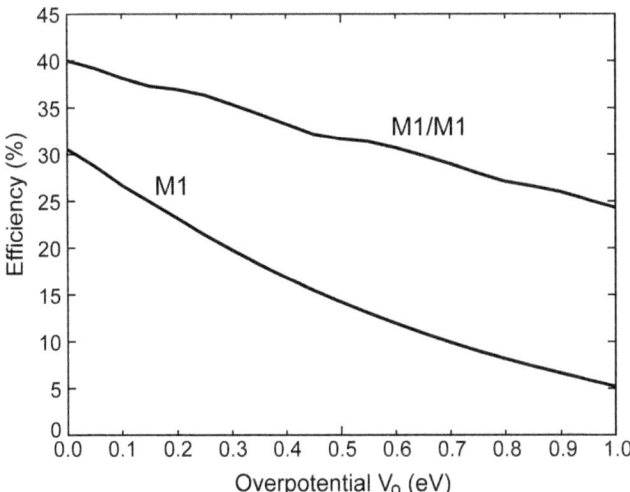

Figure 13.8 Maximum PCE values *vs.* overvoltage of a conventional photoelectro-lysis cell using a single M1 photoelectrode compared to a cell with two M1 photoelectrodes (M1/M1). The optimum band gaps for the two cases are shown in Figures 13.5 and 13.6 (from reference 42).

characteristic is a linear function (not the staircase function) and n indicates the value of $h\nu_{th}/E_g$.

The staircase function can be recovered from the linear function by considering the dynamics of hot exciton MEG rates, r_{MEG}, *vs.* cooling rates, r_{cool}. In this treatment,[74] the threshold photon energy is given by

$$h\nu_{th} = \left(2 + \frac{r_{cool}}{r_{MEG}^{2h\nu_{th}}}\right) E_g \qquad (13.11)$$

where $r_{MEG}^{2h\nu_{th}}$ is the MEG rate at $h\nu = 2h\nu_{th}$ and r_{cool} is the cooling rate (assumed to be independent of photon energy). In this treatment, cooling is defined as the process restricted to electron (or hole)-phonon scattering and hence heat; MEG also results in electrons or holes being relaxed to a thermalized, cold state, but this process is distinguished from cooling that involves phonons. As $r_{MEG}^{2h\nu_{th}}$ increases or r_{cool} decreases, $h\nu_{th}$ approaches the energy conservation limit of $2E_g$. In terms of the *e-h* pair creation energy,[71,74] ε_h (the excess energy required to produce one additional e^--h^+ pair) is given by $h\nu_{th} = E_g + \varepsilon_h$. The perfect staircase function can be obtained from the analysis when $r_{MEG}^{2h\nu_{th}} / r_{cool} > 10,000$; *i.e.* when the MEG rate is 10,000 times faster than the hot exciton cooling rate.[74]

Various possible MEG characteristics are shown in Figure 13.9; the staircase characteristic is labeled as M_{max}, and the linear characteristics are labeled $L(n)$ where n is the value of $h\nu_{th}/E_g$. The results of the S-Q thermodynamic PCE calculations at 1 sun made for various MEG characteristics are shown in Figure 13.10 and are reproduced from a previous analysis.[74] The maximum efficiency occurs for the staircase MEG characteristic (M_{max}), with a peak PCE of 44–45% for band gaps ranging from 0.7 to 1 eV. The $L2$ and $L2.5$ characteristics also produce significant PCE gains, even at band gaps below the optimum value, but when the threshold is $> 2.5E_g$ the PCE efficiency gain at 1 sun is marginal. Thus, for 1 sun applications it is critical to optimize the MEG process by approaching the M_{max} staircase characteristic as closely as possible.

13.2.3.3   PCE Enhancement Based on Molecular Chromophores Exhibiting Singlet Fission

Singlet Fission in molecular chromophores[77,78] is the analog of MEG in semiconductors. It produces two triplet excitonic states from a singlet state created by the absorption of a single photon that excites a transition from the S_0 ground state to the first excited S_1 state. In one variation of a solar cell based on SF, the molecular chromophores are bound to nanocrystalline TiO_2 particles that form a photoactive thin (20 – 30 μm) film as in a dye-sensitized solar cell (commonly called a "Grätzel cell").[79] The photoexcited molecules inject electrons into the conduction band of the TiO_2 nanocrystalline film to produce charge separation and enable extraction of electrical power. The requirement for SF in molecules is that the lowest triplet state energy (T_1) needs to be just

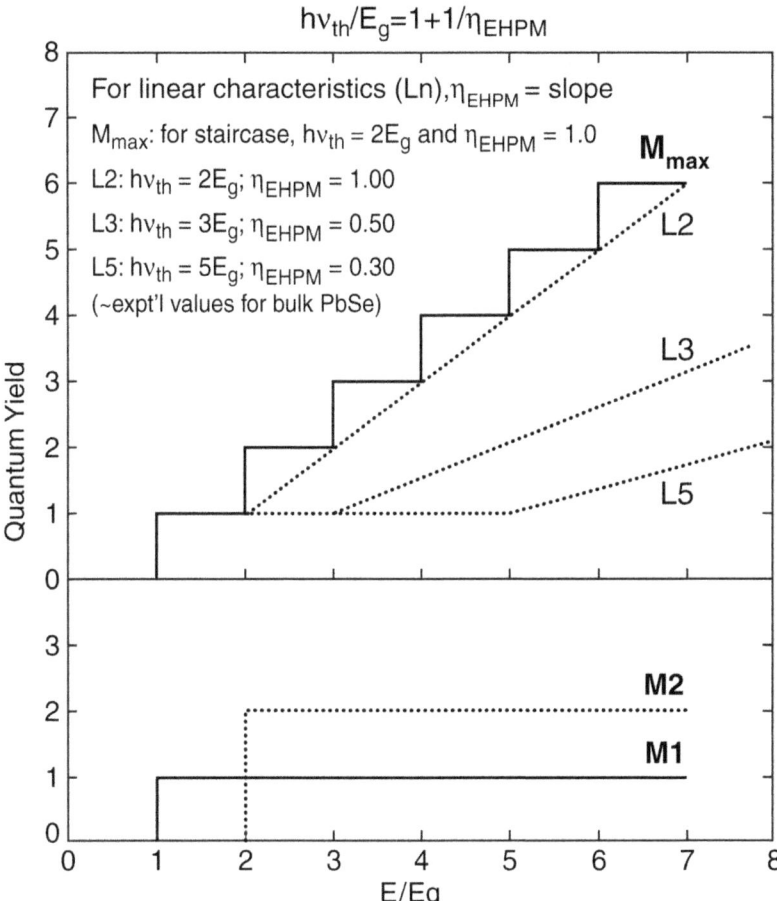

Figure 13.9 Possible characteristics of QY *vs.* absorbed photon energy (normalized to the band gap of the cell) for various MEG and SF characteristics. The efficiency of MEG is related to the normalized threshold photon energy and for the linear (*L*(n)) cases is the slope of this plot. The MEG efficiency is maximized with a threshold photon energy of $2E_g$ (from reference 74).

slightly less than one-half of the $S_1 - S_0$ energy (to prevent triplet-triplet annihilation, twice the T_1 energy needs to less than the S_1 energy); the required energy level arrangement is shown in Figure 13.11. To achieve SF, two of the molecular chromophores need to be coupled since the two triplets must be formed on separate molecules; thus, the net spin state of the two coupled molecules is a singlet, allowing the formation of 2 triplets from a singlet state. The "band gap" for a SF molecule is defined as $T_1 - S_0$ because the electrons are extracted from T_1 and the holes from S_0. Furthermore, since the direct transition from S_0 to T_1 is forbidden, a second chromophore has to be placed in optical tandem behind the first molecular chromophore to absorb the lower energy photons not absorbed from the $S_0 - S_1$ transition.

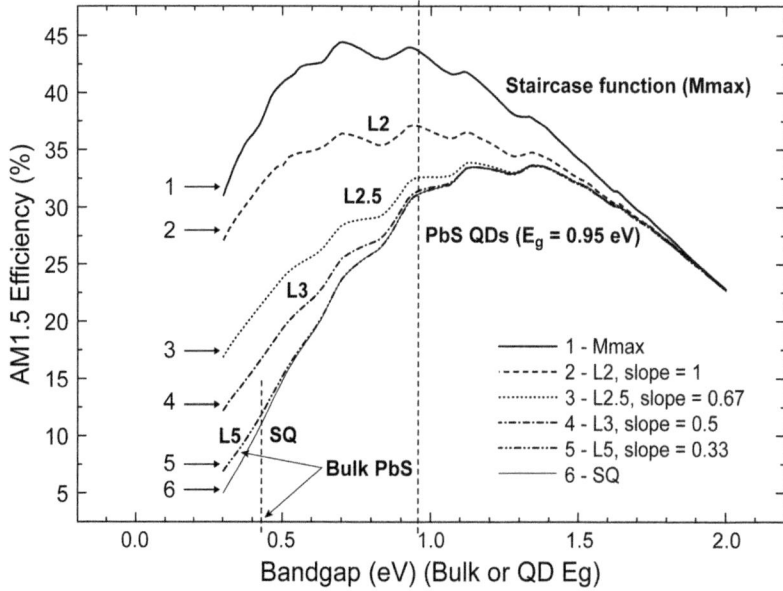

Figure 13.10 S-Q type thermodynamic calculations of the PCEs taking into account various MEG characteristics and for conventional solar cells that do not exhibit MEG (from reference 74).

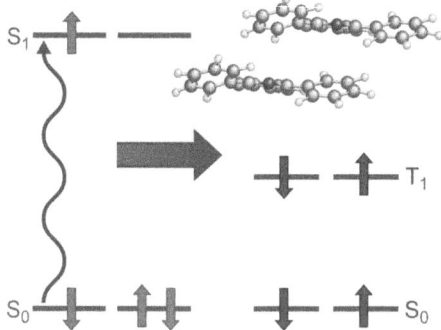

Figure 13.11 Electronic energy level scheme for molecules that can exhibit singlet fission (SF). The triplet state must be just slightly less than half the energy of the S_1 state (to inhibit triplet-triplet annihilation) and two identical chromophores must be optimally coupled to allow two triplets to be formed on each chromophore (with total spin of zero) (from reference 78).

Two schemes for SF-based solar cells are shown in Figure 13.12. In Figure 13.12a, the SF molecule (C1) is bound to nanocrystalline TiO_2 and receives the input solar flux. The 2nd chromophore (C2) produces just one EHP per photon, and it could be a small band gap semiconductor or a nanocrystalline support sensitized with either QDs or dye molecules of the appropriate

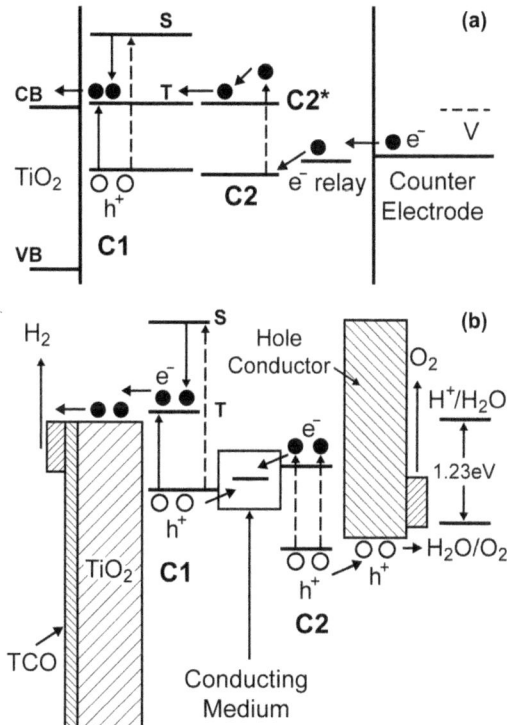

Figure 13.12 Two possible schemes for incorporating SF molecules into photoelectrolysis cells. In both cases two chromophores (C1 and C2) with different HOMO-LUMO values are needed for maximum PCE. In (a) the two chromophores are illuminated in series but are connected electrically in parallel; in (b) the two chromophores are illuminated in series and their electrical connections is also in series. The (b) scheme is analogous to the Z scheme of biological photosynthesis and the photocurrent in C1 and C2 must be equal for maximum PCE. In both cases, the C1 HOMO-LUMO (or band gap) is larger than C2, and solar photons not absorbed in C1 are absorbed in C2 (from reference 42).

band gap. A distinguishing feature of the architecture of Figure 13.12a is that while the optical absorption is in series, the photocurrents from C1 and C2 flow in parallel and need not be equal, as is required for all other tandem cells; this is a potential advantage for the cell engineering and cost. The alignment of the first excited state of C2 with T_1 and the ground state of C2 with that of C1, as shown in Figure 13.12a, produces the optimum PCE, but this is not a rigid requirement for cell operation as long as the flow of electrons and holes is energetically allowed; non-alignment only affects the upper limits of the PCE. However, the maximum voltage obtained from this cell cannot be great than one-half of S_1 - S_0, and for H_2O splitting this voltage must be at least $1.23 + V_0$. For this architecture, the maximum PCE for H_2O splitting in an ideal cell (*i.e.* zero V_0) is 33%; with a V_0 of 0.25 V it is 22%. As shown below, these theoretical PCE values are lower than those obtained by arranging the

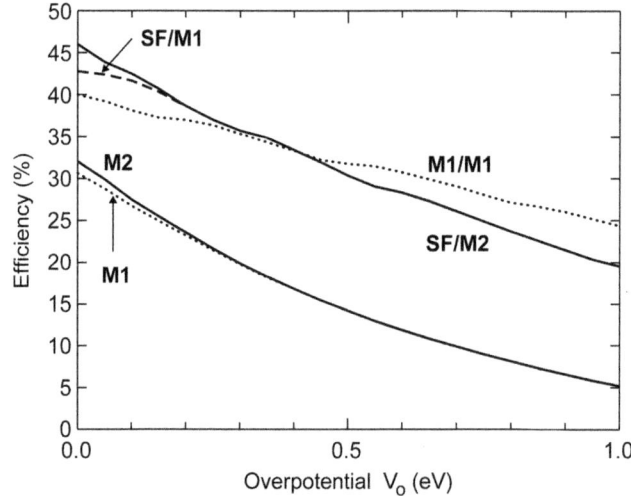

Figure 13.13   For a plot like that in Figure 13.5, the PCE values *vs.* V_0 are also added for the tandem cells SF/M1 and SF/M2, where M2 produces two excitons/photon. This plot shows adding SF to a Z-scheme type tandem architecture only improves efficiency for very low values of V_0 (from reference 42).

2 chromophores in the tandem structure shown in Figure 13.12b, but the design is simpler. In Figure 13.12b, C(1) and C(2) are arranged in a Z-scheme as discussed above. As a result, the photopotentials of C1 and C2 are additive, and the photocurrents from C1 and C2 must be equal. The PCE values for this architecture are higher than the values for the parallel connection shown in Figure 13.12a for all values of V_0.

   Figure 13.13 shows a plot like Figure 11.8 but with the results added for the tandem cell based on SF in the series connection option (Figure 13.12b). It can be seen from Figure 13.13 that for H_2O splitting, while the SF system is much better than a single photoelectrode (M1), its use of the two chromophore architectures of Figure 13.12b only improves the PCE over the M1 photo-materials (that produce just 1 EHP/photon in a Z scheme) from 40% to 46% at $V_0 = 0$. Furthermore, this advantage decreases with increasing V_0, becoming a disadvantage at $V_0 > 0.40$ V.

13.2.3.4   *Effect of Solar Concentration Coupled with Carrier Multiplication*

S-Q analyses of MEG or SF that is with combined with solar concentration show that the PCE for PV cells increases dramatically with increasing con-centration.[80] The effect is greatest for PV cells where the band gap is allowed to vary with concentration; the band gaps that yield the highest PCE shift to much smaller values. Thus, for example, the PCE of a single junction PV cell that exhibits nearly ideal MEG (L2) can increase to 65% for a quantized band gap

Table 13.1 PCE and Optimum E_gs for Tandem Cells with Different Configurations and Parameters.

Tandem Cell Configuration	Overvoltage V_0 (V)	E_1 or $T_1 - S_0$ (eV)	$S_1 - S_0$ (eV)	E_2 (eV)	Solar Concentration	Electrical Connection	PCE (%)
M1/M1	0.0	1.40 (E1)	N/A	0.50	1	Series	40
M1/M1	0.4	1.55 (E1)	N/A	0.75	1	Series	33
L2/L2	0.0	1.20 (E1)	N/A	0.30	500	Series	50
L2/L2	0.4	1.45 (E1)	N/A	0.55	500	Series	40
SF/M1	0.0	0.90	1.8	0.90	1	Series	42
SF/M1	0.4	1.0	2.0	1.1	1	Series	33
SF/M1	0.0	0.88	1.8	0.58	500	Series	50
SF/M1	0.4	0.95	1.9	1.0	500	Series	38
SF/M1	0.0	1.4	2.8	1.4	500	Parallel	42
SF/M1	0.4	1.8	3.6	1.9	500	Parallel	22

of 0.4 eV under a concentration of 500 suns, while a conventional PV cell at 500 suns will have a maximum possible efficiency of 38%. However, for H_2O splitting, the need to provide a large photovoltage ($1.23\,V + V_0$) lowers the maximum, PCE. Table 13.1 tabulates the PCE results for cells having overvoltages of 0.0 V and 0.4 V with concentrations of 1 sun and 500 suns and for different tandem configurations with different MEG characteristics for the photoelectrodes. M1 is the usual 1 exciton/photon; L2 is the linear MEG characteristic with a threshold of $2E_g$, and SF is the sensitized nanocrystal singlet fission system with SF chromophores; for SF systems, results for both series and parallel connections of the two photoelectrodes are presented.

With solar concentration, the total cell photocurrent will be increased linearly with the concentration factor. To avoid exceptionally large overvoltage losses at high current densities, it would be necessary to transport the charge carriers to large area electrodes to conduct the oxidation/reduction reactions of H_2O splitting (see Figure 13.14 for a possible design of such a cell under solar concentration). If the electrode area is increased by the same factor as the solar concentration, then it could be possible to limit the total cell overvoltage to 0.4 eV (for 1 sun intensity, the cell overvoltage is $< 0.25\,V^{80}$). Higher overvoltages will reduce the efficiencies in Table 1, but the conditions and values presented there are nevertheless useful for comparing the benefits of solar concentration to no concentration. Thus, Table1 shows that the best benefit for a tandem cell with $V_0 = 0.4\,V$ and a concentration of 500× concentration would be obtained from an SF/M1 configuration (38% maximum PCE) and an L2/L2 configuration (40% maximum PCE); for a conventional M1/M1 tandem cell at 500× concentration, the maximum PCE is 33%.

13.3 Buried Junctions and Other Novel Approaches

13.3.1 Buried p-n Junctions in a Tandem Cell

Two vital advantages of solar water-splitting and fuel-producing cells with buried junctions are: (1) the photoactive materials are appropriately

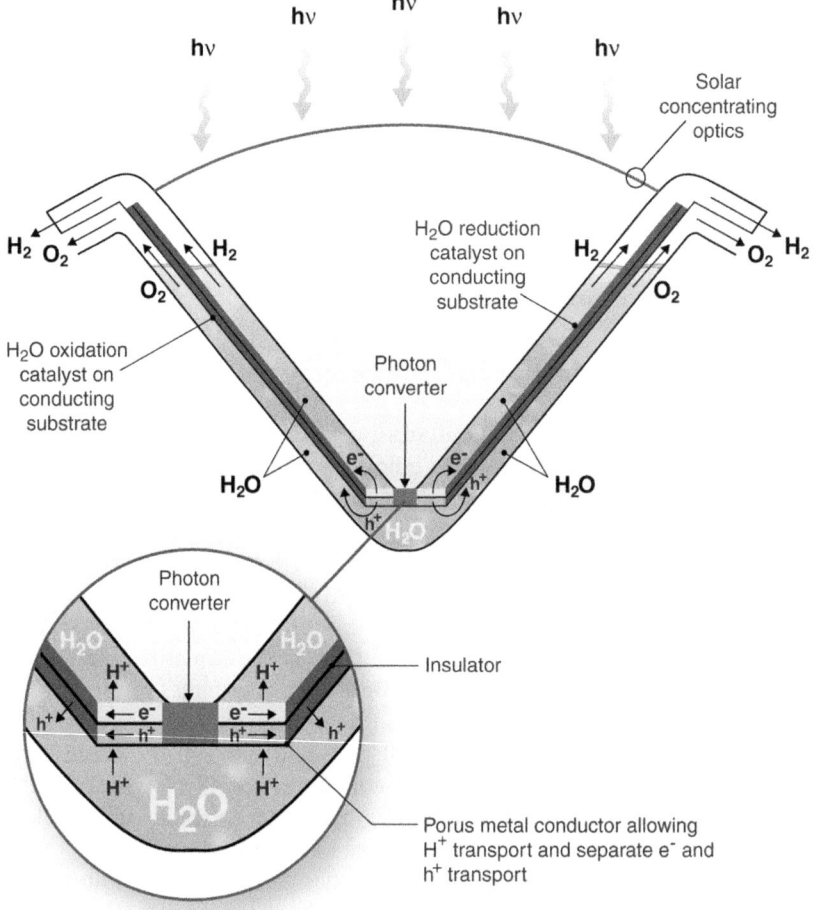

Figure 13.14 Schematic of a photoelectrolysis cell that shows MEG or SF under solar concentration where the photocurrent is distributed over a large dark electrode surface area to reduce the photocurrent density and hence the corresponding V_o in order to take advantage of the great increase in PCE possible when MEG or SF is combined with solar concentration.

encapsulated and hence are not in direct contact with the aqueous electrolyte so that photocorrosion is avoided; and (2) the electrocatalytic surfaces are not those of the photoactive and charge-separating materials, but rather the photo-generated current is transferred to optimized catalytic, dark metal electrode surfaces. This later feature means that the energy levels of the electrons and holes in the photomaterial (set by the flatband potential of the semiconductor electrodes) do not need to be appropriately aligned with the redox potentials of the two oxidation and reduction half-reactions of water splitting or CO_2 reduction, as is required in pure photoelectrochemical (PEC) systems (*i.e.* those with semiconductor/aqueous electrolyte interfaces). Provided that a sufficient

photovoltage is generated in the photomaterial to electrolyze water, the electrode potentials of the catalytic surfaces in electrical contact with the photomaterial adjust automatically and instantaneously through the redistribution of
charge in the Helmholtz layers at the catalytic anode and cathode surfaces; this
is exactly what happens in the case of metal electrodes during the dark electrolysis of H_2O or during CO_2 reduction to fuel.

Figure 13.15 shows an energy band diagram for two buried p-n junctions in
series in a tandem cell for water splitting. The photoactive semiconductor
materials are encapsulated to isolate them from the aqueous solution and thus
prevent photocorrosion. A transparent window is used for either the anode or
cathode side of the cell; either the larger or smaller band gap can be the anode
or cathode depending upon the specific materials of the junctions. The photocurrents generated in each p-n junction must be equal to balance charge and
optimize efficiency; this is done by adjusting the absorber thicknesses depending
upon their absorption coefficients. Following absorption of two photons, one
in each junction, one electron in injected into the electrolyte from the n-type
region of the smaller band gap p-n junction to drive the reduction reaction, and
one hole is injected from the p-type region of the larger band gap p-n junction
to drive the oxidation reaction. Thus, in the buried 2-junction configuration
shown in Figure 13.15, the device acts as a majority carrier device. This is the
opposite to the conventional PEC structure, where the semiconductor-liquid
junction results in electron injection from the p-type semiconductor and the

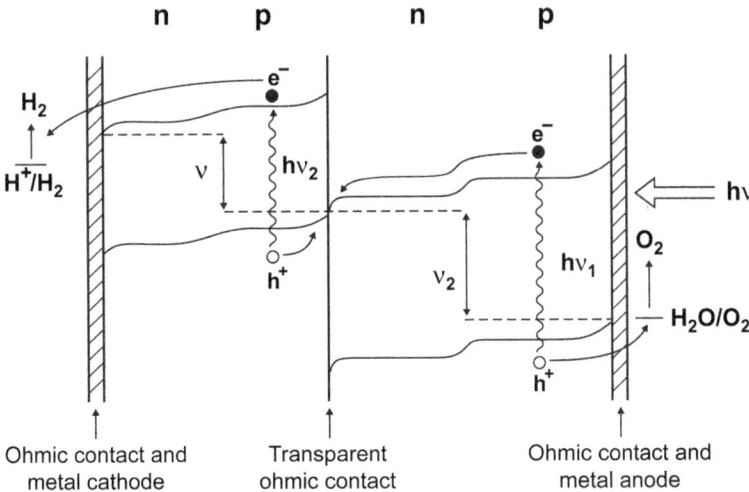

Figure 13.15   Energy level scheme for a tandem photoelectrolysis cell based on two
buried p-n homojunctions with different band gaps (their PCE values
can be determined from Figure 13.5). Photoelectrolysis cells with buried
junction do not suffer photocorrosion and eliminate the need to match
flatband potentials with the redox potentials of water oxidation-
reaction reactions (from reference 25).

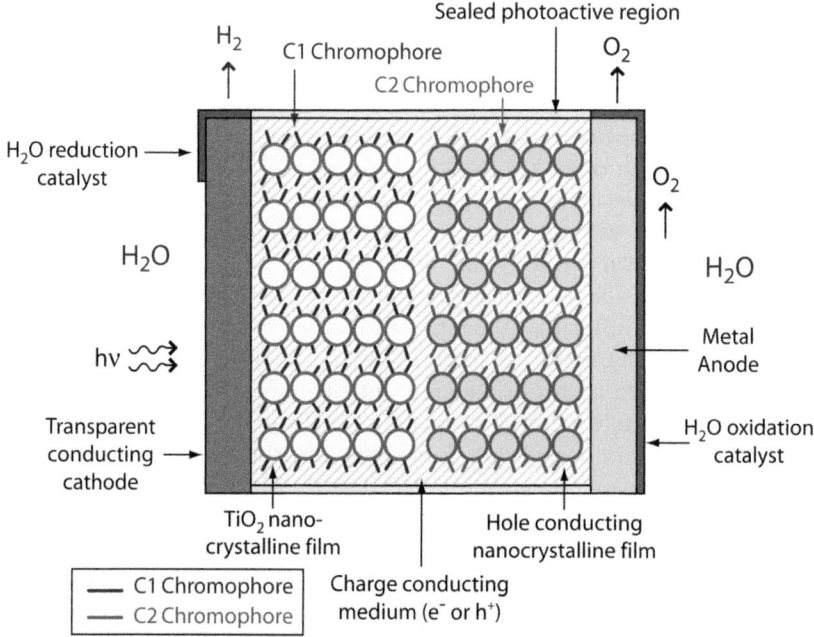

Figure 13.16   A schematic diagram of a tandem photoelectrolysis cell with buried junctions that is based on dye or QD or SF-sensitized nanocrystalline films, as in a dye-sensitized (Grätzel) cell.

hole is injected from the n-type semiconductor, making the PEC cell a minority carrier device (see Figures. 13.2 and 13.4).

The remaining electron and hole in the two p-n junction regions of the buried cell structure of Figure 13.15 recombine at the interface between the two p-n junctions *via* a transparent ohmic contact, which can be a tunnel junction produced by thin layers of heavily doped n-and p-type layers in the adjacent n- and p-type semiconductors, thereby maintaining charge balance in the device. Figure 13.16 shows a schematic of a buried junction tandem cell based on the dye or QD-sensitized nanocrystalline solar cell. A record 12.4% for water-splitting PCE in a tandem structure with half-buried junctions was reported by Khaselev and Turner;[81] the cell consisted of a GaAs p-n junction buried behind a p-type $GaInP_2$ semiconductor/aqueous electrolyte junction. Long-term photoinstability of the p-$GaInP_2$ PEC electrode is a problem with this system.

13.3.2   Buried p-n Junctions in a Texas Instruments System for H_2O Splitting

Another approach to H_2O splitting using buried junctions is a variation of a previously successful Texas Instruments (TI) system for splitting liquid HBr into H_2 and Br_2 that was based on 250 μm crystalline Si spheres processed into 250 micron-sized core-shell p-n junctions.[82] The spheres were produced by

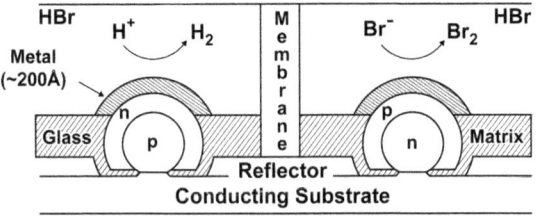

Si spheres: 10-15 mil diameter
2-3 radii apart

Figure 13.17   A schematic of the buried junction configuration of the Texas Instruments (TI) approach (late 1970 s) to solar H_2 production based on p-n
Si microspheres that have two Si p-n junctions between two microspheres (one having an n-core and p-shell and the other a p-core and
n-shell) coupled in series; a multitude of microsphere pairs operate in
parallel between two large scale panels that are spatially separated but
electrically connected by a common metal support to carry out the
oxidation and reduction reactions in separate panels. Failure of one or
more of the microshperes does not affect the operation of the system.
HBr is used as the electrolyte.

simply spraying molten Si through an atomizing nozzle to produce droplets
that subsequently solidified into small spheres. The cell structure is shown in
Figure 13.17. Each spherical micron-sized p-n junction is buried by metal and
glass coatings to protect the Si against photocorrosion; the metallic coating at
the Si-HBr interface is thin layer of Pt (20 nm). Two panels consisting of a high
density of microspheres consisting of n-Si cores with p-Si shell were dispersed
on one panel and microspheres of p-Si cores with n-Si shells were dispersed on a
second panel. The core-shell structures were processed such that all cores were
contacted by a common metal support, thus producing two Si p-n junctions in
series. Under solar illumination, H_2 was evolved at the panel with p-cores and
n-shells, and Br_2 was evolved at the panel with n-cores and p-shells. The PCE of
this system was 7–10%. One difference between the buried junctions of
Figure 13.15 and Figure 13.18 is that the p-n junctions are illuminated in
tandem in the former and in parallel in the latter; thus, the former has a
higher PCE.

A H_2O splitting system based on the general TI concept could also consist of
core-shell Si microspheres prepared both with n-type cores and p-type shells
and vice-versa, forming microsized p-n junctions. Such a possible H_2O splitting
cell is shown in Figure 13.18. It would consist of four panels illuminated simultaneously in parallel, with two panels consisting of a random distribution of
n-on-p microspheres and two panels consisting of p-on-n microspheres. The
two types of panels containing the two types of microspheres in each panel
would alternate side by side with each other. The microspheres would be embedded in an insulating matrix (*e.g.* SiO_2 or plastic), but the top and bottom
sections of the microspheres would be exposed to allow processing for the
specific types of electrical and electrochemical contacts shown in the figure.
The density of the microspheres in the panels would be about $2000/cm^2$.

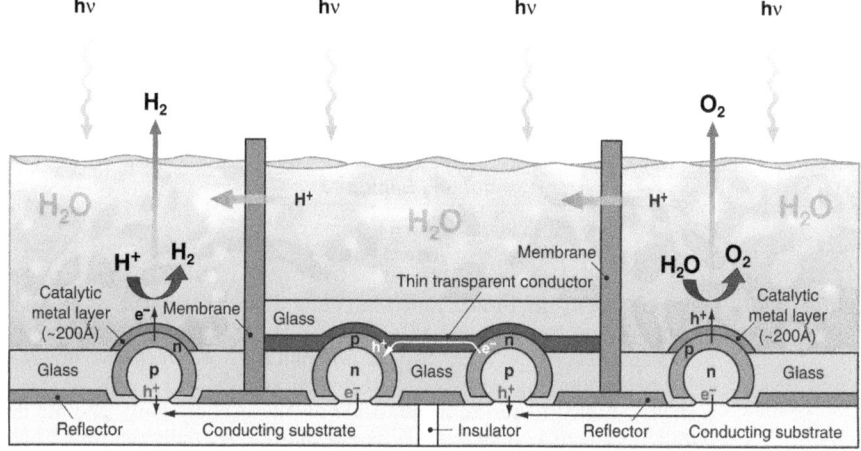

Figure 13.18   Modification of the TI approach to an H_2O splitting system by using 4 p-n Si microspheres in electrical series as shown.

The proposed configuration in Figure 13.18 would result in the effective formation of 4 Si p-n junctions through the series connection of 4 core-shell Si microspheres, giving a photovoltage of about 2 V, which will be sufficient to split H_2O at high rates. Each of the 4 p-n junctions created from 4 microspheres would be independent and operating in parallel; this means failure of any sphere(s) in the system would not affect any of the remaining microspheres. Two panels with planar p-n junctions connected in series can be substituted for the middle two panels consisting of microspheres if warranted by cost and engineering considerations. Further simplifications and modifications could substitute Si microcubes for microspheres, produced by patterning of planar Si p-n junctions.[83]

13.3.3   Buried p-n Junctions using Nanocrystals (Quantum Dots, Quantum Wires, and Nanorods)

Finally, a new approach to highly efficient H_2O-splitting cells with buried junction involves taking advantage of quantization effects in semiconductors to modify their band gaps to produce a double p-n junction tandem cell derived from just one semiconductor photomaterial. The band gaps of a given semiconductor nanocrystal can be controlled *via* their size to produce two different band gaps of optimum value (as shown in Section 13.2.3.2) that are arranged in a tandem cell. The two nanocrystalline p-n junctions can be buried and protected from the aqueous electrolyte as discussed in Section 13.3. The principle of this approach is shown in Figure 13.19 for two nanocrystalline p-n junctions made from quantum dots arrays arranged in optical and electrical series. Other variations are possible: (1) nanorods (with diameters not in the quantization regime) containing two axial or radial p-n junctions of different materials with optimum band gaps can be arranged optically in parallel or in tandem; (2) quantum wires of the same semiconductor but with different diameters to

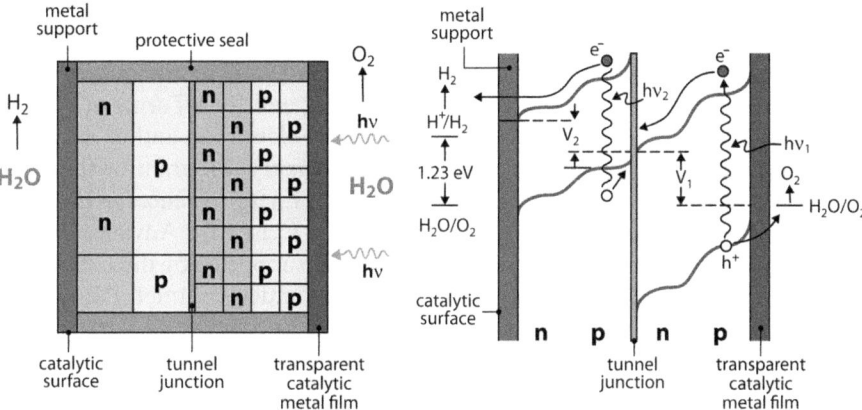

Figure 13.19 A photoelectrolysis cell using two buried junctions based on nanocrys-talline arrays of n-and p-type nanocrystals that form nanocrystalline p-n junctions. The semiconductors are the same material but of differ-ent size in their size-quantization regime; they thus have two different band gaps which can be optimized according to the required values of Figure 5 by controlling their nanocrystal size.

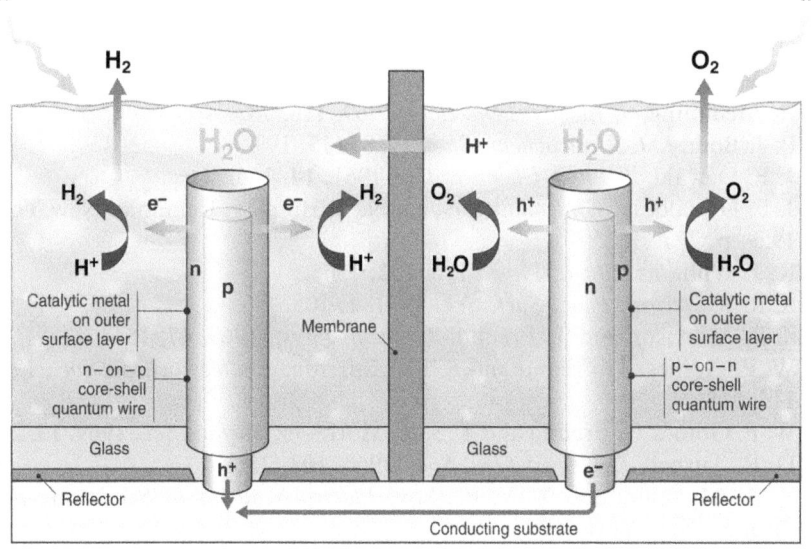

Figure 13.20 A buried junction photoelectrolysis cell based on quantum wires with axial p-n junctions with different band gaps due to size quantization. The arrangement is analogous to the initial TI system except the cathode and anode consist of arrays of quantum wires with n-shells and p-cores and p-shells and n-cores, respectively.

control their band gaps can also be arranged with two axial or radial p-n junctions optically in series or in parallel. Figure 13.20 shows a quantum wire system arranged optically in parallel and electrically in series. It is a variation of the TI approach discussed above in Section 13.3.2.

Acknowledgements

Very critical contributions to the research reviewed and reported here were made by Mark Hanna, Matt Beard, Justin Johnson, and John Turner at NREL and Josef Michl at the University of Colorado, Boulder. The author acknowledges early support from the Solar Photochemistry Program within the Division of Chemical Sciences in the Office of Basic Energy Sciences of the U.S. Department of Energy, and more recently from the Center for Advanced Solar Photophysics, an Energy Frontier Research Center supported by the U.S. DOE Office of Science. DOE funding to NREL was under contract DE-AC36-086038308.

References

1. A. Fujishima and K. Honda, *Nature*, 1972, **238**, 37.
2. W. H. Brattain and C. G. B. Garrett, *Bell Syst. Tech. J.*, 1955, **34**, 129.
3. H. Gerischer, *Z. Phys. Chem.*, 1960, **26**, 223.
4. H. Gerischer, *Z. Phys. Chem.*, 1961, **27**, 48.
5. H. Gerischer, in *Advances in Electrochemistry and Electrochemical Engineering*, Vol. 1, ed. P. Delahay, Interscience, New York, 1961, p. 39.
6. H. Gerischer, in *Physical Chemistry: An Advanced Treatise*; Vol 9 A, ed. H. Eyring, D. Henderson and W. Jost, Academic, New York, 1970, p. 463.
7. R. Memming and G. Schwandt, *Electrochim. Acta*, 1968, **13**, 129.
8. R. Memming, *J. Electrochem. Soc.*, 1969, **116**, 785.
9. P. J. Boddy, *J. Electrochem. Soc.*, 1968, **115**, 199.
10. J. F. Dewald, *J. Phys. Chem. Solids*, 1961, **14**, 155.
11. J. F. Dewald, in *Semiconductors*, ed. N. B. Hannay, Reinhold: New York, 1959, p. 727.
12. R. J. Williams, *Chem. Phys.*, 1960, **32**, 1505.
13. S. R. Morrison, *Prog. Surf. Sci.*, 1971, **1**, 105.
14. S. R. Morrison and T. Freund, *J. Chem. Phys.*, 1967, **47**, 1543.
15. W. P. Gomes, T. Freund and S. R. Morrison, *J. Electrochem. Soc.*, 1968, **115**, 818.
16. W. P. Gomes, T. Freund and T. S. R. Morrison, *Surface Sci.*, 1969, **13**, 201.
17. D. R. Turner, *J. Electrochem. Soc.*, 1956, **103**, 252.
18. V. A. Myamlin, and Y. V. Pleskov, *Electrochemistry of Semiconductors*, Plenum, New York, 1967.
19. A. K. Vijh, *Electrochemistry of Metals and Semiconductors*, Dekker, New York, 1973.
20. M. Green, M, in *Modern Aspects of Electrochemistry*, ed. O'M. Bockris, Butterworth, London, 1959, p. 343.
21. A. J. Nozik, *Ann. Rev. Phys. Chem.*, 1978, **29**, 189.
22. A. J. Nozik, *Philos. Trans. R. Soc. London*, 1980, **5**, 453.
23. A. J. Nozik, *Faraday Discussions of the Royal Society of Chemistry*, 1980, **No. 70,** 7.
24. F. Williams and A. J. Nozik, *Nature*, 1984, **312**, 21.
25. A. J. Nozik and R. Memming, *J. Phys. Chem.*, 1996, **100**, 13061.

26. M. D. Archer and A. J. Nozik, *Nanostructured and Photoelectrochemical Systems for Solar Photon Conversion*, Imperial College Press, London, 2008.
27. R. J. D. Miller, G. McLendon, W. Schmickler, A. J. Nozik and F. Willig, *Surface Electron Transfer Processes*, VCH Publishers, Weinheim, Germany, 1995.
28. R. Memming, *Semiconductor Electrochemistry*, Wiley-VCH, Weinheim, 2001.
29. Y. V. Pleskov, *Semiconductor Electrochemistry*, Consultants Bureau, New York, 1986.
30. A. J. Bard and M. A. Fox, *Acc. Chem. Res.*, 1995, **28**, 141.
31. C. A. Koval and Howard, *J. N. Chem. Rev.*, 1992, **92**, 411.
32. N. S. Lewis, *Acc. Chem. Res.*, 1990, **23**, 176.
33. K. Rajeshwar, *J. Appl. Electrochem.*, 1985, **15**, 1.
34. A. J. Nozik, *J. Cryst. Growth*, 1977, **39**, 299.
35. A. J. Nozik, *Nature*, 1975, **257**, 383.
36. A. J. Nozik, *Appl. Phys. Lett.*, 1976, **29**, 150.
37. A. J. Nozik, *Appl. Phys. Lett.*, 1977, **30**, 567.
38. M. G. Walter, E. L. Warren, J. R. McKone, S. W. Boettcher, Q. Mi, S. A. Stanton and N. S. Lewis, *Chem. Rev.*, 2010, **110**, 6446.
39. C. Chen, S. Shen, L. Guo and S. S. Mao, *Chem. Rev.*, 2010, **110**, 6503.
40. W.-J. Yin, H. Tang, S.-H. Wei, M. M. Al-Jassim, J. Turner and Y. Yan, *Phys. Rev.*, 2010, **B82**, 45106.
41. W. Shockley and H. J. Queisser, *J. Appl. Phys.*, 1961, **32**, 510.
42. M. Hanna and A. J. Nozik, *J. Appl. Phys.*, 2006, **100**, 74510.
43. ASTM G-173-3 Reference Solar Spectrum in http://rredc.nrel.gov/solar/spectra/am1.5/.
44. W. Spirkl and H. Ries, *Phys. Rev.*, 1995, **B 52**, 11319.
45. R. E. Blankenship, D. M. Tiede, J. Barber, G. W. Brudvig, G. Fleming, M. R. Gunner, W. Junge, D. M. Kramer, A. Melis, T. A. Moore, C. C. Moser, D. G. Nocera, A. J. Nozik, D. R. Ort, W. W. Parson, R. C. Prince and R. T. Sayre, *Science*, 2011, **332**, 805.
46. M. A. Green, K. Emery, Y. Hishikawa, W. Warla and E. D. Dunkop, *Prog. Photovolt. Res. Appl.: Solar Cell Efficiency Tables (V. 39)*, 2012, **20**, 15.
47. O. D. Miller, E. Yablonovitch and S. R. Kurtz, *IEEE J. Photovoltaics*, 2012, **2**, 303.
48. R. T. Ross and A. J. Nozik, *J. Appl. Phys.*, 1982, **53**, 3813.
49. F. Williams and A. J. Nozik, *Nature*, 1978, **271**, 137.
50. D. S. Boudreaux, F. Williams and A. J. Nozik, *J. Appl. Phys.*, 1980, **51**, 2158.
51. M. C. Hanna, Z. Lu, and A. J. Nozik in *Future Generation Photovoltaic Technologies,* ed. R. D. McConnell, AIP Conference Proceedings No. 404; American Institute of Physics: Woodbury, NY, 1997, p. 309.
52. G. Cooper, J. A. Turner, B. A. Parkinson and A. J. Nozik, *J. Appl. Phys.*, 1983, **54**, 6463.
53. M. A. Green, *Third Generation Photovoltaics*, Bridge Printery, Sydney, 2001.
54. A. Marti and A. Luque, *Next Generation Photovoltaics: High Efficiency through Full Spectrum Utilization*, Institute of Physics, *Bristol*, 2003.

55. D. König, K. Casalenuovo, Y. Takeda, G. Conibeer, J. F. Guillemoles, R. Patterson, L. M. Huang and M. A. Green, *Physica E*, 2010, **42**, 2862.
56. Y. Takeda, T. Ito, T. Motohiro, D. König, S. Shrestha and G. Conibeer, *J. App. Phys.*, 2009, **105**, 74905.
57. G. J. Conibeer, D. König, M. A. Green and J. F. Guillemoles, *Thin Sol. Films*, 2008, **516**, 6948.
58. P. Würfel, A. S. Brown, T. E. Humphrey and M. A. Green, *Prog. Photovol. Res. Appl.*, 2005, **13**, 277.
59. M. Wolf, R. Brendel, J. H. Werner and H. J. Queisser, *J. Appl. Phys.*, 1998, **83**, 4213.
60. K. J. Bude and K. Hess, *J. Appl. Phys.*, 1992, **72**, 3554.
61. H. K. Jung, K. Taniguchi and C. Hamaguchi, *J. Appl. Phys.*, 1996, **79**, 2473.
62. D. Harrison, R. A. Abram and S. Brand, *J. Appl. Phys.*, 1999, **85**, 8186.
63. O. J. Christensen, *Appl. Phys.*, 1976, **47**, 689.
64. A. J. Nozik, M. C. Beard, J. M. Luther, M. Law, R. J. Ellingson and J. C. Johnson, *Chem. Rev.*, 2010, **110**, 6873.
65. A. J. Nozik, *Ann. Rev. Phys. Chem.*, 2001, **52**, 193.
66. A. J. Nozik, *Physica E*, 2002, **14**, 115.
67. R. J. Ellingson, M. C. Beard, J. C. Johnson, P. Yu, O. I. Mićić, A. J. Nozik, A. Shabaev and A. L. Efros, *Nano Lett.*, 2005, **5**, 865.
68. A. Shabaev, A. L. Efros and A. J. Nozik, *Nano Lett.*, 2006, **6**, 2856.
69. A. J. Nozik, *Nano Lett.*, 2010, **10**, 2735.
70. R. Schaller and V. Klimov, *Phys. Rev. Lett.*, 2004, **92**, 186601.
71. V. I. Klimov, *Nanocrystal Quantum Dots*, 2nd Edition, CRC Press, Taylor and Francis, Boca Raton, FL, 2010.
72. M. C. Beard and R. J. Ellingson, *Laser Photonics Rev.*, 2008, **2**, 377.
73. H. W. Hillhouse and M. C. Beard, *Curr. Opin. Colloid Interface Sci*, 2009, **14**, 245.
74. M. C. Beard, A. G. Midgett, M. C. Hanna, J. M. Luther, B. K. Hughes and A. J. Nozik, *Nano Lett.*, 2010, **10**, 3019.
75. O. E. Semonin, J. M. Luther, S. Choi, H.-Y. Chem, J. Gao, A. J. Nozik and M. Beard, *Science*, 2011, **334**, 1530.
76. A. G. Midgett, H. W. Hillhouse, B. S. Hughes, A. J. Nozik and M. C. Beard, *J. Phys. Chem. C*, 2010, **114**, 6873.
77. M. B. Smith and J. Michl, *Chem. Rev.*, 2010, **110**, 6891; and *Ann. Rev. Phys. Chem.*, 2013, **64**, 361.
78. J. C. Johnson, A. J. Nozik, J. Michl, *Acc. Chem. Res.*, 2013, **46**, 1290.
79. B. O'Regan and M. Graetzel, *Nature*, 1991, **353**, 737.
80. M. C. Hanna, M. C. Beard and A. J. Nozik, *J. Phys. Chem. Lett.*, 2012, **3**, 2857.
81. O. Khaselev and J. A. Turner, *Science*, 1998, **280**, 425.
82. E. L. Johnson, Development of a Solar Energy System. National Technical Information Service Report No. DE83014236; Texas Instruments Incorporated, Dallas, 1983.
83. F. Toor, T. G. Deutsch, J. W. Pankow, W. Nemeth, A. J. Nozik and H. M. Branz, *J. Phys. Chem C*, 2012, **116**, 19262.

CHAPTER 14

Light Harvesting Strategies Inspired by Nature

EVGENY OSTRUMOV, CHANELLE JUMPER AND
GREGORY SCHOLES*

University of Toronto, Department of Chemistry, 80 St. George Street,
M5S3H6, Ontario, Toronto, Canada
*Email: greg.scholes@utoronto.ca

14.1 Introduction

The primary processes of photochemical reactions begin with the event of
absorption of light. The resulting excitation energy is transferred to a trap
where it can be used for charge separation. Through different mechanisms, the
separated charges are used in subsequent redox reactions to oxidize water and
produce a proton gradient. Thus the light harvesting apparatus uses a natural
energy resource (sunlight) to provide electric power for driving necessary
reactions. Using artificial light-harvesting devices, the energy of sunlight can be
stored in various solar fuels (*e.g.* hydrogen, methanol or other carbon-based
compounds), or it can be used for direct generation of electricity in solar cells.[1]
The efficiency of the primary energy conversion in such devices is one of the key
elements in successful utilization of the sunlight energy.

Natural systems can teach us strategies for the capture, transfer and transformation of solar energy, and improve our understanding of the principles of
these primary photoreactions in order to provide the necessary platform for
building efficient artificial light-harvesting devices. After 3.5 billion years of
evolution, photosynthetic organisms have mastered these processes, achieving

RSC Energy and Environment Series No. 9
Photoelectrochemical Water Splitting: Materials, Processes and Architectures
Edited by Hans-Joachim Lewerenz and Laurence Peter
© The Royal Society of Chemistry 2013
Published by the Royal Society of Chemistry, www.rsc.org

>90% efficiency of exciton energy transfer within light-harvesting antennae proteins and >99% efficiency of charge separation in reaction centers.[2–5] This remarkable efficiency is achieved *via* two mechanisms: modification of chromophore molecular structure and intelligent organisation of the photosynthetic apparatus.[6,7] A large variety of pigments allows organisms to increase the absorption cross-section and optimize inter-chromophore spectral overlap. For instance, in deep waters only the green part of solar spectrum is accessible, and organisms inhabiting deep waters (cyanobacteria, cryptophytes, diatoms *etc.*) express larger amounts of pigments – carotenoids and phycobilins – that absorb in this spectral region.[8] In contrast, red-absorbing chlorophylls constitute the dominant pigments in terrestrial plants and shallow water algae.[9] The second factor, the structure of proteins and their organisation in the photosynthetic unit, plays a key role in the energy transfer reactions.[10] It is known that chlorophylls in free solution with concentrations comparable to those in light-harvesting proteins (0.1 M) exhibit such strong quenching that essentially no energy transfer is possible.[11] In contrast, chlorophyll molecules embedded in proteins exhibit extraordinarily high fluorescence yield and intermolecular exciton transfer rates.[4] This phenomenon is due to the unique protein scaffold that encases the pigments in a very specific way and prevents the concentration quenching (see Figure 14.1).[12] Within the protein scaffold, the orientations and distances between pigments are optimized to keep the inter-molecular interaction strong, while the energy dissipation is already low. This interaction is described in terms of pigment-pigment and pigment-protein couplings, which are the central parameters in the theoretical description of the exciton dynamics within the photosynthetic unit and determine the inter-chromophore energy-transfer rates.[13] Several different mechanisms, as determined by pigment-pigment coupling strengths, can jointly contribute to the overall energy transfer

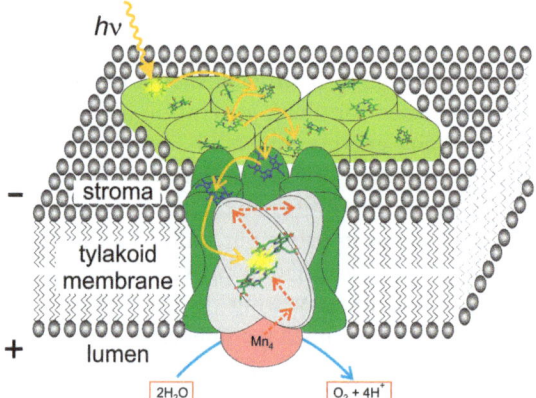

Figure 14.1 Photosystem II of higher plants. Orange arrows indicate an exciton transfer pathway between antennae chlorophylls. Red arrows indicate the electron transfer chain from the oxygen-evolving complex at the lumen to the quinone pool at the stroma.

during the migration of the exciton from the initially excited chromophore to the reaction center.

This chapter explores the main mechanisms of energy transfer employed within antenna proteins of various photosynthetic organisms. First, the principles of Förster Resonance Energy Transfer (FRET) within a system of weakly coupled pigments are discussed. The Förster theory provides the essential information on energy migration in the Donor-Acceptor system. However, in systems with strongly interacting pigments, where the excitation energy is delocalized over a group of several molecules, conventional Förster theory fails. An improved description is provided by the generalized Förster theory, which we introduce using light-harvesting proteins of purple bacteria as an example. The generalized Förster theory accounts for situations in which the pigments are grouped in weakly interacting aggregates with strong intra-aggregate coupling. Following evolutionary progression, we consider more complex antenna proteins of algae and higher plants within which both the density of pigments and their number per aggregate is increased. In recent years, theoretical and experimental studies have revealed coherent behavior in such systems, leading to the proposal of wave-like energy migration as an alternative to the classical hopping mechanism assumed in Förster theory. We critically examine the quantum coherent effects in pigment-protein complexes and discuss their possible influence on the efficiency of energy transfer. Finally the strategies of photoadaptation and photoprotection are considered. Sunlight intensity varies by 2–3 orders of magnitude during the day, and protein degradation can easily occur under these conditions. We discuss how the regulation mechanisms – developed in plants on molecular level – prevent photodamage and allow for a substantial increase in the overall efficiency.

14.2   Weakly-Coupled Chromophores: Förster Resonance Energy Transfer Theory

Experiments on unicellular algae in the early 1930s revealed that production of one oxygen molecule results from the cooperative action of a large number of chlorophyll molecules.[14] Thus, it was concluded that a single photosynthetic unit consists of ~ 300 light-harvesting chromophores and one reaction center.[15] After absorption of a light photon, the generated exciton will experience a large number of energy transfer events – 'hops', from one chromophore to another before reaching the reaction center.[16] Traditionally, the exciton transfer is described by random hop model.[17] The energy transfer mechanisms between two isolated chromophore molecules, donor (D) and acceptor (A), can be classified according to their coupling strength. In the weak coupling regime, *i.e.* when the distance between A and D is large enough, energy transfer can be described by the dipole-dipole approximation of the Coulomb interaction. This approximation is the basis of the Förster Resonant Energy Transfer theory, according to which the electronically excited donor molecule D^* returns to the ground state, while the acceptor molecule A becomes excited from its ground state (see

Figure 14.2A). This process is nonradiative and is solely due to Coulomb interaction between molecules A and D. The amplitudes of the donor de-excitation and acceptor excitation transitions depend on the transition dipole moments of each of the molecules, d_D and d_A. The Coulomb interaction couples these two transitions and the coupling strength in the point-dipole approximation takes shape

$$V_{DA} \approx \frac{d_A d_D}{R_{DA}^3} - 3\frac{(d_A R_{DA})(d_D R_{DA})}{R_{DA}^5} \tag{14.1}$$

where R_{DA} is the distance between centers of donor and acceptor molecules. Within the weak coupling limit, if the inter-chromophore distance becomes comparable with the size of the chromophores $R_{DA} \sim r_A, r_D$, the coupling term cannot be approximated by the point dipole approach, and it is therefore calculated more accurately using a quantum chemical calculation of the transition density of the donor and acceptor molecules. Using the calculated transition densities the coupling can be obtained by the so-called Transition Density Cube (TDC) method.[18] In that approach each of the A and D molecules is treated as a sum of a large number of space elements (cubes) that have much smaller dimensions than the inter-cube distances. Integrating over these small cubes gives the coupling term.

When the inter-chromophore distance becomes short enough for the wave-functions of the interacting molecules to overlap, the FRET is dominated by an alternative energy transfer mechanism – Dexter electron exchange.[19] In this mechanism, an electron is transferred between the ground and excited states of the interacting molecules as shown in Figure 14.2B.

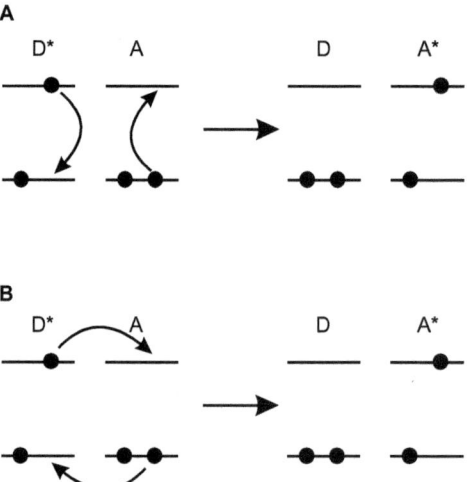

Figure 14.2   Energy transfer mechanisms between Donor and Acceptor chromophore molecules. A – long range Förster nonradiative energy transfer, B – Dexter electron exchange.

The rate of the energy transfer between any two quantum states is determined by Fermi's Golden Rule:

$$k_{DA} \approx \frac{2\pi}{h} V_{DA}^2 \delta(E_D - E_A) \qquad (14.2)$$

where E_D and E_A are energies of the donor and acceptor states. In case of energy transfer between chromophore molecules, the δ-function is replaced by a spectral density function (or spectral overlap integral) J_{DA}:

$$J_{DA} = \int S_D(E)S_A(E)dE = \frac{1}{\hbar} \int S_D(\omega)S_A(\omega)d\omega \qquad (14.3)$$

Here, $S_D(\omega)$ and $S_A(\omega)$ are the spectra of donor emission and acceptor absorption, respectively. Combining equations (14.1) to (14.3) and averaging over all possible orientations of the dipoles d_D and d_A, the final expression for the Förster transfer rate can be obtained:

$$k \approx \frac{4\pi}{3\hbar} \frac{d_D d_A}{R_{AD}^6} J_{DA} \qquad (14.4)$$

The spectral overlap integral J_{DA} is a very important quantity in the Förster theory. Together with the coupling parameter (V_{DA}), it controls the magnitude of the energy transfer, and it is used in nature to tune each particular ET pathway between different pigments to optimize the overall energy transfer efficiency. For example, the light-harvesting complex II (LHCII) of higher plants contains 8 Chl-a and 6 Chl-b molecules. However, due to interaction with the protein scaffold, instead of two absorption lines corresponding to Q_y electronic states of Chl-a and Chl-b, the LHCII complex exhibits a broadened spectrum consisting of 14 strongly overlapped exciton absorption bands.[20] This tuning of the site energy of each chromophore within the protein generates an intelligent electronic 'spider-web', where multiple energy transfer pathways optimally funnel the excitation energy to the reaction center without the exciton being 'lost' among 300 chlorophylls of the photosynthetic subunit.

While the electronic structure of an individual chromophore controls the spectral overlap, the distance between chromophores and their mutual orientation determine the coupling term.[21] Thus, the chromophores within a light-harvesting complex perform the act of light absorption and store the exciton until it is transferred to another chromophore or is dissipated. Chromophores are therefore a mere construction block. It is the protein that assembles all the necessary cofactors together, tunes their properties, aligns and orients them; creating a complete unit of the photosynthetic machinery which can optimally function in a given environment, delivering excitation energy to the reaction center.

The Förster theory provides a good estimate of the energy transfer rate and the characteristic distance between chromophores in a photosynthetic unit – the so-called Förster radius.[22] However, in its simplest form, (equation (14.4)) it does not account for another factor: the dynamics of chromophore-protein (system-bath) interaction, which can substantially alter energy transfer. In the following section, we discuss the role of the system-bath interaction and the interplay between pigment-pigment and pigment-protein couplings.

14.3 Multi-Subunit Protein Complexes: Generalized Förster Theory

Förster theory is formulated with two assumptions, which are often summarized as the 'weak coupling limit'. First, the coupling of the system of interest (the donor and acceptor states) to the bath is considered to be much stronger than the intrinsic coupling between donor and acceptor. Second, the characteristic relaxation time of the bath is much faster than the rate of FRET, and therefore the bath is considered to be completely equilibrated during the resonance energy transfer. Under these assumptions, the bath plays the role of a sink for the excitation energy, and the energy transfer has an incoherent and irreversible character. However these approximations are not always applicable, and one of the best examples of the breakdown of the weak coupling limit is the LH2 complex of purple bacteria. This complex is one of the most well studied photosynthetic proteins. Its crystal structure reveals a cylindrical shape with 9-fold symmetry (or 8-fold depending on the organism species) where each subunit consists of two helices, which scaffold four chromophores: three bacteriochlorophyll-a and one carotenoid molecules (see Figure 14.3A). The LH2 complex is optimized to absorb light in a broad spectral range, from UV (mostly by carotenoids) to IR (by Q_y excited state of BChls). The absorbed excitation energy is funnelled downhill to the lowest excited state of BChls, from where it is transferred to the reaction center (RC-LH1 complex). The geometry of the BChl 'aggregate' within the LH2 complex has been found to be most efficient in light-harvesting, performing perfect energy transfer. The BChl aggregate in LH2 complex is organized in two rings, B850 and B800 – the notation

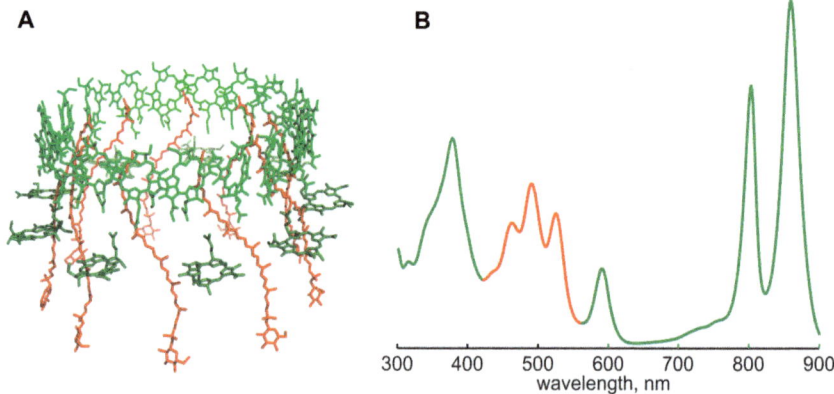

A   **B**

wavelength, nm

Figure 14.3   A. Molecular structure of LH2 complex of *Rds. acidophila*. B850 ring bacteriochlorophylls are shown in light green, B800 ring bacteriochlorophylls are shown in dark green, carotenoids are shown in red. B. Absorption spectrum of the LH2 complex of *Rds. acidophila*.

refers to the maximum wavelength of the absorption peaks–(Figure 14.3B). The B800 ring contains 9 BChl molecules separated by ~ 2 nm. The primary role of this ring is to collect light at 800 nm and forward the energy to the B850 ring. Due to the large separation distance of BChls, the energy transfer within the B800 ring and between B800 and B850 rings can be well described by Förster theory (weak coupling limit). In contrast, the properties of the B850 ring are very different. Due to a very short spacing (~ 0.9 nm) of the 18 BChls within the B850 ring, the resulting coupling ($\sim 300 \, \mathrm{cm}^{-1}$) between nearest neighbour BChls is strong enough to cause formation of 18 exciton states.

In other words, the interaction of BChls within B850 exceeds the energetic disorder represented by absorption lineshapes of isolated BChl molecules, which leads to sharing of the excitation energy among the coupled chromophores, forming so-called Frenkel excitons.[23,24] In a perfect case, only 2nd and 3rd exciton states would be optically bright, and the rest of the exciton manifold should be optically forbidden.[25] However, due to disorder in the energies of the excited states, the lowest exciton state gains substantial dipole transition moment. In addition, the disorder decreases the delocalization to typically 4 BChls sharing a single exciton.[26–29] To sum up, the arrangement of chromophores in aggregates with strong coupling serves several functions. The absorption cross-section is increased by creating exciton states that differ from the original electronic states of the isolated chromophores. The lowest energy exciton state stimulates a down-hill energy transfer cascade from the localized states, thereby increasing the efficiency and speed of energy transfer (by a factor of 10 relative to estimated values based on Förster theory). Finally, the exciton states show better properties in protecting excitation energy and directing it to the reaction center.

The key feature of the excitonic aggregate is the domination of the inter-chromophore coupling over the chromophore-bath coupling. In that case the delocalized exciton states $|M\rangle$ are given as coherent superposition of the localized excited states $|m\rangle$

$$|M\rangle = \sum_n c_m^{(M)} |m\rangle \qquad (14.5)$$

where $c_m^{(M)}$ are elements of the eigenvector matrix of the system Hamiltonian H_S, consisting of localized state energies E_m and couplings V_{mn}

$$H_S = \sum_m E_m |m\rangle\langle m| + \sum_{mn} V_{mn} |m\rangle\langle n| \qquad (14.6)$$

In contrast to the FRET, the energy transfer in a system of strongly coupled chromophores occurs *via* excited states rather than through molecular units. In a system of strongly interacting chromophores, energy transfer will occur *via* delocalized exciton states with energies defined as the eigenvalues of the Hamiltonian H_S. The calculation of the transfer rate between two exciton states is usually performed under two approximations summarized by the Redfield theory.[30] Within the Born approximation, any change in the vibrational structure during relaxation is neglected, and the system is treated perturbatively. Within the Markov approximation, the system-bath interaction is assumed to decay on a very short timescale and the relaxation (Redfield) tensor is assumed to be time-independent. Under these conditions, the rate constant between two delocalized states depends on the spectral density $J(\omega)$, the energy between the two states $\hbar\omega_{MN}$ and electronic 'overlap' factor γ_{MN}:

$$k_{MN} = 2\pi\gamma_{MN}\omega_{MN}^2(J(\omega_{MN})(1 + n(\omega_{MN})) + J(\omega_{MN})n(\omega_{MN})) \qquad (14.7)$$

where $n(\omega_{MN})$ is a Bose-Einstein distribution function for the vibrational quanta.[31] The electronic factor γ_{MN} depends on the correlation of energy fluctuations of localized states. If the fluctuations are strongly correlated, no transfer between exciton states takes place. In the opposite case, when the energies of localized states are completely uncorrelated, the relaxation rate is maximal. The spectral density term $J(\omega_{MN})$ describes how strong the dissipation of the system is.

The perturbative approach utilized in Redfield theory has been used to calculate coherent evolution in a number of protein complexes.[32–34] However, the perturbation approach is not always valid in natural systems where strong inter-chromophore coupling within aggregates is combined with weak coupling between aggregates. In some cases, only a part of the protein pigments have strong coupling and form aggregates, whereas the rest of the pigments are weakly coupled. In other proteins, all the pigments are strongly coupled, making the whole protein a single aggregate. In both scenarios, interaction between weakly coupled aggregates can be well described by Förster theory. Therefore in order to describe complete dynamics of a complex system, both strong inter-chromophore coupling and weak inter-aggregate coupling have to be taken into account simultaneously. This is addressed in the generalized Förster theory.[35–38] This generalized form operates with the effective donor and effective acceptor rather than localized states. The effective donor states $|M>$ are calculated from the interacting donor molecules $|m>$ in the absence of acceptor molecules, and effective acceptor states $|N>$ are calculated from the interacting acceptor molecules $|n>$ in the absence of donor molecules. In this way, the delocalized donor states and delocalized acceptor states are determined. In the next step, the effective spectral density J_{MN} and effective couplings V_{MN} between pairs of effective donor and effective acceptor states are estimated. The final energy transfer rate between donor exciton state $|M>$ and acceptor exciton state $|N>$ takes form analogous to equations (14.2) and (14.3)

$$k_{MN} = 2\pi\frac{|V_{MN}^2|}{\hbar^2}\int_{-\infty}^{\infty} S_M(\omega)S_N'(\omega)d\omega = 2\pi\frac{|V_{MN}|^2}{\hbar_2}J_{MN} \qquad (14.8)$$

where the effective spectral density J_{MN} contains both excitonic and exciton-vibrational contributions and the effective excitonic coupling V_{MN} is a sum of weighted intra-aggregate excitonic couplings V_{mn}:

$$V_{MN} = \sum c_m^{(M)}c_n^{(N)}V_{mn} \qquad (14.9)$$

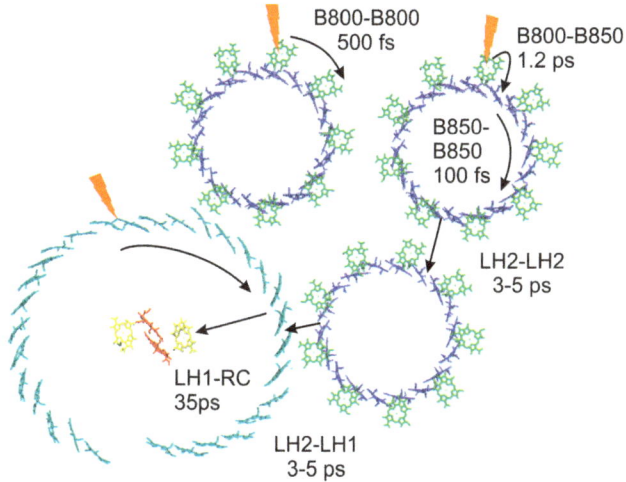

Figure 14.4 Photosynthetic unit of purple bacteria consisting of LH1-RC and LH2 complexes. Only BChl molecules are shown: strongly coupled B850 bacteriochlorophylls of LH2 complex (blue), weakly coupled B800 bacteriochlorophylls of LH2 complex (green), strongly coupled B875 bacteriochlorophylls of LH1-RC complex (cyan), special pair bacteriochlorophylls of reaction center (red), accessory bacteriochlorophylls of the reaction center (yellow). Black arrows indicate energy transfer pathways.

The scheme of the energy transfer in a photosynthetic unit of the purple bacteria is summarized in Figure 14.4. The rates shown were measured experimentally by various techniques and represent typical values for different organisms.[39,8,40–42]

14.4 Quantum Coherence in LH Proteins

The sharing of excitation energy among several chromophores arising from the strong coupling of BChls within LH2 complexes is a quantum effect well-known from condensed matter physics.[43,44] This quantum phenomenon substantially increases energy transfer rates in light-harvesting complexes, enhancing the efficiency of primary photoreactions.[45] Despite the fact that exciton sharing is a quantum coherence effect, the dephasing is so fast that the mechanism of energy transfer remains incoherent and only the magnitude of the transfer rate is affected. However, within the last decade a completely different quantum effect has been reported for several light-harvesting proteins (Figure 14.5). Oscillations (quantum beats) of significant amplitude have been observed in the excitation dynamics of these proteins indicating presence of a coherent energy transfer.[46,47] The experimental observation of quantum coherence in light-harvesting proteins appeared to be a surprising phenomenon and attracted much attention from the scientific community.[48–52]

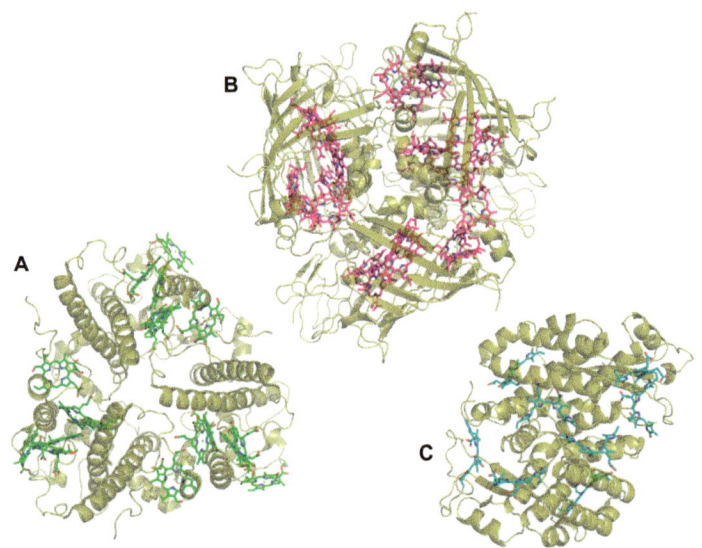

Figure 14.5   Light harvesting proteins of various organisms: chlorophyll containing LHCII from higher plants[56] (A), bacteriochlorophyll containing FMO from green sulfur bacteria[57] (B) and bilin containing PC645 from cryptophyte algae[58] (C).

The distinctive feature of these proteins is that the interacting chromophores have an intermediate coupling strength comparable with the coupling to the bath.[53,54] In other words, the characteristic line-shape width is similar to the disorder in the excited state energies of the chromophores. In molecular networks with intermediate inter-chromophore coupling, coherent quantum dynamics is generally possible; however, it is not expected to occur in natural proteins which function at high temperatures (>273 K) in the presence of water and experience strong interactions with the environment. Indeed, due to these factors, decoherence is considered to be very fast and the excitation dynamics in natural systems is traditionally described using laws of classical physics. However, a new advanced spectroscopy method, two-dimensional electronic spectroscopy (2DES)[55] developed within the last decade has provided clear evidence of coherent dynamics in light-harvesting proteins and allowed direct observation of the pathways of the energy flow.

Applying this technique to light-harvesting antenna complexes from green sulphur bacteria (FMO) and cryptophyte algae (PC645), it was found that under laser excitation a long-lived coherent beat (<600 fs) is present even at ambient temperatures.[46,47,59,60] This beat is observed as oscillations (Figure 14.6B) of the cross peak in the two-dimensional spectrum (Figure 14.6C) as a function of the population time between the pump pulse sequence and the probe pulse (Figure 14.6A). The cross peak of the 2D spectrum represents a pathway of the field-matter interaction (Liouville pathway), and the beating of the cross peak is observed only when two excited states coherently share the exciton. The beat period is defined by the energy gap between the two states, and the position of the

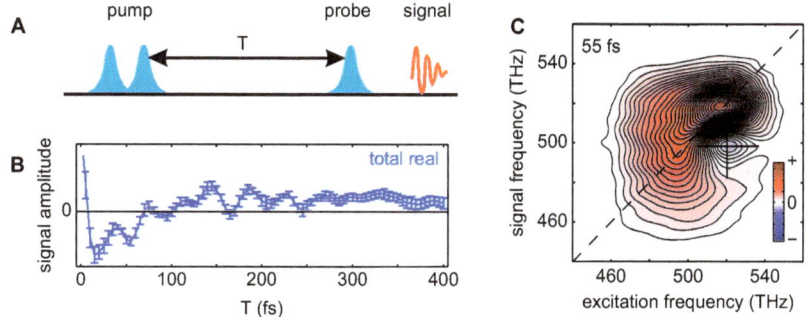

Figure 14.6 2D electronic spectroscopy: scheme of excitation pulses (A), example of coherent beat of the cross-peak of the 2D spectrum (B), and example of 2D spectrum (C).[68]
The data was kindly provided by D. B. Turner.

cross peak in the 2D spectrum is determined by the energies of the two electronic states that participate in the coherent superposition.[61] The observation of quantum beating in photosynthetic complexes stimulated theoretical studies of the possible mechanisms that may lie beneath the observed coherent signals and which could play a role in improving photosynthetic transport.[62–64] One of the proposed mechanisms of quantum energy transfer is the 'quantum walk'.[65] In a classical random walk, the excitons take steps from chromophore to chromophore in randomly chosen directions (Brownian-like motion) and eventually reach the reaction center. In contrast, according to the quantum walk model the exciton takes steps in all directions simultaneously, and the resultant path is determined by the destructive/constructive interference between all possible paths.[66] The quantum walk has significant advantages since the distance of exciton travel is proportional to the travelling time t, whereas a classical random walk it scales with $\sqrt{t}$. The critical factor in the quantum walk efficiency is the interplay between the interaction of the system with the environment and the destructive interference of possible pathways. Remarkably, the optimal transport occurs at room temperature, where the two mechanisms cancel each other.[67] Due to a generally strong interaction with the environment, the decoherence is always substantial, but the inter-chromophore distance and their relative orientation can change the electronic coupling between molecules by orders of magnitude. Therefore, it would not be surprising if during evolution nature had found an excellent alignment of pigments within the protein scaffold that allows compensation of decoherence effects.

The contribution of the quantum physics to light harvesting in natural systems is well established. The quantum phenomena of sharing excitation energy between neighboring chromophores increase both the spectral overlap and the coupling strength and result in increased energy transfer rates.[45] These phenomena are solely due to the properties of the system and do not depend on the properties of the incident light. In contrast, the quantum coherent oscillations (beats) are observed only under coherent laser excitation. However, natural solar radiation is incoherent and is of much lower spectral intensity. Therefore, despite clear evidence of coherent dynamics in light-harvesting complexes, it is not yet clear whether the quantum coherent effects contribute to the exciton dynamics *in vivo*. Nevertheless, the quantum coherence phenomena can certainly improve efficiency of the energy transfer, and the optimal arrangement of chromophores is the key to that technology.

14.5 Strategies of Photoadaptation and Photoprotection

Environmental conditions can change dramatically over time and space. These changes can cause substantial damage to the organism if no prompt readjustment is introduced. Photosynthetic organisms have developed remarkable mechanisms to survive and perform efficient photochemistry in different, often extreme conditions. For example certain algae species can live in hot springs at temperatures above 70 °C,[69] in very acidic environments (pH 0.05–4),[70] at the snow surface,[71] and under high levels of UV radiation.[72] Some species can even survive in the deep-sea where no sunlight can reach and utilize geothermal radiation for photosynthesis.[73] The incident light (either low or high levels) is one of the main stress factors, and organisms must not only show highly efficient light-harvesting under various illumination levels but also provide good protection against photodamage.

Changes of the intensity of sunlight during the day can reach several orders of magnitude within a few minutes.[74] When the rates of absorption of light and generation of excitons exceeds the rate of photochemical reactions, the excited chlorophylls have a high probability of triplet state formation, followed by the formation of singlet oxygen or other reactive species. These species can easily oxidize proteins and can severely damage photosynthetic apparatus. In order to prevent photodamage, organisms have developed a number of molecular mechanisms that allow adaptation to new light conditions within minutes, seconds or even faster.[75,76] A pH gradient across the membrane serves as a sensor of the radiation level. Under excess light, low pH triggers the reorganisation of light-harvesting proteins as well as changes in the molecular structure of the proteins, namely their carotenoid composition. These processes are often unified under the heading of non-photochemical quenching (NPQ), since they compete with photochemical charge separation in the deactivation of chlorophyll excitation energy. One of the NPQ mechanisms is related to the interconnectivity of the antenna proteins: under high light conditions, antenna proteins detach from the reaction center, decreasing the excitation flow.[77,78] This is a clear advantage of a multicomponent antenna organisation, allowing fast assembling/disassembling of the antenna complexes and thereby changing the energy transfer *vs.* excitation quenching ratio. Another mechanism of NPQ involves carotenoids, which are long known to contribute to photoprotection and which act as antioxidants in all living organisms.[79] Their most famous role is in quenching triplet chlorophylls as well as singlet oxygen, thus removing the reactive species from organisms. To increase efficiency of that process, carotenoids are embedded in protein next to the chlorophylls allowing them to quench reactive species before they can damage the protein. During the last two decades, the role of carotenoids in quenching of chlorophyll singlet states has been very extensively studied. This carotenoid-involving NPQ mechanism provides a remarkable tool for fast switching of the antenna complex from the light-harvesting to the quenching state,[80] thus attracting a significant interest from the scientific community. With minimal structural change, the machinery decreases the efficiency of light harvesting by >80%. Such a tool would be invaluable for any light-harvesting device, since it will allow adaptation to changing conditions with minimal effort. In natural proteins, it is not a surprise that only a small structural change can so drastically affect energy dynamics. Indeed, as was noted above, with the high concentration of chlorophylls (>0.6M) it is remarkable that any energy transfer reactions can take place at all, since excitation of chlorophylls is completely quenched in free solution.[11] The exact mechanism of singlet chlorophyll quenching by carotenoids is still under debate.[81] In one hypothesis, the quenching is achieved by direct contact of chlorophyll and carotenoid molecules, either *via* formation of a carotenoid cation[82] or *via* direct energy transfer from chlorophylls to carotenoids.[83] Carotenoids have excited states with extremely short lifetime, making them a perfect sink for excess excitation. According to a second hypothesis, carotenoids, along with other factors, stimulate aggregation of neighboring light-harvesting proteins (LHCII) and as the result a chlorophyll-chlorophyll charge transfer states are formed, which again act as a sink for the excess excitation,[84] competing with the energy flow to reaction center.

14.6 Summary of the Strategies Adopted by Natural Systems

Light-harvesting complexes of photosynthetic organisms provide an example of perfect devices for collecting and transporting radiation energy. The photosynthetic antenna not only fulfills these two functions but is able to adapt to various environments and changing conditions, while keeping the overall efficiency of the light-harvesting above 90%. The key to this successful operation is the multi-chromophore and multi-protein organization. It allows adjustment of sunlight absorption to the accessible spectral range, transfer the absorbed energy to the reaction centre with negligible losses and matching of the energy flow to the photochemical conversion rate of the reaction center. The latter mechanism keeps the photochemical reactions working without idling and at the same time prevents photodamage of the photosynthetic apparatus due to over-excitation of the antenna. The most striking achievement of natural light-harvesting systems is the utilization of quantum physics in the energy transfer processes by means of intelligent arrangement of chromophores within the protein scaffold as well as by adjustment of the interaction between different pigment-proteins complexes within the photosynthetic unit.

To summarize, there are two main concepts of light harvesting. First, the system has to acquire heterogeneous multi-subunit structure. As nature demonstrates, heterogeneity can be achieved even using the same type of chromophores. By assembling the chromophore in subunits with variable inter-chromophore coupling, the optimal diversity of optical properties can be produced. Second, organisation of the subunits within the system determines the mechanisms of energy transfer. Here, special attention has to be paid not only to inter-chromophore interactions, but to chromophore-environment interactions as well. In modern artificial light-harvesting devices, most emphasis is placed on the synthesis of the optimal chromophores (dyes, quantum dots *etc.*). However, the example of natural systems demonstrates that the design of the light-harvesting unit is of equal or even higher importance than the chromophore itself.

Acknowledgements

The support from the Natural Sciences and Engineering Research Council of Canada and DARPA (QuBE) are acknowledged.

References

1. D. Gust, T. A. Moore and A. L. Moore, *Acc. Chem. Res.*, 2009, **42**, 1890–1898.
2. G. T. Oostergetel, H. van Amerongen and E. J. Boekema, *Photosynth. Res.*, 2010, **104**, 245–55.
3. S. Vasil'ev and D. Bruce, *Plant Cell*, 2004, **16**, 3059–3068.
4. R. Van Grondelle, J. P. Dekker, T. Gillbro and V. Sundstrom, *BBA*, 1994, **1187**, 1–65.
5. B. Gobets, I. H. Van Stokkum, M. Rögner, J. Kruip, E. Schlodder, N. V. Karapetyan, J. P. Dekker and R. Van Grondelle, *Biophys. J.*, 2001, **81**, 407–424.
6. G. Britton, *The biochemistry of natural pigments*, Cambridge University Press, 1983.
7. J. P. Dekker and E. J. Boekema, *BBA*, 2005, **1706**, 12–39.
8. B. R. Green and W. W. Parson, Light-Harvesting Antennas in Photo-synthesis. *Advances in Photosynthesis and Respiration, Vol 13*, Kluwer Academic Publishers, Dordrecht, 2003.
9. G. C. Papageorgiou and Govindgee, *Chlorophyll a Fluorescence: A Signature of Photosynthesis. Advances in Photosynthesis and Respiration, Vol. 19*, Kluwer Academic Publishers, Dordrecht, 2004.
10. P. Joliot and A. Joliot, *Photosynth. Res.*, 2003, **76**, 241–5.
11. G. S. Beddard and G. Porter, *Nature*, 1976, **260**, 366–367.
12. T. Barros and W. Kühlbrandt, *BBA*, 2009, **1787**, 753–772.
13. T. Renger, in *Primary Processes of Photosynthesis*, ed. G. Renger, The Royal Society of Chemistry, Cambridge, 2008, pp. 39–98.
14. R. Emerson and W. Arnold, *J. Gen. Physiol.*, 1932, **16**, 191–205.

15. D. C. Mauzerall and N. L. Greenbaum, *BBA*, 1989, **974**, 119–140.
16. H. van Amerongen, L. Valkunas, and R. van Grondelle, *Photosynthetic excitons*, World Scientific Pub. Co., Singapore, 2000.
17. D. Abramavicius, L. Valkunas and R. van Grondelle, *PCCP*, 2004, **6**, 3097–3105.
18. B. P. Krueger, G. D. Scholes and G. R. Fleming, *J. Phys. Chem. B*, 1998, **102**, 5378–5386.
19. D. L. Dexter, *J. Chem. Phys.*, 1953, **21**, 836–850.
20. V. I. Novoderezhkin, M. A. Palacios, H. Van Amerongen and R. Van Grondelle, *J. Phys. Chem. A*, 2005, **109**, 10493–10504.
21. M. Sener, J. Strümpfer, J. A. Timney, A. Freiberg, C. N. Hunter and K. Schulten, *Biophys. J.*, 2010, **99**, 67–75.
22. S. E. Braslavsky, E. Fron, H. B. Rodriguez, E. S. Roman, G. D. Scholes, G. Schweitzer, B. Valeur and J. Wirz, *Photochem Photobiol Sci*, 2008, **7**, 1444–1448.
23. M. Kasha, *Rad. Res.*, 1963, **20**, 55–70.
24. R. Monshouwer, M. Abrahamsson, F. van Mourik and R. van Grondelle, *J. Phys. Chem. B*, 1997, **101**, 7241–7248.
25. X. Hu and K. Schulten, *Physics Today*, 1997, **50**, 28.
26. O. Kühn and V. Sundström, *J. Chem. Phys.*, 1997, **107**, 4154–4164.
27. J. A. Leegwater, *J. Phys. Chem.*, 1996, **100**, 14403–14409.
28. J. T. M. Kennis, A. M. Streltsov, H. Permentier, T. J. Aartsma and J. Amesz, *J. Phys. Chem. B*, 1997, **101**, 8369–8374.
29. M. Chachisvilis, O. Kühn, T. Pullerits and V. Sundström, *J. Phys. Chem. B*, 1997, **101**, 7275–7283.
30. V. May and O. Kühn, *Charge and energy transfer dynamics in molecular systems*, Wiley-VCH Verlag GmbH & Co. KGaA, Weinheim, Germany, Third., 2011, vol. 6.
31. T. Renger, V. May and O. Ku, *Physics Reports*, 2001, **343**, 137–254.
32. V. Novoderezhkin and R. Van Grondelle, *J. Phys. Chem. B*, 2002, **106**, 6025–6037.
33. V. I. Novoderezhkin, M. A. Palacios, H. Van Amerongen and R. Van Grondelle, *J. Phys. Chem. B*, 2004, **108**, 10363–10375.
34. V. Novoderezhkin, M. Wendling and R. Van Grondelle, *J. Phys. Chem. B*, 2003, **107**, 11534–11548.
35. H. Sumi, *J. Phys. Chem. B*, 1999, **103**, 252–260.
36. K. Mukai, S. Abe and H. Sumi, *J. Phys. Chem. B*, 1999, **103**, 6096–6102.
37. G. D. Scholes and G. R. Fleming, *J. Chem. Phys. B*, 2000, **104**, 1854–1868.
38. M. Yang, A. Damjanović, H. M. Vaswani and G. R. Fleming, *Biophys. J.*, 2003, **85**, 140–158.
39. H. Bergstrұm, R. Van Grondelle and V. Sundstrұm, *FEBS Lett.*, 1989, **250**, 503–508.
40. V. Sundstrұm, T. Pullerits and R. Van Grondelle, *J. Phys. Chem. B*, 1999, **103**, 2327–2346.
41. Y. Z. Ma, R. J. Cogdell and T. Gillbro, *JPC B*, 1998, **102**, 881–887.

42. Y.-Z. Ma, R. J. Cogdell and T. Gillbro, *J. Phys. Chem. B*, 1997, **101**, 1087–1095.

43. V. M. Agranovich, *Chem. Phys.*, 2000, **112**, 8156–8162.

44. D. P. Craig and S. H. Walmsley, *Excitons in molecular crystals*, Benjamin, New York, 1968.

45. J. Strümpfer, M. Sener and K. Schulten, *J. Phys. Chem. Lett.*, 2012, **3**, 536–542.

46. G. S. Engel, T. R. Calhoun, E. L. Read, T.-K. Ahn, T. Mancal, Y.-C. Cheng, R. E. Blankenship and G. R. Fleming, *Nature*, 2007, **446**, 782–786.

47. E. Collini, C. Y. Wong, K. E. Wilk, P. M. G. Curmi, P. Brumer and G. D. Scholes, *Nature*, 2010, **463**, 644–647.

48. G. D. Scholes, *J. Phys. Chem. Lett.*, 2010, **1**, 2–8.

49. P. Ball, *Nature*, 2011, **474**, 272–274.

50. R. J. Sension, *Nature*, 2007, **446**, 740–741.

51. R. V. Grondelle and V. I. Novoderezhkin, *Nature*, 2010, **463**, 614–615.

52. G. R. Fleming and G. D. Scholes, *Nature*, 2004, **431**, 256–257.

53. G. D. Scholes and G. R. Fleming, *Adv. Chem. Phys.*, 2006, **132**, 57–129.

54. R. Jimenez, S. N. Dikshit, S. E. Bradforth and G. R. Fleming, *J. Phys. Chem.*, 1996, **100**, 6825–6834.

55. T. Brixner, T. Mancal, I. V. Stiopkin and G. R. Fleming, *J. Chem. Phys.*, 2004, **121**, 4221–4236.

56. Z. Liu, H. Yan, K. Wang, T. Kuang, J. Zhang, L. Gui, X. An and W. Chang, *Nature*, 2004, **428**, 287–292.

57. Y. F. Li, W. Zhou, R. E. Blankenship and J. P. Allen, *J. Mol. Biol.*, 1997, **271**, 456–471.

58. A. B. Doust, K. E. Wilk, P. M. G. Curmi and G. D. Scholes, *J. Photochem. Photobiol. A-Chem*, 2006, **184**, 1–17.

59. G. Panitchayangkoon, D. Hayes, K. a Fransted, J. R. Caram, E. Harel, J. Wen, R. E. Blankenship and G. S. Engel, *PNAS*, 2010, **107**, 12766–12770.

60. D. B. Turner, K. E. Wilk, P. M. G. Curmi and G. D. Scholes, *J. Phys. Chem. Lett.*, 2011, **2**, 1904–1911.

61. Y.-C. Cheng and G. R. Fleming, *Annu. Rev. Phys. Chem.*, 2009, **60**, 241–262.

62. M. B. Plenio and S. F. Huelga, *New J. Phys.*, 2008, **10**, 113019.

63. P. Rebentrost, M. Mohseni and A. Aspuru-Guzik, *J. Phys. Chem. B*, 2009, **113**, 9942–9947.

64. P. Rebentrost, M. Mohseni, I. Kassal, S. Lloyd and A. Aspuru-Guzik, *New J. Phys.*, 2008, **11**, 7.

65. J. Kempe, *Contemp. Phys.*, 2003, **44**, 307–327.

66. E. Farhi and S. Gutmann, *Phys. Rev. A*, 1998, **58**, 915–928.

67. M. Mohseni, P. Rebentrost, S. Lloyd and A. Aspuru-Guzik, *J. Chem. Phys.*, 2008, **129**, 174106.

68. D. B. Turner, R. Dinshaw, K.-K. Lee, M. S. Belsley, K. E. Wilk, P. M. G. Curmi and G. D. Scholes, *PCCP*, 2012, **14**, 4857–4874.

69. R. T. Papke, N. B. Ramsing, M. M. Bateson and D. M. Ward, *Env. Microb.*, 2003, **5**, 650–659.

70. C. Ciniglia, H. S. Yoon, A. Pollio, G. Pinto and D. Bhattacharya, *Mol. Ecol.*, 2004, **13**, 1827–1838.

71. M. Fujii, Y. Takano, H. Kojima, T. Hoshino, R. Tanaka and M. Fukui, *Microb. Ecol.*, 2010, **59**, 466–475.

72. M. Stibal, J. Elster, M. Sabacká and K. Kastovská, *Fems Micriobiol. Ecol.*, 2007, **59**, 265–273.

73. J. T. Beatty, J. Overmann, M. T. Lince, A. K. Manske, A. S. Lang, R. E. Blankenship, C. L. Van Dover, T. A. Martinson and F. G. Plumley, *PNAS*, 2005, **102**, 9306–9310.

74. S. P. Long, S. Humphries and P. G. Falkowski, *Annu. Rev. Plant Physiol. Plant Mol. Biol.*, 1994, **45**, 633–662.

75. P. Muller, X. Li and K. K. Niyogi, *Plant Physiol.*, 2001, **125**, 1558–1566.

76. P. Horton, A. V. Ruban and R. G. Walters, *Annu. Rev. Plant Physiol. Plant Mol. Biol.*, 1996, **47**, 655–684.

77. A. R. Holzwarth, Y. Miloslavina, M. Nilkens and P. Jahns, *Chem. Phys. Lett.*, 2009, **483**, 262–267.

78. S. Kereïche, A. Z. Kiss, R. Kouril, E. J. Boekema and P. Horton, *FEBS Lett.*, 2010, **584**, 759–764.

79. G. Britton, in *Carotenoids Vol 4 Natural Functions*, eds. G. Britton, S. Liaaen-Jensen, and H. Pfander, Birkhauser Verlag, 2008, vol. 4, pp. 189–212.

80. B. Robert, P. Horton, A. A. Pascal and A. V. Ruban, *Trends Plant Sci.*, 2004, **9**, 385–390.

81. A. V. Ruban, M. P. Johnson and C. D. P. Duffy, *BBA*, 2011, **1817**, 167–181.

82. N. E. Holt, D. Zigmantas, L. Valkunas, X.-P. Li, K. K. Niyogi and G. R. Fleming, *Science*, 2005, **307**, 433–436.

83. A. V. Ruban, R. Berera, C. Ilioaia, I. H. M. van Stokkum, J. T. M. Kennis, A. a Pascal, H. van Amerongen, B. Robert, P. Horton and R. van Grondelle, *Nature*, 2007, **450**, 575–8.

84. Y. Miloslavina, A. Wehner, P. H. Lambrev, E. Wientjes, M. Reus, G. Garab, R. Croce and A. R. Holzwarth, *FEBS Lett.*, 2008, **582**, 3625–3631.

CHAPTER 15

Electronic Structure and Bonding of Water to Noble Metal Surfaces

HIROHITO OGASAWARA* AND ANDERS NILSSON

SLAC National Accelerator Laboratory, 2575 Sand Hill Rd, Menlo Park, CA 94025, USA
*Email: hirohito@slac.stanford.edu

15.1 Introduction

Photoelectrochemical (PEC) cells are often designed with semiconductors as photoelectrodes and noble metals as counter electrodes. The counter reaction on noble metal surface should be fast to avoid any performance limitations. At the interface, water interacts both with the noble metal surfaces and *via* hydrogen (H-) bonding with other water molecules.[1,2] Here, we will focus on the interaction of water on metal surfaces with a detailed discussion about the electronic structure and the resulting bonding mechanism. We anticipate that in particular the electronic structure of the water-surface interaction could have some similarity for water on semiconductor surfaces.

The interaction of water on metal surfaces has been the center of an extended debate during the last decade due to the multitude of bonding and overlayer structures water assumes on different single crystal surfaces. On metal surfaces, water forms two-dimensional hexagonal, or pseudo-hexagonal, H-bond networks in the first contact layer,[1,2] where the H atoms not involved in the two-dimensional hydrogen-bond network are either directed toward vacuum (H-up)

RSC Energy and Environment Series No. 9
Photoelectrochemical Water Splitting: Materials, Processes and Architectures
Edited by Hans-Joachim Lewerenz and Laurence Peter

or towards the surface (H-down). The detailed structure of the first contacting layer at a metal surface will affect barriers to dissociation and surface reactivity, including interaction with additional water layers.[3,4]

15.2 H-up, H-down and Partially Dissociated Water Layers

In the traditional structural model proposed by Doering and Madey,[5] water was considered to bind to metal surfaces exclusively *via* an oxygen *lp* orbital. The structure of the internally H-bonded water contact layer on metal surfaces was consequently considered to be the H-up structure where only every second water binds to the metal through the metal-oxygen (M-O) bond (H-up, Figure 15.1) while the other half is displaced towards vacuum and pointing the non-hydrogen-bonded OH group away from the surface towards vacuum. Recent work utilizing x-ray photoemission spectroscopy (XPS) and x-ray absorption spectroscopy (XAS) to investigate the structure of the contact layer[6–8] has, however, showed that an H-down layer, where all water molecules in the layer bind directly to the surface through alternating either M-O or metal-hydrogen (M-HO) bonds, is favored for intact adsorption of water on these close-packed metal surfaces. However, the details vary for different surfaces in terms of the long-range order in the H-up and H-down configurations.[9,10]

X-ray photoelectron spectroscopy (XPS) is based on the creation of a core hole *via* ionization and provides a method to study the geometric, electronic and chemical properties of a sample. The XPS binding energies depend very strongly on the elements involved; even for first-row elements the core-level binding energies differ on the order of 100 eV making XPS a highly

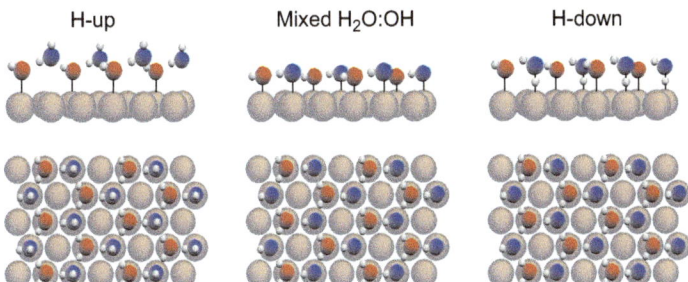

Figure 15.1   Illustrations of three proposed structural models for water adsorption on metal surfaces. All models contain a flat-lying O-bonded water (red/lighter) but differ in orientation and chemical nature of the second molecule (blue/darker). H-up: Traditional "ice-like" bilayer with the blue/darker water molecule having one of its H's pointing toward vacuum, Mixed H_2O:OH: partially dissociated water layer with flat-lying H_2O (red/lighter) and flat-lying OH (blue/darker), and H-down: non-O-bonded water molecule has one of its H's pointing toward the metal surface.
Reprinted from reference 8. Copyright 2010, with permission from Elsevier.

element-specific technique. In the case of water, XPS monitors the binding energy of O*1s* core-level state. The binding energy of the O*1s* state is affected by the valence electrons, which in turn are sensitive to the local environment. We can therefore expect that the core-levels will be chemically shifted depending on chemical nature (intact or dissociated), adsorption site, distance to the surface and molecular orientations. In x-ray absorption spectroscopy (XAS), a core electron is resonantly excited into unoccupied atomic or molecular orbitals at or above the Fermi level *via* a dipole-induced transition.[11] XAS provides element-specific information on the density and the energy level of unoccupied states, local atomic structure including molecular orientation, the nature, orientation, and length of chemical bonds (*via* bonding-antibonding orbital splitting[12]) as well as the chemical nature. Using s- and p- polarized x-ray light field, XAS measurements determine the orientation of molecular adsorbates and the directionality of bonding in an adlayer, including proton orientation, information often unattainable by other spectroscopic probes.[11]

The work on Pt(111)[6] and Ru(0001)[7] employed XPS to address the co-ordination of atoms in the molecularly intact monolayer to the surface (see Figure 15.2) and XAS to determine the orientation and coordination of the internal OH-groups in water with respect to the surface. While the H-up structure (Figure 15.1, H-up) would give only 33% surface coordination

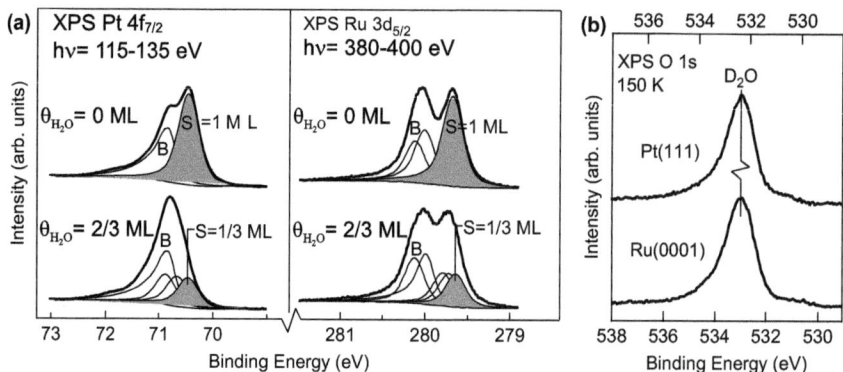

Figure 15.2   XPS spectra for clean and water-covered Pt(111) and Ru(0001), The essential results showing low concentration of non-coordinated surface Pt and Ru atoms (a) and non-dissociated water (b) can be seen directly in the experimental spectra. A qualitative curve fitting analysis (black lines) in a is indicated to guide the eye. (a): (Top) (Left) Summed Pt $4f_{7/2}$ spectra taken at three excitation energies (115, 125 and 135 eV) and (Right) summed Ru $3d_{5/2}$ spectra taken at three excitation energies (380, 390 and 400 eV) to average out photoelectron diffraction effects. The bulk (B) and uncoordinated surface (S) spectral components are indicated. (Bottom) Water coordinated to the surface atoms quenches intensity of uncoordinated surface atoms (S) and introduces core-level shifts (new spectral components) compared to the clean surface. The uncoordinated surface peak (S) is lowered in intensity by more than 60%. Reprinted from reference 8. Copyright 2010, with permission from Elsevier.

through M-O bonds, the H-down structure (Figure 15.1, H-down) would give 67% surface coordination through the combination of M-O *and* M-HO bonds. The surface coordination number can be experimentally determined by exploiting the surface core-level shift in XPS.[6,7,13–15] For clean metals, the lower coordination of atoms at the surface leads to a different core-level binding energy compared to the bulk.[14] On Pt(111), this splitting is 0.4 eV for the Pt 4*f* photoemission peak[6] (Figure 15.4a, left panel). The introduction of water on Pt(111) shifts the Pt 4*f* surface state towards the bulk value for the atoms that now coordinate to water. The change in XPS intensity of the Pt 4*f* state for the adsorbate system compared to the clean surface indicates that more than 60% of the surface Pt atoms become coordinated to water molecules. This directly and strongly indicates the H-down layer for Pt(111), where all water molecules in the contact layer bind to the surface.

On Ru(0001), the non-dissociated water contact layer forms a hexagonal two-dimensional hydrogen-bond network similar to the case of Pt(111). The surface coordination number was determined using the surface core level shift in the Ru 3*d* photoemission peak[7] (Figure 15.2a, right panel), which is sensitive to changes in local coordination similar to the Pt 4*f* case. As on Pt(111), more than 60% of the surface Ru atoms become coordinated, directly showing that all water molecules in the monolayer bind to atoms at the Ru(0001) surface. Based on the same coordination of the water layer to Ru(0001) as for the water layer on Pt(111) and near identical O 1*s* XPS spectra (see Figure 15.2b), an H-down model is suggested also for the non-dissociated water contact layer on Ru(0001).[7]

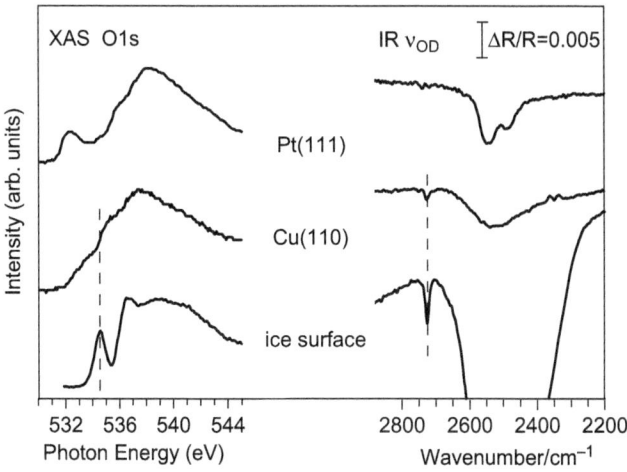

Figure 15.3   XAS (Reprinted from reference 8. Copyright 2010, with permission from Elsevier.) and IR spectra[24] for D$_2$O ice surface and D$_2$O water monolayer on Pt(111) and Cu(110). The light field oscillates in the direction normal to the surface (p-polarized).

The question emerges whether the H-down model is valid in general for water-metal monolayer adsorbate systems, or does the overlayer structure vary depending on surface structure? On the open and corrugated (110) surface of copper, it was found that a mixed monolayer with 2/3 ($\pm$15%) of the outer-layer water molecules in H-down configuration, and 1/3 with hydrogen pointing up toward the vacuum. Thus we find an H-down:H-up ratio of 2:1 for the water monolayer on Cu(110).[16] The spectroscopic indication of a mixed H-up/H-down layer is consistent with the large (7×8) unit cell for monolayer water on Cu(110),[16] indicating that the hexagonal adlayer is rather distorted with respect to the open substrate which can be expected to lead to a range of adsorption sites.

The orientation of the uncoordinated OH is confirmed by XAS, which can selectively probe the local unoccupied orbital structure in different directions in the layer by using s- and p-polarized x-rays, whose x-ray light field is either parallel or orthogonal to the surface. The interaction between water and metal surface, with the possible formation of M-HO bonds, is probed in the out-of-plane XAS while in-plane XAS is related to the formation of the two-dimensional hydrogen-bond network in the contact layer. There is, indeed, a strong anisotropy in XAS between aligning the E-vector along the surface normal (out-of-plane) and parallel to the surface (in-plane) (see reference 17. for the Pt(111) case). If we had the H-up situation, in which non-hydrogen bonded OH group of water, free OH, are present on the surface, we should an

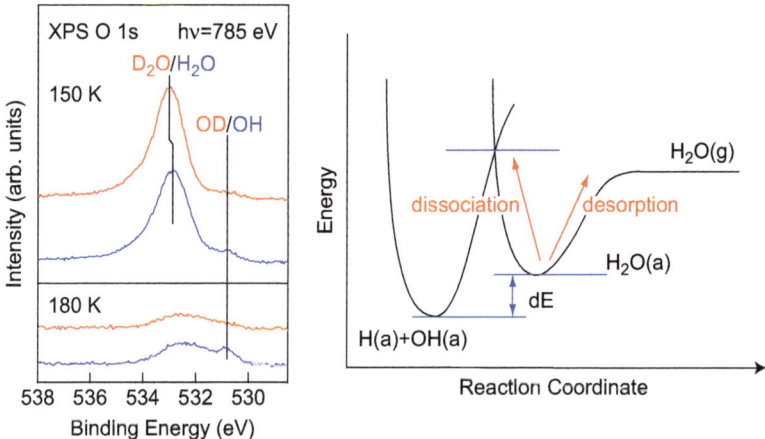

Figure 15.4   XPS for water (H_2O and D_2O) adsorbed on Ru(0001)[7] (left) and schematic illustration of desorption and dissociation barriers for H_2O on Ru(0001) (right). Chemical species on Ru(0001) are identified through chemical shifts in the O 1s XPS. The formation of hydroxyl at 180 K is characterized by the appearance of an O 1s XPS peak at *ca.* 531 eV, whereas that for intact water appears at ca. 532–533 eV. The decomposition is observed for H_2O but not for D_2O due to zero-point energy differences.
Reprinted from reference 8. Copyright 2010, with permission from Elsevier.

XAS signature of free OH. There are abundant amount of free OH on the surface of ice water, which gives rise to the peak at 535 eV assignable to orbitals localized at free OH groups. Another way to probe the orientation of the non-hydrogen-bonded OH group is *via* vibrational excitation of the OH vibrational mode. Several techniques have been utilized to probe the O-H stretch vibration of adsorbed water: IR absorption, electron energy loss and sum frequency generation.[16,18–20] The fingerprint of non-hydrogen-bonded OH groups pointing towards the vacuum is an isolated high-frequency "free OH" band at 3680 cm^{-1} (2730 cm^{-1} for deuterated water). XAS and IR studies on ammonia terminated ice surface indicate that these states almost exclusively reside at the ice surface.[21,22] In Figure 15.3, we compare p-polarized XAS and IR results for the surface of ice water[23] and the water monolayer on Pt(111) and Cu(110);[6,16,24] the polarization shown is with the light field oscillating normal to the surface (z-direction), *i.e.* in the direction of the free OH-groups. On Pt(111), OH groups are no longer uncoordinated resulting in broadened feature and the loss of intensity. The strong feature at 532 eV is attributed to molecules binding through oxygen (Pt-O) and the feature at 534 eV to the binding through hydrogen (Pt-HO), which will be discussed in the next section. On the other hand, a notable amount of free OH feature remains in XAS for the water monolayer on Cu(110) with H-down:H-up ratio of 2:1, The IR absorption of free OH is intense for water adsorbed on Cu(110)[16] but negligible on both Pt(111),[18,25,26] which corroborates with the presence of free OH on the mixed H-down:H-up water layer on Cu(110), but not on the H-down water layer on Pt(111) .

15.3 Competition Between Thermal Dissociation and Desorption

Under certain circumstances, the O-H bond of adsorbed water dissociates. Compilations of studies of water on metal surfaces[1,2] show that Ru(0001) and Cu(110) are on the border between active and inactive metal surfaces with respect to dissociation of water. The dissociation of water on these surfaces is supported by the appearance of two different O 1s XPS peaks assignable to, respectively, water and hydroxyl, see Figure 15.4 (left). Although a dissociated layer is thermodynamically favorable on Ru(0001) and Cu(110), the dissociation must overcome activation barriers. The relative heights among activation barriers for different reaction play in determining the dissociation probability. The structure of water on Ru(0001) was an issue of debate.[7,19,20,27–34] Feibelman[34] found that a partially dissociated layer consisting of a near-planar hexagonal mixed network of adsorbed water and hydroxyl (Mixed H$_2$O:OH in Figure 15.1). There is, however, an activation barrier that impedes the decomposition of water, as depicted in the schematic potential energy diagram in Figure 15.4 (right). It was found on Ru(0001) that dissociation and desorption of water occur with very similar barriers, and the probability of dissociation is thus ruled by the balance between desorption and dissociation kinetics.[7]

An indication of this delicate balance between dissociation and desorption is found from the anomalous isotope effect and kinetics in the thermal desorption spectra of water on Ru(0001).[35-39] The isotope effect arises from differences in the zero-point vibrational energy contribution to the dissociation barrier between H_2O and D_2O. The dissociation pathway involves elongation of an O-H or O-D bond. H_2O has a 0.1 eV higher zero-point vibrational energy compared to D_2O in the dissociative pathway. The bonding to the surface, on the other hand, is equivalent for the two isotopes giving similar barriers to desorption. The zero-point contribution directly affects the barrier to dissociation, which will be only slightly higher than desorption barrier for H_2O but significantly more so for D_2O. No dissociation is observed on the surfaces of neighboring elements to the right in the periodic table, *e.g.* Ni(111), Cu(111), Rh(111) and Pt(111), for which the barrier to dissociation thus becomes significantly larger than the desorption barrier.[1,2]

15.4 Electronic Structure and Bonding Mechanism

Photoelectron spectroscopy (PES) has been used to probe the valence electronic structure of water. PES determines the binding energy and character of the different occupied molecular orbitals can be determined. In the PES spectra of gas phase water,[40,41] see Figure 15.5, peaks in the valence region were assigned

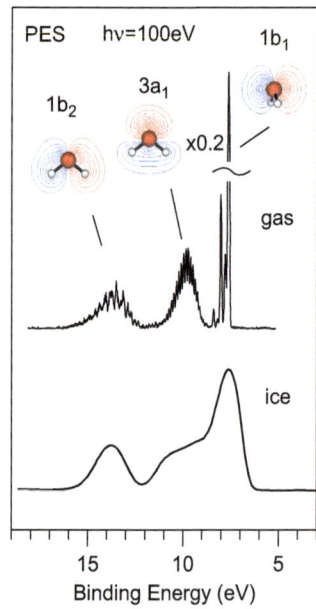

Figure 15.5 PES spectrum of gas phase water measured at a photon energy of 100 eV[40] and PES spectra of ice at photon energies of 100 eV.[43] Reprinted with permission from reference 43. Copyright 2005, American Institute of Physics.

to $1b_2$, $3a_1$ and $1b_1$ states of water. The highest occupied state ($1b_2$) has a non-bonding character and highly localized on the oxygen atom, as it has a lobe pointing away from the two hydrogen atoms. The third lowest occupied state ($1b_1$) has a lobe pointing toward two hydrogen atoms, which bonds the O and H atoms (*O-H*). Though the second lowest occupied state ($3a_1$) also has an *O-H* bonding character, it has a lobe pointing away from two hydrogen atoms, which can be thought as a *lp* character.

X-ray emission spectroscopy (XES) has also been used to determine the binding energy and character of the different occupied molecular orbitals. One of the main advantages of XES to PES is that it provides element specificity. In the case of water the XES process involves the projection of the valence electronic state onto the oxygen 1s core state of water. Since the spatially localized character of oxygen 1s orbital on an atom, it provides a tool to probe the molecular orbitals of water selectively and no contribution comes from the substrate.

In an aqueous electrolyte solution, water makes hydrogen bonds to surrounding water and ions. Though the strength of hydrogen bond is weak, generally ~ 0.25 eV per bond, the valence states of water is altered upon hydrogen bonding. The effect of hydrogen bonding to the valence states is demonstrated by comparing gas phase and ice water PES results in Figure 15.5. The $3a_1$ state undergoes the most prominent change of substantial broadening, which have also seen for PES study in liquid water.[40,42] This is due to the electron re-distribution inside the water molecules to minimize the Pauli repulsion upon hydrogen bonding.[43] PES [40,42] and XES [44,45] studies of the liquid phase have produced similar results.

Metal and semiconductor materials have valence and conduction electrons. The valence electrons are bound to individual atoms, as opposed to conduction electrons, which move freely within the material. We now consider the role of these electrons in the bonding mechanism of water to the metal through XES studies.

X-ray emission spectroscopy (XES) studies on water on Pt(111) showed the Pt-O and Pt-HO bonding mechanism of water in a site-specific way.[6] For XES measurements, a core-hole is created in the 1s orbital of water through the x-ray absorption process. The core-hole will subsequently be filled through the dipole transition of valence electronic states resulting in emission of x-rays; therefore the intensity of the x-ray emission peak corresponds to the *p* component of the valence electronic states. The s-polarized x-ray emission spectra, where the x-ray light is emitted in the surface plane, correspond to the oxygen 2p components projected in the surface plane,[46] and the electronic states involved in the two-dimensional hydrogen-bond network of the contact layer are probed. These spectra[8] are very similar to those of bulk ice.[47,48] The p-polarized x-ray emission spectra, whose x-ray light field is parallel to the surface normal, corresponds to the oxygen 2p components projected along the surface normal and a component parallel to the surface plane. On Pt(111), water molecules are alternately Pt-O and Pt-HO bonding to the surface. By tuning the excitation photon energy, we can selectively excite either Pt-O or Pt-HO bonding water, projecting the occupied electronic states on the oxygen atom of respectively the Pt-O and Pt-HO bonding species (Figure 15.6). The bonding mechanism shown in the insert of Figure 15.6 is proposed based on the analysis of spectral features in XES combined with electronic structure calculations.[6] The interaction of the O lp orbital ($1b_1$) with the valence electrons in Pt d-orbital form bonding Pt-O state and anti-bonding Pt-O* states, which appears in the vicinity the Fermi level. Here the bond strength is predicted by the "d-band model";[49] in which the degree of d-band population and the position of d-band center are important. In the case of water on Pt, a partially unfilled nature of Pt 5d-band makes the Pt-O* state partially unfilled. The depopulation is seen as a peak at 532.5 eV in the out-of-plane XAS spectrum in Figure 15.7.

The closed-shell d^{10} configuration of Cu surfaces, however, does not afford this mechanism.[50] The decomposition, and comparison with water on Cu(110)[16] (Figure 15.7, highlights the influence on the bonding by the d-band position with respect to the Fermi level as depicted in the inset. Let us now consider the M-O bonding channel. The valence electrons of Cu occupy the 3d

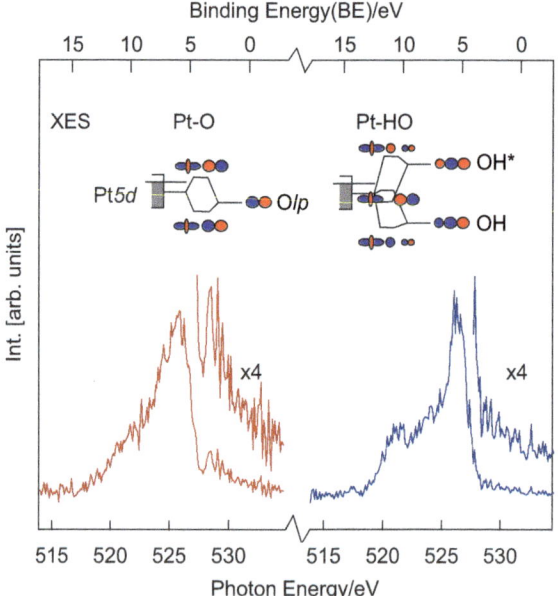

Figure 15.6   X-ray emission spectra (p-polarized) from Pt-O and Pt-HO bonding water showing the occupied orbital structure out-of-plane (p_z components).[6] The Pt-O bonding water has an x-ray absorption threshold at lower energy than that for the Pt-HO species bonding through hydrogen. This allows separated XES spectra for the Pt-O and Pt-HO bonding water to be obtained by using two excitation energies (532 eV and 538 eV) and a subtraction procedure. The inset shows schematic diagrams of the Pt-O and Pt-HO bonds.
Reprinted from reference 6. Copyright 2002 by the American Physical Society.

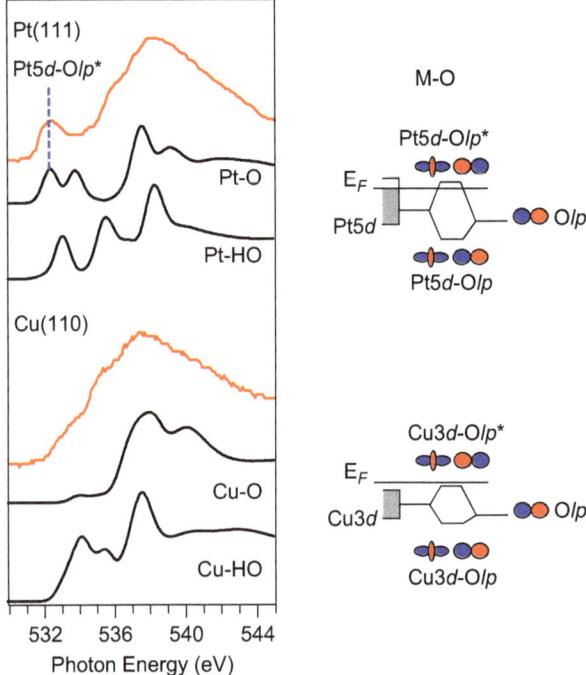

Figure 15.7 X-ray absorption spectrum (p-polarized) for water on Pt(111) (top) and Cu(110) (bottom) and computed x-ray absorption spectra for M-O and M-HO bonded water on each surface corresponding to the p_z component.[50]
Reprinted from reference 8. Copyright 2010, with permission from Elsevier. The inset shows schematic diagrams of the Pt-O and Cu-O bonds.

orbitals more spatially contracted than the 5*d* orbitals of Pt. Moreover, high density of conduction electron in Cu causes the Pauli repulsion with the electrons of water, which will inhibit the approach of O *lp* to the 3d orbitals of Cu. These effects give rise to a smaller splitting between the bonding and anti-bonding states on Cu compared to on Pt as illustrated in Figure 15.7. In the interaction with the closed-shell water lone pair, both bonding and anti-bonding states become fully occupied leading to Pauli repulsion. Accordingly, no such O-bonding related peak appears in the XAS for Cu surfaces resulting in no net attractive interaction in the M-O bonding channel. The *s*-electrons are much more mobile and can easily move away from the bonded metal atom towards neighboring atoms to minimize the overall repulsion. This can be described in a simplistic way as that the water lone-pair "digs a hole" in the *s*-band[50] as shown schematically in Figure 15.8. Since there is now a partial positive charge on the metal atom, the lone pair orbital will be stabilized through electrostatic interaction, often denoted dative bonding. This provides the main surface bonding mechanism for water and describes general lone-pair

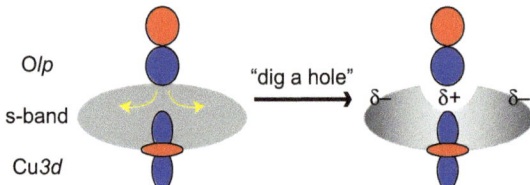

Figure 15.8   Schematic illustration of the water O lp orbital "digging the hole" in the metal sp-band to open for water-metal dative bonding.

interactions on surfaces.[50] While no chemisorbed water layer is observed on Cu(111),[50] it is observed on corrugated Cu(111) in which the depleted density of the s-electrons at atomic rows on Cu(110) lowers the energy costs to dig the hole in the s-band.

15.5   Conclusions

The structure of water at noble metal surfaces results from a complex interplay of a number of effects, including the balance between water-water and water-metal bond strength, which will directly affect barriers to dissociation, desorption and other catalytic reactions. The localized valence electrons and the mobile conduction electrons play important roles in the bonding. Localized valence electrons facilitate the covalent bonding to water forming a bonding and anti-bonding combination. The population and position of valence band are parameter to tune nature of anti-bonding state in the vicinity of the Fermi level. The Pauli repulsion by the conduction electron hampers the approach of water to the valence electrons, which can be reduced by the geometric effects. The mechanistic picture of bonding of water to noble metals is anticipated to be applicable also to semiconductors in PEC materials.

Acknowledgements

This work was supported by Office of Basic Energy Sciences, US Department of Energy under contract DE-AC02-76SF00515. Portions of this research were carried out at the Stanford Synchrotron Radiation Lightsource (SSRL), a division of SLAC National Accelerator Laboratory and an Office of Science user facility operated by Stanford University for the U.S. Department of Energy. The results discussed in this review have naturally been obtained in collaboration with a large number of extraordinary scientists, and we like to in particular thank Klas Andersson, Lars GM Pettersson and Theanne Schiros.

References

1. P. Thiel and T. Madey, *Surf. Sci. Rep.*, 1987, **7**, 211.
2. M. Henderson, *Surf. Sci. Rep.*, 2002, **46**, 1.

3. G. A. Kimmel, N. G. Petrik, Z. Dohnálek and B. D. Kay, *Phys. Rev. Lett.*, 2005, **95**, 166102.
4. G. A. Kimmel, N. G. Petrik, Z. Dohnálek and B. D. Kay, *J. Chem. Phys.*, 2007, **126**, 114702.
5. D. L. Doering and T. E. Madey, *Surf. Sci.*, 1982, **123**, 305.
6. H. Ogasawara, B. Brena, D. Nordlund, M. Nyberg, A. Pelmenschikov, L. G. M. Pettersson and A. Nilsson, *Phys. Rev. Lett.*, 2002, **89**, 276102.
7. K. Andersson, A. Nikitin, L. G. M. Pettersson, A. Nilsson and H. Ogasawara, *Phys. Rev. Lett.*, 2004, **93**, 196101.
8. T. Schiros, K. J. Andersson, L. G. M. Pettersson, A. Nilsson and H. Ogasawara, *J. Electron Spect. Rel. Phen.*, 2010, **177**, 85–98.
9. A. Hodgson and S. Haq, *Surf. Sci. Rep.*, 2009, **64**, 381–451.
10. J. Carrasco, A. Hodgson and A. Michaelides, *Nat. Mater.*, 2012, **11**, 667–674.
11. J. Stöhr, *NEXAFS spectroscopy*, Springer-Verlag, Berlin Heidelberg, 1992.
12. J. Stöhr, F. Sette and A. L. Johnson, *Phys. Rev. Lett.*, 1984, **53**, 1684.
13. O. Björneholm, A. Nilsson, H. Tillborg, P. Bennich, A. Sandell, B. Hernnäs, C. Puglia and N. Mårtensson, *Surf. Sci.*, 1994, **315**, L983.
14. A. Nilsson, A. Stenborg, H. Tillborg, K. Gunnelin and N. Mårtensson, *Phys. Rev. B*, 1993, **47**, 13590.
15. N. Mårtensson and A. Nilsson, *J. Electron Spect. Rel. Phen.*, 1996, **75**, 209.
16. T. Schiros, S. Haq, H. Ogasawara, O. Takahashi, H. Öström, K. Andersson, L. G. M. Pettersson, A. Hodgson and A. Nilsson, *Chem. Phys. Lett.*, 2006, **429**, 415–419.
17. T. Schiros, L.-Å. Näslund, K. Andersson, J. Gyllenpalm, G. S. Karlberg, M. Odelius, H. Ogasawara, L. G. M. Pettersson and A. Nilsson, *J. Phys. Chem. C*, 2007, **111**, 15003.
18. S. Haq, J. Harnett and A. Hodgson, *Surf. Sci.*, 2002, **505**, 171–182.
19. C. Clay, S. Haq and A. Hodgson, *Chem. Phys. Lett.*, 2004, **388**, 89–93.
20. D. N. Denzler, C. Hess, R. Dudek, S. Wagner, C. Frischkorn, M. Wolf and G. Ertl, *Chem. Phys. Lett.*, 2003, **376**, 618–624.
21. P. Wernet, D. Nordlund, U. Bergmann, M. Cavalleri, M. Odelius, H. Ogasawara, L. Å. Näslund, T. K. Hirsch, L. Ojamäe, P. Glatzel, L. G. M. Pettersson and A. Nilsson, *Science*, 2004, **304**, 995–999.
22. H. Ogasawara, N. Horimoto and M. Kawai, *J. Chem. Phys.*, 2000, **112**, 8229–8232.
23. D. Nordlund, H. Ogasawara, P. Wernet, M. Nyberg, M. Odelius, L. G. M. Pettersson and A. Nilsson, *Chem. Phys. Lett.*, 2004, **395**, 161–165.
24. H. Öström, A. Nillson, M. Kawai and H. Ogasawara, unpublished.
25. H. Ogasawara, J. Yoshinobu and M. Kawai, *Chem. Phys. Lett.*, 1994, **231**, 188.
26. M. Nakamura, Y. Singaya and M. Ito, *Chem. Phys. Lett.*, 1999, **309**, 123.
27. D. Menzel, *Science*, 2002, **295**, 58.
28. D. N. Denzler, S. Wagner, M. Wolf and G. Ertl, *Surf. Sci.*, 2003, **532–535**, 113.

29. A. Michaelides, A. Alavi and D. A. King, *J. Am. Chem. Soc.*, 2003, **125**, 2746.

30. J. Weissenrieder, A. Mikkelsen, J. N. Andersen, P. J. Feibelman and G. Held, *Phys. Rev. Lett.*, 2004, **93**, 196102.

31. S. Meng, E. G. Wang, C. Frischkorn, M. Wolf and S. Gao, *Chem. Phys. Lett.*, 2005, **402**, 384.

32. G. Materzanini, G. F. Tantardini, P. J. Lindan and P. Saalfrank, *Phys. Rev. B*, 2005, **71**, 155414.

33. N. S. Faradzhev, K. L. Kostov, P. Feulner, T. E. Madey and D. Menzel, *Chem. Phys. Lett.*, 2005, **415**, 165.

34. P. J. Feibelman, *Science*, 2002, **295**, 99.

35. D. N. Denzler, S. Wagner, M. Wolf and G. Ertl, *Surf. Sci.*, 2003, **544**, 348–348.

36. D. N. Denzler, S. Wagner, M. Wolf and G. Ertl, *Surf. Sci.*, 2003, **532–535**, 113–119.

37. W. Hoffmann and C. Benndorf, *Surf. Sci.*, 1997, **377–379**, 681–686.

38. G. Held and D. Menzel, *Surf. Sci.*, 1995, **327**, 301–320.

39. P. J. Schmitz, J. A. Polta, S. L. Chang and P. A. Thiel, *Surf. Sci. Lett.*, 1987, **186**, 283–284.

40. B. Winter, R. Weber, W. Widdra, M. Dittmar, M. Faubel and I. V. Hertel, *J. Phys. Chem. A*, 2004, **108**, 2625–2632.

41. S. Myneni, Y. Luo, L. Å. Näslund, M. Cavalleri, L. Ojamäe, H. Ogasawara, A. Pelmenschikov, W. Ph, P. Väterlein, C. Heske, Z. Hussain, L. G. M. Pettersson and A. Nilsson, *J. Phys.-Condens. Mat.*, 2002, **14**, L213.

42. M. Faubel, B. Steiner and J. P. Toennies, *J. Chem. Phys.*, 1997, **106**, 9013–9031.

43. A. Nilsson, H. Ogasawara, M. Cavalleri, D. Nordlund, M. Nyberg, P. Wernet and L. G. M. Pettersson, *J. Chem. Phys.*, 2005, **122**, 154505–154509.

44. J. H. Guo, Y. Luo, A. Augustsson, J. E. Rubensson, C. Såthe, H. Ågren, H. Siegbahn and J. Nordgren, *Phys. Rev. Lett.*, 2002, **89**, 137402.

45. S. Kashtanov, A. Augustsson, Y. Luo, J. H. Guo, C. Såthe, J. E. Rubensson, H. Siegbahn, J. Nordgren and H. Ågren, *Phys. Rev. B*, 2004, **69**, 024201.

46. A. Nilsson and L. G. M. Pettersson, *Surf. Sci. Rep.*, 2004, **55**, 49.

47. B. Brena, D. Nordlund, M. Odelius, H. Ogasawara, A. Nilsson and L. G. M. Pettersson, *Phys. Rev. Lett.*, 2004, **93**, 148302.

48. T. Tokushima, Y. Harada, O. Takahashi, Y. Senba, H. Ohashi, L. G. M. Pettersson, A. Nilsson and S. Shin, *Chem. Phys. Lett.*, 2008, **460**, 387.

49. B. Hammer and J. K. Nørskov, *Surf. Sci.*, 1995, **343**, 211–220.

50. T. Schiros, O. Takahashi, K. J. Andersson, H. Öström, L. G. M. Pettersson, A. Nilsson and H. Ogasawara, *J. Chem. Phys.*, 2010, **132**, 094701.

CHAPTER 16

New Perspectives and a Review of Progress

HANS-JOACHIM LEWERENZ*[a] AND LAURENCE PETER*[b]

[a] Joint Center for Artificial Photosynthesis, California Institute of Technology, 1200 E. California Blvd, Pasadena, CA 91125, USA;
[b] Department of Chemistry, University of Bath, Bath BA2 7AY, United Kingdom
*Email: lewerenz@caltech.edu; l.m.peter@bath.ac.uk

16.1 Introduction

This final chapter reviews a range of topics that could advance the field of light-induced energy conversion, in particular, of photoelectrochemical approaches, beyond current research and development activities. This compilation represents a subjective view with data and results considered from the fields of photonics, electronics, electrochemistry and life sciences. Our view of the relevance of these topics to the content of this book is given in short notes, and suggestions are outlined about how to incorporate the concepts and findings into the next generation of solar fuel generating structures and devices. The chapter concludes with a brief survey of progress towards the ultimate goal of generating solar fuels.

16.2 Advanced Photonics

16.2.1 Surface Plasmons

The excitation of volume and surface plasmons is well known in metal physics, and the reader is referred to a review by Raether and references therein for

RSC Energy and Environment Series No. 9
Photoelectrochemical Water Splitting: Materials, Processes and Architectures
Edited by Hans-Joachim Lewerenz and Laurence Peter
Published by the Royal Society of Chemistry, www.rsc.org

further details.[1] In simple terms, the collective electron interaction is based on the long-range part of the Coulomb interaction and determines substantially the optical properties of metals and semiconductors.[2] The dispersion relations for volume and surface plasmons and for photons ($\omega = cK$) are displayed in Figure 16.1. Three main features can be distinguished: the branch starting at ω_P denotes the dispersion of longitudinal volume plasmons, the branch starting at zero energy approximating the frequency $\omega_P/\sqrt{2}$ denotes the surface plasmon dispersion. In the shaded region, density fluctuations are damped exponentially, but excitation by light is possible, as can be seen from the figure where the photon dispersion intersects with the damped modes. The volume plasmon dispersion can be deduced from the properties of a nearly free electron gas, where the effective mass m* contains the energy band structure terms.[3] The volume plasmon resonance is given by

$$\omega_p = \sqrt{\frac{n_e q^2}{\varepsilon_0 m^*}} \qquad (16.1)$$

and its dispersion for small K values is

$$\omega = \omega_p \left(1 + \frac{3v_F^2}{10\omega_p^2} K_{//}^2\right) \qquad (16.2)$$

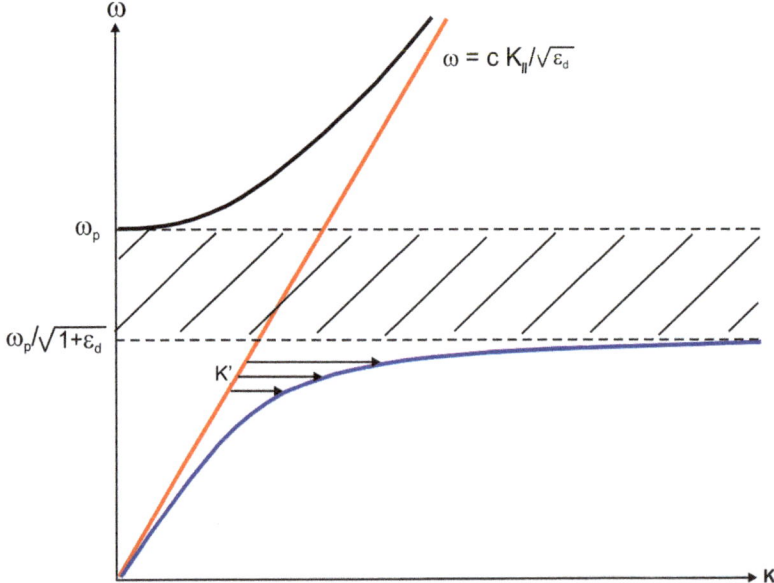

Figure 16.1 Dispersion relations for volume and surface plasmons of a nearly free electron gas; c, velocity of light; $K_{\|}$, wave vector parallel to the surface; ω_p, volume plasmon resonance frequency; ε_d permittivity of dielectric; the shaded region shows ω - $K_{\|}$ relations where no real solutions exist (damped modes); note the additional K' vectors from surface roughness that connect the light line to the surface plasmon dispersion (see text).

The wave vector $K_\parallel$ is parallel with the bounding surface, and one sees that additional k vectors are needed to excite surface plasmons with light. These can be provided by surface roughness[4] or by specific geometries.

The role of surface plasmons[5] at the solid-electrolyte interface is an interesting topic, and in the present context one can ask whether surface plasmons can be used to induce photoelectrochemical reactions away from equilibrium conditions. First, the excitation and the decay of (delocalized) surface plasmons on electrochemically roughened Ag films is considered. In this case, the roughness adds a K'-spectrum to the parallel wave vector dispersion that allows surface plasmon excitation, as indicated in Figure 16.1. A perfect test of the surface plasmon decay is provided by photoemission into electrolytes.[6] This method allows one to investigate the energy relations of solids at photon energies below the vacuum work function, since the threshold for yield photoemission at the solid-acidic electrolyte contact is ~ 3.1–3.2 eV.[7] Figure 16.2 shows the photoemission signal from the surface plasmon of roughened Ag (111), where the resonance energy lies at 3.5 eV, *i.e.* only slightly above the photoemission threshold. A distinct signal is noted at the surface plasmon energy. It follows that the surface plasmon can decay by emission of hot electrons with energy centered at the plasmon resonance energy. This is a first indication that such excitations can be used to induce photoelectrochemical reactions that otherwise need a large overpotential, such as CO_2 reduction, where the first step requires an overpotential of 1.9 V.[8] However, the quantum yield of electron emission is less than 10^{-3}, which can be attributed to the following effects. Firstly, the emission cone for photoemission is very narrow for photoelectrons with kinetic energy only slightly above the threshold energy, and therefore a large fraction of the excited electrons will not be detected in the experiment (see Figure 16.3). Secondly, the generated roughness defines largely

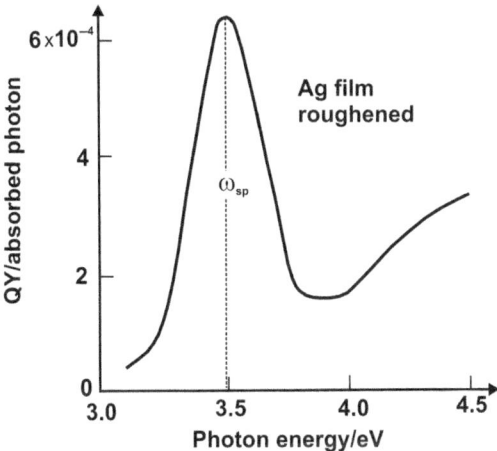

Figure 16.2 Photoemission-into-electrolyte spectrum of a roughened Ag film. Note the onset of the photoemission near 3.1 eV. Electrolyte, 1N H_2SO_4, counter electrode Pt, reference electrode NHE (1 bar H_2) blackened Pt; electrode potential -0.2 V (NHE) (after reference 8).

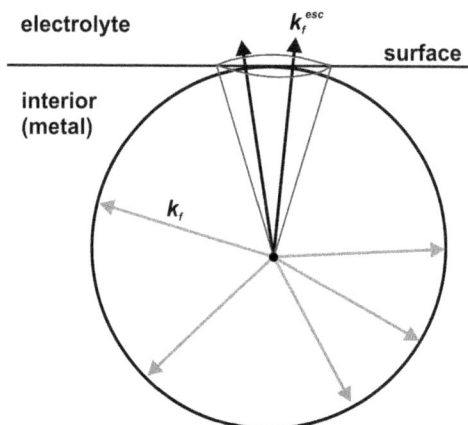

Figure 16.3 Schematic for the photoelectron emission cone in a yield experiment plotting the electron momenta in the final state, k_f, and indicating the surface (see text). The superscript 'esc' denotes the escaping photoelectrons that have sufficient momentum perpendicular to the surface to overcome the work function.

the spectrum of K' vectors, and from the photoemission experiment it was not possible to assess which roughness distribution would be particularly suited for a high yield of photoelectrons from surface plasmon decay.

The escape cone is defined by the electron momentum $\hbar \cdot k_f$ perpendicular to the surface, which is given by

$$\left|k_f^\perp\right| = \frac{1}{\hbar}\sqrt{2m^*(E_f - \Phi)}$$

where the energetic difference between the final state energy E_f and the work function Φ determines the width of the escape cone:

$$\theta = \cos^{-1}\left(\frac{|k^\perp|}{|k|}\right)$$

One observes that (i) hot electrons are injected into the electrolyte by surface plasmon decay and (ii) the quantum yield for the process is rather low, which can be related to the small energetic difference between the electron excess energy and the work function at the Ag-electrolyte contact. Therefore, in principle, it should be possible to induce reactions relatively far from equilibrium as shown in a schematic for two CO_2 reduction steps in Figure 16.4. The first step occurs at -1.9 V *vs.* NHE and the second step (leading to the formate ion) at -1.5 V.[8]

The role of localized surface plasmons in catalysis and electrochemistry has already been discussed to some extent. The resonance energy is typically lower than that of the surface plasmons shown in Figure 16.2, although it can be

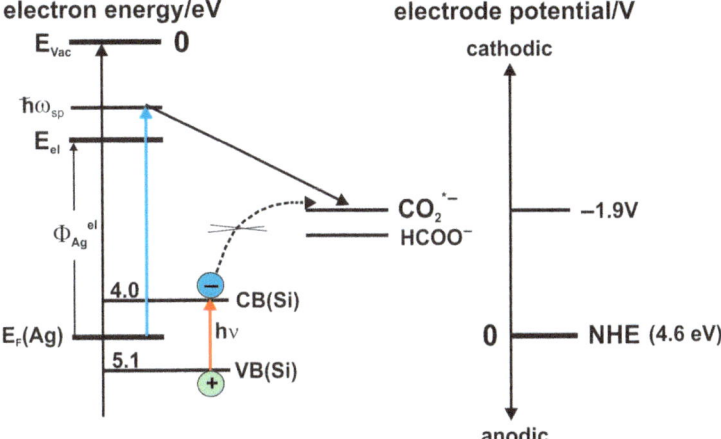

Figure 16.4 Energy and potential scheme for intermediate formation in carbon dioxide reduction; left: energy scheme related to the physics vacuum scale (E_{Vac}); the work function for Ag in acidic electrolyte, Φ_{Ag}^{el}, is labeled E_{el} and is approximately 3 eV. Also shown are the band edges of Si (electron affinity χ) located at 4 eV (conduction band, CB) and at 5.1 eV (valence band). The photoexcitation of p-Si is shown to produce excess minority electrons that are not energetic enough to overcome the first step in CO_2 reduction (note crossed dotted line); the excitation of a Ag surface plasmon (delocalized) can overcome the threshold. Also shown are the potentials for the CO_2^- radical and for $HCOO^-$ relative to NHE (see text).

tuned by size and shape of the nanoparticles that show the resonance. The resonance of localized surface plasmons (LSP) is found from the expression for the photon scattering cross section σ_{SC} in conjunction with the formula for the particle polarizability α

$$\sigma_{SC} \propto \left(\frac{2\pi}{\lambda_{ph}}\right)^4 |\alpha|^2 \text{ or with } k_{ph} = \frac{2\pi}{\lambda_{ph}}, \ \sigma_{SC} = k_{ph}^4 |\alpha|^2 \tag{16.3}$$

$$\alpha \propto V_{NP} \frac{\varepsilon_{NP} - \varepsilon_M}{\varepsilon_{NP} + 2\varepsilon_M} \tag{16.4}$$

where the photon wavelength, the nanoparticle volume V_{NP}, and the permittivities of the metal nanoparticle and of the host, ε_{NP}, ε_M, respectively enter the expressions. A typical result of the enhancement of the absorbance of Au nanoparticles with different sizes is shown in Figure 16.5.

One also observes a small shift in the resonance frequency towards lower wavelength with decreasing size. The surface plasmon resonance is given by the Froehlich condition $\varepsilon_{NP} = -2\varepsilon_M$. The resonance energy in Figure 16.5 is about 2.3 eV, which would be enough to induce non-equilibrium photoelectrochemical reactions. However, a direct effect based on LSPs has not yet been demonstrated. LSP excitation is also accompanied by a strong increase in the electrical field strength,[9] a fact that has been used in application of surface

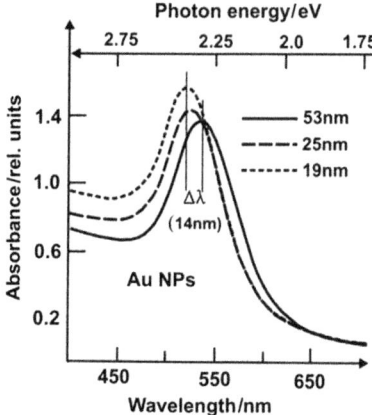

Figure 16.5   Spectral absorption of Au NPs of different sizes in the region near the LSP resonance (see text).

enhanced Raman scattering in biomedical sciences, for instance.[10] It is also known that this resonance excitation results in temperature increases.[11] Therefore, besides reactions involving energetic hot electrons, increases in reaction rates due to local heating and non-linear field effects might be envisaged. It will be difficult, however, to disentangle the contributions of the various photonic effects from the size- and shape dependence of the electrocatalytic activity of nanoparticles. Hitherto, convincing electrocatalysis based on LSP excitation has not yet been demonstrated, despite rather intense research in the field.[12–16]

16.2.2   Coupled Förster Excitation Energy Transfer

Carrier-free transfer of photonic excitation energy is known from photosynthesis, where the light absorption and the reactive centers are spatially separated.[17] The transfer of the excitation energy to the reaction centers takes place in nature *via* multichromophoric Förster transfer,[18,19] (see also chapter 14). In principle, this transfer follows the Hertzian dipole interaction and is described for a donor-acceptor molecular transfer by the transfer rate

$$k_T = k_D \left(\frac{R_0}{R}\right)^6 \tag{16.5}$$

where the Förster radius R_0 is given by

$$R_0^6 = 8.8 \cdot 10^{17} \frac{\kappa^2}{n^4} J \tag{16.6}$$

In equations (16.5) and (16.6), k_D denotes the donor fluorescence decay rate, κ is the dipole orientation factor, n the refractive index, and J the overlap integral, given by the integral over energy of the product of the donor emission spectrum $F(v)$ and the acceptor absorption spectrum $A(v)$

$$J = \int \frac{F(v) \cdot A(v)}{v^4} dv \qquad (16.7)$$

A simplified scheme showing the spectral overlap is presented in in Figure 16.6 for a single Förster transfer.

For photocatalytic reactions, it is of interest to know over what distance excitation energy can be transferred *via* the Förster mechanism. The Förster radius, R_0, is defined as the distance at which the energy transfer is 50% efficient. Typical values for R_0 lie in the range between 5 and 10nm. For Dexter energy transfer, this range is smaller, being ~ 1 nm.[20] For the construction of a photoelectrocatalytic device where the location of absorption is clearly separated from that of the catalytic reaction (at a molecular catalyst for CO_2 reduction, for instance), efficient operation would demand a highly porous surface texture, similar to that of dye sensitized solar cells.[21,22] The complication and the challenge would be to achieve even distribution of absorption centers in very similar proximity (*i.e.* in the range of 5 nm) to the catalyst surface. Although this goal might be achievable, it is more desirable to spatially separate light absorption and catalytically active sites, since then only catalyst stability is required if the remote absorber has no interface with the electrolyte. Non-radiative Förster transfer is limited to energy transfer over distances less

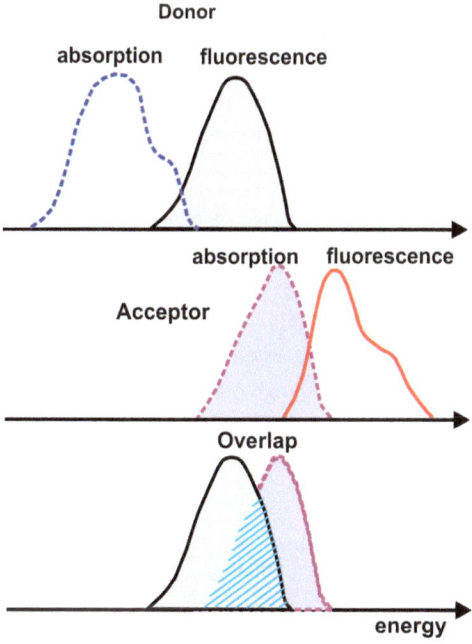

Figure 16.6 Schematic of the spectral overlap of donor fluorescence and acceptor absorption of a single molecule Förster transfer (see text).

than 10 nm, and it lacks controllable directionality. An approach that allows mitigation of these problems was introduced about a decade ago.[23] It is based on a coupled Förster transfer – the surface plasmon polariton process. Surface plasmon polaritons (SPP) can be viewed as mixed electromagnetic-mechanical waves that propagate along interfaces between metals and dielectric media. Their propagation lengths can reach 100 μm. The coupling of the Förster process to SPP occurs *via* the coupling of the donor molecule dipole moment to the SPP mode. SPP excitation results in efficient dipole emission to acceptor molecules.

Figure 16.7 shows the arrangement for the coupling experiment. Donor and acceptor doped dielectric layers were separated by Ag films with varying thicknesses d that are too large for classical Förster transfer. We select here $d = 60$ nm. As donor dye, Alq$_3$, dispersed in polymethylmetacrylate (PMMA) was used, and the acceptor dye was rhodamine 6 G in PMMA. The donor system was deposited onto silica, and Ag films were successively evaporated on the donor layer. The acceptor layer was deposited on top of the Ag layer.

Figure 16.8 shows a major result from the study for a 60 nm thick Ag film, where three spectra are displayed. The control spectra show the photoluminescence of the Alq$_3$, which overlaps only weakly with the acceptor spectrum (blue line), and the directly excited emission from the R6G acceptor through the Ag film (red line). One sees that the donor control signal at $\lambda_{res} = 520$ nm is larger than that of the acceptor ($\lambda_{res} = 565$ nm). This is attributed to the fact that the excitation wavelength used (408 nm) matches the Alq$_3$ absorption but is off resonance with the R6G absorption and, furthermore, that the acceptor signal is attenuated by the Ag film.

The black line in Figure 16.8 displays the characteristic features of the donor and acceptor. However, one observes that the emission from the R6G dye is enhanced by a factor of ~ 10 compared to the control sample (red line).

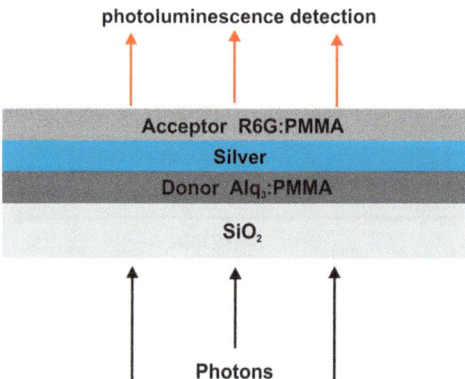

Figure 16.7   Experiment schematic for coupled Förster-SPP excitation energy transfer; laser excitation continuous wave diode laser, pump wavelength 408 nm, *i.e.* minimum of acceptor absorption (R6G); PMMA thicknesses: 80 nm, Ag thickness 60 nm (see text).

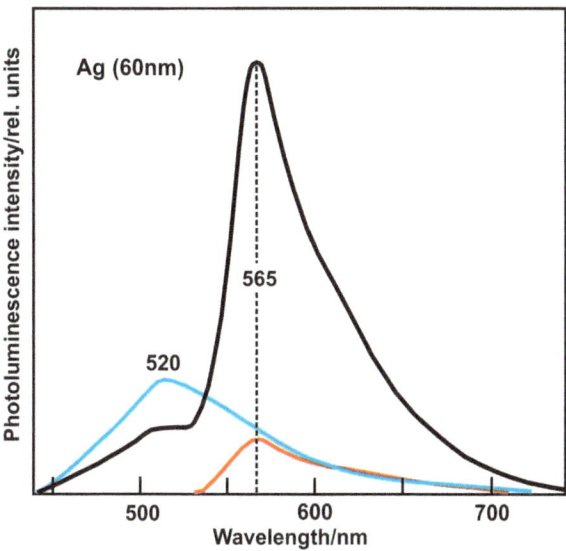

Figure 16.8 Photoluminescence spectra upon excitation of the structure shown in Figure 16.7 by laser light with 408 nm wavelength. Blue line, control spectrum for donor only, red line acceptor control spectrum, black line signal from samples with, both, donor and acceptor layers (see text).

This enhancement of the acceptor emission demonstrates that the excitation energy is transferred efficiently from the donor to the acceptor molecules by mediation of the Ag SPP over a distance that is much too large for classical Förster transfer. In principle, this effect could be used to design new architectures for synthetic light harvesting systems, where the SPP modes channel photonic excitation energy from light absorbing sites to catalytic reaction centers. So far, such designs have not yet been realized in practice.

16.2.3 Lévy Processes

In this subsection, the implications of the diffusive properties of light on energy conversion processes are summarized briefly, particularly in the context of the recent realization of a so-called Lévy glass.[24,25] Models and theories of diffusion date back many years. They include the understanding of Brownian motion[26] and Einstein's formulation in 1905 of Brownian random walk, actually used as proof for the existence of atoms.[27,28] In optics, diffusion has proven to be a useful descriptor of systems in which randomly positioned particles scatter light in independent scattering events such that possible interference processes are cancelled out by the system's disorder. This "diffusion of light" was initially investigated in astrophysics in order to analyze interstellar light that has passed through dust and nebulae.[29]

Although light is described in classical electrodynamics as a wave, light scattering by inhomogeneities with multiple scattering paths can be described as a random walk process. The random walk picture allows one to describe, for instance, the propagation of light through opaque media such as clouds. If the successive scattering events inside such a medium are independent, the multiple scattering and the disorder smoothen out, resulting in isotropic Gaussian transport that manifests itself by the smooth white color of clouds of different altitudes and composition, although each water droplet in the clouds is transparent.

The fact that diverse phenomena such as heat-, sound- and light-diffusion can be described by Brownian random walk results from the Central Limit Theorem.[30] This states that, for a large enough sample from a group of events (here light scattering) with a finite variance, the mean of all samples from that group will be approximately equal to the mean of the group of events and that all such samples show an approximate normal distribution pattern and all variances are approximately equal to the variance of the group of events divided by the size of each sample taken. In classical diffusion, the average squared displacement increases linearly with time

$$\langle x^2 \rangle = D \cdot t \tag{16.8}$$

having the same form as the well-known relation for diffusion length L of minority carriers in semiconductors.[3]

$$L = \sqrt{D \cdot \tau} \tag{16.9}$$

In systems that exhibit very strong fluctuations such as the spectral fluctuations in random lasers, the average step length is described by a power law $(\gamma > 1)$

$$\langle x^2 \rangle = D \cdot t^\gamma \tag{16.10}$$

indicating that extremely long jumps can occur and that the normal diffusion breaks down. This is the case for the so-called Lévy flights[31] and has been termed superdiffusion, where the average squared displacement increases superlinearly with time. Classical diffusion is therefore a limiting case of Lévy flights. For $\gamma = 2$, one has ballistic motion. Lévy random walks are a modification of the Lévy flights and, to compensate for the divergence of infinite jumps, long steps are penalized by coupling space and time. Classical and Lévy random walks are contrasted in Figure 16.9.

One sees that Lévy random walks are characterized by larger jumps and a type of loop near the end or beginning of the next jump. These features have been proven advantageous, for instance, in foraging. As another example, surface adsorbed polymers having a finite number of contact points with the surface can be modeled as a Lévy flight. As the figure shows, Lévy flights generally spread faster, leading to superdiffusion. Figure 16.10 compares the probability distribution functions for Gaussian and Lévy statistics. The latter is

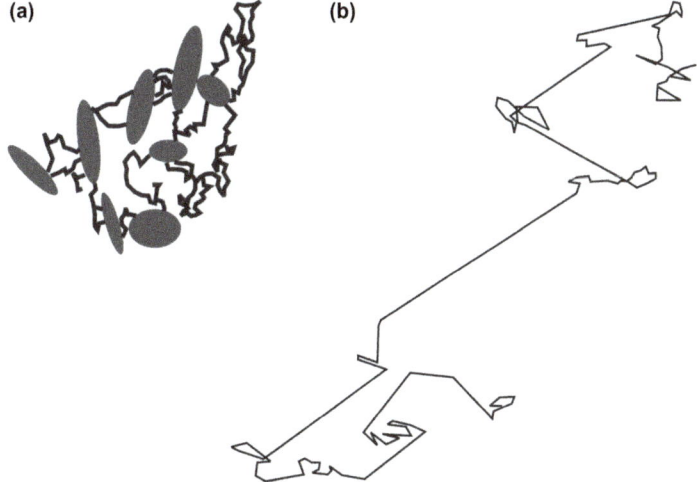

Figure 16.9 Schematic of classical Brownian walk (a) and of a Lévy random walk (b) for light scattering by a set of scattering elements with different sizes; in (a), the ellipsoidal and circular shapes indicate areas of high density of curling paths that are displayed for simplicity as filled areas; indeed, these areas are highly interconnected by random paths.

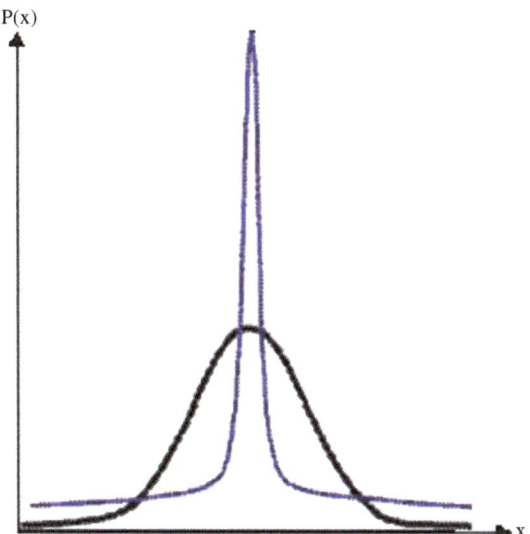

Figure 16.10 Comparison of the probability distribution function $P(x)$ of a Lévy flight (blue line) with that of a Gaussian distribution (black line) (see text).

characterized by a sharper and higher distribution near the center, *i.e.* stronger localization and, for *x*-values where the Gaussian distribution is $\sim 1/5^{th}$ of its maximum value, the Lévy tail becomes substantially larger compared to the

Gauss distribution. The heavy tail (blue line in the figure) decays according to the inverse power law

$$P(x) = \frac{1}{x^{\alpha+1}} \quad 1 < \gamma \leq 2 \tag{16.11}$$

The value of α can be related to the superdiffusion coefficient γ (see equation (16.10)): $\gamma = 3 - \alpha$ for $1 \leq \alpha < 2$, hence $1 < \gamma \leq 2$. For superdiffusion of light, this means that the transmission profile, $T(d)$ can appear as shown in Figure 16.11, which has been realized by the fabrication of a so-called Lévy glass.[24,32] The heavy tail follows the form

$$T \propto 1 / d^{\alpha/2} \tag{16.12}$$

where d denotes the thickness of a slab-like sample, assuming that absorption is zero and only scattering occurs. $\alpha = 2$ describes classical diffusive behavior, whereas superdiffusion occurs according to the above inequality for α.

An intriguing aspect of superdiffusive light propagation is that the properties can be tuned by changing the parameters that determine the step length distribution *via* density variations arising from the chosen distribution of glass microsphere diameters. In the context of solar energy converting structures, the

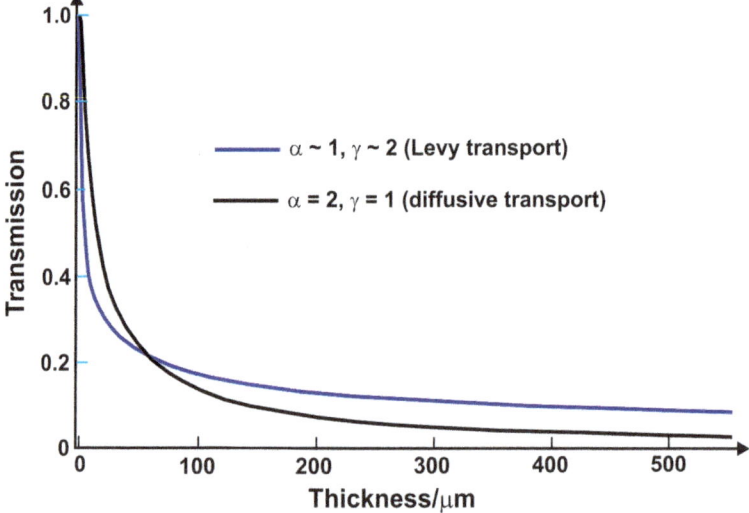

Figure 16.11   Total transmission *vs.* thickness for a fabricated Lévy glass with $\alpha \sim 1$ (corresponding to a Lorentzian Lévy flight–Cauchy distribution) compared to classical diffusive transport. The superdiffusive glass was fabricated by a suspension of titania nanoparticles in sodium silicate including an accurately chosen distribution of glass microspheres that have different diameters resulting in a modification of the density of scattering elements. The titania concentration was adjusted to achieve one scattering event occurring in the titania-filled spaces between adjacent glass microspheres (which themselves do not participate in scattering).

possibility to design strongly altered and spatially extended excess minority carrier profiles is hitherto unexplored. A first approach in this direction is a process where the randomly moving particles are minority carriers of a semi-conductor.[33,34] The process, which is referred to as photon recycling, consists of photon assisted hopping of minority carriers as a result of radiative re-combination that produces a photon that is re-absorbed at a distance *via* interband excitation. This presupposes that the material is a direct band gap semiconductor with a high rate of radiative recombination. The effect has been observed in moderately doped n-InP, where the reduced free carrier absorption ensures that the photon recycling process continues for ~100 times before a light generated minority hole recombines non-radiatively with a majority electron. It is the emission spectrum from the radiative recombination in combination with the probabilities for interband absorption and for photon propagation that produce the randomness of the process and which define the probability distribution of the walk of photons. Photons emitted near the absorption threshold of the semiconductor travel long distances before re-absorption takes place, resulting in the divergent variance typical for Lévy flights that leads to long-range Lévy walks for the excess carriers in semi-conductors. Figure 16.12 shows the experimental arrangement for lumi-nescence measurements and Figure 16.13 the results for a moderately doped n-InP wafer.

It was found that the observed luminescence intensity was proportional to the intensity of the exciting light and that the spectral behavior of the lumi-nescence was identical for all positions along the long axis of the slab. Based on comparison with model calculations, the proportionality of the local lumi-nescence intensity with the excess carrier concentration could be shown. Ac-cordingly, the observed intensity profile can be correlated to the local excess carrier concentration that is only a function perpendicular to the edge side.

Considering equation (16.10), where the average *step length* has been defined, the actual *width* of the distribution is described by $L \sim (Dt)^{1/\gamma^*}$ where $\gamma^* = 0.5$ denotes classical diffusive transport (L_0) as indicated in equation (16.9). For

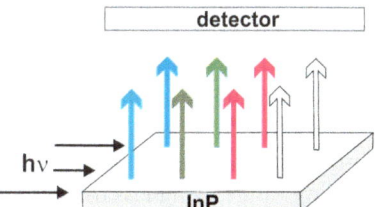

Figure 16.12 Experimental geometry for detection of long-range walks in a lumi-nescence experiment; illumination: 808 nm laser (photon energy 1.53 eV, *i.e.* above the band gap energy of InP of 1.35 eV) at the 7 mm long edge face of an InP slab (length 20 mm, thickness 350 μm). The luminescence spectra and luminescence intensity are detected per-pendicular to the illumination direction as function of distance from the edge. The colored arrows mark different position from the edge.

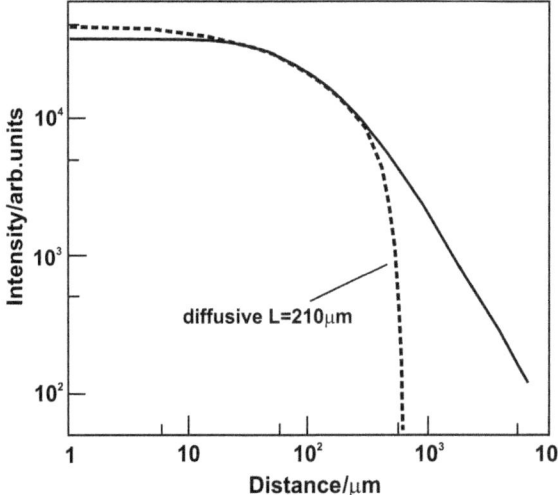

Figure 16.13   Luminescence intensity (excess carrier concentration) *vs.* distance from
the illuminated edge side for moderately $(N_D = 3 \times 10^{17} \text{cm}^{-3})$ doped
n-InP plotted on a log-log scale; full line: experimental result; dashed
line: calculation for classical diffusion assuming the untypically large
diffusion length of $L = 210$ µm (see text).

values $\gamma^* < 2$, $L^{1/\gamma^*} \gg L_0$. The tail of the solid line in Figure 16.13 is described by
an exponent $\gamma^* = 0.7$, which corresponds to a very pronounced spread beyond
that expected for classical diffusion. The data thus demonstrate that, under
specific circumstances, one can generate excess minority carriers much further
away than the classical diffusion length from the light absorption profile. An
initial interesting approach would be to analyze carefully the optical properties
of nanoparticulate light absorbers as photoanode or photcathode materials
with regard to the nanoparticle scattering properties and the size and shape
dispersion of the particles. Another consideration would be to analyze the resulting excess minority carrier profiles – particularly in nanoscopic materials
with a direct energy gap – and to assess under which conditions carrier
transport can occur that extends beyond classical diffusive motion or carrier
hopping. In microwire arrangements for solar fuels generators,[35] the use of
light scattering elements positioned between the wires at the bottom of the
structure in close proximity to metallic electrocatalysts has been realized, and
enhancements of the scattering range using less material can be envisaged. This
would also allow better access of electrolyte to the catalysts and allow for enhanced bubble removal. In thin film solar cells with direct gap semiconductors,
the losses due to superdiffusion should be considered explicitly. The spectral
distribution at the front side will contain more light with band gap energy due
to photon recycling. Also, the detailed balance calculations for theoretical efficiency of solar cells have to be modified if photon recycling is significant. This
contribution can become substantial in materials where thermalized electrons

and holes recombine radiatively. In systems where carrier transport poses a problem, carrier generation by superdiffusion of light could be used to generate carriers at a distance from the absorption site near collecting parts of a structure.

16.3 Electrodes – Structural Aspects

The concept of obtaining an optical Lévy flight by using scatterers that have self-similar fractal structures cannot be realized experimentally because of the size-dependence of the light scattering; larger particles would lead to Mie scattering, whereas small particles would almost not scatter in the Rayleigh limit. For electrodes, however, the surface structure and the total accessible surface area for catalysis are of decisive importance for catalytic activity and, besides metal electrodes, also semiconductor electrodes with fractal geometry have been realized.[36–38] A large surface-to-volume ratio is thus generally desirable for electrochemical energy conversion devices such as combined photo-electrodes and electrocatalysts, batteries or fuel cells. Interestingly, a major input on this topic originates from biology, where allometry, *i.e.* the measurement of size or mass of body parts in relation to the whole body, has led to revealing insights[39,40] regarding the allometric scaling law, which is formulated as

$$Y = Y_0 M^b \qquad (16.13)$$

Here, Y is a biological variable (metabolic rate or life span, for example), M the body mass, b the scaling exponent and Y_0 depends on the kind of organism. In Euclidean geometric scaling, reflecting simple geometric constraints, the exponent b should be a multiple of 1/3. Scaling of area (l^2) or volume (l^3), for example, leads to a scaling exponent of 2/3. It is found, however, that the exponent b follows a 1/4 behavior for properties as different as cellular metabolism, heartbeat, maximal population growth (all $b = -1/4$), blood circulation, life span, embryonic development ($b = 1/4$) and cross sectional areas of mammalian aortas, metabolic rates of entire organisms, cross-sectional areas of tree trunks ($b = 3/4$). This behavior has been connected with three unifying biological principles that obviously govern evolution by natural selection: (i) a space-filling branching structure which is fractal-like is needed to allow transport to the entire volume of an organism, (ii) the energy that is required to distribute resources down to the cellular level is minimized and (iii) the final branches of the fractal-like network is a size-invariant fixed unit. According to these principles, nature has developed methods to maximize metabolic rates by optimizing the surface areas where exchange of resources occur (for instance in organs such as the lung) and, also, the internal transport efficiency has been maximized by minimizing transport times *via* reduction of the distances within the network. Fractals are an optimal geometry for minimization of energy losses due to the transfer network while maximizing the effective surface area. Figure 16.14 illustrates a topological network representation adapted from

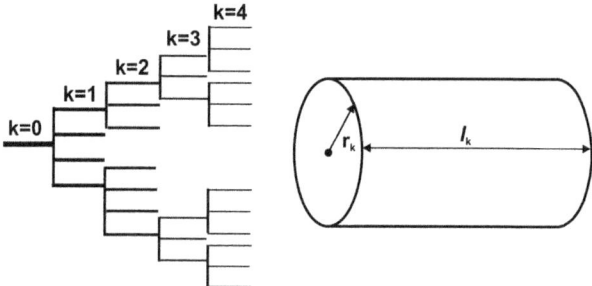

Figure 16.14   Topological representation of a biological network; k: order of the respective level from large ($N = 0$) to the final fixed size (capillary, for example) at $N = k$; the right hand side shows the dimensions of a tubular element at the kth level (see text).

ref. 38. It has been shown that, in order to obey the principle of minimized energy dissipation in the system, the network has to be a self-similar fractal where the number of branches, n, increases according to $N_k = n^k$ when their size decreases from $N_0 = 0$ to N as displayed in the figure.

Electrodes operating on the above principles would be ideal for many applications in electrochemistry and – in the context of this book – in photoelectrochemistry. This has been recognized and specifically addressed more recently.[38] The case has been made that space-filling fractal structures provide the means for efficient charge transfer from an extensive surface area to a current collector. If one considers operation of electrodes, however, an additional parameter enters that has, so far, been neglected in the biological view of allometry. Besides needing to fulfill the above-mentioned three criteria for transport, the network also has to provide efficient paths for current transport and collection, which is a non-trivial problem due to the increase of specific resistivity with decreasing diameter of the current transporting element. A further challenge is the synthetic generation of such networks, which is difficult. Attempts to prepare such networks have been made using chemical vapor deposition,[38] polishing,[41] solid state reaction (calcination)[42] or metal electrodeposition,[43] to name only a few. Synthesis of fractal architectures can be achieved by "building" the structures, as in CVD, or by generating porosity (material removal) that follow a scaling law. The latter concept has been applied for light-induced fabrication of fractal structures at silicon electrodes, which show distinct azimuthal features related to the crystal orientation used.[36,37] Figure 16.15 shows a typical result obtained by illumination of the silicon electrode with about 15% of 1 sun (white light, W-I lamp) for 10 min at anodic potentials in concentrated ammonium fluoride where oxygen evolution occurs. The structure is considerably extended over several mm, as can be seen from the 1×1 mm display of Figure 16.15. In principle four processes compete in the slightly alkaline NH_4F solution:[44] (i) anodic light-generated hole induced oxidation, O_2 evolution; (iii) oxide etching and (iv) silicon dissolution. A typical topography of such a structure (measured by a dactylograph because the depth

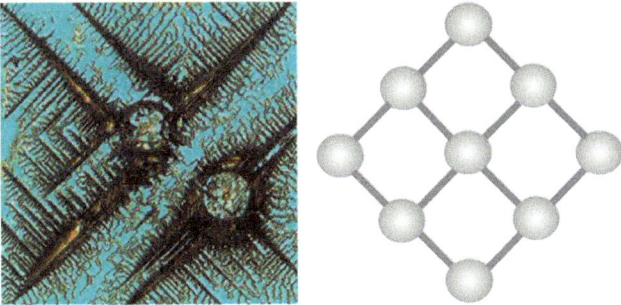

Figure 16.15 Left: light microscope image of n-Si(100) after extended photoelectrochemical etching in 40% NH$_4$F at anodic potential of V = 6V(SCE).
Light intensity between 4 and 8 mWcm^{-2} (white light). Scale: 1×1 mm;
right: crystal surface orientation of a (100) fcc lattice indicating the
azimuthal symmetry of the fractal structures (see text).

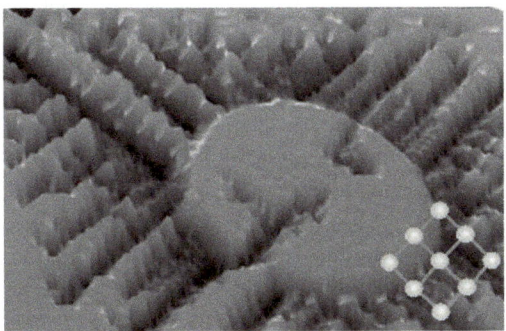

Figure 16.16 Dactylogramm of a fractal structure obtained on Si(100) (see crystal
orientation insert) that shows depressions that are up to 6 μm deep; the
circular, less corroded structure in the center is about 100 μm in diameter, the total picture frame is ∼300×200 μm.

is too large for AFM measurements) shows depressions that are a result of
crack propagation and silicon dissolution (see Figure 16.16).

The images Figures 16.15 and 16.16 show a striking resemblance to the many
pictures existing of snowflakes and, particulary, of pictures of frost on a win-
dowpane, although, instead of assembling particles from a random Brownian
walk, here a dissolution pattern with rather deep trenches is observed. These
features have been described using the diffusion-limited aggregation model
(DLA).[45] DLA is a process where the particles that are are added successively
(one at a time) to a growing entity (cluster, aggregate, condensate) arrive by a
random Brownian walk (classical diffusion, see also above under Lévy flights).
Simulations in two dimensions show that highly ramified structures are formed
that exhibit fractal geometries. For 2D and 3D clusters, the fractal exponent (or
fractal dimension) is 1.7 and 2.5, respectively.[46] The similarity between crack
propagation and DLA becomes more visible in the dielectric breakthrough in
insulators and the corresponding observed branching.[47] The basic similarity of

crack growth to non-equilibrium processes such as DLA[48] is that the probability for growth in both cases depends on the local value of a field close to the surface of the growing entity. This scalar field can be the electric potential or the concentration in the case of dielectric breakthrough or DLA, respectively. The growth probability is a function of the local value of the field or of its gradient. A complication of theoretical considerations arises from the fact that the field at each point is a function of the entire structure, which typically has a complex (fractal) geometry.

For analysis of the data on silicon electrodes, the fractal cracking approach has been followed, which reproduces the basic features by considering normal and shear stress in the surface plane assuming that Hooke's law holds. The physical origin of the stress is attributed to the well-known volume mismatch between Si and its oxide since it has been shown by photoelectron emission microscopy (PEEM) that the structures formed are partially oxidized (see Figure 16.17.[49] The figure demonstrates that the tip and the surrounding areas exhibit different surface conditions; whereas the tip shows a signal from silicon (right hand side of Figure 16.17), the surrounding areas consist largely of oxidized surfaces. For applications in photoelectrochemical water splitting, this results in a spatially inhomogeneous distribution of electrocatalysts, where the deposition occurs on the tips but not in the surrounding areas.

In the model, the oxide-induced interfacial stress and strain results in crack propagation and development of fractal structures. It has been assumed that near edges or at tips of the structures the influence of the stress results in larger strain as atoms at tips, kink sites and edges are less coordinated to the Si lattice. Strain at a surface-near region is known to facilitate crack propagation.[36,37] In the photoelectrochemical experiment for fractal structure formation, the cracking is accompanied by solvolytic attack of backbonds by water and, in the presence of holes, oxide and hydroxide formation (pH ~ 8) with resulting dissolution and three-dimensional etching at the cracks. The trench formation is a result of the deflection of holes to the sites where the electric field is largest and

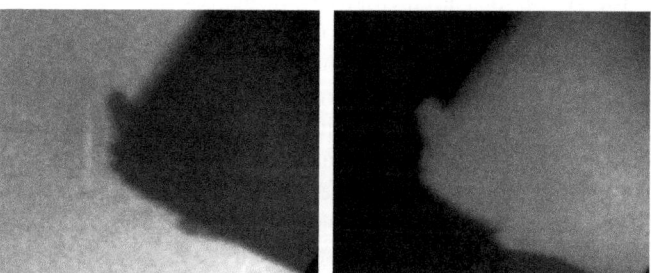

Figure 16.17   PEEM data from the inner part (a tip) of a fractal structure as shown in Figures.16.14 and 16.15. The contrast is defined as follows: left, photoelectron yield from silicon oxide, measured at a XPS bindingenergy of 104 eV; right, photoelectron yield from silicon measured for a binding energy of 99 eV. The displayed tip length is about 15 μm (see text).

where holes are consumed predominantly. Nevertheless, considering the lateral and perpendicular extension of the structures, L_p, L_s, respectivly the ratio L_p/L_s is $\sim 10^{-3}$ indicating that a 2D description is a good approximation.

One notes in Figures 16.15 and 16.16 that a large central area exists and that the structure development appears to initiate predominantly at the boundaries of the structure. Since substantial O_2 evolution takes place in the photoelectrochemical experiment, it has been assumed that particularly pronounced oxide formation occurs at the three-phase boundary at the circumference of a not yet detached oxygen bubbles where O_2 (inside the bubble), electrolyte and silicon coexist. It is known the Si oxidizes particularly efficiently in the presence of both oxygen and water, generating silicon oxide with a high rate and maximum thickness. The oxide formed is thought to be the seed for the spread of the fractal structure that can be seen in the figures. Figure 16.18 shows a schematic of the present model of the origin of the structure development. The arrows indicate the compressive stress due to the volume expansion of the oxide ($V_{SiO2} \sim 2.2 \; V_{Si}$).

For applications in water splitting structures, it would be necessary to extend the fractal geometry further into the z-direction (perpendicular to the surface) in order to generate well-defined fractal geometries with extremely large internal surface areas. The preparation process is scalable and occurs at room temperature; however, a method to deepen the structures has not yet been developed. Based on the experience with thin film silicon, it suffices if one creates trenches that are in the range of those observed in Figures 16.15 and 16.16.[50] Actually, since some thin film Si systems operate with thicknesses down to 1.5µm, one has to be consider leaving an intact Si backbone for electrode support.

Deposition of electrocatalysts can be achieved more uniformly if the material is etched to remove surface oxide. In the view of recent attempts to fabricate photovoltaic "cores" of solar fuel generators using Si-based tandem structures, the method shown here also holds promise for subsequent deposition of interlayers and a surface transition metal oxide as photoanode for realization of large-area photoelectrodes with defined geometry. A geometry that exhibits a fractal-type structure has been introduced by the Grätzel group (see also

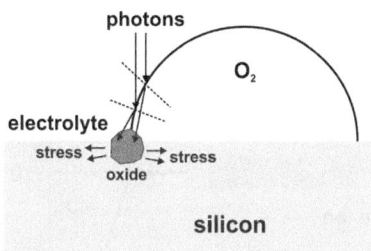

Figure 16.18 Schematic of the formation of an oxide seed for crack propagation at the circumference of an oxygen bubble; note that the bubble is an optically less dense medium compared to the electrolyte (see text).

chapter 4), where hematite (Fe_2O_3) has been prepared with a "broccoli-type" structure[51] because the carrier diffusion lengths are so low (in the low nm range) that excess minority carrier transport to the surface is restricted to the near-surface region (some reasoning for this electronic behavior will be given below in the next sub-section on transition metal oxides). In a ramified structure, carriers have a larger probability of reaching the surface for charge transfer than in a planar one. Since, the performance of $BiVO_4$ is also believed to be transport limited,[52] this points to (i) the need to control mesoscopic transport properties across nanoparticular boundaries and (ii) possible complexities of the energy band structure of transition metal oxides. The latter topic is addressed in the following section.

16.4 A Note on Electronic Properties of Transition Metal Oxides

Transition metal oxides (TMOs) are a materials class that is currently being investigated intensively for application as photoanodes in solar fuel generation systems (see chapters 4 and 5). Recent experiments on TMOs show some peculiarities of their band structure that are outlined here since they influence transport, Fermi level position, contact behavior, carrier lifetimes and optical properties, including the excess minority carrier profiles after photoexcitation. Much work has been focused on the origin of the mobility gap, particularly in the context of understanding perovskite high temperature superconductors.[53,54] The origin of the energy gap can be envisaged in a simplicistic schematic as shown in Figure 16.19 for the case of a ground state configuration of the five d orbitals in a crystal field of cubic symmetry. The splitting of the levels is such that their center of gravity with respect to energy is not affected by the perturbation; hence $3E_{t2g} + 2E_{eg} = 0$. Historically, the energy of the free ion ground state was set to zero, and the energetic difference between the three t_{2g} and the two e_g levels was defined as 10 Dq.[55] The splitting of the levels is then, as shown in Figure 16.19:

$$E_{t2g} = -4Dq \quad E_{t2g} = -4Dq \quad E_{eg} = +6Dq \tag{16.14}$$

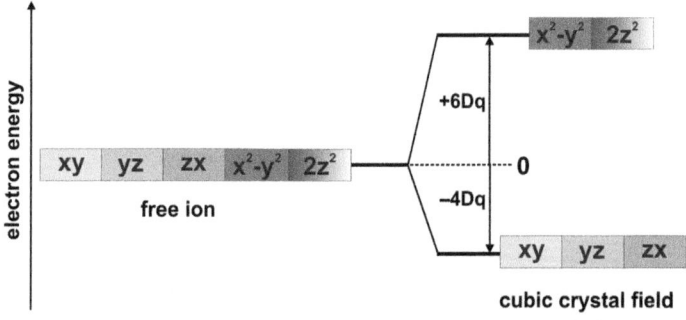

Figure 16.19 Schematic of d-band splitting in a solid with cubic crystal field (see text).

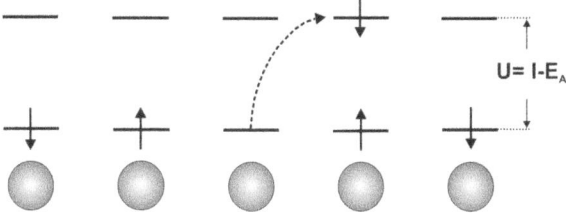

Figure 16.20   Schematic on Coulomb repulsion resulting in Mott-Hubbard gap, where the Hubbard interaction is expressed by the difference between ionization energy I and electron affinity E_A (see text).

Rather recently, it has been recognized that the crystal field splitting (which is larger for tetragonal symmetries) and the influence of electron repulsion, which leads to localization (see Figure 16.20), can play a role in the determination of the energy band structure of TMOs.[56] Here, we restrict ourselves to the Coulomb repulsion and related effects for the analysis of band structure properties. As this subsection is written as a note, it emphasizes specific findings without claiming generality, and the focus here is on the localization and delocalization of electrons in strongly correlated systems. Consider the simple case of a one dimensional arrangement of d-atoms, originally introduced by Hubbard, Kanamori and Gutzwiller,[57–59] with Coulomb electron repulsion for placing two electrons at an atomic site within a solid (Figure 16.20). The energy balance is written

$$U = E(d^{n+1}) + E(d^{n-1}) - 2E(d^n) \tag{16.15}$$

where, $E(d^n)$ is the total energy of the system of n electrons in a given atom and in a given d-shell. The above considerations are valid for materials where the energy gap arises between the d-states of the transition metal having an intra-metallic d-d band gap. This is the case only for the so-called early TMOs. In that case, the occupied oxygen related p-bands are energetically well below the Hubbard-type d-bands. For late TMOs in the periodic table, the situation changes: the d-states are located closer to the O p bands such that $E(d) - E(p) < U$. As result, the lower lying Hubbard band is located energetically below the top of the O p band as shown for both situations in Figure 16.21.

In the latter case, the lowest excitation energy corresponds to a transition from occupied oxygen orbitals to empty d states (the unoccupied upper Hubbard band) which, in the notation of Zaanen, Sawatzky and Allen[60] is called a charge transfer insulator (as indicated in the figure) because the lowest excitation corresponds to an electron transfer from an O to a metal atom within the compound. For a single band approximation, the interaction is typically described by the Hamilton operator in the framework of second quantization[61,62]

$$H = -t \cdot \sum_{\langle i,j \rangle, \sigma} c_{i\sigma}^{+} c_{j\sigma} + U \cdot \sum_{i} n_{i\uparrow} \cdot n_{i\downarrow} \tag{16.16}$$

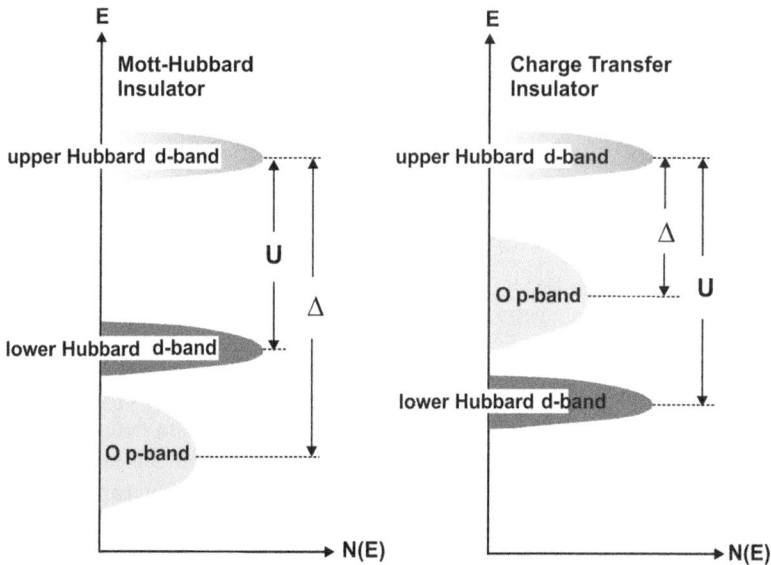

Figure 16.21 Schematic of the energy relations of d- and p- bands in Hubbard (left) and charge transfer insulators(right); U: Hubbard parameter; Δ: charge transfer parameter (see text).

Here, t denotes the so-called transfer integral in the framework of tight binding theory for conduction[63] and U is the Hubbard parameter; both are distance dependent. The symbol c stands for the creation (c^+) and annihilation operator (c) of an electron at position i or j, respectively, with spin σ. The bracket with i, j denotes the nearest neighbor interaction in the lattice. The number operators $n_{i\uparrow}, n_{i\downarrow}$ describe the doubly occupied sites and the second term in equation (16.16) represents the energy for the doubly occupied single sites of the system. The charge transfer energy (first term) and the Coulomb interaction (including exchange, second term) describe the basic properties of the system. Figure 16.22 shows a highly simplistic visualization of the situation for a 1 D chain of atoms in the tight binding approximation.

The schematic above refers to transport in a system where the low energy localized states are formed by a charge transfer band. Accordingly, the extended states (see upper level) form Bloch-type bands with delocalized electrons. The inner shells are described by localized states where intersite tunneling can occur which delocalizes these electrons to some extent. In the lowest part of Figure 16.22, the inner structure of the atom and the orbitals is neglected and electrons are considered that hop from site to site. Hopping between more distant sites can be neglected in this approximation, where the intersite tunneling depends on the distance and the localized shell's radial wave function decays exponentially with distance. The corresponding Hamiltonian can be described by (see also equation (16.16))

$$H = - \sum_{\langle i,j \rangle, \sigma} t_{ij} c_{i\sigma}^+ c_{j\sigma} \qquad (16.17)$$

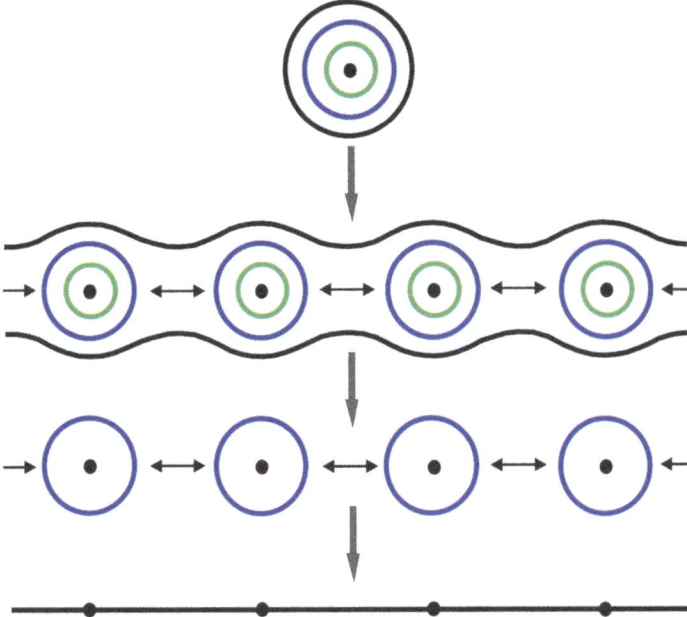

Figure 16.22   Simplified representation of tight binding descriptions of bonding. Upper level: single atom with core (black) and electrons in different (black and color) orbitals. Below: upon decreased distance in a solid, outer electrons (black circles) become itinerant, those in inner shells (dark blue, green) remain localized at the site of the atom, but, as indicated by the arrows, can tunnel with a certain probability to another adjacent colored orbital; third level. One considers only the behavior of the electrons in the more localized shells only and their role on the low energy electronics of the system. Lowest level: the electron behavior is given by localized electrons (at the position of the lattice sites) that move by hopping from on site to another (see text).

where $t_{ij} = t_{ji}$, and $t_{ii} = 0$. The matrix element for tunneling, t_{ij}, describes the wave function overlap and the tunneling probability of an electron from site i to site j:

$$t_{ij} = \int \varphi^*(r - R_i) \cdot \varphi(r - R_j) d^3r \qquad (16.18)$$

The hopping Hamilton operator in equation (16.17) sums over all hopping processes: the operators c_i^+ and c_j destroy an electron at site j and recreate the electron at the adjacent site i (see also above text).

The energy band structure of the TMOs can be assessed by resonant photoemission,[64] inverse photoemission[65] and x-ray absorption spectroscopy.[66] For the hematite (Fe_2O_3) and titania (TiO_2), which are investigated intensively for solar fuel and solar photovoltaic applications, the energy band structure resulting from recent measurements[67] is presented schematically in Figures 16.23 and 16.24.

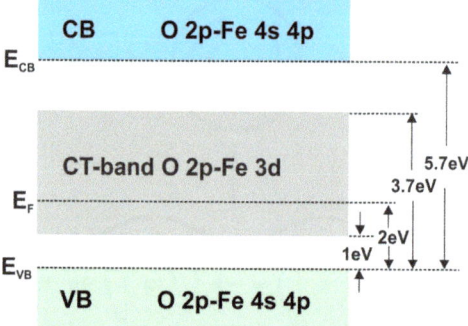

Figure 16.23   Energy band structure of hematite (α-Fe$_2$O$_3$) as determined from resonant photoemission.[64] CT-band: charge transfer band; VB, CB: conduction- and valence band; E_{VB}, E_{CB}: valence band maximum and conduction band minimum, respectively; E_F, Fermi level; the optical gap of 2 eV is located between Bloch-like valence band and charge transfer states (see text).

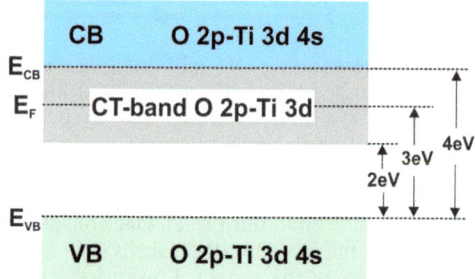

Figure 16.24   Energy band structure for TiO$_2$; symbols as in Figure 16.22. Note that in both cases shown in the above figures, the Fermi level is positioned within the charge transfer band (see text).

For TiO$_2$, the following transitions have been identified by optical methods. Photoluminescence shows transition energies of 2.0 eV and 2.2–2.4 eV. Ellipsometry shows transitions at 3.2 and 4.2 eV; this is further supporting evidence of the energy scheme shown above. For hematite, the 2 eV transition is mostly known and considered as the energy gap.[68] The presence of Hubbard- or charge transfer bands within a Bloch-state energy gap explains the problems in electron transport encountered in these materials. It has to be shown in future work how one can control the position of the Fermi level position, the formation and extension of space charge layers and how to improve transport properties. Since the development of stable and efficient photoanodes is a crucial component of light-induced water splitting, such research will be particularly beneficial for this goal.

16.5 Looking Back

It is always difficult to assemble a book like the present one in such a way that the coverage is comprehensive and strongly interlinked. Therefore, in the concluding paragraphs of this final chapter, we try to pull some of the ideas together by looking at trends that can be discerned from the preceding chapters as well as from the current literature. The first feeling that the editors gained from colleagues working on this book is a palpable sense of excitement about the science involved, a feeling that this time we are really getting closer to the final goal. At the same time, enthusiasm for the basic science has to be tempered by an increasing awareness that as research in the laboratory is translated into working devices, considerations of cost and competitiveness need to guide the selection of technologies for scale up and testing. In Chapter 1, Parkinson and Turner take a hard look at the potential contribution of photoelectrochemical energy conversion and storage to the global energy future. They argue convincingly that technoeconomic analysis indicates that the rather comfortable assumption that 10% solar to hydrogen efficiency is adequate for a viable technology is not justified. Instead a much more demanding efficiency of 15% or more, combined with a lifetime in excess of 20 years, will be required. These are challenging targets that deserve more widespread discussion. Parkinson and Turner also take the – possibly controversial – view that the photoelectrochemical reduction of CO_2 is really an approach of last resort; one that only makes sense when virtually all carbon-based power generation has been replaced by renewable alternatives. Of course, this conclusion should not deter us from continuing to search for ways of mimicking photosynthesis, and in chapter 11, Rajeshwar and his colleagues give an excellent overview of some of the exciting science involved in trying to reduce CO_2, while Schouten and Koper give a very comprehensive account of theoretical aspects of CO_2 reduction in chapter 12. The point about Parkinson and Turner's discussion is that is draws attention to the importance of realistic assessments of the potential of different technologies. All too often, in the introduction to scientific papers one reads bland statements about a supposedly "low cost technology" that are not backed up by any citations of a corresponding technoeconomic analysis.

Although the majority of the book deals with basic science, the importance of engineering in the design of practical water splitting systems is highlighted by McKone and Lewis in chapter 3. Many of the electrochemical engineering issues discussed in this chapter are similar to those involved in the design of commercial electrolyzers for hydrogen or chlorine generation: mass transport, bubble formation and safe separation of the products. It certainly makes sense to involve electrochemical engineers at an early stage of device development in order to be able to define a complete system that can be subjected to technoeconomic analysis. The overall design becomes more complex when a photoelectrochemical cell is combined with an integrated solar cell in the kind of tandem system advocated by Sivula and Grätzel in chapter 4. The matching of the two active light absorbing components will clearly be critical, and of course all of the preceding engineering considerations apply as well. This leads us to consider the trade-off between complexity and performance. If Parkinson and Turner are right, many of the photoelectrochemical systems currently being studied may not make it past a critical cost analysis. Only ultra-low cost systems can afford lower efficiencies, and it is an open question whether the kind of particulate systems described by Maeda and Domen in chapter 5 may ultimately fall into this category. If, on the other hand, very high efficiency systems really are the only option, then perhaps the group II-nitride systems discussed by Collazo and Dietz in chapter 8 or the III-V thin film structures reviewed in chapter 9 by Hannappel *et al.* may prove to be the most viable way forward. Nevertheless, materials science aspects of water splitting are likely to remain important. Some idea of the richness of the existing materials landscape is given by Fiechter in chapter 7, and the potential of rapid screening methods to identify promising new materials is clear from the survey given in chapter 6 by Bard and co-workers. At the same time, it is often helpful to return to Nature for new ideas and inspiration, as is clear from chapter 14, where Ostroumov *et al.* examine different light harvesting strategies.

The role of computational methods in the development of water splitting systems has already been mentioned. The general relevance of computational studies in energy materials is nicely illustrated by a recent overview by Catlow *et al.*[68], and more specific examples are given in the present book by Hellman in chapter 10. Sometimes chemical instinct is not enough to ensure progress in such a complex field as light-driven water splitting, and it seems likely that computational studies will make and increasingly important contribution to a range of relevant topics, including new materials, catalyst as well as aspects of kinetics and mechanisms. Here, as always, the link between theory and experiment will be crucial. Many of the recent suggestions from computational work for routes to performance enhancement remain to be tested in practice, and in the future will be essential to establish much closer cooperation between theoretical and experimental groups in order to pinpoint key issues and to guide research directions.

16.5   Progress Towards Light Driven Water Splitting

In these final sections, we want to take stock of the current situation and review progress. Forty years have passed since the publication of Fujishima and Honda's *Nature* much-cited paper,[69] but in spite of a resurgence of interest, we are still some way from realizing the goal of a stable low cost system for splitting water efficiently using sunlight. One has to ask why progress has been so slow. The answer is largely connected with the complexity of the water splitting reaction, particularly the half reaction leading to molecular oxygen. The four electron/4 proton oxidation of two molecules of water to form O_2 is achieved in higher plants by Photosystem II, in which the four electrons are transferred to two plastoquinone molecules *via* the manganese centre, which contains 4 manganese atoms, a bound calcium ion and a tyrosine radical.

The P680$^+$ formed by the first electron transfer quenching of the excited P680* state is a strong oxidant, and plants repair of the Photosystem II centre several times a day. This places into context the extraordinarily ambitious goal of designing a water splitting system that will be stable for decades. Even oxides like TiO$_2$ undergo slow photocorrosion, particularly in acidic solutions, and generally other compound semiconductors are much less stable than transition metal oxides. The difficulty of preventing normal corrosion of a twenty year time scale is a fact of everyday life, and water splitting systems are no exception. The concept of self-repair is a feature of the cobalt-based oxygen evolving catalyst discovered by Nocera *et al.*[70] and it may be possible to develop self-repairing components for water splitting systems as a way of overcoming basic issues of thermodynamic instability.

The complexity of the light-driven oxygen evolution reaction presents an enormous challenge to theory and practice. It is clear that the 4-hole reaction at n-type photoanodes must proceed *via* a series of surface bound intermediates, but even in the case of model photoanode systems such as rutile, details of the oxygen evolution mechanism remain controversial[71–73] in spite of spectroscopic evidence (FTIR) for the formation of peroxo or superoxo species. In the case of hematite photoanodes, UV-visible spectroscopic evidence also suggests formation of a surface peroxo or superoxo intermediate, but the exact identity of the species is unclear.[74] In both cases, the first step is generally thought to involve the trapping of a hole at surface site. The progressive refinement of computational approaches[75–78] is providing useful guidance regarding possible pathways for the oxygen evolution reaction, but much remains to be done to obtain reliable quantitative results that can be compared with experiment. Generally, computational approaches deal with well-defined surfaces corresponding to particular crystal orientations, but as far as experimental studies are concerned, little is known about the influence of surface structure and orientation on the mechanism and kinetics of light-driven water splitting. One of the problems in this respect is that some of the best photoanode materials such as hematite have been nanostructured to enhance performance. The increase in complexity makes it difficult to apply conventional models for the semiconductor/electrolyte interface, and as a consequence extraction of kinetic information from photoelectrochemical experiments can be difficult. Probably the best way forward for the quantitative study of such systems is to use time-resolved in situ spectroscopic techniques.[79]

The search for new materials for solar water splitting is being pursued using combinatorial[80–84] and also computational[84–87] approaches. Here one of the main problems is that optimization of a large range of solid-state, optical and chemical properties is generally necessary in order to obtain an efficient and stable[88] photoanode or photocathode material. The pathway from identifying promising candidates to development of a viable stable water splitting electrode is generally time-consuming, but the necessary research input is now becoming more readily available as water splitting moves towards centre stage, partially displacing other heavily researched areas such as dye-sensitized solar cells. A measure of the enormous increase in research effort is provided by figures for the Joint Center for Artificial Photosynthesis, the world's largest research programme dedicated to the development of an artificial solar fuel generation technology: a budget of $122M over 5 years, a workforce of over 130, 8 research divisions and 5 scientific user facilities. In his State of the Union Address in 2011, Barack Obama put it in these terms: "We're issuing a challenge. We're telling America's scientists and engineers that if they assemble teams of the best minds in their fields and focus on the hardest problems in clean energy, we'll fund the Apollo projects of our time ... At the California Institute of Technology, they're developing a way to turn sunlight and water into fuel for our cars ... We need to get behind this innovation." The signs are that the next decade will take us much closer towards the ultimate goal of splitting water to produce solar fuels, and we hope that this book will be a source to inspire new ideas about how to achieve this.

Acknowledgements

This material is based upon work performed by the Joint Center for Artificial Photosynthesis, a DOE Energy Innovation Hub, supported though the Office of Science of the US Department of Energy under Award No. DE-SC0004993. HJL acknowledges additional financial support from DFG, grant No. Le 1192-4. The authors are grateful to C. Pettenkofer and M. Lublow for experiments regarding PEEM and fractal Si formation.

References

1. H. Raether, Surface Plasmons on Smooth and Rough Surfaces and on Gratings, *Springer Tracts in Modern Physics*, Springer, Berlin, New York, 1988.
2. N. J. M. Horing, in *Introduction to Complex Plasmas*, Springer Ser. Atomical, Optical and Plasma Phys., Vol. 59, ed. M. Bonitz, N. Horing, P. Ludwig, Springer Heidelberg, 2010, pp. 109–134.
3. S. M. Sze, *Semiconductor Devices*, Wiley and Sons, 1980.
4. R. H. Ritchie, *Surf. Sci.*, 1973, **34**, 1–19.
5. R. Kötz, H. J. Lewerenz and E. Kretschmann, *Phys. Lett. A.*, 1979, **70**, 452–454.
6. J. K. Sass and H. J. Lewerenz, *J. de Physique*, 1977, **38**, 277–284.
7. H. Neff, J. K. Sass, H. J. Lewerenz and H. Ibach, *J. Phys. Chem.*, 1980, **84**, 1135–1139.
8. Y. Hori, H. Wakebe, T. Tsukamoto and O. Koga, *Electrochim. Acta*, 1994, **39**, 1833–1839.
9. K. A. Willets and R. P. Van Duyne, *Annu. Rev. Phys. Chem.*, 2007, **58**, 267–297.
10. T. Vo-Dinh, *Trends in Analytical Chem.*, 1998, **17**, 557–582.
11. D. K. Roper, W. Ahn and M. Hoepfner, *J. Phys. Chem. C*, 2007, **111**, 3636–3641.

12. M. Lublow, K. Skorupska, S. Zoladek, P. J. Kulesza, T. Vo-Dinh and H. J. Lewerenz, *Electrochem. Comm.*, 2010, **12**, 1298–1301.
13. R. Solarska, A. Krolikowska and J. Augustynski, *Angew. Chem. Internat. Ed.*, 2010, **49**, 780–783.
14. X. Zhang, Y. L. Chen, R.-S. Liu, D. P. Tsai and *Rep. Prog, Phys.*, 2013, **76**, 046401.
15. D. B. Ingram and S. Linic, *J. Am. Chem. Soc.*, 2011, **133**, 5202–5205.
16. K. Awazu, M. Fujimaki, C. Rockstuhl, J. Tominaga, H. Murakami, Y. Ohki, N. Yoshida and T. Watanabe, *J. Am. Chem. Soc.*, 2008, **130**, 1676–1680.
17. H. J. Lewerenz: *Photons in Natural and Life Sciences - an Interdisciplinary Approach*, Springer Heidelberg, New York, 2012.
18. S. Jang S, M. D. Newton and R. J. Silbey, *Phys. Rev. Lett.*, 2004, **92**, 218301.
19. G. D. Scholes and G. R. Fleming, *J. Phys. Chem. B*, 2000, **104**, 1854.
20. D. L. Dexter, *J. Chem. Phys.*, 1953, **21**, 836.
21. B. O'Regan and M. Grätzel, *Nature*, 1991, **353**, 737–740.
22. M. Graetzel and J. Photochem., *Photobio. C: Photochem. Rev.*, 2003, **4**, 145–153.
23. P. Andrew and W. L. Barnes, *Science*, 2004, **306**, 1002–1005.
24. P. Barthelemy, J. Bertolotti and D. S. Wiersma, *Nature*, 2008, **453**, 495–498.
25. M. Burresi, V. Radhalakshmi, R. Savo, J. Bertolotti, K. Vynck and D. S. Wiersma, *Phys. Rev. Lett.*, 2012, **108**, 110604.
26. R. Brown, *Phil. Mag.*, 1828, **4**, 161–173.
27. A. Einstein, *Ann. d. Phys.*, 1905, **17**, 549.
28. A. Einstein: *Investigations of the Theory of Brownian Movement*, Dover, 1956.
29. J. Bertolotti, Ph D thesis, University of Florence, 2007.
30. G. Polya, *Math. Zeitschr.*, 1920, **8**, 171–181.
31. Z. Cheng and R. Savit, *J. Math. Phys.*, 1987, **28**, 592.
32. R. Patel and R. Mehta, *J. Nanophoton.*, 2012, **6**, 069503.
33. S. Luryi and A. Subashiev, *Internat. J. High Speed Electr. Syst.*, 2012, **21**, 1250001.
34. S. Luryi, O. Semyonov, A. Subashiev and Z. Chen, *Phys. Rev. B*, 2012, **86**, 201201.
35. E. L. Warren, J. R. McKone, H. A. Atwater, H. B. Gray and N. S. Lewis, *Energy & Environm. Sci.*, 2012, **5**, 9653.
36. M. Lublow and H. J. Lewerenz, *Electrochem. Solid-State Letters*, 2007, **10**, C51–C55.
37. M. Lublow and H. J. Lewerenz, *Electrochim. Acta*, 2009, **55**, 340–349.
38. B. Y. Park, R. Zaouk, C. Wang and M. J. Madou, *J. Electrochem. Soc.*, 2007, **154**, P1–P5.
39. G. B. West, J. H. Brown and B. J. Enquist, *Science*, 1997, **276**, 122–124.
40. G. B. West, J. H. Brown and B. J. Enquist, *Science*, 1999, **284**, 1677–1679.

41. V. Lakshminarayanan, R. Srinivasan, D. Chu and S. Gilman, *Surf. Sci.*, 1977, **329**, 44–51.
42. A. Eftekhari, *Electrochim. Acta*, 2002, **47**, 4347–4350.
43. C.-P. Chen and J. Jorné, *J. Electrochem. Soc.*, 1990, **137**, 2047.
44. M. Letilly, K. Skorupska, H. J. Lewerenz, *J. Phys. Chem. C*, in print.
45. T. A. Witten and L. M. Sander, *Phys. Rev. B*, 1983, **27**, 5686–5697.
46. O. Malcai, D. A. Lidar, O. Biham and D. Avnir, *Phys. Rev. E*, 1997, **56**, 2817–2828.
47. L. Niemeyer, L. Pietronero and H. Wiesmann, *Phys. Rev. Lett.*, 1984, **52**, 1033–1036.
48. Y. Termonia and P. Meakin, *Nature*, 1986, **320**, 429.
49. M. Lublow, W. Bremsteller and C. Pettenkofer, *J. Electrochem. Soc.*, 2012, **159**, D333–D339.
50. K. Yamamoto, A. Nakajima, M. Yoshimi, T. Sawada, S. Fukuda, T. Suezaki, M. Ichikawa, Y. Koi, M. Goto, T. Meguro, T. Matsuda, M. Kondo, T. Sasaki and Y. Tawada, *Solar Energy*, 2004, **77**(2004), 939–949.
51. A. Kay, I. Cesar and M. Graetzel, *J. Am. Chem. Soc.*, 2006, **128**, 15714–15721.
52. F. F. Abdi, N. Firet and R. v.d. Krol, *ChemCatChem*, 2013, **5**, 490–496.
53. A. P. Goncalves, I. C. Santos, E. B. Lopes, R. T. Henriques, M. Almeida and M. O. Figueiredo, *Phys. Rev. B*, 1988, **37**, 7476–7481.
54. E. Dagotto, *Rev. Mod. Phys.*, 1994, **66**, 763–84.
55. F. J. Morin, *Bell Syst. Tech. Journ.*, 1958, **07**, 1047–1084.
56. A. I. Poteryaev, M. Ferrero, A. Georges and O. Parcollet, *Phys. Rev. B*, 2008, **78**, 045115.
57. J. Hubbard, *Proc. R. Soc. London A*, 1963, **276**, 238.
58. J. Kanamori, *Prog. Theor. Phys.*, 1963, **30**, 275.
59. M. C. Gutzwiller, *Phys. Rev. Lett.*, 1963, **10**, 159.
60. J. Zaanen, G. A. Sawatzky and J. W. Allen, *Phys. Rev. Lett.*, 1985, **55**, 418.
61. H. P. Paar: *An Introduction to Advanced Quantum Physics*, Wiley & Sons, 2010.
62. F. H. L. Essler, H. Frahm, F. Goehmann, A. Kluemper, V. E. Korepin: *The One Dimensional Hubbard Model*, Cambridge Univ. Press, Cambridge, New York, 2005.
63. P. A. Cox: *Transition Metal Oxides: An Introduction to Their Electronic Structure and Properties*, Clarendon Press, Oxford, 1992.
64. A. E. Bocquet, T. Mizokawa, K. Morikawa, A. Fujimori, S. R. Barman, K. Maiti, D. D. Sarma, Y. Tokura and M. Onoda, *Phys. Rev. B*, 1996, **53**, 1161–1170.
65. R. Zimmermann, P. Steiner, R. Claessen, F. Reinert, S. Hüfner, P. Blaha and P. Dufek, *J. Phys, Condens. Matter*, 1999, **11**, 1657.
66. T. Saitoh, A. E. Bocquet, T. Mizokawa, H. Namatame, A. Fujimori, M. Abbate, Y. Takeda and M. Takano, *Phys. Rev. B*, 1995, **51**, 13942–13951.
67. D. Schmeisser, to be published.

68. C. R. A. Catlow, Z. X. Guo, M. Miskufova, S. A. Shevlin, A. G. H. Smith, A. A. Sokol, A. Walsh, D. J. Wilson and S. M. Woodley, *Philosophical Transactions of the Royal Society a-Mathematical Physical and Engineering Sciences*, 2010, **368**, 3379–3456.
69. A. Fujishima and K. Honda, *Nature*, 1972, **238**, 37–38.
70. M. W. Kanan and D. G. Nocera, *Science*, 2008, **321**, 1072–1075.
71. R. Nakamura and Y. Nakato, *J. Am. Chem. Soc.*, 2004, **126**, 1290–1298.
72. P. Salvador, *Prog. Surf. Sci.*, 2011, **86**, 41–58.
73. R. Nakamura, T. Okamura, N. Ohashi, A. Imanishi and Y. Nakato, *J. Am. Chem. Soc.*, 2005, **127**, 12975–12983.
74. C. Y. Cummings, F. Marken, L. M. Peter, K. G. U. Wijayantha and A. A. Tahir, *J. Am. Chem. Soc.*, 2012, **134**, 1228–1234.
75. J. Rossmeisl, Z. W. Qu, H. Zhu, G. J. Kroes and J. K. Norskov, *J. Electroanal. Chem.*, 2007, **607**, 83–89.
76. H. Dau, C. Limberg, T. Reier, M. Risch, S. Roggan and P. Strasser, *Chemcatchem*, 2010, **2**, 724–761.
77. A. Valdes, Z. W. Qu, G. J. Kroes, J. Rossmeisl and J. K. Norskov, *Journal of Physical Chemistry C*, 2008, **112**, 9872–9879.
78. P. Liao and E. A. Carter, *Chem. Soc. Rev.*, 2013, **42**, 2401–2422.
79. A. J. Cowan and J. R. Durrant, *Chem. Soc. Rev.*, 2013, **42**, 2281–2293.
80. M. Woodhouse and B. A. Parkinson, *Chem Soc Rev*, 2009, **38**, 197–210.
81. W. Liu, H. Ye and A. J. Bard, *Journal of Physical Chemistry C*, 2010, **114**, 1201–1207.
82. S. P. Berglund, H. C. Lee, P. D. Nunez, A. J. Bard and C. B. Mullins, *PCCP*, 2013, **15**, 4554–4565.
83. J. M. Gregoire, C. Xiang, S. Mitrovic, X. Liu, M. Marcin, E. W. Cornell, J. Fan and J. Jin, *J. Electrochem. Soc.*, 2013, **160**, F337–F342.
84. K. J. Young, L. A. Martini, R. L. Milot, R. C. Snoeberger, III, V. S. Batista, C. A. Schmuttenmaer, R. H. Crabtree and G. W. Brudvig, *Coord. Chem. Rev.*, 2012, **256**, 2503–2520.
85. I. E. Castelli, D. D. Landis, K. S. Thygesen, S. Dahl, I. Chorkendorff, T. F. Jaramillo and K. W. Jacobsen, *Energy & Environmental Science*, 2012, **5**, 9034–9043.
86. I. E. Castelli, T. Olsen, S. Datta, D. D. Landis, S. Dahl, K. S. Thygesen and K. W. Jacobsen, *Energy & Environmental Science*, 2012, **5**, 5814–5819.
87. A. Valdes, J. Brillet, M. Graetzel, H. Gudmundsdottir, H. A. Hansen, H. Jonsson, P. Kluepfel, G.-J. Kroes, F. Le Formal, I. C. Man, R. S. Martins, J. K. Norskov, J. Rossmeisl, K. Sivula, A. Vojvodic and M. Zach, *PCCP*, 2012, **14**, 49–70.
88. S. Chen and L.-W. Wang, *Chem. Mater.*, 2012, **24**, 3659–3666.

Subject Index

Page numbers in *italics* refer to index entries in figures or tables.